Progress in Botany/Fortschritte der Botanik 55

55 Progress in Botany

Structural Botany Physiology
Genetics Taxonomy Geobotany

Fortschritte der Botanik
Struktur Physiologie Genetik
Systematik Geobotanik

Editors
H.-Dietmar Behnke, Heidelberg
Ulrich Lüttge, Darmstadt
Karl Esser, Bochum
Joachim W. Kadereit, Mainz
Michael Runge, Göttingen

Springer-Verlag
Berlin Heidelberg New York
London Paris Tokyo Hong Kong
Barcelona Budapest

With 23 Figures

ISBN-13:978-3-642-78570-2 e-ISBN-13:978-3-642-78568-9
DOI: 10.1007/978-3-642-78568-9

Softcover reprint of the hardcover 1st edition 1994

Library of Congress Catalog Card Number 33-15850

Typesetting: M. Masson Scheurer, Homburg/Saar
31/3145- 5 4 3 2 1 0 - Printed on acid-free paper

Contents

A. Structural Botany

I. Cell Architecture and Cellular Morphogenesis of Eukariotic Algae (Chlorophyta)

B. Physiology

D. Taxonomy

E. Geobotany

List of Editors

Section A:

Professor Dr. H.-D. BEHNKE, Zellenlehre, Universität Heidelberg, Im Neuenheimer Feld 230, 69120 Heidelberg, FRG

Section B:

Professor Dr. U. LÜTTGE, TH Darmstadt, Fakultät für Botanik, FB Biologie (10), Schnittspahnstraße 3–5, 64287 Darmstadt, FRG

Section C:

Professor Dr. Dr. h.c. mult K. ESSER, Lehrstuhl für Allgemeine Botanik, Ruhr Universität, Postfach 10 21 48, 44780 Bochum, FRG

Section D:

Professor Dr. J. W. KADEREIT, Institut für Spezielle Botanik und Botanischer Garten, Universität Mainz, Saarstraße 21, 55099 Mainz, FRG

Section E:

Professor Dr. M. RUNGE, Lehrstuhl für Geobotanik, Systematisch-Geobotanisches Institut der Universität, Untere Karspüle 2, 37073 Göttingen, FRG

Professor Dr. Dr. h.c. Hubert Ziegler
33 years of Fortschritte and Progress

Fortschritte der Botanik or *Progress in Botany* is deeply indebted to Hubert Ziegler, one of its most faithful advocates. He joined the series' endeavour to present a wide readership year after year a digest of the progress made in the whole field of botany or the plant sciences in 1960 with Volume 22, when he took over from Bruno Huber the chapter on *Wasserumsatz und Stoffbewegungen*. Subsequently, starting with Volume 29 in 1967, he assumed the responsibility for editing the section *Physiology* from Erwin Bünning. He remained devoted to this task until Volume 53, i.e., for 25 volumes or years, or just a bit more, because *Physiology* in Volume 54 was still jointly edited by him, together with his successor, before his retirement. *Fortschritte der Botanik/Progress in Botany* warmly thanks Hubert Ziegler for the invaluable contributions he has made to this series. We hope that his spirit of inquiry, his imparting of knowledge, friendship, and his help to those studying problems and their solutions will continue and will keep *Progress in Botany* thriving for many years and volumes to come.

A. Structural Botany

I. Cell Architecture and Cellular Morphogenesis of Eukariotic Algae (Chlorophyta)

By Diedrik Menzel

1. Introduction

The chlorophyta comprise a phylogenetically coherent assemblage of algae characterized by their possession of chloroplasts surrounded by only two membranes and containing chlorophylls a and b as the principal light-capturing pigments. Further genuine features of subcellular architecture, developmental patterns, and modes of reproduction set them well apart from other classes of algae (for a recent discussion of this phylum, see Bold and Wynne 1985 and Margulis et al. 1990, Chaps. 9, 32). Evolutionary advanced taxa of the class Charophyceae share sufficient similarities with archegoniate land plants to support the concept that just one ancestral lineage of green algae has founded all higher forms of modern plant life (Graham 1984), a conclusion that is also backed by recent molecular evidence (Manhart and Palmer 1990). It may, therefore, not come as a surprise that principles of cellular architecture in the higher plants, ferns, and mosses exhibit considerably lower diversity compared with the wealth of variations seen in their ancestral stocks, the green algae. For the same reason, the chloroplast genomes from higher and lower plants are extraordinarily similar, whereas those from green algae taxa are extremely divergent (Palmer 1985).

Textbooks and reviews some 10 to 15 years ago could leave the reader with the impression that the major trends in the evolutionary history of the green algae, as well as the principles of their cellular architecture, were already resolved to a reasonable degree of clarity (see Mollenhauer 1978). Such optimism, however, has proven to be delusive. In fact, the past decade has seen the redefinition of major lineages of green algae. Instrumental in this endeavor has been the finding that the composition and three-dimensional organization of the flagellar rootlet system in the phytoflagellates as well as in the flagellated generative cells of most thallic algae represent extremely conserved structural features. Cataloguing and comparing this information proved invaluable for the assessment of natural congeneal relationships among the green algae (for further details, see Melkonian 1989). The turbulence created in the process of applying new fine structural markers to the problem of understanding phylogenetic relationships among algal groups is clearly mirrored in Deason's (Deason and Floyd 1989) review on the Chlorosarcinales. Currently, further refinements in the construction of evolutionary trees are in progress in many laboratories by the employment of modern recombinant DNA technology and PCR-based sequence analysis of algal genomes (Olsen 1990; Melkonian 1991).

Progress in Botany, Vol. 55

Concurrent with the gathering of fine structural data throughout the past decade, improvements in cell fractionation, biochemical and immunological analysis have also been achieved. Furthermore, new methods of specimen preparation and three-dimensional light microscopical image acquisition techniques probing subcellular structures and whole tissues have provided a wealth of information which, combined, have allowed us to explore structure-function relationships operating in algal growth, morphogenesis, cell differentiation, locomotion, and other vital processes. Some of this development is reflected in recent multiauthor volumes such as C Lobban et al. (1988), Coleman et al. (1989), Round and Chapman (1990, 1992), Wiessner et al. (1990), Melkonian (1992b), Menzel (1992), Reisser (1992).

Since the last overview of algal cell structure and morphogenesis appeared in this Series (Leedale 1982), countless articles have been published in this field, including supplementary material on the influence of light, salinity, metal ions, inhibitors, and growth substances on green algal morphogenesis. New species have been described and details of their intracellular organization have been revealed. It would be an insurmountable task, to attempt a complete coverage of all this material. This review will consider a necessarily subjective and limited number of topics that have been raised in publications over the past decade. As a method of selection, modes of multicellular morphogenesis have not been covered. When the architecture of cells is considered, whether monad or part of a tissue, where-ever possible emphasis has been placed on the use of experimental model systems for the indepth study of structure-function relationships.

It has been known for a long time, and many new examples have been described in the past decade, that cell and thallus morphology may be subject to considerable variation depending on growth conditions, population dynamics and gene dosage. These aspects of morphogenetic plasticity and the occurrence of ecomorphs, however, will not be discussed here (more indepth discussion of this subject may be found in Ettl 1985; Hoshaw et al. 1985, 1987; Vidyavati 1985; McCourt and Hoshaw 1990; Trainor and Egan 1990).

2. Endomembrane Systems

Three major endomembrane compartments (not counting their derivatives) exist in every plant cell including the green algae: the endoplasmic reticulum (ER) usually present in two variants, rough/cisternal and smooth/tubular, the Golgi apparatus (GA) consisting of one to many dictyosomal stacks, and the vacuole in the form of either an inconspicuous population of vesicles, a single central vacuole, or a dynamic, highly structured osmoregulatory organelle "termed contractile vacuole" (Robinson 1985; Harris 1986). Communication between these compartments and the plasmamembrane is mediated by an unknown number of vesicle types of which the coated vesicles are the best known (Coleman et al. 1988, Robinson and Depta 1988). In the green algae, structure of the endomembrane system, principle functions of its components, and interaction between the various compartments have recently been comprehensively treated by Domozych (1991a). The following chapter will summarize some of the background information as well as considering recent findings in greater detail.

a) Endoplasmic Reticulum (ER) and Derivatives

In general, the ER is thought to be the compartment of secretory protein synthesis, processing, and storage, whose products can either be directly exported to the cell surface or sent to the GA for further processing. However, specific functions of the ER in a given cell type are very difficult to assess. Often, intriguing spatial relationships are reported between the ER and other cell organelles, and theories on ER function may be built on the basis of this information. One such example suggests an involvement of the ER in phycoplast-mediated cytokinesis. In the volvocalean flagellate *Carteria*, ER membranes proliferate at and closely associate with the edge of the incipient cleavage furrow without actually fusing with it. Domozych (1987a) interprets this intimate contact as an indication for a role of the ER in the delivery and regulation of calcium ions at the furrow edge for optimal perfomance of the cytokinetic apparatus. In another example, the cisternal compartments actually fuse, thus bypassing the GA to provide the extracellular space for cell wall assembly (*Cylindrocapsa geminella*, Sluiman 1984).

A few studies have also dealt with a biochemical analysis of the ER. Zhang et al. (1989) have isolated ER and dictyosomes from *Chlamydomonas* by isopycnic sucrose density centrifugation and discovered O-glycosylation activity, arabinosyl transferase in particular, in the ER fraction. This has later been confirmed by immunocytochemistry (Zhang and Robinson 1990).

Coated vesicles (CVs) of several size classes have been sighted on many occasions in the green algae. They are associated with vesicle shuttling, exocytosis from the GAs trans-face and endocytosis from the plasmalemma. They often occur in conjunction with a post-Golgi apparatus (GA) membrane compartment coined "partially coated reticulum" (PCR, Pesacreta and Lucas 1984), and are prominent during cytokinesis when intense membrane recycling takes place. However, most of these reports were based on conventional fine structural analysis, leaving it open as to whether these vesicles represent the classic clathrin-coated type or whether different types of coates are awaiting discovery.

In recents studies, a membrane compartment similar to the PCR in *Nitella* has been observed by applying rapid freeze fixation and freeze substitution in the coccoid green algae *Friedmannia israelensis* (Lokhorst et al. 1989). The PCR is suggested to be involved in membrane recycling processes during cytokinesis of this alga. This was even more clearly depicted in the ulvophycean alga *Gloeotilopsis planctonica*, where the PCR formed a continuous network in the vicinity of the forming cleavage furrow (analogous to the localization of the ER at the cleavage furrow as discussed above; Sluiman 1991). The author suggests that large numbers of vesicles typically seen in chemically fixed ulvophycean algae could be the remnants of an ill-preserved PCR, and he concludes that the PCR-mediated membrane recycling could be more typical for cytokinesis in the Ulvophyceae than previously known.

One example for an unusual type of CV has been described in *Acetabularia.* It was named "cytotomic cleavage vesicles" by Franke and coworkers (1976) because of its extreme abundance at a stage when the cell performs multiple cytoplasmic compartmentation during cyst formation. The vesicles originate from the GA, bear distinct bristle coats, and seem to deliver their content into large vacuolar cavities, which likewise become bristle-coated. This, and the fact the vesicles are very heterogenous in size, speak against a classic clathrin type. The only documented case of the occurrence of classic CV, as defined by biochemical evidence for the presence of clathrin, is that of *Chlamydomonas* (Denning and Fulton 1989). While much of the activity of CVs during cytokinesis is attributed to endocytosis, actual experimental evicence supporting this function is rare. One example where this has been done is the protoplast fragmentation process in the giant siphonoclad green alga *Boergesenia*, where uptake of exogenous cationized ferritin by CVs has been demonstrated (O'Neil and La Claire 1988).

It has become a widely held view that the ER and the chloroplast envelope are both part of the extracytoplasmic (also referred to as extracellular) space, and claims have been made that they are, in fact, physically connected with each other; however, it has been difficult to provice artifact-free evidence. By employing quick-freeze and freeze-fracture techniques, which currently provide perhaps the best available method to approach this problem, McLean et al. (1988) have recently presented the first convincing support for the existence of such a continuity between the chloroplast envelope and the ER in *Chara* as well as in *Equisetum*.

In order to study membrane behavior in vitro, Higashi-Fujime (1988) managed to produce membranous tubular networks in vitro from endoplasmic droplets of *Nitella* internodal cells. She observed elongation and branching of these tubes live under dark field illumination when a solution of muscle actin and ATP was added. The membrane structures were reminiscent of polygonal systems of cortical tubular ER found in many plants, including algae, but it was not clear whether the network was a remnant of the droplet's plasma membrane or represented tubular ER that had survived disruption of the droplet. Since the network interacted with external actin and underwent shape changes after the addition of ATP, it is most likely of ER origin and it is conceivable that the interacting molecule is ER-bound endogenous myosin. Despite some uncertainties, Higashi-Fujime's approach could prove to be of importance, simply because experimental conditions can be manipulated much more easily in a cell-free system than in the intact cell.

b) Golgi Apparatus (GA)

In green algae, as in most other plants, much of the synthetic activity of the GA is dedicated to cell wall precursor production. The discovery, three decades ago, that most species in the Prasinophyceae produce delicately structured scales in the cisternal stacks of their GA has provided an early opportunity to study the organization of the GA on the fine structural level. Since several of the prasinophyte species are easy to grow axenically in large quantities, and because scale formation can be experimentally induced, they have subsequently lent themselves to more rigorous experimental approaches involving refined cytochemical, immunological as well as biochemical techniques (Melkonian et al. 1988; Melkonian et al. 1991a). Mainly species of the four genera *Pyramimonas, Scherffelia, Tetraselmis*, and *Mesostigma* have been studied in detail.

Scales are produced entirely in the cisternal lumen of the GA, and since they conveniently serve as internal markers, the behavior of the cisternal compartments can be easily determined by fine structural analysis. Through such studies it has become clear that maturation of cisternal compartments occurs by progression through the stack rather than by vesicle shuttling between stationary cisternae. It is argued that cisternal progression is not just a special adaptation restricted to scale-producing cells, but a much more general mode of activity than was earlier recognized (Melkonian et al. 1988). This, of course, does not preclude the possibility that vesicle shuttling also contributes to the biosynthetic activity of the GA. In fact, all the studies have revealed the existence of small vesicles at the periphery of the cisternal stacks, but scales have never been found in these vesicles (Domozych 1991a).

During interphase, the GA in the prasinophytes appears to express residual activity in hair scale processing or can even be completely idle. However, the synthetic machinery of the GA can be stimulated by an external signal to rapidly resume activity. For instance, reactivation can be experimentally provoked by deflagellation, and is most likely linked to a transient rise in intracellular free calcium (Melkonian et al. 1991b). Another signaling system must be operating that monitors flagellar regeneration and causes an inactivation of the GA after flagellar regeneration is completed.

The mechanisms regulating GA activity are complicated, since the GA can operate in at least two different modes depending on whether it is engaged in the production of body scales or flagellar scales. In *Scherffelia*, both modes are temporally separated and involve slightly different activities of cisternal compartments (McFadden et al. 1982; McFadden and Melkonian 1986). Body scales depart from the GA within the outermost trans-cisterna destined to fuse as a whole with the plasma membrane in the furrow zone or the flagellar grove. On the other hand, flagellar scales appear to become presorted at the rim of the trans-most cisterna, and undergo a final sorting on their way to the plasma membrane with the effect that each type of scale is found in a separate vesicle (McFadden and Melkonian 1986). Many prasinophytes possess a scale reservoir within which scales seem to become sorted and sequentially layered for export to the flagellar pit through a membrane ductus. How sorting actually works in any of the stages of scale export is not known. An additional problem is that during exocytosis of flagellar scales, a surplus of membrane is incorporated in the plasma membrane; this material must be immediately retrieved in order to maintain cellular integrity. In *Scherffelia*, a more elaborate post-trans face membrane reticulum and a coated polygonal vesicle at the trans-side of the GA seem to be involved in membrane retrieval (McFadden and Melkonian 1986; McFadden et al. 1982).

The chlamydomonad flagellate *Gloeomonas* has also been extensively studied with respect to GA morphology and membrane flow. Up to 18 dictyosomal stacks are arranged radially around the central nucleus oriented with the trans-faces towards the nuclear envelope. Each stack is surrounded by an ER pocket. Outside the perinuclear GA field is a layer of large peripheral vacuoles which could possibly be "the main autophagic locus of the cell, aiding in bulk degradation and recycling of membranes" (Domozych 1989a). Domozych further suggests that the ultimate destination for trans-face dictyosome vesicles carrying wall precursor material is the apically located contractile vacuole, which itself opens up to the cleavage furrow during cell division. In an extensive cytochemical study, Domozych (1989b) provides evidence for a strict polar compartmentalization of several marker enzymes typically associated with carbohydrate processing in the dictyosome, such as nucleotide mono-, di-, and triphosphatases, acid phosphatase, and alkaline phosphatase. It has hence become possible to monitor changes in the compartmentalization of these marker enzymes upon treatments with various inhibitors in greater detail than by employing fine structural analysis alone.

Despite this progress, the GA remains enigmatic. Specific factors that control GA polarity, intracellular orientation, dynamics of the membrane flow, shape and elaboration of the cisternal compartments have not yet been identified.

Based on the effects of a number of pharmaca on the integrity of the GA in a prasinophyte species (Domozych 1987b) and a chlamydomonad alga (Domozych 1989a, b, 1991b), several assumptions can be made. The maturing face requires a balanced ionic environment for proper function and maintaining integrity. Both K^+/Na^+-ionophores nigericin and monensin cause extensive curling of the trans-face. Calcium is also important, because disturbing calcium ions leads to the formation and buildup of large vesicles from the trans-face. The methylxanthine derivative caffeine – a potent blocker of cell plate formation in higher plants and inhibitor of phosphodiesterase in animals – prevents release of material from the trans-face and causes hypertrophy of the entire cisternal stack. This suggests that the potential targets of the inhibitor, i.e., phosphodiesterases, Ca-dependent ATPases, and Ca-homeostasis in general (for references, see Domozych 1989b), are involved in the processing of the departing vesicles from the trans-face. Finally, intracellular orientation and apportioning of cisternal stacks could be controlled by MTs, as has been shown for animal cells (Kreis 1990). This conclusion has been drawn from several studies employing anti-MT drugs. *Tetraselmis convolutae* possess two polarly arranged cisternal stacks in undisturbed cells. After application of colchicine or the anti-MTOC drug IPC (isopropyl N-phenyl carbamate), the number of stacks increased to up to 20 and their polar orientation was lost (Domozych 1987c). In *Gloeomonas kupfferi*, IPC also had a colchicine-like effect upon short-term treatment, supporting an involvement of MT-dependent processes; however, prolonged treatments with IPC massively altered the physiological status of the GA reflected by a changed pattern of marker enzyme localization within the cisternal compartments and ultimately by a vesicular disintegration of the GAs (Domozych 1991b). Although an MTOC involvement in dictyosome positioning is an interesting aspect that deserves further attention in future studies, all these effects are more reminiscent of the disturbances caused by the ionophores and caffeine, and hence may not reflect MT or MTOC functions.

The majority of conclusions on GA dynamics and function have been derived from structural observations made on fixed cells. Little progress has yet been made to actually visualize the dictyosomal membrane stack in the light microscope, for instance by video-enhanced microscopy and by using vital fluorescence dyes. Such approaches could contribute to questions of GA repositioning and division (dictyokinesis) in preparation for cell division. Domozych (1987c) mentioned that in *Tetraselmis convolutae*, division of the GA starts from the trans-face and terminates at the cis-face. The direction of division through the cisternal stack, however, does not seem to be fixed, since in *Gloeomonas*, division starts at the cis-face (Domozych 1989a), as has long been described for *Oedogonium* (Pickett-Heaps 1975). In this context, it is noteworthy to learn that Tschermak-Woess (1988) has been able to image dictyosomes by conventional DIC microscopy in *Chlorella* strains, even without digital image manipulation.

During the development of the desmid *Closterium ehrenbergii*, the dictyosomes go through a differentiation cycle producing characteristic stage-specific morphologies (Noguchi 1988). The resting configuration (prodictyosome) consists of only two stacked cisternae lacking any visible sign of proliferation. Dictyosomes do not form de novo, upon germination the stacks grow to seven to ten members and subsequently double by dictyokinesis once every day for 4 days, before they start producing low and high density secretory vesicles carrying wall precursor to the plasma membrane.

3. Seminautonomous Organelles

a) Chloroplasts

The chloroplasts of the green algae have been the subject of numerous studies (O'Kelly 1982; Setlik and Zachleder 1984; Przibilla and Galling 1987; Rochaix 1987; Goyal et al. 1988; Kajikawa et al. 1988; Leible et al. 1989; Ehara et al. 1990; Okabe et al. 1991; Wagner and Grolig 1992). Some of the issues currently under investigation concern the eyespot of phytomonads and the DNA nucleoids.

In a survey of about 100 species of eukaryotic algae, some 25 green algae were examined by staining with DNA-specific fluorochromes (Coleman 1985). In all of them the nucleoids were morphologically ill-defined, often several to many foci were present in one plastid and an association with pyrenoids was sometimes seen but was most typically depicted in *Spirogyra*. It was also noted that in species of the Dasycladales a larger proportion and in *Volvox nagariensis* a small proportion of chloroplasts lacked nucleoids (see also Lüttke 1988). This was interpreted as a sign of terminal differentiation of the organelles that do not participate in the continued growth of the organism. Another reason for a relaxed evolutionary constraint on the maintenance of at least one genome per plastid could be that only a fraction of the organelle population will eventually be segregated into the gametes and thus take part in fertilization.

However, the same argument would apply for the Caulerpaceae, but here a 1:1 ratio of nucleoid per plastid was observed and the nucleoids are firmly embedded in the pyrenoids of those caulerpalean plastid types that possess a pyrenoid (Miyamura and Hori 1991). The plastids in the Caulerpaceae are typically dimorphic, chloroplasts and amyloplasts, and both types have a highly asymmetric morphology (Grant and Borowitzka 1984). The dorsal side bulges out and the ventral side is flat, supported on the inner side of the plastidal double membrane by a sheath of fibrils, a feature not seen in any other green algal chloroplasts (Menzel 1985). The tip of the plastid is formed by a concentric lamellar body. Although the functional roles of the morphological features just described are not known, it has been speculated that they are a consequence of evolutionary "streamlining" in the literal sense, because these plastids are subject to fast unidirectional movement along MT bundles and it has been observed that the direction of movement is always in the direction of the pointed tip (Menzel 1985, 1987). With respect to nucleoid position in caulerpalean plastids which lack pyrenoids, it may be noteworthy that they are situated between the large posterior starch grain and the anterior concentric lamellar body (see Fig. 14 in Miyamura and Hori 1991).

A recent study on *Chlamydomonas* demonstrated that the shape of platidal nucleoids change with the stages of the cell cycle from a dispersed multiple aggregate to a compact aggregate around the pyrenoid and followed by a thread-like configuration which seems to represent the predivision condition. After division of the chloroplast, nucleoids disperse again (Ehara et al. 1990). It might be just a strange coincidence that in the volvocalean alga *Pseudocarteria* the cytoplasmic nucleus is situated inside the chloroplast (Suda et al. 1990).

In the green algae, as in most other plants, only the maternal chloroplasts are inherited; this is achieved by selective elimination of the male chloroplast (cp) nucleoids. This phenomenon was originally observed in *Chlamydomonas* by DAPI staining of the chloroplast nucleoids in postfusion zygotes (Kuroiwa et al. 1982) and was subsequently seen in several other algae from Ulvohyceae (Kuroiwa and Hori 1986) to Charophyceae (Sun et al. 1988). It proved that in taxa with anisogamous and oogamous fertilization, chloroplast nucleoids were destroyed during gametogenesis, whereas in the isogamous species this process was delayed until fusion of gametes had occurred.

The mechanism of chloroplast division is still elusive. After the discovery in *Cyanidium caldarium* by Mita and Kuroiwa (1988) that the cyanobacterial endosymbiont divides via an actin ring situated in the plastid's constriction, there have been several similar but less conclusive reports in mosses (Tewinkel and Volkmann 1987) and higher plants (Hashimoto 1986; Oross and Possingham 1989), and a further example has been recently described among the green algae (Hashimoto 1992).

The eyespot of phytoflagellates, gametes, and zoospores has been of obvious interest because of its role in light perception and phototactic behavior. An exhaustive review by Melkonian and Robenek (1984) drew together all the current structural, physiological, and behavioral data. Since then, progress has been made with respect to reflective properties of the eyespot, the actual photoreceptor, and with modeling the signaling chain from perception of the light to the change in flagellar behavior.

Chlamydomonas flagella beat with a frequency of 40–60 Hz. The slightly asymmetric beating pattern produces a helical path from which the cell deviates, depending on the direction and intensity of an incident light beam. Rotation of the cell causes a rhythmic exposure of the photoreceptor to a perpendicular light beam. The question remaining is, does the eyespot provide an efficient shield from light coming through the cell body, so that the photoreceptor will be excited only once per revolution, or does it function as a focusing device for detecting maximum intensity when the eyespot faces the light source? Recent work by Kreimer and Melkonian (1990) using confocal laser scanning microscopy in the reflection mode indicate that the eyespot acts as a quarter-wave interference reflector with the inherent property that the light signal is focused back to the photoreceptor patch in the membrane overlaying the eyespot. The proposed enhancement effect was also confirmed since the reflection efficiency was increased in species with multilayered eyespots.

Nultsch (1988) proposed a model in which the photoreceptor in the plasma membrane patch overlaying the eyespot is excited and causes neighboring Ca^{2+} channels to transiently open. The Ca^{2+} influx somehow causes a depolarization of the membrane that transmits to the flagellar surface where further Ca^{2+} channels become activated. This increase in the axonemal Ca^{2+} concentration could then modulate the dynein-based motility machinery of the flagellum. A phosphoprotein has recently been identified as candidate to mediate the Ca-dependent photophobic response in *Chlamydomonas* (Segal and Luck 1985).

The nature of the photoreceptor in *Chlamydomonas* has recently been identified as a rhodopsin (Foster et al. 1984). Harz and Hegemann (1991) have been able to experimentally confirm the above hypothesis by measuring ion currents with a suction pipette specifically in the area of the eyespot (eyespot inside the pipette) or in the area of the flagella (flagella inside the pipette). According to these data, the alga *Chlamydomonas* has a visual system whose basic mechanism of light perception works just like in amimals.

Kreimer et al. (1991a) have begun to analyze the protein and pigment composition of eyespots isolated from *Spermatozopsis similis* and found that the extracted eyespot contained two retinal derivatives, most likely representing the chromophoric groups of the rhodopsin photoreceptor, which itself has not yet been identified in this alga (Kreimer et al. 1991b).

b) Mitochondria

It has been known for more than 20 years that mitochondria in some green algae can coalesce to form a single reticulum per cell. This has recently been redescribed for *Pyramimonas* (McFadden et al. 1982) and for *Chlamydomonas* (Morris et al. 1985), where the shape of the mitochondrion appears to depend on the osmotic pressure. Taking advantage of the affinity of the lipophilic dye $DiOC_6$ towards mitochondria, which has also been used as an in vivo marker dye for the ER, Hatano and Ueda (1988) discovered that the mitochondria form an extended reticulum in the large coenocytic cells of the net-like alga *Hydrodictyon reticulatum* (Chlorococcales). The reticulum gradually breaks down during zoospore formation, and finally fragments into single spherical units in the zoospores. However, soon after commencement of daughter net formation by association of zoospores, the mitochondrial reticulum reforms in each cell and expands as the multicellular net expands. The study did not reveal a specific funtional advantage of the mitochondrial reticulum over a population of individual smaller units, and no attempts have been made to disturb this process. Therefore, it remains to be seen whether the organization of the reticulum depends on the distribution and action of cytoskeletal elements or is an intrinsic property of the mitochondria. In the coccoid alga *Chlorococcum infusionum*, a similar stage-dependent reticulation of the mitochondrion has been observed (Chida and Ueda 1986).

4. Interphase Nucleus

Van den Hoek et al. (1988) recognized the various modes by which green algae can divide as significant for taxonomic classification. These include closed, partially closed and open mitosis, referring to the appearance and behavior of the nuclear envelope during mitosis.

One recent example of such variation is provided by studies on the conjugatophycean mitosis. Abhayavardhani and Ueda (1986) have reported that the nuclear envelope in *Triploceras gracile* recondeses prematurely around the anaphase nuclei and that an unusual amount of nucleolar material is carried along with the chromosomal masses. According to earlier observations cited by these authors, this could be a specific feature of the Conjugatophyta (see also Ueda et al. 1986).

The evolution of these different modes most likely has a cytoskeletal basis. It is known in animal cells that the ability to go through cycles of nuclear envelope breakdown and recondensation is directly dependent on the presence of cytoskeletal proteins of the intermediate filament class, called lamins (Gerace 1986). They form a thin fibrous layer, the lamina, on the inside of the nuclear envelope, which begins to break down when lamin becomes phosphorylated by the cell cycle kinase $P34^{CDC2}$ (Nigg 1992). It is still a controversial issue as to whether intermediate filament proteins occur in plants (Menzel 1993), and there has indeed been little structural evidence for the existence of a lamina. In the green algae, however, two examples exist. In an early publication by Burr and West (1971), attention was drawn to the giant primary nuclei in the *Bryopsis* sporophyte. Using conventional TEM, the authors described a densely stained fibrous reticulum on the inside of the *Bryopsis* nuclear envelope. In a recent study on hand-isolated nuclei of *Acetabularia* using whole mount and metal shadowing techniques Tischendorf et al. (1992) presented evidence for the existence of a cytoskeletal structure reminiscent of the *Xenopus* oocyte nuclear lamina. Certainly, this structure requires additional characterization and the search for similar structures needs to be extended to other algal taxa. It would be of particular interest to reinvestigate the lamina-like structure in the *Bryopsis* nucleus. From such work, it would be possible to examine the cytoskeletal basis for the behavior of the nuclear envelope in the various modes of mitosis of the green algae.

It may also be considered that nuclear integrity could be supported by an exonuclear skeleton, particularly if nuclei are subject to shear forces resulting from rapidly streaming cytoplasm. Wasteneys and Williamson (1991) described a ring of actin filaments following the long edges of elliptical nuclei in the internodal cells of *Nitella*. Since the nuclei perform an unusual tumbling rotation about their longest diameter rather than spinning around their small diameter, it is suggested that the actin ring could be involved in the mechanism of rotation. The phenomenon is reminiscent of the rotation of polygonal actin rings that occur in extruded *Nitella* cytoplasm (for a recent review on motility phenomena in extruded characean endoplasm, see Higashi-Fujime 1991). Involvement of the actin ring in the motility mechanism is further supported by the finding that nuclei in another characean species, *Chara corallina*, are nonrotational and do not possess actin rings. However, it is not clear whether *Nitella* actually benefits from rotational nuclei. It is unlikely, that the actin ring assists in nuclear division since it is oriented at right angles to the cleavage plane of the nucleus.

Although the giant *Acetabularia* nucleus has attracted attention for a long time because of its large size and its usefulness for studying nucleocytoplasmic interactions (for review, see Berger et al. 1987), its karyological status has only recently been clarified. De and Berger (1990) provide the first complete karyotype consisting of 20 bivalents in the diploid primary nucleus. Shihira-Ishikawa (1984) observed that in the course of premeiotic maturation the distribution of chromosomes within the nucleus changes and that both nucleoli and chromosomes undergo stage-specific shape changes. The same author has also studied chromosomal behavior through the remaining phases of the cell cycles of both sporophytic and gametophytic stages (Shihira-Ishikawa and Kuroiwa 1984).

An additional aspect of interest is the lampbrush morphology of chromosomes in *Acetabularia* (Spring et al. 1975). Recently, Berger et al. (1993) have been able to study loop fine structure in these chromosomes in unsurpassed detail under the confocal laser scanning microscope by employing an antibody against nuclear phosphoprotein epitopes. One of the most unusual features of the nucleus in the Dasycladales is the presence of perinuclear dense bodies, which on the basis of affinity binding of gold-labeled RNase and DNase and by other indirect evidence are suggested to contain RNA and DNA (Berger and Schweiger 1986). If these findings can be further substantiated, the current dogma on the regulation of gene expression might require revision, and a role for the extrachromosomal DNA needs to be found. The question could also be addressed whether the perinuclear dense bodies fit the role of the species-specific morphogenetic substances which have long been postulated to be present in *Acetabularia* (for a summary, see Berger et al. 1987).

5. Cytoskeleton

It has become almost a trivial statement that cell architecture and morphogenesis ultimately depend on the cytoskeleton. This can be deduced from conventional TEM observations. However, it has only recently become possible to fully appreciate the three-dimensional image of the cytoskeleton, because immunocytochemistry became applicable to algal cells. Even more important, it is now possible to visualize and document the changes that occur as a consequence of pharmacological treatments or during normal growth and development. Many alga taxa still seem to resist this approach due to indigestible walls and voluminous mucilage layers or other adverse factors.

The disadvantage of conventional immunofluorescence is that cells must be fixed and permeabilized in order to have access to the cytoskeleton. A more elegant method is the rapid permeabilization of live cells without fixation under conditions that stabilize cytoskeletal components. Nevertheless, in both cases, the cells are dead when they are examined under the microscope. Ultimately, one would like to follow cytoskeletal changes directly in the living organism. This goal has come into immediate reach for plant cell biologists by a technique called analog fluorochrome cytochemistry (Wang and Sanders 1990). It entails microinjection of fluorescently labeled monomeric cytoskeletal precursors such as tubulin and actin into the cell or labeled phalloidin that will then bind to actin. The fate of the injected material can then be directly followed by enhanced video fluorescence microscopy or confocal laser scanning microscopy (CLSM). Recent advances have been summarized by Hepler and coworkers (Hepler et al. 1993), who have been poineering in the application of this technique for plant and algal cells.

a) Phytoflagellates. Zoospores, Gametes

The phytoflagellate cytoskeleton is absolutely dominated by the basal body-associated rootlet apparatus with its MTs and assortment of fibers, connectives, and plates. The ultrastructural analysis of these structures has continued to be one the most fruitful enterprises in algal cell biology during the past decade (see Melkonian 1980 for an early review). Possibly up to 100 new publications have appeared since. The majority have been made from a phylogenetic or general cytological perspective; however, there has also been substantial progress in our understanding of structure-function relationships and developmental aspects of flagellar formation and regeneration. This has been recently reviewed competently and exhaustively by the protagonist workers in this field. For flagellar motility and structure and molecular components of the axoneme, the reader may consult Kamiya (1992). An update of the work on flagellar rootlets is presented by Lechtreck and Melkonian (1991), Melkonian (1991), and O'Kelly (1992). Significance of the recently discovered contractile proteins centrin, assemblin, and other molecular components is discussed in Melkonian (1992b) and Salisbury (1992). The phenomenon of surface motility of the flagellar membrane has been reviewed by Menzel (1990), transmembrane signaling through the flagellar membrane is covered in the review by Bloodgood (1991) and a monograph has also appeared dealing with all aspects of the flagellar membrane (Bloodgood 1990).

In the following, some additional aspects dealing with basal body centriole dualism will be discussed and several examples of unusual cytoskeletal structures will be mentioned. Problems inherent to cell shape maintenance of naked cells will be dealt with in Section 8a.

Green algae, with a few exceptions, are centric organisms possessing either centrioles in vegetative stages or basal bodies (BBs) in the gametes and zoospores. The exceptions are the Dasycladales and Caulerpales (class Ulvophyceae) and Charales (class Charophyceae), featuring acentric mitotic and meiotic spindles in the vegetative stages but normal centrioles/BBs in gametes, and all taxa of the Zygnematophyceae lacking centrioles as well as basal bodies. BBs and their associated rootlet structures duplicate in a semiconservative fashion. The consequence of this is that each cell possesses a parent and a daughter BB, each with a different developmental history. This explains why flagella can display different axonemal growth behavior (recently reviewed by Beech et al. 1991) and slightly different beating patterns (Kamiya and Witman 1984), which ist the reason why a cell spins during forward swimming. An analysis of the BB rootlet apparatus structure of *Chlamydomonas* demonstrated that despite the overall rotational symmetry of the BB rootlet complex as viewed from the top, plus and minus mating types differ with respect to the position of their mating structures. In the plus gametes it is associated with the two-stranded MT rootlet on the trans-side (opposite the eyespot) and in the minus type with the two-stranded MT rootlet on the cis-side (Holmes and Dutcher 1989). The cis-side itself is determined by the eyespot, which is always associated with the four-stranded MT rootlet of the daughter basal body (Huang et al. 1982).

An observation by Luck and coworkers that the genes of the uni-linkage group in *Chlamydomonas* whose mutations affect only one flagellum (formed by the daughter basal body; Holmes and Dutcher 1989) reside on a circular DNA and behave unlike nuclear chromosomes (Luck 1984), revived the long-standing issue of the existence of extra-chromosomal DNA in the basal bodies (for review, see Goodenough 1989). Indeed, after in situ hybridization with a uni-locus gene sequence the two basal bodies were distinctly labeled, indicating that the circular DNA could reside in the BBs (Hall et al. 1989). However, careful reinvestigations with DNA-specific fluorochromes (Kuroiwa et al. 1990) and antibodies against DNA Johnson and Rosenbaum (1990) failed to substantiate the therory of a semiautonomous basal body organelle (Bermudes et al. 1987).

In the coenocytic coccoid alga *Dictyochloris fragrans*, the rare situation exists that the centrioles still have a virtually complete cruciate basal body rootlet complex attached to them and with each round of mitosis the rootlets become faithfully duplicated (Kouwets 1991). As the number of nuclei increases, the cell becomes distinctly layered, with chloroplasts situated in the cortex, dictyosomes following next, and nuclei with their centrioles facing a central dense mass of cytoplasm filled with ER profiles and vesicles. It is conceivable that this internal organization is a direct consequence of the existence of the rootlet system running parallel along the nuclear flanks toward the cortex. It is also implicated that rootlet components could take part in prophase movement of centrioles. In a more general way, Kouwets suggests that the retention of the basal body rootlet complexes in this soilborne alga, together with its coenocytic cell architecture, reflects an adaptation to its habitat and allows instantaneous compartmentation into swarmers after sudden rain showers.

Persistence of a rootlet template, consisting of a connecting band and a short MT rootlet, has been encountered in a filamentous member of the Chlorococcales, *Cylindrocapsa geminalla* (Sluiman 1985). The two pairs of centrioles migrate into the equatorial plane after mitosis and seem to assist the formation of a transverse MT system in the cleavage plane. The role of the rootlet template is not determined.

In the coccoid green alga *Friedmannia israelensis*, ill-defined electrondense projections and some MTs at the proximal end of the centrioles have also been interpreted as rootlet templates. In this alga, the mitotic spindle is flat and acentric, because the centrioles stay at an equatorial position throughout mitosis, marking the prospective cleavage site (Lokhorst et al. 1989).

The effect of MT inhibitors in phytoflagellates on GA morphology and function has already been considered (see Sect. 2 b). In addition to these effects, Domozych et al. (1983) reported that MTs in *Mesostigma viride*, especially those subtending the central flagellar pit membrane, become depolymerized. Subsequently, the pit disappears and the cell loses its distinct disk-like shape (see also Sect. 8 a). Furthermore, MT proteins reorganize into crystalline bands consisting of 25 nm hexagonally packed subunits surrounded by ER cisternae. They are reminiscent of crystals formed after depolymerization of MTs in a variety of organisms including plants (Apostolakos et al. 1990), and may even be naturally occurring structures which may indeed have a function in MT dynamics (Golz and Hauser 1986). Since MT depolymerization in phytoflagellates does not regularly lead to paracrystal formation, it may be concluded that in *Mesostigma* specific factors with an affinity to MTs also become liberated by the treatment and associate with the surplus of tubulin subunits in an unusual way. MT associated proteins (MAPs) are still enigmatic molecules in the plant cell, so that it might be worthwhile to reexamine *Mesostigma* as a potential source for plant/algal MAPs.

Another kind of paracrystalline structure has been discovered in the male gametes of *Atractomorpha echinata* (Hoffman 1984). The structure is a trilamellar flat stack of electron-dense dots right underneath and parallel to the basal bodies. In all other respects, the cells appear quite normal, including the existence of a cruciate rootlet system. Surprisingly, gametes from other sphaeropleacean species, even female gametes of the same species, do not possess this structure, suggesting that although perhaps not crucial for survival, it may still be a male-specific structure important in an unknown way for fertilization or postfertilization development of this species.

In the chlamydophycean brackish water alga *Brachiomonas submarina*, which also provides an example of shape changes involving modifications of the wall (see Sect. 8 b), a so-called layered structure reminiscent of the paracrystalline structure in *Atractomorpha* is found close to each basal body and in association with the four-membered MT rootlet. Unfortunately, nothing is known about the role and composition of this structure (Watanabe et al. 1989).

Many, but by no means all, green algal gametes possess a mating structure which is often an electron-dense oval plaque in the subapical plasma membrane of the female gamete and a similar, more compact structure in the male, respectively. The latter can produce a fertilization tubule during mating by formation and subsequent extension of an actin filament core much like the acrosomal reaction in many coelenterate sperms (Detmers 1985; Tilney 1985), which eventually fuses with the female mating structure. Many of the phylogenetically advanced taxa lack a mating structure or only the female gametes have one. A recent study on gametic fusion in the siphonous green alga *Bryopsis maxima* (Hori 1988) has stressed the point that the mating structure in females is most probably functionally irrelevant, since anisogamous fusion occurs anywhere along the cell surface of the female gamete.

This prompts speculations as to why mating structures should be advantageous in the first place. The answer could be that they are important for spatially controlling the addition and integration of two separate cytoskeletons to one entity. Only in cases where the zygote does not immediately rely on internal spatial order is coordination during mating unnecessary. In *Bryopsis*, for instance, postfusion processes lead to a degradation of both male and female flagellar rootlet systems, concurrent with a rounding up and settling of the cell. In other cases, however, where the zygote remains motile and usually behaves in a negatively phototactic way, both rootlet skeletons need to become integrated in a highly controlled fashion so that functionality is maintained. In the latter case, a mating structure could provide spatial information during the mating process, by which the positioning of basal bodies and flagellar rootlet structures becomes predetermined. Unfortunately, very little ist known about the regulation of dynamic changes occurring in early postfusion zygotes.

b) Walled Cells and Giant Coenocytes

In cells preparing for the next cell division, interphase cells, or terminally differentiating cells, control over the position of organelles relative to each other is a vital prerequisite for optimal physiological performance and faithful expression of genetically programmed developmental patterns. With structural reinforcement by a wall, cells can become large, and with an increase in size the numbers of organelles necessarily increase, and this in turn has consequences on the design of the cytoskeleton. The following section will explore some of the strategies that have evolved in the green algae to control movement and positioning of organelles.

α) Motility Phenomena with Uncertain Mechanism

Intracellular movements have continued to attract the curiosity of cell researchers during the past decade. Consequently, the output of original publications as well as review articles has been comparatively high. The majority of this work has focused on the traditional model system such as the characean internodal cell and the giant coenocytic sipohonoclads and siphonous green algae, in which the motility mechanisms are either well understood or at least testable hypotheses have been developed (see below; for reviews, see Kamiya 1986; Williamson 1986, 1992; Tazawa and Shimmen 1987; Shimmen 1988, 1992; LaClaire 1989; Kuroda 1990; Wagner and Grolig 1992), but other intriguing motility phenomena have also been described which have remained largely unexplained.

Since Marchant (1977) rediscovered chloroplast rotation around the bases of setae (long stiff hair-like projections protruding from the dorsal cell face) in the charophycean alga *Coleochaete scutata*, no further attempts have been made to analyze the mechanism of this movement in detail. According to Marchant's description, a ring of MTs encircles the base of the seta and the chloroplast holds on to the MT ring by a finger-like extension called diverticulum. However, unexpectedly, colchicine did not stop chloroplast movement, although it did cause depolymerization of the MTs. Cytochalasin B was also not completely effective in stopping movement, so that there appears to be no clue as to the nature of this movement. The suggestion that chloroplast rotation in *Coleochaete* could be driven by a novel mechanism involving neither MT- nor actin-dependent motors, but could nevertheless be guided by MTs, provides a challenging model to reexamine. A similar phenomenon has not been reported in any other green alga.

Ca-dependent chloroplast translocation in the chlorococcalean green alga *Eremosphera* has also remained unresolved. The organism seems to be particularly well suited for experimentation, because the motility phenomenon can be triggered: chloroplasts become translocated to the cell center upon blue light illumination or after raising the intracellular Ca-concentration with the ionophore A 23187 (Weidinger and Ruppel 1985). Superficially, the phenomenon resembles light-induced aggregation of chloroplasts in the archegoniate water plant *Egeria densa* (Witztum and Parthasarathy 1985), which has been described as an actin-based phenomenon, but as yet there is no evidence for the presence of actin in *Eremosphera.*

β) Actin-Based Organelle Transport

In the plant kingdom, the majority of intracellular movements are based on actin, either as a structural polymer providing the tracks along which organelles become transported, or as networks that become relaxed or contracted providing less directional movement. This is not an exclusive phenomenon of the green algae (as discussed in this section), although some of the best studied examples of actin-based motility come from this group.

The attractiveness of *Nitella* as a model system for actin-based organelle transport is not at all surprising to anyone who has watched the restless large-scale rotational movement of the endoplasmic layer along the gently spiraling rows of stationary chloroplasts in the cortex. *Nitella* was the first green plant cell within which actin filaments were positively identified (Williamson 1974; Palevitz and Hepler 1975). It has since served as a successful model system for actin-based organelle motility. Manipulative techniques have been perfected on the giant characean internodes to a degree that just about everything seems to be possible. Cells can be cut open, perfused, permeabilized, and reactivated. Foreign organelles and beads with myosin attached to them can be introduced and reactivated. The endoplasm can be extruded and manipulated in vitro to display a variety of actin-based phenomena (for an overview of all these techniques, see Shimen 1992 and Higashi-Fujime 1991). In the latest advancement, Wasteneys et al. (1993) have been able to microinject fluorochromed probes and proteins into *Nitella* to study cytoskeletal dynamics in vivo.

Now that the basics of the mechanism, i.e. organelles driven along actin bundels by myosin-like motors, have been revealed, the improved spectrum of experimental techniques will help to meet the current challenges in understanding the regulation of intracellular movements. It is already beginning to emerge that calcium ions and subsidiary signaling steps such as protein phosphorylation lie at the center of the regulatory events (Williamson 1992), but we are still far from understanding these events in detail. One step in this direction has been taken by McCurdy and Harmon (1992) on the identification of calcium-regulated protein kinase in *Nitella* and its immunofluorescence localization at the actin bundles. The labeling pattern suggests that the kinase binds to myosin, and hence is a candidate for regulating myosin function, possibly in conjunction with a matching protein phosphatase acting on the same target epitopes.

Another problem, relevant to the much more dynamic actin bundle system in other algal systems (see below) and higher plants, concerns the processes that cause actin filaments to assemble to higher order structures. In *Nitella*, the parallel bands of actin bundles running alongside the chloroplast rows are exceedingly stable, but since they stretch along the entire length of the cell it is obvious that they must keep pace with cell elongation. They apparently also have the capacity to regenerate after being severed, as elegantly demonstrated by Williamsom et al. (1984; Williamson and Hurley 1986).

Analysis of the cytoskeleton and organelle motility in *Nitella* has begun to resolve to the molecular level. A library of monoclonal antibodies has been produced against various components of the detergent-extracted cytoskeleton (Williamson et al. 1986) and actin has been isolated and characterized by one- and two-dimensional electrophoretic analysis (Williamson et al. 1987). This latter study identified a single cytoplasmic actin and a second polypeptide of 58 kDa which also cross-reacted with the same actin antibody, but which was surprisingly localized in the chloroplasts. An actin-related protein has also been detected in pea chloroplasts (McCurdy and Williamson 1987). The significance of this finding is not at all clear. There have been many reports of actin-like proteins in prokaryotes (see Barnett and Cunningham 1992), and hence there is a possibility that similar proteins might occur in an organelle. This would explain its phylogenetic origin but not its function.

The dasyclad green alga *Acetabularia* is one of the best-known experimental model systems for examining dynamic rearrangements of the actin bundle system in a background of diverse organelle behavior. Many investigators have described the lively multistriate streaming in this alga (see Kamiya 1986), but it was not until Koop and Kiermayer (1980a, b) provided microcinematographic documentation of the organelle motility in *Acetabularia* that the flexibility and manifold variations of the motility mechanism existing in this system were appreciated. Since it has become possible to visualize the cytoskeleton in this alga by immunofluorescence, we know that the parietal layer of cytoplasm is interlaced with a network of roughly longitudinally oriented actin bundles of various thickness (Menzel 1986). Although the shear mass of the actin cytoskeleton is certainly comparable to that in *Nitella*, the order of the bundles ist not. In contrast to the usual type of actin-based organelle movement, chloroplasts in *Acetabularia* remain tied together by flexible as well as contractile links and move along in more or less orderly rows. Unfortunately, the cytoskeletal basis for these links has not been clarified. Other organelles, however, move individually at speeds typical for each type of organelle. The most enigmatic motility type in *Acetabularia* is the headed streaming band, so named by Koop (Koop and Kiermayer 1980b) because it consists of a flexible, structureless cytoplasmic sack that sits in front of a striated cytoplasmic band. The whole structure moves like a miniature roller coaster with speeds of ca. 15 µm/s through the cytoplasm. The head structure labels heavily with actin antibodies and the posterior band is nothing more than a particularly wide actin bundle (Menzel and Elsner-Menzel 1989a). It is not yet understood what mechanisms power the head and there is no clear concept to explain the behavior of the chloroplasts. It is, however, clear that all motility phenomena encountered in *Acetabularia* must be based on actin as the sole structural cytoskeletal component and that regulatory components must be sought among the actin-binding proteins or the elements acting on this class of proteins (Menzel et al. 1992). The more conventional type of actin-based motility, typical for higher plant cells, can, for instance, be observed in the conjugatophyte *Spirogyra* (Grolig 1990).

Rotational reorientation movement of the plate-like chloroplasts in *Mougeotia* is also an actin-binding process, albeit an unusual one in which the mechanism relies on a physical connection between the margins of the chloroplast and the plasma membrane. This connection is thought to be provided by actin filaments which bind with their filament ends to actin-binding proteins distributed in the membrane by a a phytochrome-dependent gradient (a recent summary of this hypothesis is given in Grolig 1992; Wagner and Grolig 1992).

γ) Actin-Based Contraction and Wound Response in Giant Green Coenocytes

Cells can survice injury of their walls and plasma membrane, and even loss of cytoplasm, if a minimum of essential components is still available. This tolerance against mechanical deformation and impacts at the cellular level is the reason why cells tolerate micromechanical manipulations such as microinjection, protoplast formation, and the like. This property does not seem to have consequences for small cells, because recovery of a single small cell can have only a negligible effect on the survival of the entire large population of cells. The ability, however, to survive mechanical wounds becomes a major advantage if a cell is large. It is, therefore, conceivable, that the evolution of cytoskeleton-based wound closure mechanisms from an inherent tendency to close holes in the membrane and relocate or retract cytoplasm from affected cell regions has guaranteed survival of large-celled algae. In the marine Dasycladales, Caulerpales, and Siphonocladales, this machinery is based on the ability of actin networks in the cytoplasm to contract when cytoplasmic free calcium exceeds a threshold concentration. Since cellular injuries involve transient leaks in the plasma membrane, marine organism will experience an instant rise in internal calcium ion concentration upon wounding. Although there are clear differences in the details, the general principles of the wound responses are the same in the three orders of green algae mentioned above (Menzel and Elsner-Menzel 1989b; LaClaire 1992). In addition, the Dasycladales and Caulerpales rely on protection by an extracellular cap or plug that quickly forms by gelation of a vacuolar mixture of polysaccharides and glycoproteins extruded upon wounding (Menzel 1988). Giant-celled characean algae usually do not survive transversal cuts as the siphonous green algae do; however, for smaller punctation wounds they possess a similar mechanisms of plug formation (Foissner 1987).

δ) MT-Based Organelle Transport

Perhaps the most impressive type of long range MT-dependent organelle movements studied so far occurs in all members of the green algal order Caulerpales. Some of these giant coenocytes, such *as Caulerpa cactoides*, can have a stolon diameter of up to 1 cm and reach many meters in length. Organelles, except for the majority of chloroplasts, are transported along meandering tracks through the stolon and up into the assimilatory blades visible in the binocular at low magnification. Although organelle traffic occurs in all directions, transport along individual tracks is unidirectional. The tracks have massive MT bundles at their center, from which smaller bundles branch off and run up into the cortical layer, where chloroplasts remain localized and are seen as long and wavy chains (Menzel 1987). They can only be dislodged by stress or injury (Menzel and Elsner-Menzel 1989c). In all the caulerpales such as *Caulerpa* (Manabe and Kuroda 1984), *Halimeda* (Drew and Abel 1990), *Chlorodesmis* (Menzel 1985), and even the distantly related *Dichotomosiphon* (Maekawa and Nagai 1989), the mechanism of transport is based solely on MTs (Kuroda and Manabe 1984), most probably carried by a homologue of cytoplasmic dynein (Kuroda 1990), although further study is needed to substantiate this. The removal of chloroplasts from the cortex, however, is an actin-based process (Menzel and Elsner-Menzel 1989c).

ε) Nuclear Rotation in Constricted Desmids

The nucleus in the Desmidiaceae is known to perform several types of postmitotic movements (for a recent summary, see Meindl 1992a). In contrast to the conventional forward movement of the nucleus into the growing half-cell normally followed by a return towards the istmus, as, for instance, in *Micrasterias*, the nucleus in the species *Pleurenterium tumidum* passes through a gap between the lobes of the parietal chloroplast, attaches to the isthmus, which is considerably wider than in other constricted desmids, and begins to follow the rim of the isthmus in a slow rotational movement. Fine structural analysis of the isthmus region shows that the plasma membrane is drawn out into a fold in the plane of the isthmus so that an inwardly directed collar is formed. The collar lies deep in an invagination of the nuclear envelope. The isthmus band of MTs, which is also a typical feature in constricted desmids, lines the outer rim of the collar and is apparently involved in the motility mechanism, because depolymerization of MTs stops nuclear rotation. An additional striking element of nuclear behavior in *Pleurenterium* is that the direction of movement reverses several times in the course of some 12 to 70 turns, before the nucleus returns to its position at the center of the isthmus (Meindl 1986). Except for the developmentally timed occurrence of this movement, there is no apparent function associated with it, and there is no theory to describe the possible molecular motor that drives this MT-dependent bidirectional translocation. It may be noted that the current concept of molecular motors may require revision to accommodate bidirectional movements. In the freshwater amoeba *Reticulomyxa*, a bidirectional dynein-like motor has been discovered (Schliwa et al. 1991). It remains to be seen as to whether *Pleurenterium* provides another example.

ζ) Mixed Mechanism for Chloroplast Transport

In *Bryopsis*, yet another mechanism for translocation of chloroplasts appears to be operating because it requires MTs and actin filaments in equal parts. This conclusion is based on a careful comparison of motility tracks, motile behavior, and distribution of MTs and actin bundles under a variety of treatments (Menzel and Schliwa 1986a, b). This is an interesting model, and has prompted others to seek for mutualism between MTs and actin filaments elsewhere. Undoubtedly, the close alignments between MTs and actin filaments in the preprophase band (McCurdy and Gunning 1990) and the phragmoplast (Kakimoto and Shibaoka 1987) of higher plants, as well as in the cortical cytoskeleton (Kengen and Degraaf 1991; Lancelle and Hepler 1991) of a variety of organisms, indicate interdependence between the two cytoskeletal elements. This concept has recently been further corroborated by the finding that actin and MTs copurify from plant extracts (Abe et al. 1992). A molecular basis for such an interaction is only vaguely identifed in the actin-binding function of the MT binding protein MAP1 (Arakawa and Frieden 1984; Correas et al. 1990). Interestingly, movements of mitochondria in *Brypopsis* rely exclusively on MTs. This has been shown by employing a cytoplasmic exudate system which allows the relocation of a previously video-recorded motility track in a sample processed for immunofluorescence. In each of the recorded cases, the tracks coincided with MTs and movement of organelles along the MT tracks was not inhibited by cytochalasoin D (Menzel and Elsner-Menzel 1989d). Actual physical links between mitochondria and MTs have been described in the desmid *Micrasterias crux-melitensis* (Nakamura and Ueda 1982).

6. Cell Division

Cell elements such as the mitotic spindle and the cytokinetic apparatus serve indispensable functions for the cell's survival and the perpetuation of a species. Hence, it is reasonable to assume that they have been placed under constraints against evolutionary changes. It is, therefore, commonly agreed that their structural features provide useful supplementary information, and it has proved that by applying this information to existing schemes, a more realistic picture of phylogenetic relationships among the green algae has been achived. A great number of original publications have been dedicated to this subject. These will not be discussed in detail here, since they have recently been excellently reviewed (van den Hoek et al. 1988; Sluiman et al. 1989; Margulis et al. 1990).

Much less attention has been given to the study of the actual mechansims underlying these processes. Some new structures that have been identified or suggested as important elements will be briefly listed below.

In phytoflagellates, several additional features of cytokinesis have been revealed since the concept of phycoplast has been introduced (see Pickett-Heaps 1975 for review). In *Brachiomonas*, a particularly prominent role of the four-membered MT rootlets in mitotic segregation of basal bodies is suggested (Segaar and Gerritsen 1989). According to Segaar's scheme, both of the four-membered MT rootlets come to lie next ot each other in an antiparallel orientation after basal bodies have duplicated. Detailed fine structural serial section analysis further suggests that they slide past each other in opposite directions so that the two basal body complexes become pushed apart, and situated at the two opposite spindle poles. At mid-metaphase, the four-membered roots have slid apart and sit astride the anterior rim of the cell. By this time, the cytoplasmic MTs have disappeared and a new set of MTs is initiated from the four-membered roots and is sent downwards along both flanks of the cell in arrays following along the cleavage furrow. It is thus becoming clear that flagellar rootlet complexes have at least two more functions besides (1) anchoring the basal body in interphase cell, (2) spatial control of organelle positioning, and (3) shock absorbance during flagellar motility, these being active MT-sliding and acting as MTOCs for the cytokinetic MTs.

In coccoid algae, a number of variations and special features have been described, including those that have already been employed for classification (van den Hoek et al. 1988). One of the sparsest cytokinetic mechanisms so far encountered in a green alga, in terms of the participating structural components, operates in the chlorococcan microalga *Marvania geminata* (Sluiman and Reymond 1987). Before cytokinesis starts, the cell elongates on one side in a bud-like fashion, breaking through the old wall and covering the bud with a new wall. The position of the prospective cleavage site is predetermined by the neck of the bud, where the old cell wall ends. This is followed by division of the single dictyosome and mitochondrium and positioning of both pairs of organelles at the opposite sides of the cell. In order to accomplish cytokinesis, the alga requires just a single cytoplasmic depression (not furrow!) next to the MT. The depression ends in a membrane pit with a diameter of no more than 50 nm. Both the MT and the pit sink side by side inwards into the cell, impinge on the acentrically dividing nucleus whose intranuclear spindle also gets by with just a few MTs, succeed in getting past the dividing chloroplast and eventually through to the other side of the cell. During this process, exocytosis of what could be mucilaginous material into the cleavage invagination might help to build up some colloidal pressure from the outside to keep the newly formed membranes separated.

It is, of course, obvious that from this kind of descriptive data the actual mechanism of cytokinesis in *Marvania* cannot yet be extrapolated; however, it is also quite clear that this most amazing reduction in the number and elaboration of essential functional elements poses an ideal model for further studies on the role of each of the individual elements in cytokinesis.

Aphanochaete magna has been examined by indirect immunofluorescence and conventional transmission electron microscopy to determine the type of mitosis and cytokinesis in the chaetophoralean green algae (Segaar and Lokhorst 1988; Segaar 1989a). Mitosis is of the closed intranuclear type but at telophase the long interzonal spindle body collapses and as the daughter nuclei approach each other again, a star-like MT system develops between them and aids in cell plate formation. A similar cytokinetic MT system is also found in *Cylindrocapsa* (Sluiman 1985) an *Fritschiella* (Segaar 1989b); in the latter case it has been argued that the center of the star-like system could be considered an MTOC.

Segaar and Lokhorst (1987a) focus on the cytokinetic MTs in *Ulotrix palusalsa*, which they detect at the leading edge of the cleavage furrow by conventional TEM and alongside the entire furrow by immunofluorescence. Since the latter images did not correspond to the TEM images, they suggest that the fluorescent material represents stores of tubulin subunits. They also discuss whether the term phycoplast should be maintained (Segaar and Lokhorst 1987b).

Mitosis and cytokinesis in the Zygnematales have been examined at the fine structural (Bakker and Lokhorst 1987) and, more recently, the light microscope levels using indirect immunofluorescence procedures (Galway and Hardham 1991). Both studies have confirmed and extended earlier observations by Pickett-Heaps and Fowke (1970), including the existence of a cleavage furrow intersecting the persistent telophase spindle and postmitotic migration of daughter nuclei with the aid of a pointed cone-like MT array.

These features and that of vesicle guidance at telophase have recently been discussed in detail by Grolig (1992). Usually, if cleavage is said to be accomplished by a furrow, it is actually not clear whether the furrow is caused by a constricting ring or by localized membrane growth guided by other mechanisms. This problem seems to have been resolved in the Zygnematales, since a constricting ring built from actin filaments has been conclusively demonstrated (Goto and Ueda 1988).

Considering other types of cytokinesis, for instance those involving phycoplast MTs as guiding elements, it may prove that even in these cases actin plays a role. A recent immunofluorescent study on *Chlamydomonas* demonstrated a clear labeling of the advancing cleavage furrow with an antibody against actin (Harper et al. 1992). Multiple cleavage of cyst domains in *Acetabularia* also involves both MTs and actin (Menzel 1986).

Mitosis and cytokinesis in *Urospora* (Acrosiphonales) are unique, because nuclei become presorted and positioned in the equatorial plane of the large multinucleate cells before mitosis commences synchronously. The cleavage furrow that cuts through the two rows of recently divided nuclei seems to be guided by a circumferentially oriented MT system positioned in front of the furrow; but neither this configuration nor any of the phenomena involved in the preceding steps, such as nuclear movement and positioning, are understood in any detail (Lokhorst and Star 1983).

A cytoplasmic septum forms in the filamentous green alga *Oedogonium* between the recently divided nuclei, and extends across the central vacuole. Its growth is exclusively aided by a sheath of MTs which also impinges on the parietal chloroplasts prompting them to divide. Wall precursor vesicles become transported into the cytoplasmic septum, possibly guided by the sheath of MTs, and fuse to form a separating cross wall. Whereas growth of the septum and division of the chloroplasts are blocked by the MT-inhibitor vinblastine, vesicle fusion is blocked by cytochalasin B, suggesting that the final step in cross-wall formation is an actin-dependent step (Nakagawa et al. 1990).

The terrestrial, packet-forming green alga *Chlorokybus atmophyticus* (Charophyceae) is remarkable in several ways (Lokhorst et al. 1988). Although its zoospore ultrastructure and some features of cytokinesis and mitosis are typically charophycean, other details are reminiscent of ulvophycean traits. Particularly striking are the extra pyrenoid that seems to aid in chloroplast division, and the prominent peroxisome that is associated with the centrioles and the nuclear envelope in interphase and mitosis. Since centrioles duplicate and migrate to opposite cell poles in prophase, the microbody becomes stretched along the entire length of the mitotic spindle, to which it remains laterally associated before it is split by the incipient cleavage furrow. The functional significance of peroxisomal behavior is not known, and remains unclear also in other cases where a close association to the spindle or spindle poles has been observed, such as in taxa of the Prasinophyceae and Ulvophyceae. Lokhorst and coworkers (1988) note that other features of the cleavage process are certainly not charophycean. A furrow forms early during prophase as a shallow depression on the side of the cell opposite the position of the chloroplasts, but begins to advance only at late telophase aided by a transverse MT system. The authors suggest that transverse MT systems could be a more general feature of cytokinesis than presently acknowledged, but might be more labile than other MT systems so that they are only preserved by rapid freezing fixation, as in this study. If this holds true, the term phycoplast would require redefinition and the presence of a phycoplast-like structure might have a lower significance in terms of algal classification than originally thought.

A phragmoplast has been discovered in the endophytic green alga *Cephaleuros parasiticus* (Trentepoliophyceae) that is referred to as "higher plant-like", indicating a close alliance of the Trentepoliophyceae with the Charophyceae (Chapman and Henk 1986). However, the authors express concern that the Trentpoliophycea could withstand classification by classic features such as phragmoplast versus phycoplast and multilayered structure of the zoospores versus cruciate rootlet system, because both features appear to be anomalous in this species. As far as the phragmoplast is concerned, it is noted that it is an unusually dense structure stretching up to the telophase nuclei; in fact, it is clearly a remnant of the telophase spindle. In higher plants, the phragmoplast is well separated from daughter nuclei and it is not suspected of being a remnant of the telophase spindle, but a spearate MT structure that could assembles de novo (Lambert 1993). Recent studies employing molecular phylogenetic tools such as 18S ribosomal RNA gene sequencing have clearly shown that the Trentepoliophyceae are closer allies of the Ulvophyceae than of the Charophyceae (Zechman et al. 1990).

7. Cell Wall

Cell coverings of green algae range from spurious amounts of glycoprotein or polysaccharides spread over the cell surface, a condition usually referred to as naked, to massive multilayered walls impregnated and coated with phenolic material or covered with thick layers of mucilage. Many walls bear elaborate reinforcements, ornamentations, and appendages. Chemical composition and structure of algal walls have been extensively studied during the past decade. Reviews summarizing this field include Roberts et al. (1982), Kristen (1985), Brown (195), Kloareg and Quatrano (1988), Neville (1988). The property of the wall has an immense influence on cell shape, but it almost never provides the original architectural blueprint. As discussed below, this function is exerted solely by the protoplast, either by providing the case upon which the wall is molded or by locally modifying the existing wall. Current interest in wall morphogenesis centers on the question of how the cell is able to direct wall precursor material to specific locations on the cell surface, how this material assembles and becomes secondarily processed in subsequent stages of cell development, and how polymer synthesis can be specifically oriented.

a) Scales and Thecae

Organic scales are a typical feature of prasinophyte flagellates but are also found in zooids and gametes of the charophycean and ulvophycean lineages, whose vegetative thallic cells form regular cellulosic or mixed polysaccharide cell walls. Scales occur in countless shapes and sizes, often half a dozen different types on the same species and often separable in two groups, flagellar scales and body scales. They may be rectangular, polygonal, disk- or hair-like, or even dorsoventrally differentiated, featuring a base plate and a more or less delicate crown made of reticulate threads and appendages. The dimensions of scales are usually within the nanometer range, but some, such as the elaborate three-dimensional basket scales of *Mesostigma viride*, fall just within the range of light microscope resolution (Domozych et al. 1991). Each different type of scale produced by one cell somehow finds its way to predestined areas of the cell surface and into specific layers as subunit component of the thecal casing or the flagellar covering (see Sect 2.b).

Since thecal casings are shed from parental cells in the course of cell division, they can be conveniently harvested, allowing determination of their chemical composition (Becker et al. 1989). Likewise, flagellar scales can be obtained in large quantities by induction of flagellar shedding (Becker et al. 1990). Scales are also typically found in chrysophycean algae, but in contrast to these, the prasinophyte scales do not possess cellulose but a polymeric substance with an unusual carbohydrate composition (for recent review, see Melkonian et al. 1991b) The major components, up to 80% of the carbohydrate fraction, are 2-keto-sugar acids, otherwise unknown in algae (Becker et al. 1991), but more frequently encountered in higher plants (Stevenson et al. 1988) and bacterial walls (Lindberg 1990). The polymer status of the 2-keto-sugar acids is not yet determined and the reason for their preferred occurrence in scales is not known. In some of the more elaborate types of scales, such as in the basket scales in *Mesostigma*, the organic matrix is mineralized with calcium phosphate and trace amounts of sulfur (Domozych et al. 1991). The existence of a mineral type containing phosphate would be quite intriguing, since most mineral deposits in green algal walls consist of calcium carbonate. The way in which the unusual sugar acid and mineralization influence scale shaping and assembly properties remains to be elucidated. Scales also contain glycoproteins, the role of which, at least of the flagellar underlayer scales, seems to be a little clearer. Several glycoprotein species are present predominantly in the underlayer scales, e.g., in the pentagonal scales of *Tetraselmis striata*, presumably mediating contact to the plasma membrane surface (Becker et al. 1990; Becker and Melkonian 1992). Other scales, such as the flagellar hair scales, apparently insert directly in the membrane (Melkonian 1982). Upon excretion, scales become rapidly incorporated into the thecal casing by self-assembly processes, and it has been suggested that an excess of free 2-keto-sugar acids and galacturonic acid monomers, presumably exocytosed alongside with prefabricated scales act as a stabilizing cement (Becker et al. 1991) and calcium could play a role as complexing agent (Melkonian 1982; Melkonian et al. 1991b). Recently, polyclonal antibodies have been raised against isolated *Mesostigma* basket scales and used to stain the theca by indirect immunofluorescence (Domozych et al. (1991). Individual scales have been clearly identified by confocal microscopy so that it should soon be possible to directly study the scale release mechanism and thecal assembly.

b) Glycoprotein Walls in the Chlamydophyceae

The walls of *Chlamydomonas* and other volvocalean green algae are also excellent examples for the formation of intricate three-dimensional structures by self-assembly of its subunit components. It has long been known from electron microscope studies including conventional TEM, freeze fracture, negative staining, and metal shadowing techniques that the *Chlamydomonas* wall consists of several paracrystalline layers (Roberts et al. 1982). Biochemical analysis has revealed that the major macromolecular components of the wall are glycoproteins with strong self-association properties (Roberts et al. 1985).

Goodenough and Heuser (1985a) have demonstrated that the quick-freeze deep-edge technique is a far superior EM technique, providing detailed information about the three-dimensional architecture of the wall layers of cells subjected to various experimental conditions. By a combination of perchlorate extraction and autolysin digestion, they were able to extract the major glycoproteins and reassemble selected combinations of them by dialysis back into paracrystalline sheets. Comparing the paracrystalline patterns obtained from those combinations, they were able to identify the glycoprotein complement typical for each of the seven wall layers occurring in *Chlamydomonas* and in *Volvox* (Monk et al. 1983; Goodenough et al. 1986; Goodenough and Heuser 1988a, b). As additional reinforcement, the organic matrix, may become encrusted with mineral such as calcite in the Phacotaceae (Gierig et al. 1990), and colonial Volvocales can possess considerable amounts of sulfated polysaccharides in the extracellular matrix (Crayton 1982).

c) Noncellulosic and Mixed Polysaccharide Walls

The overall property of the wall is determined by the nature of the structural polymer and by the mixture of branched, heterogeneous matrix polysaccharides alongside other minor components such as glycoproteins and secondary metabolites (for review, see Kloareg and Quatrano 1988). Instead of cellulose, $1 \rightarrow 4$-linked linear homopolymers of β-D-mannose can be a prominent if not the sole linear homopolymer serving as the skeletal framework in some groups of algae. Often a fraction of nonfibrillar mannan is found in the matrix of such walls as well, but more frequently β $1 \rightarrow 3$ glucans serve as matrix material. Examples of mannan walls occur in members of the siphonous green algae Dasycladales (Chanzy et al. 1984). In other siphonous green algae such as the Caulerpales, xylan is the dominant skeletal polymer (Fukushi et al. 1988). Structurally, the microfibrils of mannans and cellulose (β $1 \rightarrow 4$ glucans) form flat ribbons upon crystallization, whereas xylans form single or triple helices in the higher-order crystalline conformation (Saito et al. 1991). Interestingly, within the siphonous green algae, the same species may hold the genetic capacity to encode the synthetic machineries for two or even more types of linear biopolymeres in its genome, sometimes coexistent within the same wall, but often expressed sequentially in different stages of the life history. In the *Derbesia/Bryopsis* group (Caulerpales), the sporophytic stage features an almost exclusive mannan wall, whereas the gametophytes known as *Halicystis* possess cellulose and some xylan (Huizing et al. 1979). More recent studies have shown that even those species that do not have a *Halicystis* phase, such as *Bryopsis maxima*, have cellulose in their gametophytic stage although as a minor component and in a noncrystalline conformation, the major component being xylan (Maeda et al. 1990). In the genus *Acetabularia*, the stage-specific difference is similarly pronounced. Here, the sporophytic phase presents mannan and galactan walls, whereas the gametophytic phase, usually referred to as cysts, has an almost pure, highly cystalline cellulosic wall (Nisizawa et al. 1974). In contrast, the coenocytic, multichambered Siphonocladales, which in many textbooks are considered close relatives of the siphonous green algae, have classic cellulosic cell walls found in the Charophyceae and higher plants. The rare coexistence of chitin with cellulose has been described for *Pithophora oedogonia* (Pearlmutter and Lembi 1980). Chitin remains restricted to the cross wall layers that are the first to be deposited after cytokinesis, whereas subsequent cross wall layers and the rest of the cell wall is composed of cellulose.

d) Cellulosic Walls

With the exception of the Prasinophyceae, Chlamydophyceae, some Chlorophyceae, and the siphonous species of the Ulvophyceae, most other green algae bear cellulosic walls ranging from noncrystalline to crystalline forms, highly ordered in crossed fibrillar layers such as those found in the Siphonocladales, Cladophorales, and some Oedogoniales and Chlorococcales. The matrix material in this case is made of β 1 $\rightarrow$ 3 glucan and a variable fraction of highly branched sulfated heteropolysaccharides (for review, see Kristen 1985, Kloareg and Quatrano 1988).

α) Terminal Cellulose Synthetase Complexes (TCs)

The concept that the cellulose microfibril is produced by a transmembrane multisubunit synthetase complex that is fed with UDP glucose from the cytoplasmic face of the plasma membrane releasing the polymer on the external face has become widely accepted since it was first proposed by Roelofson in 1958. By the crystallization process, in the course of which a variable number of single elementary polymer fibers are transformed into a microfibril, the synthesizing complex is propelled forward in the plane of the plasma membrane so that the cell becomes wrapped in a meshwork of such fibrils (Delmer 1987; Preston 1988). The supramolecular appearance of TCs may differ. Rosettes are found in the plasma membrane of the Conjugatophyceae and Charophyceae either as single units during primary wall formation or as hexagonal arrays during secondary wall formation (summary in Emons 1991). Linear TCs, on the other hand, have been discovered in Chlorococcales, Eremosphera, Siphonocladales, and Glaucocystis (summary in Quader 1991). Since the initial report by Brown and Montezinos (1976), much of what we have learned about microfibril orientation and TC fine structure stems from freeze fracture studies (reviewed in Brown 1985) and other EM techniques involving metal shadowing (i.e., Mizuta and Wada 1981). Only recently, Kudlicka et al. (1987) have also been able to depict TCs in ultrathin sectioned material demonstrating that the TC is not just a transmembrane complex but, in fact, protrudes into the cytoplasm for more than one span of the membrane.

Rosettes are probably not directly assembled in the plasmamembrane. It was first discovered by Kiermayer in *Micrasterias* (summarized in Kiermayer 1981) and has subsequently been shown by several others that rosettes occur embedded in the membranes of Golgi cisternae and trans-Golgi vesicles (Noguchi and Ueda 1985; Haigler and Brown 1986 for higher plants), suggesting that the enzyme subunits enter the GA via transit vesicles from the ER, become processed and assembled within the GA and reach the plasma membrane through exocytosis of trans-Golgi vesicles. In the moss *Funaria*, rosettes have even been sighted in the ER membrane (Rudolph 1987). It also seems to be of relevance in this context to note that some organisms do produce cellulose in the GA (*Pleurochrysis*, Chrysophyceae, see for instance Herth et al., 1975), although this has not been shown for the green algae. Though the general pathway should be the same, linear TCs may not be preassembled to the mature complex in the GA but arrive at the plasma membrane as subunits and assemble there (Itoh and Brown 1988).

The turnover of cellulose synthesizing complexes is not known for green algae, but from calculations in *Funaria* (Schnepf et al. 1985), it is reasoned to be in the order of minutes, which might also be a realistic assumption for green algae.

Despite intense research on the cellulose deposition, two key questions have remained unresolved: what is the biochemical identity of the cellulose synthesizing enzyme, and how is the orientation of microfibrils achieved?

β) Cellulose Microfibril Orientation

When it was discovered that microfibrils and microtubules (MTs) often run parallel to each other and that anti-MT drugs cause the cells to lose their shape, the idea was conceived that MTs control the orientation of microfibril deposition (Heath and Seagull 1982; Robinson and Quader 1982; Brown 1985; Preston 1988). One early model assumed that the orientation of movement of the cellulose synthase complexes was directly guided by MTs via bridges (MT-dependent motors) between the TCs and the MTs. Another, more popular model stated that adjacent MTs cross-linked to the membrane provided the boundaries within which cellulose synthase complexes could be restricted in their freedom of movement in the plane of the membrane; the latter could be circumscribed as the fluid channel model. However, neither model appears completely satisfactory because they do not appear to accommodate all the observed phenomena. For instance, there are too many cases where MTs do not run parallel to recently deposited microfibrils (for reevaluation, see Emons et al. 1992).

These inconsistencies demand an explanation. One possibility may be that more than one mechanism for the control of microfibril aligment has evolved, and those cases where parallelism does exist would indeed fit the current model. In the desmid *Closterium*, for instance, Giddings and Staehelin (1988) came to the conclusion that their observations of the plasma membrane fracture faces were perfectly consistant with the fluid channel model but not with direct guidance as far as primary wall formation is concerned. On the other hand, during secondary wall formation when rosettes form larger hexagonal arrays whose synthetic products are flat parallel ribbons of cellulose rather than single fibrils, there was absolutely no correlation with MTs, and there was no preferred orientation of the ribbons either. This exemplifies that some stages do and others do not require spatial control of cell wall deposition.

The chlorococcalean alga *Oocystis*, with its crossed fibrillar layers, has been advocated as a prime example where strict parallelism exists and where destruction of MTs abolishes the regular alternations of orientation from one layer to the next (Robinson and Quade 1982). Mizuta (1992) has compared the situation in *Oocystis* with that in the giant Siphonocladales and came to the conclusion that a different mechanism must be active in the latter. Here, the MTs are oriented axially, whereas the orientation of microfibrils switches regularly between layers to form a more elaborate pattern of crossed fibrillar texture than that seen in *Oocystis*. Nevertheless, MTs are still important in this system, because their depolymerization causes an interruption of the regular sequence of crossed fibrillar layers and rounding up of cells. This dependence of cell shape on MTs is seen in various green algal groups including Chladophorales (Mizuta et al. 1989; Nishioka et al. 1990), Conjugales (Hogetsu 1992), and Charales (Wasteneys 1992).

For those systems which lack a general parallelism between cellulose fibrils and MTs, it would be a plausible explanation if MTs only governed that part of the alignment mechanism that achieves the switching of orientation rather than providing physical guidance all the way through. If there is a structural basis for the switching mechanism, rapid freezing should be sufficiently fast to capture this event. In *Oocystis solitaria*, each wall layer takes about 1 h to be synthesized, and upon completion of the currently deposited layer a new complement of TCs is inserted in the membrane and aligned roughly perpendicular to the previous orientation before new synthesis commences (Robinson and Quader 1982). Unfortunately, this process is not synchronized in culture, so that each cell when fixed or frozen will be captured in its individual cycle. Quader (1986) exposed the cells to a pulse of Congo Red, which structurally alters the "old" TCs so that they can be distinguished from the new ones. This made it possible to tell that after treatment with the anti-MT herbicide APM, the new ones were not oriented regularly as in untreated cells. This implies that TCs become aligned before they actually start moving. Quader suggested, in the sense of the fluid channel model, that they still require guidance as they move. Unfortunately, the orientation of MTs was not shown simultaneously in this study, so that some doubt remains as to the actual causes of the misalignment after APM treatment.

In the siphonoclad system, Mizuta and collaborators have been able to routinely visualize both microfibrils and MTs in metal shadowed thin sections (see Mizuta and Okuda 1987). Recently, Okuda et al. (1990) have examined the fine structure of the polylamellated cell wall texture in the tip of *Chamaedoris orientalis*, and confirmed their previous assumptions that new wall layers are initiated from the tip and that their deposition runs basipetally in a sequentially repeating order of orientations: meridian, right-handed helix, and left-handed helix, meridian, etc. They further suggest that only the meridian layer might be spatially controlled by MTs, because only this layer runs parallel to the MTs. However, whether this applies to the apex region as well needs to be determined. Judging from the few immunofluorescence images published so far, the orientation of MTs in the apex appears to be random (Okuda et al. 1990). One possibility that needs to be explored in the future is whether it would be sufficient for maintaining a given orientation of the moving front of TCs on its way down to the base, if they were just pointed into that direction at the moment of their release from the tip, or if they need to be guided all the way down. Okuda et al. (1990) suggest the operation of an as yet undiscovered vectorial component. As far as the helicoidal patterns are concerned, it has been argued that these do not require a special guiding mechanism once the sign of the helix is determined (for review, see Neville 1988).

The idea that TCs receive directional information by initial alignment with an MT only, easily accommodates those cells where overall parallelism exists but MTs remain relatively short as opposed to the Siphonocladales. The *Nitella* internode would be an example (Wasteneys 1992). If one would then make the additional assumption that the TCs are relatively short-lived, so that they would be disengaged after, say, one full turn around the cell circumference, there would be statistically significant numbers of microfibrils running in an overall parallel course to whatever the direction of the MTs is. It is indeed striking to observe that microfibril imprints in siphonocladalean walls can be visually followed over exceedingly long distances, whereas those in *Nitella, Closterium* primary walls, or higher-plant primary walls quickly become obscure. This could simply reflect a much greater number of starts of new fibrils and hence a larger degree of tangential crossings. This model neither requires that MTs are long and straight nor is it necessary that they are cross-linked to the membrane throughout their entire length, and last but not least, they do not have to be present all the time.

In the effort to understand the mechanism that controls TC orientation, Legge and Brown (1988) have explored the influence of the physical status, i.e., viscosity of the membrane on the proposed movements of TCs. They used the chemical cholesteryl hemisulfate in the *Boergesenia* protoplast system to perturb membrane fluidity. The results indicate that membrane fluidity was indeed reduced leading to patchwork deposition of cellulose rather than regular helical arrays as assayed by calcofluor staining and freeze fracture analysis. They suggest that reduced mobility of the synthesizing complexes was responsible for the patchwork pattern.

Cellulose is a commercially important polymer for biotechnology and paper industry and its production in vitro could alleviate current exploitation of natural resources. However, as Delmer pointed out in her review on cellulose synthesis (1987), it has been impossible so far to preserve the plant TC in a native functional state in fractionated cell extracts. One of the problems seems to be that the supramolecular assembly of subunits is a rather labile state and that proper embedment in a membrane is a prerequisite for the TC to remain functional. While recombinant DNA technology could eventually help to reveal the subunit structure of the TC, cell biological approaches could help to approach the requirements necessary for assembly and interaction with the membrane. As a promising first step in this direction, Mizuta and Harada (1991) have been able to work out conditions that preserve membrane sheets derived from *Boergesenia* protoplasts and maintain embedded TCs in an operational state.

e) Wall Appendages

When the idea of a membrane-resident cellulose synthetase was first conceived, a theory was put forward which required a stationary lattice of synthetase complexes for production and orientation of microfibrils ("ordered granule hypothesis", see Preston 1988 for historical background). This theory was dismissed soon after it was formulated because it was not globally applicable to cellulose wall formation, and has long been replaced by the "mobile terminal complex model" as discussed above. However, it would be worthwile to reexamine some of the cell wall appendages seen in the Chlorococcales, because these structures seem to be formed by a mechanism of cellulose microfibril alignment that is reminiscent to the ordered granule hypothesis. An overview of this group of algae with their variously shaped appendages has been given by Hegewald and Schnepf (1984). Additional taxa bearing amazingly eccentric wall structures have been described by Reymond (1985; Reymond and Hegewald 1988, 1990). These include the "horse tail" appendage of *Ankyra inermis*, the "double leaf" structure of *Ankyra ancora*, and a delicate cage-like structure made up of an assemblage of interconnected radial threads in *Desmatractum indutum*. From a detailed developmental study on the cellulose spikes of *Acanthosphaera zachariasi* by Schnepf and coworkers (1982), we know more about the formation of these wall structures. According to this study, *A. zachariasi* possesses several pit fields in the plasmamembrane within which TCs are anchored, one per pit. A cellulose fiber emerges from each pit, points away from the cell, and merges at a distance with other fibers into a broad crystalline bundle, many microns in length. Only after these spikes are completed does the pit field disappear and the cell forms a regular wall.

Two more variations of the same theme can be discovered in Reymond's fine structural description of *Treubaria setigera* (Chlorococcales, Reymond 1985). Firstly, instead of a pit field, a circular electron-dense collar sits in the plane of the membrane at the base of each of the three cone-shaped appendages that are formed by this organism. The protoplast protrudes through the collar into the base of the cones during the period of cellulose synthesis and retracts when the cones have reached their final length. Secondly, in young cells, the primordial cones lie side by side, but at a later stage begin to move apart and assume equidistant position while a meshwork of cellulose microfibers is formed between the cones. This peculiar behavior suggests that in the concluding phase of wall formation the orientation of the putative TCs of the collar is altered to such an extent that the newly formed fibrils now move along the surface of the cell until they interdigitate with fibers from the neighboring collars. This could provide the force by which the cones are pushed apart. It may be noted that these interpretations are still highly speculative and require more detailed study especially on the nature of the collar region, but it is quite obvious that chlorococcalean taxa display an intriguing evolutionary variation in the theme of cellulose fiber orientation. Studying these variations could significantly contribute to our understanding of TC function and orientation.

Appendages of a proteinaceous nature have also been described in the Chlorococcales, notably in the genera *Micractinium, Pediastrum, Polyedriopsis,* and *Scenedesmus* (Schnepf 1983), which superficially resemble the cellulosic spikes. The exact mode of formation of these appenages is not yet known, but apparently the protein precursors are excreted through the wall and assemble on the surface into long paracrystalline spikes.

Long, straight, or curled cell wall appendages termed "cirri" have been discovered in ornamented zygote walls of several sphaeropleacean species (Hoffman and Buchheim 1989). Cirri are composed of organic material, but neither their chemical nature, nor their mode of formation are known. These structures apparently have no precedents although zygotes with similarly thick and ornamented walls have been described from other green algae such as Oedogoniales, Zygnematales, and Charales.

8. Morphogenesis

Shaping of a cell could possibly involve four hierarchical levels of cell activity: (1) active shaping of the protoplast, (2) spatially controlled deposition of wall components, (3) localized or general processing and modification of the wall to accommodate further shape changes and (4) permanent fixation of cell shape by structural reinforcement of the wall (covalent cross-linking and excretion of cementing matrix components). It is important to note that, although each phase builds upon the previous one, the succession of phases may not need to be completed, depending on the developmental stage of the cell as well as on the evolutionary advancement of the organism. It must also not be forgotten that as soon as a wall is laid down (stage 3), the cell can utilize hydrostatic pressure as a general driving force for expansion (discussed in Harold 1990 and see Sect. 8. c below).

We are presently a long way from understanding any of these processes in detail. There is ample evidence that MTs have a role in these processes. Some of this evidence has already been discussed in the context of recent work on TC orientation (see Sect. 7.d. β) and more recent evidence will be discussed below. More attention should now be paid to structural links between the cytoskeleton and the membrane. The following examples present some of the recent progress that has been made in the field of green algal morphogenesis.

a) Morphogenesis by Shaping the Protoplast

It has long been known that some of the naked phytoflagellates display asymmetrical cell shapes which need to be actively maintained, or that shaping precedes the deposition of a wall (see, for instance, Pickett-Heaps 1975). Although it is likely that MTs must have an instrumental function in cell shaping, as was suggested early on (Marchant 1982), detailed studies focusing on the underlying mechanisms have not been carried out. Perhaps the best-characterized MT cytoskeleton in a naked flagellate is that of the crescent-shaped, more or less twisted chlamydophycean alga *Spermatozopis*. In the quadroflagellate species *S. exsultans*, a secondary cytoskeletal MT (SCMT) system was seen as two parallel arrays running underneath the plasma membrane along both flanks of the twisted cell (Melkonian et al. 1987). In the biflagellate species *S. similis*, both the flagellar rootlet MTs and the SCMTs terminate in a curved, fibrous band (rhizosyndesmos) at the posterior cell pole, a structure that is only visible in detergent extracted whole-mount preparations (Lechtreck et al. 1989). Apart from the various basal body-associated fibrous rootlets, this is the first non-MT cytoskeletal structure discovered in a naked green alga, and provides an ideal model system for testing shape control. It is conceivable that the rhizosyndesmos simply locks the MTs in place and that asymmetric length distribution in the microtubular system and asymmetric pulling or pushing forces transmitted or mediated by the MTs between both cell poles create the twisted crescent shape.

The employment of detergent-extracted cytoskeletons is not new (Hyams and Borisy 1978; Marano et al. 1985). It has primarily been used for the study of flagellar beating and basal body behavior (Kamiya and Witman 1984, Kamiya 1992). In extension of these earlier studies, extracted cytoskeletons present the possibility of subfractioning structural components of the basal body rootlet complex allowing analysis of their biochemical composition. Using this technique, the Ca-binding protein centrin has been isolated from the rhizoplast (system II fiber) of the prasinophyte *Tetraselmis* and the phosphoprotein assemblin from the striated MT-associated system I fiber of *Spermatozopsis similis* (see Sect. 5. a). This approach also holds great promise for the study of other cytoskeletal features in naked phytoflagellates.

Another extreme cell shape is created by the formation of a posterior tail, displayed in several green algal gametes and zoospores. The zoospores of *Urospora penicilliformis* are an example (Sluiman et al. 1982). The formation of the tail takes place at the onset of swarming, presumably by a long evagination of the chloroplast pushing into posterior direction. No particular cytoskeletal structure has been identified in this species that could produce or support the tail, but it is unlikely that the chloroplast by itself can deform to such an extent. MTs might have been overlooked or not sufficiently preserved, since the situation is very similar in the gametes of the close relative *Spongomorpha duriuscula*, except that here MT do run down the projecting posterior tail (Miyaji and Hori 1984). The stephanokont zoospores of *Pseudobryopsis hainanensis* also possess a tail with MTs in the inner core (Hori and Kobara 1982).

Other examples of posterior body extension are present in the Prasinophyceae. *Pyramimonas longicauda* possesses a very long, thin caudate appendage which can quickly be withdrawn or extended. Unlike *U. penicilliformis*, the chloroplast is excluded from the appendage, but again as in the former, no obvious cytoskeletal elements have been identified that might lend this structure its shape and dynamics (Inouye et al. 1984).

b) Morphogenesis by Spatial Control of Wall Deposition

The desmid *Closterium* has proved to be an excellent example among the green algae to demonstrate that localized wall extension in an area of the cell, which exhibits a roughly cylindrical shape, is dependent on a spatially controlled insertion and restriction of cellulose synthetase complexes in the segment which is about to expand. During cell expansion, the coincidence between MTs and wall fibrils is very striking indeed. After expansion has ceased, only a circular MT band remains, separating the old and new semicells from each other, and the cellulose synthesizing machinery switches to the criss-cross band pattern as discussed in Sect. 7.d. β. Yet, cellulose synthesis remains restricted to the new half of the cell. Cellulose deposition spreads into the old semicell only after the MT ring has been depolymerized. Hogetsu (1983) proposed the question of whether this was because TCs are physically restrained, for instance by the ring of MTs forming a boundary in the membrane between the old, nongrowing semicell and the new, growing semicell. Interestingly, he found that the MT ring was not a physical barrier for the TCs, because cellulose microfibril imprints ran across it into the old half of the cell. However, soon after they passed over it they terminated. He suggested therefore that it must be due to another membrane-bound factor restrained by the boundary, and that this factor controls either TC activity or its state of assembly (see a summary in Hogetsu 1992).

During lid formation in the *Acetabularia* cyst, a narrow band of MTs coincides with a line demarcating the prospective lid position. Although microfibril imprints have not yet been studied in this example, it has been proposed that TCs may not cross this line from either side, so that a narrow circular gap results in the cell wall which is filled with matrix material but devoid of cellulose polymer. This produces a structurally weak, circular zone in the cyst wall that enables the lid to be opened when gametes begin to swarm (Menzel and Elsner-Menzel 1990b).

Protoplasts prepared from the zygnematalean green alga *Mougeotia* have recently been empolyed as experimental model to study the causal relationship between MT distribution, wall deposition, and cell expansion. Directly after protoplast production, the cortical MTs appear random, but they soon reorganize into symmetrical arrays centered on two foci. Patchy cell wall synthesis starts and the cell remains spherical. As the MT array gradually changes into a broad band running through the two poles previously localized by the position of the two foci, cellulose synthesis becomes restricted to the region of the band and expansion starts perpendicular to it resulting in a cylindrical shape (Galway and Hardham 1986). This is an example where a change in cell shape is exclusively dependent on the presence of a wall.

c) Morphogenesis by Modification of the Wall

Without the necessary chemical and structural means to modify an existing wall, a cell would be trapped in its own shell unable to reproduce and divide. Loricae impregnated with lime, such as those in the chlamydophycean alga *Phacotus*, overcome this problem by a preformed fracture line in the shell that breaks under a certain turgor threshold (Gierig et al. 1990). A simple way to get rid of an existing wall to allow for a change in cell shape is to digest the wall internally to the extent that the protoplast can leave it and regenerate. This principle is realized in many prasinophycean, chlamydophycean, and chlorococcalean green algae, and has best been studied in *Chlamydomonas*. The active agents extruded by the cells are very specific proteases called cellolysins (Spessert and Waffenschmidt 1990)

In all other cases where it is necessary to induce a change in shape after a wall has been laid down, such as generalized cell growth, tip growth, predivision elongation, budding, side branch formation, some kind of wall processing is required, with the effect that the existing wall is softened to the point where it yields to the internal pressure. At the same time, new wall material must be incorporated to limit expansion.

If a mechanism were to be engineered that would be capable of regulating all these factors, it would have to (1) determine the area to be affected, (2) perform localized extrusion of softening agents, (3) sense strain and (4) limit expansion. Some examples will be discussed below concerning localized extrusion and sensing of strain.

α) Spatial Control of Exocytosis

Cells that are programmed to modify their shape in the process of wall deposition need to control the biosynthetic machinery of the GA temporarily and spatially, the latter would involve the determination of preferential sites for the extrusion of wall material. Any tip growing cell has only one preferetial site for extrusion that would by determined by the apicobasal polarity (Schnepf 1986). In cells with more complex morphologies such as desmids and dasyclads, a mechanism with a higher degree of sophistication should be operating to achieve localized exocytosis (for an introduction to morphology and morphogenesis in both cell systems, see: Dasycladales – Berger and Kaever 1992; Desmidiaceae – Pickett-Heaps 1983). For the desmids, Kiermayer (1981) has postulated the existence of recognition sites for exocytotic vesicles. This hypothesis has been substantiated in several ways. In *Micrasterias*, Hauser and Herth (1983), as well as Meindl (1982), have been able to localize calcium ion accumulation exactly at the sites of future wall growth before any deformation of the wall has actually occurred. In the dasyclad system, Harrion and coworkers (1988) have also localized calcium ions at the prospective branching sites, and in both systems, disturbance of calcium homeostasis results in perturbation of morphogenesis (Goodwin et al. 1983; McNally et al. 1983; Lehtonen 1984; McNally and Swift 1986). The detection of localized calcium ion currents at these sites are in good agreement with this line of reasoning (Troxel and Scheffey 1991; Nawata 1992). At the structural level, it is conceivable that future growth points are marked by a clustering of calcium channels, which could bring in more calcium at the same sites until a threshold concentration is achieved which, in turn, triggers further structrual changes resulting in the assembly of fully operational exocytosis sites. In a similar way, pores could be inserted in the secondary wall of *Micrasterias*. As discussed in Kiermayer and Meindl (1989), special large pore vesicles are directed to specific sites arranged in the plane of the plasma membrane in a species-specific pattern. Insertion of the pore vesicles into the secondary wall prevents cellulose being deposited at these spots. It is thus suggested that the same recognition mechanism for determining cell shape during primary wall formation could also be responsible for producing pore patterns in the secondary wall.

Although mathematical algorithms have been calculated to simulate morphogenetic pattern formation in desmids and dasyclads (Harrison and Kolar 1988), it has been difficult to identify a structural representation of the mathematical process in the cell. Goodwin and Briere (1992) considered the situation in the dasyclad system and speculated that the cytoskeleton may play a role in these formative processes. However, a direct influence of the cytoskeleton has not been substantiated by experimental evidence (Menzel and Elsner-Menzel 1989a). A little more is known in the desmid system. Long ago, it was suggested that MTs were not involved in *Micrasterias* morphogenesis (Kiermayer and Fedtke 1977), but rather there was some preliminary evidence for the role of actin (Ueda and Noguchi 1988). What this function of actin might be is not yet clear. A rather too simple explanation would be that lobe extension in *Micrasterias* could be directed by polarized particles streaming into the prospective growth sites. This certainly is an important prerequisite and preliminary localization of myosin on dictyosome vesicles (Lehtonen and Suoronta (1990) seems to support this notion, but the key for an understanding of localized exocytosis must be sought for in the supramolecular structure of the submembrane cytoskeleton. A recent fine structural study employing modern pressure freezing and freeze substitution technology has not revealed any particular cytoskeletal structures, although the highly ordered arrangement of exocytotic vesicles lining up in front of fairly regularly spaced exocytosis sites, strongly suggests the existence of structural organization of some sort (Meindl 1992b).

Localized wall expansion by an unknown mechanism and of no obvious function has been described in the chlamydophycean alga *Brachiomonas submarina* (Watanabe et al. 1989). In the process of the formation of four unequal lateral protrusions and a posterior process, the cytoplasm first moves forward as the wall expands and later retracts to leave matrix filled cell wall horns behind.

β) Sensing the Direction of Strain

A possible mechanism of controlled expansion in cylindrical cells such as the *Nitella* internode or *Mougeotia* has recently been discussed by Williamson (1990). He considers the following: once the first wall layer is deposited and begins to bear stress, the cell should be able to control the direction and degree of yield at and by which the wall deforms during overall or localized expansion. This could be achieved by a positive feedback mechanism which magnifies a small asymmetry of an existing cell shape by reinforcing the mechanism that originally caused the asymmetry. The components of a sensing mechanism could involve MTs and a transmembrane protein capable of sensing the direction of maximum strain in the expanding wall. This directional information would then be transmittet to the MT-aligning and/or MT-organizing mechanism, which in turn leads to a preferred deposition of cellulose microfibrils along the line of maximum strain. The actual response of the aligning mechanism could further be modified in direction and degree, depending on developmentally expressed proteins which possess different properties. This would enable the cell to interpret the directional information coming from the sensor in a development-dependent manner.

Experiments by Galway and Hardham (1989) on *Mougeotia* protoplasts treated with MT depolymerizing agents during cell wall regeneration indicate that there must be some sort of memory stored in the membrane which tells MTs re-appearing after recovery from drug treatment where the previous axis of orientation was. This could be a first experimental indication for the existence of a transmembrane protein of the kind suggested by Williamson (1990)

Acknowledgements. I thank Geoffrey Wasteneys for discussions, Julia Willingale for critically reading the manuscript and Prof. Peter Traub for his kind support. Funding by the Deutsche Forschungsgemeinschaft is much appreciated.

References

Abe S, Ito Y, Davies E (1992) J Exp Bot 43:941–949. **Abhayavardhani P**, Ueda K (1986) Caryologia 39:409–417. **Apostolakos P**, Galatis B, Katsaros C, Schnepf E (1990) Protoplasma 154:132–143. **Arakawa T**, Frieden C (1984) J Biol Chem 259:11730–11734.

Bakker ME, Lokhorst GM (198) Protoplasma 138:105–118. **Barnett LA**, Cunningham MW (1992) Infect Immun 60:3932–3936. **Becker B**, Hard K, Melkonian M, Kamerling JP, Vliegenthart FG (1989) Eur J Biochem 182:153–160. **Becker B**, Becker D, Kamerling JP, Melkonian M (1991) J Phycol 27:498–504. **Becker D**, Melkonian M (1992) Eur J Cell Biol 57:109–116. **Becker D**, Becker B, Satir P, Melkonian M (1990) Protoplasma 156:103–112. **Beech PL**, Heimann K, Melkonain M (1991) Protoplasma 164:23–37. **Berger S**, Kaever MJ (1992) Dasycladales, an illustrated monograph of a fascinating algal order. Georg Thieme, Stuttgart. **Berger S**, Schweiger H-G (1986) J Cell Sci 80:1–11. **Berger S**, de Groot EJ, Neuhaus G, Schweiger M (1987) Eur J Cell Biol 44:349–370. **Berger S**, Menzel D, Traub P (1993) Eur J Cell Biol 60 (Suppl37):87. **Bermudes D**, Margulis L, Tzertzinis G (1987) Ann NY Acad Sci 503:187–197. **Bloodgood RA** (ed) (1990) Ciliary and flagellar membranes. Plenum, New York. **Bloodgood RA** (1991) Protoplasma 164:12–22. **Bold HC**, Wynne MJ (1985) 2nd edn, Prentice-Hall, Englewood Cliffs, NJ. **Brown RM** (1985) J Cell Sci Suppl 2:13–32. **Brown RM**, Montezinos D (1976) Proc Natl Acad Sci US 73:143–147. **Burr FA**, West JA (1971) J Phycol 7:108–113.

Chanzy HD, Grosrenaud A, Vuong R, Mackie W (1984) Planta 161:320–329. **Chapman RL**, Henk MC 1986) J Phycol 2283–88. **Chida Y**, Ueda K (1986) Phycologia 25:503–509. **Coleman AW** (1985) J Phycol 21:1–16. **Coleman AW**, Goff LJ, Stein-Taylor JR (eds) (1989) Algae as experimental systems in cell biology. Alan R Liss, New York. **Coleman J**, Evans D, Hawes C (1988) Plant Cell Environ 11:669–684. **Correas I**, Padilla R, Avila J (1990) Biochem J 269:61–64. **Crayton MA** (1982) J Phycol 18:3316–344.

De DN, Berger S (1990) Protoplasma 155:19–28. **Deason TR**, Floyd GL (1989) Crit Rev Plant Sci 8:259–272. **Delmer DP** (1987) Annu Rev Plant Physiol 38:259–290. **Denning GM**, Fulton AB (1989) J Protozool 36:334–340. **Detmers PA** (1985) Can J Biochem Cell Biol 63:599–607. **Domozych DS** (1987a) Protoplasma 136:170–182. **Domozych DS** (1987b) J Exp Bot 38:199–1411. **Domozych DS** (1987c) Bot Gaz 148:308–317. **Domozych DS** (1989a) Protoplasma 149:95–119. **Domozych DS** (1989b) J Cell Sci 93:375–384. **Domozych DS** (1991a) Int Rev Cytol 131:213–253. **Domozych DS** (1991b) Plant Cell Environ 14:391–398. **Domozych DS**, Rogers CE, Mattox KR (1983) J Exp Bot 34:1080–1088. **Domozych DS**, Wells B, Shaw PJ (1991) J Cell Sci 100:397–407. **Drew EA**, Abel KM (1990) Bot Mar 33:31–45.

Ehara T, Oagasawara Y, Osafune T, Hase E (1990) J Phycol 26:317–323. **Emons AMC** (1991) In: Haigler CH, Weimer PJ (eds) Biosynthesis and biodegradation of cellulose. Marcel Dekker, New York, pp 71–97. **Emons AMC**, Derksen J, Sassen MMA (1992) Physiol Plant 84:486–493. **Ettl H** (1985) Arch Hydrobiol Suppl 71:603–619.

Foissner I (1987) Protoplasma 142:164–175. **Forster KW,** Saranak J, Patel N, Zarilli G, Okabe M, Kline T, Nakanishi K (194) Nature 311:756–759. **Franke WW**, Kartenbeck J, Spring H (1976) J Cell Biol 71:196–206. **Fukushi Y**, Otsuru O, Maeda M (198 Otsuru O, Maeda M (1988) Carbohydr Res 182:313–320.

Galway ME, Hardham AR (1986) Protplasma 135:130–143. **Galway ME**, Hardham AR (1989) J Plant Physiol 135:337–345. **Galway ME**, Hardham AR (1991) Am J Bot 78:451–461. **Gerace L** (1986) Trends Biochem Sci 11:443–446. **Giddings TH**, Staehelin LA (1988) Planta 173:22–30. **Gierig B**, Krienitz L, Casper SJ, Peschke T, Raidi (1990) Arch Protistenkol 138:75–88. **Golz R**, Hauser M (1986) Eur J Cell Biol 40:124–129. **Goodenough UW** (1989) Cell 59:1–3. **Gooldenough UW**, Heuser JE (198, Heuser JE (1985) Cel 41:341–342. **Goodenough UW**, Heuser JE (1988a) J Cell Sci 90:717–733. **Goodenough UW**, Heuser JE (1988b) J Cell Sci 90:735–740. **Goodenough UW**, Gebhart B, Mecham RP, Heuser JE (1986) J Cell Biol 101:1550–1568. **Goodwin BC**, Briere C (1992) In: Menzel D (ed) The cytoskeleton of the algae. CRC, Boca Raton, pp 219–238. **Goodwin BC**, Skelton JL, Kirk-Bell SM (1983) Planta 157:1–7. **Goto Y**, Ueda K (1988) Planta 173:442446. **Goyal A**, Betsche T, Tolbert NE (1988) Plant Physiol 88:543–546. **Graham LE** (1984) Am J Bot 71:603–608. **Grant BR**, Borowitzka MA 1984) Bot Rev 50:267–307. **Grolig F** (1990) Protoplasma 155:29–42. **Grolig F** (1992) In: Menzel D (ed) The cytoskeleton of the alage. CRC, Boca Raton, pp 165–194.

Haigler CH, Brown RM (1986) Protoplasma 134:111–120. **Hall JL**, Ramanis Z, Luck DJL (1989) Cell 59:121–132. **Harold FM** (1990) Microbiol Rev 54:381–431. **Harper JDI**, McCurdy DW, Sanders MA, Salisbury JL, John PCL (1992) Cell Motil Cytoskel 22:117–126. **Harris N** (1986) Annu Rev Plant Physiol 37:73–92. **Harrison LG**, Graham KT, Lakowski BC (1988) Development 104:255–262. **Harrison LG**, Kolar M (1988) J Theor Biol 130:493–515. **Harz H**, Hegemann P (1991) Nature 351:489–491. **Hashimoto H** (1986) Protoplasma 135:166–172. **Hashimoto H** (1992) Protoplasma 167:88–96. **Hatano K**, Ueda K (1988) Eur J Cell Biol 47:193–197. **Hauser I**, Herth W (1983) Protoplasma 117:167–173. **Heath IB**, Seagull RW (1982) In: Lloyd CW (ed) The cytoskeleton in plant growth and development. Academic Press, London, pp 163–182. **Hegewald E**, Schnepf E (1984) Nova Hedwigia 39:297–383. **Hepler PK**, Cleary AL, Gunning BES, Wadsdworth P Wasteneys GO, Zhang DH (1993) Cell Biol Int Rep 17:127–142. **Herth W**, Kuppel A, Franke WW, Brown RM (1975) Cytobiologie 10:268–284. **Higashi-Fujime S** (1988) Protoplasma Suppl 2:27–36. **Higashi-Fujime S** (1991) Int Rev Cyto 125:95–138. **Hoffman LR** (1984) J Phycol 20:573–584. **Hoffman LR**, Buchheim MA (1989) J Phycol 25:149–159. **Hogetsu T** (1983) Plant Cell Physiol 24:777–781. **Hogetsu T** (1992) In: Menzel D (ed) The cytoskeleton of the algae, pp 113–131. CRC, Boca Raton. **Holmes JA**, Dutcher SK (1989) J Cell Sci 94:273–285. **Hori T** (1988) Jpn J Phycol 36:113–126. **Hori T**, Kobara T (1982) Sorui Jpn J Phycol 30:31–39. **Hoshaw RW**, Wang J-C, McCourt RM, Hull HM (1985) Am J Bot 27:1005–1011. **Hoshaw RW**, Wells CV, McCourt RM (1987) J Phycol 23:267–273. **Huang B**, Ramanis Z, Dutcher SK, Luck DJL (192) Cell 29:745–753. **Huizing HJ**, Rietema H, Sietsma JH (1979) Br J Phycol 14:25–32. **Hyams JS**, Borisy GG (1978) J Cell Sci 33:235–253.

Inouye I, Hori T, Chihara M (1984) Jpn J Phycol 32:113–123. **Itoh T**, Brown Jr RM (1988) Protoplasma 144:160–169. **Johnson KA**, Rosenbaum JL (1990) Cell 62:615–619.

Kajikawa H, Okada M, Ishikawa F, Okawa Y, Nakayama K (1988) Plant Cell Physiol 29:549–556. **Kakimoto T**, Shibaoka H (1987) Protoplasma 140:151–156. **Kamiya N** (1986) Bot Mag 99:444–467. **Kamiya R** (1992) In: Melkonian M (ed) Algal cell motility. Chapman and Hall, New York, pp 155–178. **Kamiya R**, Witman GB (1984) J Cell Biol 98:9–107. **Kengen HMP**, Degraaf BHJ (1991) Protoplasma 163:62–65. **Kiermayer O** (1981) In: Kiermayer O (ed) Cytomorphogenesis in plants. Springer, Vienna, New York, pp 147–189. **Kiermayer O**, Meindl U (1989) In: Coleman AW, Goff LJ, Stein-Taylor JR (eds) (1989) Algae as experimental systems in cell biology. Alan R Liss New York, pp 149–167. **Kiermayer O,** Fedtke C (1977) Protoplasma 92:163–166. **Kloareg B**, Quatrano RS (1988) Oceanogr Mar 26:259–315. **Koop HU**, Kiermayer O (1980a) Protoplasma 102:295–306. **Koop HU**, Kiermayer O (1980b) Protoplasma 102:147–166. **Kouwets FAC** (1991) Crypt Bot 2/3:104–114. **Kreimer G**, Melkonian M (1990) Eur J Cell Biol 53:101–111. **Kreimer G**, Brohsonn U, Melkonian M (1991a) Eur J Cell Biol 55:318–327. **Krei-**

mer G, Marner FJ, Brohsonn U, Melkonian M (1991b) FEBS Lett 293:49–52. **Kreis TE** (1990) Cell Motil Cytoskel 15:67–70. **Kristen U** (1985) Prog Bot 47:1–18. **Kudlicka K**, Wardrop A, Itoh T, Brown RM Jr (1987) Protoplasma 136:96–103. **Kuroda K** (1990) Int Rev Cytol 121:267–307. **Kuroda K**, Manabe E (1984) Proc Jpn Acad 59:Ser B, 131–134. **Kuroiwa T**, Hori T (1986) Protoplasma 133:85–87. **Kuroiwa T**, Kawano S, Nishibayashi S (1982) Nature 298:481–483. **Kuroiwa T**, Yorihuzi T, Yabe N, Ohta T, Uchida H (1990) Protoplasma 158:155–164.

La Claire JW (1989) Planta 177:47–57. **La Claire JW** (1992) In: Menzel D (ed) The cytoskeleton of the algae, pp 239–254. CRC, Boca Raton. **Lambert A-M** (1993) Curr Opinion Cell Biol 5:116–122. **Lancelle SA**, Hepler PK (1991) Protoplasma 165:167–172. **Lechtreck KF**, Melkonian M (1991) Protoplasma 164:8–44. **Lechtreck KF**, McFadden GI, Melkonian M (1989) Cell Motil Cytoskel 14:552–561. **Leedale GF** (1982) Prog Bot 4:32–42. **Legge RL**, Brown RM (1988) Protoplasma 143:38–42. **Lehtonen J** (1984) Plant Sci Lett 33:53–60. **Lehtonen J**, Suoronta U-M (1990) In: Peachy LD, Williams DB (eds) Proc XIIth Int Congr EM, San Francisco, p 961. **Leible MB**, Berger S, Schweiger H-G (1989) Curr Genet 15:355–361. **Lindberg B** (1990) Adv Carbohydr Chem Biochem 48:279–320. **Lobban CS**, Chapman DJ, Kremer BP (1988) Experimental phycology, a laboratory manual. Cambridge University Press, Cambridge. **Lokhorst GM**, Star W (1983) Protoplasma 117:142–153. **Lokhorst GM**, Sluiman HJ, Star W (1988) J Phycfol 24:237–248. **Lokhorst GM**, Segaar PJ, Star W (1989) Crypt Bot 1:275–294. **Luck DJL** (1984) J Cell Biol98:789–794. **Lüttke A** (1988) J Phycol 24:173–180

Maeda M, Fukushifujikura Y, Otsuru O (1990) Carbohydr Res 207:91–99. **Makawa T**, Nagai R (1989) Protoplasma Suppl I, 162–171. **Manabe E**, Kuroda K (1984) Proc Jpn Acad 60:Ser B, 118–121. **Manhart JR**, Palmer JD (1990) Nature 345:268–271. **Marano F**, Santa-Maria A, Krisnawamy D (1985) Protoplasma 127:82–92. **Marchant HJ** (1977) J Phycol 13:28–36. **Marchant HJ** (1982) Academic Press, London, pp 295–320. **Margulis L**, Corliss JO, Melkonian M, Chapman DJ (1990) Handbook of protoktista, Jones & Barlett, Boston. **McCourt RM**, Hoshaw RW (1990) Syst Bot 15:69–78. **McCurdy DW**, Gunning BES (1990) Cell Motil Cytoskel 15:76–87. **McCurdy DW**, Harmon AC (1992) Planta 188:54–61. **McCurdy DW**, Williamson RE (1987) J Cell Sci 87:449–456. **McFadden GI**, Preisig HR, Melkonian M (1986) Protoplasma 131:174–184. **McFadden GI**, Melkonian M (1986) Protoplasma 130:186–198. **McFadden GI**, Moestrup O, Wetherbee R (1982) Phycologia 21:103–111. **McLean B**, Whatley JM, Juniper BE (1988) New Phytol 109:59–65. **McNally JG**, Swift H (1986) J Cell Sci 86:305–322. **McNally JG**, Cowan JD, Swift H (1983) Dev Biol 97:137–145. **Meindl U** (1982) Protoplasma 110:1U43–146. **Meindl U** (1986) Protoplasma 129:7–87. **Meindl U** (1992a) In: Menzel D (ed) The cytoskeleton of the algae. CRC, Boca Raton, pp 133–148. **Meindl U** (1992b) Protoplasma 170:104–114. **Melkonian M** (1980) BioSystyems 12:85–104. **Melkonian M** (1982) Protoplasma 111:221–233. **Melkonian M** (1989) Prog Bot 50:214–245. **Melkonian M** (1991) Prog Bot 52:271–307. **Melkonian M** (1992a) In: Melkonian M (ed) Algal cell motility. Chapman and Hall, New York, pp 179–221. **Melkonian M** (ed) (1992b) Algal cell motility. Chapman and Hall, New York. **Melkonian M**, Robenek H (1984) Progr Phycol Res 3:193–268. **Melkonian M**, McFadden GI, Reize IB, Preisig HR (1987) Plant Syst Evol 158:47–61. **Melkonian M**, Becker D, McFadden GI, Reize IB (1988) In Cell-free analysis of membrane traffic, Alan R Liss, New York, pp 297–301. **Melkonian M**, Andersen RA, Schnepf E (eds) (1991a) Protoplasma 164:1–167. **Melkonian M**, Becker B, Becker D (1991b) J Electron Micr Tech 17:165–178. **Menzel D** (1985) Protoplasma 125:103–110. **Menzel D** (1986) Protoplasma 134:30–42. **Menzel D** (1987) Protoplasma 139:71–76. **Menzel D** (1988) Protoplasma 144:73–91. **Menzel D** (1990) Bot Acta 103:223–225. **Menzel D** (ed) (1992) The cytoskeleton of the algae, CRC, Boca Raton. **Menzel D** (1993) Bot Acta 106(4):294–300. **Menzel D**, Elsner-Menzel C (1989a) In: Coleman AW, Goff LJ, Stein-Taylor JR (eds) Algae as experimental systems in cell biology. Alan R Liss, New York, pp 71–92. **Menzel D**, Elsner-Menzel C (1989b) Bot Acta 102:164–171. **Menzel D**, Elsner-Menzel C (1989c) Protoplasma 150:1–8. **Menzel D**, Elsner-Menzel C (1989d) Bot Acta 102:241–248. **Menzel D**, Elsner-Menzel C (1990a) In: Wiessner W, Robinson DG, Starr RC (eds) Experimental phycology, Springer, Berlin Heidelberg New York, pp 105–123. **Menzel D**, Elsner-Menzel C (1990b) Protoplasma 157:52–63. **Menzel D**, Schliwa M (1986a) Eur J Cell Biol 40:275–285. **Menzel D**, Schliwa M (1986b) Eur J Cell Biol 40:286–295. **Menzel D**, Jonitz H, Elsner-Menzel C (1992a) In: Menzel D (ed) The cytoskeleton of the algae, pp 195–218. CRC, Boca Raton. **Mita T**, Kuroiwa T (1988) Protoplasma Suppl 1:133–152. **Miyaji K**,

Hori T (1984) Jpn J Phycol 32:307–318. **Miyamura S**, Hori T (1991) Protoplasma 161:192–196. **Mizuta S** (1992) In: Menzel D (ed) The cytoskeleton of the algae, pp 149–164. CRC, Boca Raton. **Mizuta S**, Harada T (1991) Bot ar 34:411–415. **Mizuta S**, Okuda K (1987) Bot Gaz 148:297–307. **Mizuta S**, Wada S (1981) Bot Mag Tokyo 94:343–353. **Mizuta S**, Kurogi U, Okuda K, Brown RM (1989) Ann Bot 64:383–394. **Mollenhauer D** (1978) Prog Bot 40:285–338. **Monk BC**, Adair WS, Cohen RA, Goodenough UW (1983) Planta 158:17–533. **Morris GJ**, Coulson GE, Leeson EA (1985) J Cell Sci 76:145–153.

Nakagawa M, Noguchi T, Ueda K (1990) Jpn J Phycol 38:353–364. **Nakamura Y**, Ueda K (1982) Cytologia (Tokyo) 47:713–716. **Nawata T** (1992) Plant Cell Physiol 33:881–888. **Neville AC** (1988) Progr Phycol Res 6:1–22. **Nigg EA** (1992) Curr Opinion Cell Biol 4:105–109. **Nishioka H**, Okuda K, Mizuta S (1990) Bot Mar 33:289–297. **Nisizawa K**, Kuroda K, Tomita Y, Shimahara H (1974) Bot Mar 17:16–19. **Noguchi T** (1988) Protoplasma 147:135–142. **Noguchi T**, Ueda K (1985) Protoplasma 128:64–71. **Noguchi T**, Ueda K (1988) Protoplasma 143:188–192. **Nultsch W** (1988) Sitzungsber Wiss Ges JW Goethe Univ Frankfurt 24:5–37.

O'Kelly CJ (1992) In: Menzel D (ed) The cytoskeleton of the algae, CRC, Boca Raton pp 315–347. **O'Kelly CJ** (1982) Bot Mar 25:133–137. **Okuda K**, Matsuo K, Mizuta S (1990) Plant Cell Physiol 31:357–364. **Olsen JL** (1990) J Phycol 26:209–214. **O'Neil RM**, La Claire JW (1988) Cytobios 53:11–125. **Oross JW**, Possingham JV (1989) Protoplasma 150:131–138.

Palevitz B, Hepler P (1975) J Cell Biol 65:29–38. **Palmer JD** (1985) Annu Rev Genet 19:325–354. **Pearlmutter NL**, Lembi CA (1980) J Phycol 11:602–616. **Pesacreta TC**, Lucas WJ (1984) J Cell Biol 98:1537–1545. **Pickett-Heaps JD** (1975) Green algae, structure, reproduction and evolution in selected genera. Sinauer, Sunderland, Ma. **Pickett-Heaps JD** (1983) In: McIntosh JR (ed) Spatial organization of eukaryotic cells, pp 241–258. Alan R Liss, New York. **Pickett-Heaps JD**, Fowke LC (1970) J Phycol 6:189–215. **Preston RD** (1988) Planta 174:67–74. **Przbilla E**, Galling G (1987) Protoplasma 139:41–50.

Quader H (1986) J Cell Sci 83:223–234. **Quader H** (1991) In: Haigler CH, Weimer PJ (eds) Biosynthesis and biodegradation of cellulose. Marcel Dekker, New York, pp 51–69.

Reisser W (ed) (1992) Algae and symbiosis. Biopress, Bristol. **Reymond OL** (198 (1985) Cryptogam Algol 6:109–124. **Reymond OL**, Hegewald E (1988) Arch Protistenkol 135:167–172. **Reymond OL**, Hegewald E (1990) Arch Hydrobiol Suppl 85:15–28. **Roberts K**, Hills GJ, Shaw PJ (1982) Academic Press, New York. **Roberts K**, Grief C, Hills GJ, Shaw PJ (1985) J Cell Sci Suppl 2:105–127. **Robinson DG** (1985 (1985) Plant membranes, endo- and plasmamembranes of plant cells. Wiley, New York. **Robinson DG**, Depta H (1988) Annu Rev Plant Physiol 39:53–99. **Robinson DG**, Quader H (1982) In: Lloyd CW (ed) The cytoskeleton in plant growth and development. Academic Press, London, pp 109–126. **Rochaix J-D** (1987) FEMS Microbiol Rev 46:13–34. **Roelofson PA** (1965) In: Preston RD (ed) Advances in botanical research, vol 2, Academic Press, London, pp 69–149. **Round FE**, Chapman DJ (1982-1992) Progress in physiological research, vols 1–8. BioPress, Bristol. **Rudolph U** (1987) Naturwissenschaften 74:439.

Saito H, Yamada J, Yoshioka Y, Shibata Y, Erata T (1991) Biopolymers 31:933–940. **Salisbury JL** (1992) In: Menzel D (ed) The cytoskeleton of the algae. CRC, Boca Raton, pp 393–410. **Schliwa M**, Shimizu T, Vale RD, Euteneuer U (1991) J Cell Biol 112:119–1203. **Schnepf E** (1983) Biol Rundsch 21:329–332. **Schnepf E** (1986) Annu Rev Plant Physiol 37:23–47. **Schnepf E**, Deichgräber G, Herth W (1982) Protoplasma 110:203–214. **Schnepf E**, Witte O, Rudolph U, Deichgräber G, Reiss H-D (1985) Protoplasma 127:222–229. **Segaar PJ** (1989a) Can J Bot 67:239–246. **Segaar PJ** (1989b) Cypt Bot 1:3–14. **Segaar PJ**, Gerritsen AF (1989) Crypt Bot 1:249–274. **Seegar PJ**, Lockhorst GM (1987a) Acta Bot Neerl 36:3–11. **Segaar PJ**, Lokhorst GM (1987b) Phycologia 26:100–110. **Segaar PJ**, Lokhorst GM (1988) Protoplasma 142:176–187. **Segal RA**, Luck DJ (1985) J Cell Biol 101:1702–1712 (1985). **Setlik I**, Zachleder V (1984) In: Wiessner W, Robinson DG, Starr RC (eds) Compartments in algal cells and their interaction. Springer, Berlin Heidelberg New York, pp 28–35. **Shihira-Ishikawa I** (1984) Protoplasma

122:27–34. **Shihira-Ishikawa I**, Kuroiwa T (1984) Jpn J Phycol 32:147–157. **Shimmen T** (1988) Bot Mag (Tokyo) 101:533–544. **Shimmen T** (1992) In: Menzel D (ed) The cytoskeleton of the algae. CRC, Boca Raton, pp 297–314. **Sluiman HJ** (1984) J Cell Sci 72:89–100. **Sluiman HJ** (1985) J Phycol 21:523–532. **Sluiman HJ** (1991) J Phycol 27:291–298. **Sluiman HJ**, Reymond OL (1987) Acta ot Neerl 36:231–245. **Sluiman HJ**, Roberts KR, Stewart KD, Mattox KR, Lokhorst GM (1982 J Phycol 18:1–12. **Sluiman HJ**, Kouwets FAC, Blommers PCJ (1989) Arch Protistenkd 137:277–290. **Spessert R**, Waffenschmidt S (1990) Eur J Cell Biol 51:17–22. **Spring H**, Scheer U, Franke WW, Trendelenburg MF (1975) Chromosoma 50:25–43. **Stevenson TT**, Darvill AG, Albersheim P (1988) Carbohydr Res 179:269–288. **Suda S**, Watanabe MM, Inouye I (1990) Phycologia 29:54–64. **Sun GH**, Uyeda TQP, Kuroiwa T (1988) Protoplasma 144:185–188.

Tazawa M, Shimmen T (1987) Int Rev Cytol 109:259–312. **Tewinkel M**, Volkmann D (1987) Planta 172:309–320. **Tilney LG** (1985) In: Metz C, Monroy A (eds) The biology of fertilization, Academic Press, New York, pp 157–213. **Tischendorf G**, Hirschmann M, Menzel D, Traub P (1992) Elecron microscopy, vol 3 EUREM 92, Granada Spain, pp 263–264. **Trainor FR**, Egan PF (1990) Phycologia 29:461–469. **Troxell CL**, Scheffey C (1991) Planta 184:218–225. **Tschermak-Woess E** (1988) Plant Systy Evol 159:123–139.

Ueda K, Abhayavardani P, Noguchi T (1986) Bot Mag Tokyo 99:301–308. **Ueda K** Noguchi T (1988) Eur J Cell Biol 46:61–67.

van den Hoek C, Stam WT, Olsen JL (1988) Helg Meeresunters 42:339–383. **Vidyavati SG** (1985) Proc Indian Acad Sci Plant Sci 95:193–198.

Wagner G, Grolig F (1992) In: Melkonian M (ed) Algal cell motility. Chapman and Hall, New York, pp 39–72. **Wang Y-L**, Sanders MC (1990) In: Foskett JK, Gristein S (eds) Noninvasive techniques in cell biology, Wiley-Liss, New York, pp 173–212. **Wasteneys GO** (1992) In: Menzel D (ed) The cytoskeleton of the algae. CRC, Boca Raton, pp 273–296. **Wasteneys GO**, Williamson RE (1991) Protoplasma 162:86–98. **Wasteneys GO**, Gunning BES, Hepler PK (1993) Cell Motil Cytoskel 24:205–213. **Watanabe S**, Tsuchimoto K, Floyd GL (1989) Phycologia 28:188–196. **Weidinger M**, Ruppel HG (1985) Protoplasma 24:184–187. **Wiessner W**, Robinson DG, Starr RC (1990) Experimental phycology, vol 1, Cell Walls and surfaces, reproduction, photosynthesis. Springer, Berlin Heidelberg New York. **Williamson RE** (1974) Nature 248:801–802. **Williamson RE** (1984) Plant Cell Environ 7:431–440. **Williamson RE** (1986) Plant Physiol 82:631–634. **Williamson RE** (1990) Aust J Plant Physiol 17:601–613. **Williamson RE** (1992) In: Melkonian M (ed) Alga Cell Motility. Chapman & Hall, London, pp 73–98. **Williamson RE**, Hurley UA, Perkin JL (1984) Eur J Cell Biol 34:221–228. **Williamson RE**, Perkin JL, McCurdy DW, Craig S, Hurley UA (1986) Eur J Cell Biol 41:1–8. **Williamson RE**, McCurdy DW, Hurley KA, Perkin JL (1987) Plant Physiol 85:268–272. **Witztum A**, Parthasarathy MV (1985) Eur J Cell Biol 39:21–26.

Priv. Doz. Dr. Diedrik Menzel
Max-Planck-Institut für Zellbiologie
Rosenhof
D-68526 Ladenburg, FRG

II. Comparative Morphology, Anatomy, and Function of the Stem and Root of the Flowering Plants

By William C. Dickison

1. Introduction

This contribution represents an extension of an earlier report covering the morphology and anatomy of the stem of flowering plants (Dickison 1992). As such, I have found it convenient to organize recent research under many of the same topic headings that were recognized previously. This chapter is different from my first contribution, in that selected literature devoted to root organization and structure is also summarized briefly. As before, this is not intended as an exhaustive literature review but a selection of current research that represents an overview on the progress made in the field of stem and root morphology and anatomy during the past few years. Also, as before, literature dealing with fossil plants, tissue culture, the phloem, the periderm, and regional, purely descriptive, or applied wood anatomy is not treated. The wide range of topics reviewed in this chapter, ranging from considerations of the cell theory, patterns of differentiation, evolution, and organismal morphology and computer modeling, reflect the current diversity of interests and research progress within the broadly based field of structural botany.

Notable books include the title *Plant Biomechanics: an Engineering Approach to Plant Form and Function* by K.J. Niklas (1992). This volume represents in the publishers words, the "first comprehensive treatment of plant biomechanics... (that) analyzes plant form and how it has evolved in response to basic physical laws... (and that) examines the ways these laws limit the organic expression of form, size and growth in a variety of plant structures and in plants as whole organisms". The book is a thoughtfully assembled collection of chapters devoted, in detail, to subjects such as the mechanical behavior of materials, the effect of geometry on mechanical behavior, plant-water relations, plant cell walls, the mechanical behavior of tissues, the mechanical attributes of organs, the plant body, fluid mechanics, and biomechanics and plant evolution. The author employed a method of direct literature citation and the text is followed by an extensive and up-to-date bibliography. This contribution is a landmark volume that establishes plant biomechanics as a discipline of considerable importance in discussions of higher plant form, function, and evolution.

The climbing habit in plants has clearly evolved multiple times among diverse plant species. *The Biology of Vines* is a multiauthored volume edited by F.E. Putz and H.A. Mooney (1991) that presents an up-to-date account of what is known about climbing plants (vines and lianas). For the structural botanist the following accounts are of special interest: a compilation of anatomical features of vines (Carlquist 1991a), biomechanical

Progress in Botany, Vol. 55

studies (Putz and Holbrook 1991), water flux and xylem structure (Ewers et al. 1991), and structural responses to stem injury in vines (Fisher and Ewers 1991).

The prominent angiosperm phylogenist A. Takhtajan summarized his view on the evolutionary morphology of flowering plants in a volume entitled *Evolutionary Trends in Flowering Plants* (Takhtajan 1991).

2. Relationship of Cells to Organisms in Plants

The long debated question of the interaction and relationship of cells to organisms in multicellular plants has once again come to the forefront. Although Sitte (1992) stressed the need for formulating a unifying synthesis of both cell and organismal theories, Kaplan and Hagemann (1991), Niklas and Kaplan (1991), and Kaplan (1992) analyzed the evidence from a broad range of studies and favored the idea that multicellularity in vascular plants is more accurately described and interpreted by the organismal theory as opposed to the cell concept (cell theory), in which cells are regarded as the fundamental building units of organisms. According to the organismal view multicellularity has arisen not by cell aggregation but rather by the secondary chambering of a unified protoplast into the subunits we now refer to as cells, and in turn the entire functional organism. These authors stress the need to address fundamental structural, developmental, and physiological questions from an organismal perspective. Accordingly, the morphology of a plant cannot be considered the product of its anatomy.

Hagemann (1992) extended the view that anatomy and morphology are independent of one another and that cellular structure in a higher plant body is a secondary phenomenon. A unique theoretical perspective is presented to explain the relationships of anatomy and morphology. According to this theory, the differentiation of form is not dependent upon cellular structure. A simple permanently growing thallus is viewed as the ideal model for a primitive but perfect plant, from which stems and roots were added in evolution.

3. Interpretations of the Shoot (Organography)

LaCroix et al. (1990) described the condition in *Leea quineensis* in which the continuity between leaf, the stipules, and shoot axis is striking at early stages of development, therefore making it difficult to delimit these different structures. The need to retain a flexible view of the structural categories such as leaf, stem, and root is pointed out. Sattler and Rutishauser (1992) also showed the existence of basic similarities between the early development of compound leaves and shoots, thus providing support to the partial-shoot theory of the leaf postulated by Agnes Arber (1950; i.e., pinnate leaves are partially homologous to shoots).

The morphological interpretation of the unusual organography of leaves and epiphyllous shoots in *Chisocheton* (Meliaceae) has remained controversial, since the foliar organs of this plant possess characteristics of both leaf and shoot in the same organ. Following a thorough description of the leaves, axillary buds, and epiphyllous shoots of this taxon, Fisher and Rutishauser (1990) concluded that the ontogenetic and structural evidence indicates that the epiphyllous buds arise by heterotopy, i.e., a change in position of bud initiation. Another unusual shoot organization is described by Posluszny and Tomlinson (1991) for the seagrass *Halophila* (Hydrocharitaceae). In this case, the organization of the erect shoot and position of flowers is shown to have systematic value. The genus shows what is perhaps the greatest degree of preformation in the shoot system among any other plant. Up to 8 branch orders and as many as 50 appendages can be recognized within a single diminutive bud. Another example of preformed shoot axes is reported in *Cytisus scoparius* (Zhou and Hara 1990).

Heteroblastic development refers to situations where profound changes are evident in leaf and stem morpholgy between early (juvenile) and late (adult) stages of development. Lee and Richards (1991) discussed this developmental phenomenon in vines and emphasized strong correlations between heteroblastic stages and microclimate. Evidence points to the fact that these variations are adapted to the environmental condition (e.g., light quantity and light quality) encountered by each stage. Ray (1990) also emphasizes that unusually elaborated patterns of heteroblastic shoot development have evolved among members of the monocotyledonous family Araceae in association with the climbing habit. This condition is also seen as directly related to varying environmental conditions.

Wilder (1992b) described orthodistichous phyllotaxy and dorsiventral symmetry on adult shoots of *Cyclanthus bipartitus* (Cyclanthaceae, Monocotyledoneae). This represents a type of phyllotaxy previously unknown for *Cyclanthus* and a growth habit unreported for the family. This species grows upright on forest-covered slopes, and the concave surface and convex surface of a shoot are slope-opposed and slope-facing, respectively. It is suggested that the form, symmetry, development, and orientation of a shoot are adaptations to unequal illumination on its slope-facing versus slope-opposed sides. A revised classification of growth habits in the Cyclanthaceae is presented.

An interesting problem of morphological interpretation is offered by the parasitic angiosperm *Ombrophytum subterraneum* (Balanophoraceae). The vegetative body is an irregularly spherical structure, called a tuber, consisting of two portions, one composed of host cells, the other of *O. subterraneum* parenchyma cells and associated vascular bundels (Mauseth et al. 1992). The "tuber" has no apical meristem, no leaves, no axillary buds, no eustele, and no epidermis. It produces slender runners that have an apical meristem, a protoderm and epidermis, but no root cap, no leaf primordia, and no root hairs. In some places the runners stele is rootlike, in other places it is stemlike. The homology of the runners is unclear, and the terms root, stem, and leaf are avoided.

4. Shoot Architecture and Biomechanics

A review of various approaches to describing and modeling the complexities of form in trees was presented by Fisher (1992). It is clear that computer models can predict the growth of small trees that are not subjected to environmental stress and other variables that produce opportunistic responses. On the other hand, stochastic and deterministic models cannot predict the response of trees or shrubs to unpredictable influences. Smith et al. (1992) developed a method utilitzing computer graphics to provide rapid and objective means of accurately describing plant architecture, that is, the framework and spatial arrangement of different plant structures.

An example of an unpredictable architectural model is provided by Tourn et al. (1992) who described a condition in three tree species of Rhamnaceae in which each meristem of a lateral bud complex has a different development potential, increasing the architectural possibilities of the trees and the resulting architectural model. The architecture of these species does not conform to that of any known model.

Examples of morphological responses of crops and weed species of different growth forms to ultraviolet-B radiation are provided by Barnes et al. (1990). Under greenhouse conditions, UV-B exposure (simulating a 20% reduction in stratospheric ozone) was found to reduce leaf blade and internode lengths and increase leaf and axillary shoot production in several species. An attempt to test and modify modeling methods in order to understand stolon growth and branching in a horizontally spreading clonal plant *(Glechoma hederacea)* was undertaken by Birch and Hutchings (1992). Growth was analyzed using a plastochron index based on the number of ramets on each stolon and internode growth. Horizontally spreading clonal plants may be viewed as groups of raments connected by spacers to form branching structures.

a) Patterns of Shoot Growth

Final shoot size of trees is determined by the number of nodes and the length of internodes during any growth period. Since the cellular components of growth contribute to final internode length, cell size and number assume primary importance in morphogenetic studies. Brown and Sommer (1992) followed shoot growth and histogenesis in five unrelated tree taxa possessing diverse patterns of shoot development. This study substantiated the significant role of subapical meristematic activity in shoot growth and development, and the need for focusing more attention on fundamental processes ultimately regulating mitotic activity in developing internodes along the shoot axis. These authors correctly point out that the sequential development of individual internodes and the orderly pattern of shoot development as a whole is, in the final analysis, the result of differences in the rates and duration of both cell division and elongation. The two processes cannot be separated.

Most tree species modify their crown architecture by controlling numbers of growing buds and small branches. Dewit and Reid (1992) investigated aspects of the abscission (self-pruning) of small branches in poplar. Branch abscission in poplar may have a number of functions, such as reducing canopy sail area and allowing for plasticity of response to unpredictable wind conditions. The self-pruning of less productive branches growing on the interior of the canopy should also be useful in allowing diversion of nutrients to other important activities.

b) Biomechanics

The hypothesis that determinate growth in an organ can reduce the mechanical liability of dynamic wind loadings was examined for the peduncles of *Allium sativum* (Niklas 1990). It is evident that neither thigmomorphogenesis (rubbing, or wind-induced bending and torsion) or determinate growth alone is an adequate design factor to ensure against buckling. When combined in mature organs, however, they confer a design factor that is resistant to stress. Nobel and Meyer (1991) investigated and discussed the complex biomechanics of cladodes (flattened stem segments) and of cladode-cladode connections for *Opuntia ficus-indica* (Cactaceae). Data resulting from wind-, drought-, and temperature-induced flexing indicate that the cladodes of *Opuntia* are quite rigid and only slightly altered by environmental influences.

The interesting constructional architecture of Indian sola wood (*Aechynomene aspera* and *A. indica*), reportedly the lightest known wood (specific gravity, 0.04), has been clarified by Nair and Mohan Ram (1992). The wood chiefly consists of fusiform-shaped cells with simple pits. The term "fusiform wood cell" is used to demarcate them from other cell types. The large-scale replacement of thick-walled fibers by thin-walled, air-filled, fusiform wood cells and the very low frequency of vessels may account for the extremely light property of sola.

Murphy and Alvin (1992) found that the degree of polymerization in the fiber cell wall of bamboos is variable and related to the position of the vascular bundle. It is evident that this variation in wall composition and structure (lamellae) probably influence the mechanical properties of the stem.

5. Stem Anatomy and Function

a) Apical Meristems

An excellent contemporary review of the subject of vegetative apical meristems, including terminology and molecular and genetic data, has been provided by Medford (1992). A classical problem of shoot growth centers on the control of the morphogenetic changes at the apical shoot meristem. Green (1992) addressed this problem using a novel approach in which minimal energy configurations were used to demonstrate a likely control mechanism. A mathematical model was formulated that resulted in three-dimensional computer-generated structures from which phyllotactic patterns emerge. One of the difficulties in understanding the complexity of the shoot apex arises from its continuous changing during the plastochron, including enzyme changes. The histochemical identification of the enzymes of various secondary compounds in shoot apices has resulted in the finding and localization of chalcone synthase, the key enzyme of flavonoid production, at different moments of the plastochron (Zobel and Hrazdina 1992). Early ontogeny of the vegetative axillary shoot apex of *Aucuba japonica* was considered by Hara (1991).

α) Shoot Regeneration

A number of recent studies have focused on shoot regeneration, from callus-derived roots (Mathur 1992), from buds formed on the hypocotyl and roots (Baird et al. 1992), from axillary bud explants (Miller et al. 1991), from cultured shoot apices (Kathal et al. 1992), from leaf petioles (Colby et al. 1991), and from root and stump buds located near or below ground level (Paukkonen et al. 1992).

b) Stem Structure

Seedlings of *Prosopis* grown in the presence of increased salinity showed structural changes in root and shoot tissue (Serrato Valenti et al. 1991, 1992). As salt concentration increases, the number of storage cells in the epidermis and cortical layers is increased, lignification is initiated earlier, and secondary xylem differentiates earlier. The earlier lignification of secondary tissues is apparently a general characterization of plants growing in habitats with low water potential and could be an adaptive strategy to facilitate water transport.

Mauseth and Sajeva (1992) reported the occurrence of cortical vascular bundles in the persistent, photosynthetic stems of 62 species in 45 genera of the cactus subfamily Cactoideae. The cortical bundles permeate in the broad, water-storing inner cortex and extend to the base of the outer, photosynthetic palisade cortex. Within the Cactoideae the cortical bundles appear to: (1) transport photosynthase from outer cortex to stele, (2) transport sugars to and from amyloplasts in the inner, nonphotosynthetic cortex, and (3) transport water throughout the entire cortex. The cortical bundles in stems of Cactoideae appear to have arisen independently and are not homologous with those of any other family.

McDonald (1992) discussed the evolutionary implications of typical and anomalous secondary growth in arborescent members of the genus *Ipomoea* (Convolvulaceae) and, in addition to describing growth patterns with concentric rings of included phloem derived from successive cambia, concluded that anatomical data provide evidence that the arborescent habit in *Ipomoea* has been derived at least twice through distantly related vining prototypes.

The diversity of recent stem morphological and anatomical studies is reflected in the following investigations. The occurrence, type, and location of calcium oxalate crystals in leaves and stems of 16 poisonous plants growing naturally in Saudi Arabia supported the view that the presence of calcium oxalate was not related to plant toxicity (Doaigey 1991). The development and structure of oil reservoirs in stem, rhizome, and root of *Solidago canadensis* (Asteraceae) were examined by Curtis and Lesten (1990). Aspects of leaf and stem anatomy of the following groups have been described: *Agalinis*, Scrophulariaceae (Canne-Hilliker and Kampny 1991), *Carex*, Cyperaceae (Standley 1990), Solanaceae (Gastizoro 1991), and Ophiopogoneae, Convallariaceae (Cutler 1992). Growth and development of the endophyte system of *Phoradendron juniperum* (Viscaceae) were the subject of study by Calvin et al. (1991). Folkerts (1989) regarded facultative rhizome dimorphism in *Sarracenia* to be an adaptation to deepening substrate. The physiological and ecological importance of abundant foliar sclerenchyma, heavy cuticle, and water-storing cells in the pseudobulb of *Stantiopea* (Orchidaceae), was emphasized by Stern and Morris (1992).

The overwintering strategy of *Diapensia lapponica* was examined by studying the seasonal variation of storage substances in the stem cells throughout different seasons (Pihakaski-Maunsbach and Walles 1990). Fiqueiredo-Ribeiro et al. (1992) studied changes in the yield and composition of hemicelluloses from the underground organs (xylopodia) of *Ocimum nudicaule* including changes in composition in dormancy and sprouting.

α) Lateral Meristems of Monocots

The importance of lateral meristems in the establishment of moncot growth habits was reviewed by Rudall (1991). The primary thickening meristem near the shoot apex is present in virtually all monocots and is regarded as a plesiomorphic feature whereas the secondary thickening meristem is confined to a few genera of Liliiflorae.

β) Dilatation Growth

Dilatation growth produces parenchymatous tissues and adjusts bark structure to the increase in the circumference of the axis resulting from secondary growth. Dilatation is the result of both cell size enlargement and meristematic activity. It is not known what mechanism(s) regulates dilatation growth although the experimental induction of dilatation meristems led to the conclusion that a plant growth regulator (probably ethylene), rather than physical force is the primary determining factor in meristematic growth or differentation of dilatation growth (Lev-Yadun and Aloni 1992b).

c) Nodal Anatomy

Variability of nodal anatomy in the Rhizophoraceae is reported by Raju (1991). It turns out the mangrove members of the family show multitraced, multilacunar nodes whereas among inland species the node is trilacunar, three-traced. Specialized split-lateral traces are also described. Lavania and Govil (1991) surveyed nodal organization in *Solanum*.

d) Vascular Differentiation

A summary of current literature and the state of knowledge on the subject of vascular differentiation was presented by Aloni (1992), including discussions of the mechanisms that control vascular differentiation in plants. Recent studies on the genetic control of apical meristem behavior have uncovered genes which are expressed in the procambium

(Medford et al. 1991; Pri-Hadash et al. 1992). It is well established that the vascular tissues and fibers are induced and controlled by longitudinal streams of inductive signals, primarily the hormones auxin, gibberellin, and cytokinin. Aloni and Peterson (1990) reported that horizontal streams of auxin between the vascular strands control the formation of phloem anastomoses in stems which operate as an emergency translocation system. The influence of cytokinin originating in the root apex on sieve tube element and vessel differentiation throughout the plant axis was the subject of investigations by Aloni et al. (1990) and Baum et al. (1991. Hormones also regulate the lignin structure of the fiber walls (Aloni et al. 1990). Ethylene is now known to induce and control vascular ray differentation, size, and spacing (Lev-Yadun and Aloni 1991a, b). The effects of ethylene on the initiation and differentiation of the ray system was discussed further by Lev-Yadhun and Aloni (1992a) and provide a basis for the development of an "ethylene diffusion hypothesis". An increase in ray number and dimensions can be experimentally altered by stem wounding or partial girdling.

The fact that auxin flow controls vessel diameter and density is confirmed by recent studies which show that low-level streams of auxin control the differentiation of wide earlywood vessels in ring-porous trees. Moderate or high auxin levels applied to ring-porous trees at the time of bud break limit the size of the earlywood vessels and result in a diffuse-porous type of wood (Aloni 1991). It is hypothesized that selective pressures in limiting environments resulted in the evolution of specialized ring-porous wood that increased the efficiency of water transport and that was accompanied by a decrease in the intensity of vegetative growth, reduced levels of growth regulators, and an increase in the sensitivity of the cambium to low levels of auxin ("limited growth hypothesis").

The vascular development and pattern of lignification were studied in *Arabidopsis*, widely used as a model system for exploring floral development, and it was proposed that the genus could make an effective model to study the biochemistry and molecular biology of lignin biosynthesis (Dharmawardhana et al. 1992). The work of de Castro et al. (1992) can be cited as an example of recent progress in understanding the spatial and temporal gene expression patterns that occur during stem development.

6. Root Architecture

A substantial amount of information on the impact of the environment on roots and root systems, including morphology and anatomy, is provided by a collection of papers edited by Barlow (1993). A number of recent contributions have dealt in various ways with aspects of root architecture (Gilman 1989; Huante et al. 1992). Although a considerable amount of information is available concerning the development of individual roots, relatively few investigations have been carried out on the development of the root system as a whole or on that of individual roots over extended periods of time. Birdsall and MacLeod (1990) suggested that onion (*Allium*) provides a suitable model for the study of a root system. Among other findings, results show that the period of maximal cell proliferaiton in apices of primary and adventitious roots of *Allium cepa* was relatively short in duration. Fitter and Stickland (1992) showed the validity and potential of the techniques of architectural analysis for the study of root systems grown under natural conditions.

Hetrick et al. (1991) found that cool-season grasses had significantly more primary, secondary root development than warm-season grasses although the diameter of primary, secondary, and tertiary roots of cool-season grasses was significantly smaller than that of warm-season taxa. Apparently, root architecture of the mycorrhizal-dependent warm-season grasses is quite plastic, allowing energy expenditure for root development to be conserved. The root architecture for the less mycorrhizal-dependent cool-season grasses appears to be fixed and does not alter to accommodate the symbiosis. Brundrett and Kendrick (1990a) found root growth and mycorrhiza formation and spread in five perennial herbaceous plants from the hardwood forest communities in Ontario, Canada, to be much slower than in other herbaceous plants in which these events have been measured.

It is also apparent that there are significant differences in root morphology among species with different life histories. For example, among 12 herbaceous species that were examined, annuals grew faster and produced larger and more branched roots than did biennials and perennials (Gross et al. 1992). The importance of understanding the potential of root systems to expand and produce new shoots under different environmental conditions is emphasized by Nadeau and Vanden Born (1989) in their study of the weedy Canada thistle (*Circium arvense*).

7. Root Anatomy and Function

a) Apical Meristem

Yamashita (1991) reviewed the evidence for and against interpreting the primary root of seed plants to be endogenous or exogenous in development, and sided against the view that all roots should be considered endogenous structures. Quiescent center ontogeny during early seed germination was studied by Bitonti et al. (1992). Establishment of the quiescent center in growing roots is related to structural changes of embryo radicle during germination. Davis et al. (1992) proposes a sequence of vacuolar ontogeny in root meristematic cells in relation to tissue germination. Hejnowicz and Karczewski (1993) modeled meristematic growth of root apices by using a complex curbilinear, orthogonal coordinate system, which resembles the patterns of periclinal and anticlinal divisions within the root apex.

Enstone and Peterson (1992a, b) utilized a rapid fluorescence technique to probe the permeability of the root apoplast (extraprotoplasmic component of the plant body). The alkaloid berberine is useful as a mobile apoplastic tracer and is readily precipitated by thiocyanate, forming bright yellow, needle-like fluorescent crystals. Chemicals do not pass the endodermal Casparian band, indicating that the system traces apoplastic pathways. By using this technique it was determined that even though root apices lack suberized structures, zones near the root tip are partially apoplastically isolated from the external environment.

b) Root Structure

Lu et al. (1991) and Rost et al. (1991) described the interesting occurrence of temperature-induced cavities and specialized parenchyma cells in the vascular cylinder of roots of cool-season legumes when grown under warm temperature conditions. When plants were grown at low temperature (below 15 °C) a normal stele formed in the primary roots. However, when grown under warm conditions (above 15 °C), cavities developed in the vascular cylinder, which, in turn, become filled by the ingrowth of specialized parenchyma cells. This is clearly a temperature-induced phenomenon that has no apparent negative effect on seedlings growth, although it is unclear why and how cavities form or how the specialized parenchyma cells develop. Fagerstedt (1991) described the development of the schizogenous spaces in roots and rhizomes of *Carex rostrata* (Cyperaceae). Although spaces form in both root and rhizome, there is no direct contact between the gas spaces of the two organs.

Previous work has reported the loss of nuclei from cells in otherwise healthy looking cortical tissue of seminal roots of wheat and barley. This loss was interpreted as an early stage in a normal nonpathological developmental sequence of cell senescence and death. By using various methods to detect nuclei, Wenzel and McCully (1991) found that cortical cells of mature corn roots remain alive with no evidence of the early loss of nuclei. Elefthesiou et al. (1993) documented the serious effects of aluminate on the morphology and ultrastructure, particularly the cell walls, of root epidermal and root cap cells.

The anatomies of roots with specialized functions have been the subjects of a number of investigations. Three genera of Cyclanthaceae (Monocotyledoneae) have a root-climbing, viny habit, and form roots of different types. The comparative morphology and anatomy of the absorbing roots (feeding roots) and anchoring roots (climbing roots) of these taxa are presented by Wilder (1992a). In the epiphyte *Tillandsia* the absorbing function of the roots no longer exists and the plants are regarded as representing the highest form of morphological and physiological adaptation in the Bromeliaceae. At maturity the roots are either free or function to anchor the plant to its support. Brighigna et al. (1990) investigated the ultrastructure of the apical zones of both free and anchored roots. Anchored roots in comparison to free roots show two features which can be related with their anchoring role: (1) the presence in the root cap of mucopolysaccharide material well suited for contact with the support, and (2) the presence of unicellular trichomes that secrete a lipopolysaccharide substance.

A survey of 34 species of 19 genera of Orchidaceae has revealed the root cortex to be partially or totally constructed of living tracheoidal cells that form an almost velamen-like tissue ("pseudovelamen") (Burr and Barthlott 1991). This region does not appear to be involved in the rapid absorption of water but rather to be a mechanical stabilizer in roots of ecologically specialized taxa, i.e., species of dry habitats as well as taxa that form debris-collecting "trash-baskets" of roots. The anatomy of roots of epiphytic Puerto Rican orchids was the subject of a study by Parrilla and Ackerman (1990). Stern et al. (1993) reported a distinctive and specialized type of amyloplast ("spiranthosome") from cortical root cells of orchids belonging to subfamily Spiranthoideae and suggested that they may be unique to orchids in general and may serve to characterize the spiranthoid orchid subfamily.

Peterson and Perumalla (1990) found that monocot species possessing roots with either a multiseriate hypodermis or epidermis also typically contain a lignosuberized hypodermal Casparian band. It is recommended that the term "exodermis" be used to refer to a hypodermis which has such a Casparian band. Brundrett and Kendrick (1990b) emphasized that the presence of the Casparian band on exodermal and endodermal cell walls may well determine the way in which fungi enter roots of plants with mycorrhizal associations. Some aspects of root structure most likely evolved as a means of regulating the efficiency of mycorrhizal association, and the efficiency of fungal exchange is perhaps one of the most important selective pressures influencing the evolution of root form and structure.

Surveys of root anatomy have been provided by French (1987) on resin canals in roots of Araceae; Brundrett et al. (1990) on root structure of Canadian trees; Raechal and Curtis (1990) on the Bambusoideae (Poaceae), and Fritsch (1992) on the genus *Allium*. Dabydeen and Sirju-Charran (1990) discussed root anatomy in yam bean (*Pachyrhizus erosus* and included an account of the development of "anomalous secondary cambia" that develop from parenchyma cells, and subsequently, "tertiary cambia" that differentiate from tissues produced by the secondary cambia. The development of the interesting tuberous roots of *Habenaria radiata* (Orchidaceae), which are initiated closely under the shoot apex of the main axis, was investigated by Yamashita and Nishikawa (1991). Lemon and Considine (1993) reported on the anatomy and histochemistry of the root system of the kiwifruit vine, *Actinidia deliciosa* var. *deliciosa*.

The term "cluster root" was used by Racette et al. (1990) to refer to a dense cluster of determinate lateral roots (rootlets), in preference to the terms proteoid root and proteoid-like root. An examination of cluster root formation by *Gymnostoma papuanum* (Casuarinaceae) in relation to aeration and mineral nutrient availability found the greatest root formation occurred in plants in aerated water cultures supplied with nitrate and with little or no phosphorus. Everard et al. (1991) showed that hypoxia (5 kPa partial pressure O_2 in nitrogen gas) stimulates mesocotyl root initiation in *Echinochloa phyllopogon* (Poaceae) and that this leads to the increased number of mesocotyl roots observed in hypoxically treated plants. A study of regeneration responses in pea roots after tip excision at different levels found that the type of regeneration response and the vascular pattern within the regenerates were correlated with the state of differentiation of cells at each excision distance (Reihman and Rost 1990).

Piper auritum is a small weedy tree of the neotropics that forms dimorphic prop roots, i.e., with differences in form, structure, and orientation between aboveground and subterranean portions. Greig and Mauseth (1991) presented detailed observations on the structure and function of these roots and found significant anatomical differences that can be related to efficiency of water conduction. Putz (1992) described a technique that can be used for the direct measurement of the force of a single contractile root. By using this method, the downward movement of a plant into the ground can be calculated.

α) Endodermis

There have been very few studies relating the development of the Casparian band to the development of the lateral root and those that have been reported present mixed observations. Peterson and LeFcourt (1990) devised a clearing and fluorochrome staining procedure for broad bean (*Vicia faba*) roots with laterals that allowed observation of the development of tracheary elements and the Casparian band in situ. In *Vicia* the Casparian band is continuous from the main root to the lateral root and encloses the xylem of the main and lateral roots. The Casparian band maturation was simultaneous with the maturation of the protoxylem elements in growing roots. In nongrowing roots, the Casparian band clearly matured distal to the protoxylem. The results illustrate the developmental variation which can occur among lateral roots of one species. The origin of lateral roots from adventitious roots in *Typha glauca* and the participation of the parent root pericycle and endodermis were described by Seago and Marsh (1990). Barnabas and Peterson (1992) examined the initiation, structure, and physiology of Casparian bands and suberin lamellae separated in onion roots by using specialized staining techniques. The Casparian band of the endodermis undergoes considerable growth after its inception and the enlargement appears to be correlated with the development of the suberin lamellae. Growth of the Casparian band during suberin lamellae development may play a role in maintaining the apoplastic barrier in this layer by enlarging the contact area between the band and the lamellae.

c) Tissue Differentiation

Despite the uniqueness and important role of the epidermis in root growth and absorption, there have been relatively few studies of root epidermis differentiation. In an attempt to fill this void, Moore and Smith, (1990) completed a morphometric analysis of epidermal differentiation in primary roots of *Zea mays*. Epidermal differentiation was characterized into six cell types, in sequence, from youngest protodermal cell to cells at the zone of elongation, based on two-dimensional cellular shape and cytoplasmic appearance, Stereology was used to quantify cellular changes associated with differentiation. Histochemical and morphometric analyses were used by Mueller (1991) to identify and define early stages of procambial differentiation in *Trifolium praetense*. By utilizing esterase activity as an early marker in both the cell wall and cytoplasm, it was possible to determine the exact location, size, and shape of the histochemically stained cells within the root apex. Esterase activity identified the proendodermis and procambial cylinder two or three cells proximal to the root cap junction and thus proved to be an earlier marker for procambial differentiation than cytological changes. It appears likely that these techniques will be useful in characterizing procambial pattern development in more complex shoot systems.

8. Wood Evolution

The value of wood anatomical evidence in the study of the phylogeny and classification of dicotyledons is well known and appreciated. In no other vegetative tissue of the plant are the trends of evolution as clearly defined. Significantly, however, these trends were recognized following extensive surveys of extant taxa, albiet without reference to existing taxonomic systems and thus without reference to the relative primitiveness or advancement of the plants in which they occur. In a major contribution, Wheeler and Baas (1991) for the first time undertook a comprehensive analysis of data for the fossil record for dicotyledonous wood in order to test the "Baileyan" model for trends of specialization, i.e., are the widely recognized primitive wood features more common in geologically older woods than younger taxa? It was found, for the most part, that the fossil record supports the widely accepted xylem trends. The incidence of "primitive" features, such as scalariform perforation plates, is much higher in the Cretaceous than in the Tertiary, while the incidence of "advanced" features such as short vessel elements and simple perforation plates, is lower. Correlations of wood anatomical features with ecology do not appear to have been constant throughout time.

a) Vessel Element Structure and Evolution

It is well known that angiosperm vessel elements have been phylogenetically derived from scalariform-pitted tracheids by the loss of pit membranes in the region of tracheid overlap. Tracheids, therefore, are imperforate elements whereas vessel elements are perforate cells since the membranes in end wall perforations undergo lysis at maturity. Vessel elements which are most tracheid-like are the most primitive type. Carlquist (1990b, 1992a, b, d) has discussed the very noteworthy occurrence of diverse types of pit membrane remnants in perforation plates of primitive dicotyledons. As viewed with the SEM, the pit membranes of tracheary elements range from nearly intact to situations where only scattered remnants of the membrane remain. These observations point out the fact that the transition between tracheids and vessel elements is not a sharp one in angiosperms and that various gradations are evident in extant dicotyledons. The presence of membrane remnants in perforations of mature vessel elements should be added to the list of features regarded as primitive in vessel elements of dicotyledons.

Carlquist (1992a) summarized research on the wood structure of Chloranthaceae, a dicotyledonous family that has attracted considerable phylogenetic interest as possible near relatives of monocotyledons. Wood anatomy of each of the four genera was summarized and similarities noted with Lactoridaceae and Piperaceae. The sympodial habit of *Chloranthus* and *Sarcandra*, and the presence of vessels in roots but not in stems of *Sarcandra*, are conditions like those thought to be basic to the origin of monocotyledons. The possibility that Chloranthaceae are close to Piperales and that these groups are close to the origin of monocotyledons should be considered. The idea of a polyphyletic origin of secondary vessellessness was discussed and rejected on the basis of the histological and ecological scenarios that must be hypothesized for these events to have occurred. Of particular significance in this regard is the often overlooked fact that vesseled wood is composed of a heterogeneous collection of cell types, as opposed to the rather homogeneous vesselless condition.

Two recent contributions addressed the taxonomic value of vessel elements; Bhat et al. (1990) for the genus *Hibiscus* and Bhat et al. (1992) for the Sterculiaceae. In the axis of the unusual parasitic dicot *Ombrophyton subterraneum* (Balanophoraceae) all vessel elements possess irregular, knobby ingrowths of unknown adaptive value on their secondary walls (Mauseth and Montenegro 1992).

b) Systematic and Ecological Wood Anatomy

Carlquist (1992f) presented a comprehensive summary of the wood anatomy of 56 sympetalous families, discussing systematic relationships and evolution of the woody habit. Within the Convolvulaceae three new types of vessel restriction patterns (patterns in which vessels are rarely in contact with rays and have a nonrandom distribution, including the absence of vessels within some parts of the secondary xylem) are reported by Carlquist and Hanson (1991). Successive cambia occur in some taxa and fragmentation of the vascular cylinder by growth from the inner cambia leads to initiation of radially oriented cambia ("ray cambia") along the ray zone where fracture occurs. A systematic presentation of the wood anatomical features of woody species throughout Europe has been presented in an atlas by Schweingruber (1990). This book (*Anatomy of European Woods*) includes a chapter on the range and variation of wood anatomy in relation to ecology and evolution (Baas 1990). Trends in the European flora are particularly suggestive of the preferential selection of advanced xylem features in dry, warm regions and in more mesic regions with high temperatures. The same author (Schweingruber 1992) has produced an extensive survey (ca. 670 species from 47 families) of annual growth rings and growth zones in woody Australian plants in areas with Mediterranean and cool, subtropical climates. The need to identify an unknown hardwood specimen and to relate it to a photograph or a published description has been satisfied by an atlas of hardwood identification (Ilic 1991).

The following list contains notable comparative, systematic, and ecological wood anatomical literature of the past 3 years. When discussed, most ecological trends conform to correlations established previously for a number of other families, although in the case of the Rosaceae the percentage of anatomical variation explained by ecological factors is claimed to be quite low (2–10%): Bombacaceae (Miller 1991); Celastraceae, *Bhesa* (Zhang et al. 1990); Cunoniaceae (Patel 1990); Eupomatiaceae (Carlquist 1992e); Lactoridaceae (Carlquist 1990a) Lamiaceae (Carlquist 1992c); Leguminosae (Wheeler and Baas 1992); Loranthaceae (Patel 1991); Meliaceae (Nair 1991; Gasson and Cheek 1992); Proteaceae (Patel 1992); Rosaceae (Zhang and Baas 1992; Zhang et al. 1992); Theaceae (Deng and Baas 1990); Ticodendraceae (Carlquist 1991b) and Ulmaceae (Zhong et al. 1992). Prior and Gasson (1990) reported on the wood anatomy of Afromontane and bushveld species from Swaziland, southern Africa.

9. Hydraulic Conductivity

Recent years have continued to provide important advances in understanding the vascular construction and functioning, or hydraulic architecture, of both stems and roots of flowering plants. The hydraulic design of trees, shrubs, and vines clearly has an immense influence on the overall movement of water (conductance) within the plant body, and is perhaps the single most important factor in determining plant size, the vulnerability of plant axes to periods of drought, the water storage capacity of tissues, and finally the geographic distribution of species. A collection of papers dealing with the topic of structural and functional responses of plants to the environmental stress of water shortage was edited by Kreeb et al. (1989). A number of chapters in this volume are of interest to persons involved with problems of hydraulic architecture. Pallardy (1989), for example, presented an overview of architecture and conductivity and concluded: "It is clear that evolution of xylem as a long-distance transport tissue was quite advantageous to terrestrial plants; however, the persistence of species of diverse xylem anatomy indicates that maximum xylem conductivity is not an attribute of overriding evolutionary value."

a) Stems

An outstanding review of how the hydraulic architecture of trees influenced the movement of water between roots and leaves is provided by Tyree and Ewers (1991). Parameters of particular importance in consideration of hydraulic architecture include: (1) the vulnerability of stems to drought-induced cavitation events, (2) the leaf-specific conductivity of stems because it determines the pressure gradients and most negative water potentials needed to sustain evaporation from leaves, and (3) the water storage capacity of tissues because this might determine the ability of trees to survive long drought periods. All of these parameters are determined by the structure and function of anatomical components of trees.

Ewers et al. (1990) surveyed vessel dimensions in stems of tropical lianas and other growth forms, followed by study of the xylem pathways in liana stems with variant or "anomalous" secondary growth patterns (Fisher and Ewers 1992). By reconstructing serial sections and studying dye movements they were able to reconstruct three-dimensional, long-distance xylem pathways in both unbranched and branched stems. Even in the most complex stems the xylem is structurally integrated by interconnections and appears not to have a particularly obvious influence on hydraulic architecture. Anomalous growth in lianas appears to be most significant in increasing flexibility of stems and as an aid in climbing, protecting the phloem, and increasing the storage parenchyma. Vessel elements in lianas are exceptionally wide and stems are relatively narrow when compared to leaf surface area. Ewers et al. (1991) suggested that water conduction through the xylem may be one of the most important factors limiting growth and survival of lianas. Cavitation is common in lianas and although positive root pressures could theoretically refill cavitated vessels, xylem refilling has been demonstrated only in the grapevine (*Vitis*). Ewers and Chiu (1992) compared wood structure and water transport in different growth forms of temperate honeysuckles (*Lonicera*) – a twiner, a scrambler, and a shrub.

Following a study of the hydraulic architecture of the rose stem (*Rosa hybrida*), Darlington and Dixon (1991) emphasized the role of the poorly vascularized abscission zone between flower and stem as a region of low conductance and high resistance to water flow owing to changes in the alignment of the xylem elements. Suzuki et al. (1991) described the structure of unusual and unique tracheids in wood of the vesselless genus *Te-*

tracentron and suggested they might function to allow rapid water movement in a radial direction. Lewis (1992) presented a method for estimating the hydraulic diameter of a pore or conduit having a noncircular opening. The concept of "hydraulic diameter" gives a convenient description of the hydraulic characterization of both circular and noncircular pores. This is important in water transport research since workers often must estimate the flow rates and other hydraulic characteristics of noncircular pores. The role of vessel occlusions in plants was hypothesized by Bonsen and Kucera (1990). The idea is presented that tyloses evolved in vessels as a barrier by which plants can withstand the spread of microrganisms and disabling embolisms. Tyloses were replaced by the production of gums in more advanced woody plants when vessel element-parenchyma cell pit size became reduced.

The probable role of water storage within the stem of arborescent palms is outlined by Holbrook and Sinclair (1992a, b). An adequate water storage potential of the stem may insure survival of leaves during prolonged periods of low soil water availability and may explain the ability of *Sabel palmetto* to grow in habitats where water is limited. The significant contribution of stored water to the plant's water requirements is apparently greatly enhanced by the occurrence of extremely low minimum leaf conductance.

b) Roots

Markhart and Smit (1990) reviewed the topic of measurement of root hydraulic conductance with the objectives of (1) discussing the nature of the resistance that limit water flux through the root, (2) evaluating methods of measuring root hydraulic conductance, (3) describing the effects of temperature and oxygen on root function, and (4) discussing the nature of the signals that may be sent from the root to the shoot.

Evlagon et al. (1990) reported the interactive effects of salinity and calcium on hydraulic conductivity, osmotic adjustment, and growth in primary roots of maize. Salinity reduces both hydraulic conductivity of the water uptake pathway and root growth, whereas additional calcium partially reversed these reductions. Meinzer et al. (1991) presented results that show that even under favorable conditions, the roots may play an important role in modulating shoot behavior since root hydraulic conductance and total stomatal conductance on a per plant basis changed in parallel during growth of sugarcane as an apparent result of changes in root xylem sap composition. The effects of temperature on the development of metaxylem in primary wheat roots and its hydraulic consequence were considered by Huang et al. (1991). The diameter of the central late metaxylem (LMX) and its porportion to the steles cross-sectional area increased in the acropteal direction in winter wheat primary roots. The diameter of the LMX and stele decreased with an increase in growing temperature. Calculations showed that relative conductivity decreased as growing temperature increased.

Saliendra and Meinzer (1992) pointed out the substantial developmental coordination between stomatal conductance, transpiration, and plant root hydraulic properties. Zhu and Steudle (1991) reported that the double pressure probe technique appears to be an attractive tool for measuring flows of water in tissues and for evaluating models of water transport in roots. In recent years, the traditional view that the bulk of the transpiration water enters the root near the apex has been questioned. In view of the fact that axil roots of maize are immature for some distance behind the apex with closed, nonconducting metaxylem the function of branch roots in water uptake needs to be reevaluated. Varney et al. (1991) presented important data relating to branch root diameter and diameters of the largest open tracheary elements. On the basis of root diameter and vascular anatomy, four classes of branch roots have been distinquished and their possible role in water uptake considered.

A number of recent studies have dealt with problems related to aspects of anatomy and water uptake or loss by roots in desert succulents (*Agave deserti, Opuntia ficus-indica, Ferocactus acanthodes*), plants subjected to prolonged droughts that are interrupted by sporadic and often light rainfall (Nobel et al. 1990; Lopez and Nobel 1991; Alm et al. 1992; North and Nobel 1992). North and Nobel (1991) examined histological and physiological responses caused by drying and rewetting of main roots. Under wet conditions, the formation of new lateral roots facilitates a high hydraulic conductance, although conductance decreases with distance from the apex and with root age. During periods of drought, hydraulic conductivity declined in young roots to an extent proportional with increases in embolism of xylem vessels, suberization of endodermal cells and periderm, formation of cortical lacunae, as well as the abscission of lateral roots, all of which tend to minimize water loss from the roots. All changes are reversible upon rewetting. Lateral roots also undergo structural changes that increase water uptake from wet soil and reduce water loss to dry soil (Huang and Nobel 1992). Roots also tend to shrink during stressful drought periods in response to lowered water potential, forming an air gap between roots and soil that further minimizes water loss from plants to the drier soil (Nobel and Cui 1992b). Additional considerations of changes in roots of desert succulents in drying soils are provided by Nobel and Cui (1992a). It is now clear that the root-soil system acts like a rectifer in response to cycles of soil water content, permitting ready water uptake under wet conditions but restricting water loss during drought. The structure and influence of anatomically complex junctions between main roots and lateral roots on water flow were investigated for *Agave* and *Ferocactus* by North et al. (1992). Under wet conditions, the junctions did not restrict xylem water flow from lateral roots to main roots. Junctions may be important in limiting water loss from main roots during initial phases of drought and hasten the recovery of water uptake by the root system when soil moisture is replenished.

The root-shoot junction was examined anatomically and experimentally in six cereal species by Aloni and Griffith (1991). It is proposed that the concept of "hydraulic segmentation", originally proposed for shoots, be expanded to include root-shoot junctions. Two types of hydraulic architecture were found: (1) a very "safe" root-vessel system, as in winter rye, in which the vessels of the roots are separated from those of the shoots by tracheids, and (2) a completely "unsafe" system, as in corngrass, where the vessels in the root are continuous with the vessels in the shoot. The hydraulic segmentation of the root-vessel system from the shoot-vessel system ("hydraulic safety zone") may represent an important vascular adaptation to dry and cold environments.

References

Alm DM, Cavelier J, Nobel PS (1992) Ann Bot 69:87–92. **Aloni R** (1991) In: Raghavendra AS (ed) Physiology of trees. Wiley, New York, pp 175–197. **Aloni R** (1992) Int J Plant Sci 153:s90–s92. **Aloni R**, Griffith M (1991) Planta 184:123–129. **Aloni R**, Peterson CA (1990) Planta 182:583–590. **Aloni R**, Tollier T, Monties B (1990) Plant Physiol 94:1743–1747. **Arber A** (1950) The natural philosophy of plant form. Univ Press, Cambridge, pp 70–92.

Baas P (1990) In: Schweingruber FH (ed) Anatomy of European woods: an atlas for the identification of European trees, shrubs, and dwarf shrubs. Haupt, Bern, pp 739–765. **Baird JH**, Dute RB, Dickens R (1992) Int J Plant Sci 153:320–328. **Barlow PW** (ed) (1993) The impact of the environment on roots and root systems. Environ Exp Bot 33:205 **Barnabas AD**, Peterson CA (1992) Can J Bot 70:2233–2237. **Barnes PW**, Flint SD, Caldwell MM (1990) Am J Bot 77:1354–1360. **Baum SF**, Aloni R, Peterson CA (1991) Ann Bot 67:543–548. **Bhat RB**, Inamdar JA, Weber DJ (1990) Angew Bot 64:113–121. **Bhat RB**, Inamdar JA, Weber DJ (1992) Angew Bot 66:121–123. **Birch CPD**, Hutchings J (1992) New Phytol 122:545–551. **Birdsall M**, MacLeod RD (1990) Can J Bot 68:747–753. **Bitonti MD**, Chiappetta A, Innocenti AM, Liso R, Arrigoni O (1992) New Phytol 121:577–580. **Bonsen KJM**, Kucera LJ (1990) IAWA Bull ns 11:393–399. **Brighigna L**, Cecchi Fiordi A, Palandri MR (1990) Caryologia 43:27–42. **Brown CL**, Sommer HE (1992) Am J Bot 79:335–346. **Brundrett M**, Kendrick B (1990a) New Phytol 114:457–468. **Brundrett M**, Kendrick B (1990b) New Phytol 114:469–480. **Brundrett M**, Murase G, Kendrick B (1990) Can J Bot 68:551–578. **Burr B**, Barthlott W (1991) Flora 185:313–323.

Calvin CL, Wilson CA, Varughese G (1991) Ann Bot 67:153–161. **Canne-Hilliker JM**, Kampny C (1991) Can J Bot 69:1935–1950. **Carlquist S** (1990a) Am J Bot 77:1498–1504. **Carlquist S** (1990b) Aliso 12:667–684. **Carlquist S** (1991a) In: Putz FE, Mooney A (eds) The biology of vines. Univ Press, Cambridge, pp 53–71. **Carlquist S** (1991b) Ann Mo Bot Gard 78:96–104. **Carlquist S** (1992a) IAWA Bull ns 13:3–16. **Carlquist S** (1992b) Am J Bot 79:660–672. **Carlquist S** (1992c) Aliso 13:309–338. **Carlquist S** (1992d) Aliso 13:447–462. **Carlquist S** (1992e) Bull Torrey Bot Club 119:167–180. **Carlquist S** (1992f) Ann Mo Bot Gard 79:303–332. **Carlquist S**, Hanson MA (1991) Aliso 13:51–94. **Colby SM**, Juncosa AM, Stamp JA, Meredith CP (1991) Am J Bot 778:260–269. **Curtis JD**, Lersten NR (1990) Nord J Bot 10:443–449. **Cutler DF** (1992) Bot J Linn Soc 110:385–419.

Dabydeen S, Sirju-Charran G (1990) Ann Bot 66:313–320. **Darlington AB**, Dixon MA (1991) Can J Bot 69:702–710. **Davis KL**, Davies MS, Francis D (1992) New Phytol 121:581–585. **de Castro LAB**, Carneiro M, Neshich D de CM, de Paiva GR (1992) Plant Cell 4:1549–1559. **Deng L**, Baas P (1990) IAWA Bull ns 11:337–378. **Dewit L**, Reid DM (1992) Int J Plant Sci 153:556–564. **Dharmawardhana DP**, Ellis BE, Carlson JE (1992) Can J Bot 70:2238–2244. **Dickison WC** (1992) In: Behnke H-D, Esser D, Kubitzki K, Runge M, Ziegler H (eds) Progress in botany, vol 53. Springer, Berlin Heidelberg New York, pp 44–62. **Doaigey AR** (1991) Am J Bot 78:1608–1616.

Eleftheriou EP, Moustalaas M, Fragiskos N (1993) J Exp Bot 44:427–436. **Enstone DE**, Peterson CA (1992a) Can J Bot 70:1493–1501. **Enstone DE**, Peterson CA (1992b) Can J Bot 70:1502–1512. **Everard JD**, Le Cain DR, Rumpho ME, Kennedy RA (1991) Am J Bot 78:462–469. **Evlagon D**, Ravina I, Neumann P (1990) Isr J Bot 39:239–247. **Ewers FW**, Chiu S-T (1992) Trees 6:216–224. **Ewers FW**, Fisher JB, Chiu S-T (1990) Oecologia 84:544–552. **Ewers FW**, Fisher JB, Fichtner K (1991) In: Putz FE, Mooney HA (eds) The biology of vines. Univ Press, Cambridge, pp 127–160.

Fagerstedt KW (1991) Nord J Bot 12:115–120. **Fiqueiredo-Ribeiro R de CL**, Isejima EM, Campos Dietrich SM, Chaves Correa JB (1992) Ann Bot 70:405–408. **Fisher JB** (1992) Int J Plant Sci 153:S137–S146. **Fisher JB**, Ewers FW (1991) In: Putz FE, Mooney HA (eds) The biology of vines. Univ Press, Cambridge, pp 99–124. **Fisher JB**, Ewers FW (1992) Bot J Linn Soc 108:181–202. **Fisher JB**, Rutishauser R (1990) Can J Bot 68:2316–2328. **Fitter AH**, Stickland TR

(1992) New Phytol 121:243–248. **Folkerts GW** (1989) Phytomorphology 39:285–289. **French JC** (1987) Bot Gaz 148:360–371. **Fritsch R** (1992) Beitr Biol Pflanz 67:129–160.

Gasson P, Cheek M (1992) Kew Bull 47:753–758. **Gastiazoro MTCD** (1991) Kurtziana 21:111–152. **Gilman EF** (1989) J Environ Hort 7:88–90. **Green PB** (1992) Int J Plant Sci 153:S59–S75. **Greig N**, Mauseth JD (1991) Bull Torrey Bot Club 118:176–183. **Gross KL**, Maruca D, Pregitzer KS (1992) New Phytol 120:535–542.

Hagemann W (1992) Int J Plant Sci 153:S38–S48. **Hara N** (1991) Phytomorphology 41:331–340. **Hejnowicz Z**, Karczewski J (1993) Am J Bot 80:309–315. **Hetrick BAD**, Wilson GWT, Leslie JF (1991) Can J Bot 69:112–135. **Holbrook NM**, Sinclair TR (1992a) Plant Cell Environ 15:393–400. **Holbrook NM**, Sinclair TR (1992b) Plant Cell Environ 15:401–410. **Huang BR**, Nobel PS (1992) J Exp Bot 43:1441–1449. **Huang BR**, Taylor HM, McMichael BL (1991) Ann Bot 67:163–166. **Huante P**, Rincon E, Savito M (1992) Trees 6:77–82.

Ilic J (1991) CSIRO Atlas of hardwoods. Crawford House, Bathurst, Australia, 525 pp.

Kaplan DR (1992) Int J Plant Sci 153:S28–S37. **Kaplan DR**, Hagemann W (1991) Bioscience 41:693–703. **Kathal R**, Bhatnagar SP, Bhojwani SS (1992) Beitr Biol Pflanz 67:45–53. **Kreeb KH**, Richter H, Hinckley TM (eds) (1989) Structural and functional responses to environmental stresses: water shortage. SPB Acad, The Hague, 308 pp.

LaCroix CR, Gerrath JM, Posluszny U (1990) Bot Gaz 151:204–209. **Lavania S**, Govil CM (1991) J Indian Bot Soc 70:87–90. **Lee DW**, Richards JH (1991) In: Putz FE, Moony HA (eds) The biology of vines. Univ Press, Cambridge, pp 205–243. **Lemon CW**, Considine JA (1993) Ann Bot 71:117–129. **Lev-Yadun S**, Aloni R (1991a) Trees 5:22–29. **Lev-Yadun S**, Aloni R (1991b) Ann Bot 68:85–91. **Lev-Yadun S**, Aloni R (1992a) Int J Plant Sci 153:348–357. **Lev-Yadun S**, Aloni R (1992b) Ann Bot 70:379–386. **Lewis AM** (1992) Am J Bot 79:1158–1161. **Lopez F**, Nobel PS (1991) J Exp Bot 42:143–149. **Lu P**, Gladish D, Rost TL (1991) Am J Bot 78:729–739.

Markhart AH, Smit B (1990) Hortscience 25:282–287. **Mathur J** (1992) Ann Bot 70:419–422. **Mauseth JD**, Montenegro G (1992) Am J Bot 79:456–458. **Mauseth JD**, Sajeva M (1992) Ann Bot 70:317–324. **Mauseth JD**, Hsiao S-C, Montenegro G (1992) Bull Torrey Bot Club 119:407–417. **McDonald JA** (1992) Bull Torrey Bot Club 119:262–267. **Medford JI** (1992) Plant Cell 4:1029–1039. **Medford JI**, Elmer JS, Klee HJ (1991) Plant Cell 3:359–370. **Meinzer FC**, Grantz DA, Smit B (1991) Aust J Plant Physiol 18:329–338. **Miller RB** (1991) Brittonia 43:88–92. **Miller RM**, Kaul V, Hutchinson JF, Richards D (1991) Ann Bot 67:35–42. **Moore R**, Smith H (1990) Am J Bot 77:727–735. **Mueller RJ** (1991) Am J Bot 78:53–62. **Murphy RJ**, Alvin KL (1992) IAWA Bull ns 13:403–410.

Nadeau LB, Vanden Born WH (1989) Can J Plant Sci 69:1199–1206. **Nair MNB** (1991) Phytomorphology 41:63–73. **Nair MNB**, Mohan Ram HY (1992) Curr Sci 62:746–751. **Niklas KJ** (1990) Am J Bot 77:762–771. **Niklas KJ** (1992) Plant biomechanics. An engineering approach to plant form and function. Univ Press, Chicago, pp 607. **Niklas KJ**, Kaplan DR (1991) In: Dudley EC (ed) The unity of evolutionary biology. Dioscorides, Portland, Oregon, pp 489–502. **Nobel PS**, Cui M (1992a) Ann Bot 70:485–491. **Nobel PS**, Cui M (1992b) J Exp Bot 43:319–326. **Nobel PS**, Meyer RW (1991) Am J Bot 78:1252–1259. **Nobel PS**, Schulte PJ, North GB (1990) J Exp Bot 41:409–415. **North GB**, Nobel PS (1991) Am J Bot 78:906–915. **North GB**, Nobel PS (1992) New Phytol 120:9–19. **North GB**, Ewers FW, Nobel PS (1992) Am J Bot 79:1039–1050.

Pallardy SG (1989) In: Kreeb KH, Richter H, Hinckley TM (eds) (1989) Structural and functional responses to environmental stresses: water shortage. SPB Acad, The Hague, pp 3–19. **Parrilla AT**, Ackerman JD (1990) Orquidea 12:105–116. **Patel RN** (1990) NZ J Bot 28:347–355. **Patel RN** (1991) NZ J Bot 29:429–449. **Patel RN** (1992) NZ J Bot 30:415–428. **Paukkonen K**, Kauppi A, Ferm A (1992) Flora 187:353–367. **Peterson CA**, LeFcourt BEM (1990) Can J Bot 68:2729–2735. **Peterson CA**, Perumalla CJ (1990) Bot J Linn Soc 103:113–125. **Pihakaski-Maunsbach K**, Walles B (1990) Nord J Bot 10:493–500. **Posluszny U**, Tomlinson PB (1991) Can J Bot 69:1600–1615. **Pri-Hadash A**, Hareven D, Lifschitz E (1992) Plant Cell 4:149–159. **Prior JAB**, Gasson PE (1990) IAWA Bull ns 11:319–336. **Putz N** (1992) Can J Bot 70:1433–1439. **Putz FE**, Holbrook M (1991) In: Putz FE, Mooney HA (eds) The biology of vines. Univ Press, Cambridge, pp 73–97. **Putz FE**, Mooney HA (eds) (1991) The biology of vines. Univ Press, Cambridge, 526 pp.

Racette S, Louis I, Torrey JG (1990) Can J Bot 68:2564–2570. **Raechal LJ**, Curtis JD (1990) Am J Bot 77:475–482. **Raju JSSN** (1991) J Indian Bot Soc 70:235–237. **Ray TS** (1990) Am J Bot 77:1599–1609. **Reihman MA**, Rost TL (1990) Am J Bot 77:1159–1167. **Rost TL**, Lu P, Gladish D (1991) Bot Acta 104:300–305. **Rudall P** (1991) Bot Rev 57:150–163.

Saliendra NZ, Meinzer FC (1992) J Exp Bot 43:1209–1217. **Sattler R**, Rutishauser R (1992) Bot Jahrb 114:61–80. **Schweingruber FH** (1990) Anatomy of European woods. Haupt, Bern, 802 pp. **Schweingruber FH** (1992) IAWA Bull ns 13:359–379. **Seago JL Jr**, Marsh L (1990) Am J Bot 7:713–721. **Serrato Valenti G**, Ferro M, Ferraro D, Riveros F (1991) Ann Bot 68:47–53. **Serrato Valenti G**, Melone L, Orsi O, Riveros F (1992) Ann Bot 70:399–404. **Sitte P** (1992) Int J Plant Sci 153:S1–S6. **Smith GS**, Curtis JP, Edwards CM (1992) Ann Bot 70:265–269. **Standley LA** (1990) Can J Bot 68:1449–1456. **Stern WL**, Morris MW (1992) Lindleyana 7:34–53. **Stern WL**, Aldrich HC, McDowell LM, Morris MW, Pridgeon AM (1993) Protoplasma 172:49–55. **Suzuki M**, Joshi L, Fujii T, Noshiro S (1991) IAWA Bull ns 12:23–33.

Takhtajan A (1991) Evolutionary trends in flowering plants. Columbia, New York, 241 pp. **Tourn GM**, Tortosa RD, Medan D (1992) Bot J Linn Soc 108:275–286. **Tyree MT**, Ewers FW (1991) New Phytol 119:345–360.

Varney GT, Canny MJ, Wang XL, McCully ME (1991) Ann Bot 67:357–364.

Wenzel CL, McCully ME (1991) Am J Bot 78:1528–1541. **Wheeler EA**, Baas P (1991) IAWA Bull ns 12:275–332. **Wheeler EA**, Baas P (1992) In: Herendeen PS, Dilcher DL (eds) Advances in legume systematics: Pt 4. Fossil Record, Royal Bot Gard, Kew, England, pp 281–301. **Wilder GJ** (1992a) Can J Bot 70:38–48. **Wilder GJ** (1992b) Can J Bot 70:1388–1400.

Yamashita T (1991) Beitr Biol Pflanz 66:371–391. **Yamashita T**, Nishikawa K (1991) Beitr Biol Pflanz 66:351–370.

Zhang S-Y, Baas P (1992) IAWA Bull ns 13:21–91. **Zhang SY**, Baas P, Zandee M (1992) IAWA Bull ns 13:307–349. **Zhang X**, Baas P, Mennga AMW (1990) IAWA Bull ns 11:57–60. **Zhong Y**, Baas P, Wheeler EA (1992) IAWA Bull ns 13:419–453. **Zhou TS**, Hara N (1990) Can J Bot 68:2576–2582. **Zhu G**, Steudle E (1991) Plant Physiol 95:305–315. **Zobel AM**, Hrazdina G (1992) Ann Bot 70:423–427.

Professor Dr. William C. Dickison
Department of Biology
The University of North Carolina
Chapel Hill, NC 27514, USA

B. Physiology

I. Electrophysiology of Stomata

By Gerhard Thiel and Dietrich Gradmann

1. Introduction

The discovery in 1968 that osmotically driven movement of guard cells is correlated with substantial transport of K^+ through the plasmalemma (Fisher 1968) has challenged electrophysiologists ever since; but only recent progress in tailoring the experimental conditions suitable for the use of conventional microelectrodes (Blatt 1987a, b) or application of patch-clamp techniques to guard cell protoplasts (Schroeder et al. 1984) has paved the way to studying electrical parameters of the membrane in greater detail. Today, guard cells are, for diverse reasons, the electrophysiologically best investigated cells of vascular plants. First, because stomatal physiology has always been a major topic in botany. Second, because guard cells are exceptional in that they have no cytoplasmic contact to adjacent cells (Wille and Lucas 1984). This provides the researcher with the desired system of an electrically isolated cell and makes guard cells most suitable for both physiological and biophysical studies. Finally, for patch-clamp investigations, guard cells offer another benefit: the small protoplasts can be easily distinguished from the mass of cells from other tissues.

Over the past years, aspects of stomatal movement have been a favorite topic of many reviews with varying emphasis (Pallas 1966; Raschke 1975; MacRobbie 1988, 1992; Raschke et al. 1988; Mansfield et al. 1990; Blatt 1991; Hedrich and Jeromin 1992; Schroeder 1992). The most recent ones demonstrate the successful application of the hormone concept adopted from animal physiology: the function of guard cells is subjected to numerous exogenous and endogenous factors via complex signal transduction cascades employing ion transport through the plasmalemma to a varying extent. The main ion transporters have, however, immediate effects and immediate dependencies on the membrane voltage. The present chapter focuses mainly on these processes on the plasmalemma. The role of electrical processes at endomembranes in guard cells will not be considered here, as there is no direct information on this issue.

In the first Section, the most important methods in the field are discussed. In the second, the rapidly growing knowledge of the properties of the main transport processes will be reviewed in a physiological context. In the last Section, we try to summarize the data in terms of a model for electrical and osmotic relations of stomata.

Progress in Botany, Vol. 55

2. Benefits and Drawbacks of Various Methods

The following paragraphs discuss the possibilities and limitations of the most important techniques which are currently used in stomatal electrophysiology.

a) Voltage Clamp on Cells

The conventional voltage clamp technique is the best approach to analyze the electrical properties of the membrane in the physiologically intact cell. The few researchers who do this have already worked out enough current-voltage data for a solid framework of the steady-state electrical membrane characteristics (Blatt 1991). It can be expected that dynamic properties in the ms range will be worked out in the near future. The setup of conventional voltage clamp with impaled micropipettes allows reliable long-term recordings of the free-running membrane voltage because of minor exchange between pipette solution and cytoplasm. On the other hand, the very process illustrates a disadvantage of the conventional voltage clamp. Other than in patch clamp, the cell interior can be perfused with well-defined media to a limited extent only (Blatt et al. 1990a; Thiel and Blatt 1991). Furthermore, only macroscopic membrane currents through an ensemble of transporters can be recorded, rather than microscopic currents through individual transporter molecules.

b) Patch Clamp on Protoplasts

Introduction of patch-clamp techniques (Hamill et al. 1981) to plants (Moran et al. 1984; Schroeder et al. 1984) has stimulated stomatal physiology tremendously. As already mentioned, the main advantages compared to the conventional voltage clamp method are the high resolution of current measurements and the chemical access to both sides of the membrane, either by using excised patches of either orientation ("right-side-out" or "inside-out"), or by equilibration of the pipet solution with the cell interior in the "whole-cell" configuration. The consequence is usually that the environment of the membrane on the cytoplasmic side is no longer physiological. Accordingly, "rundown" effects (relaxations) are observed in the whole-cell configuration, when pipette solution and cell interior equilibrate just after breaking the patch (Oleson et al. 1993). These relaxations to non-physiological steady-state conditions are disadvantageous, as such, but systematic measurements of the actual speed of relaxation can be used to estimate the molecular size of the essential factors (Pusch and Neher 1988). An experimental challenge is to minimize unwanted rundown effects by composing the pipette solution in such a way that it simulates the cytoplasmic environment, e.g., adding nucleotides effectively prevents rundown of slow and fast activating anion channels in guard cell protoplasts; (Schroeder and Hagiwara 1989; Hedrich et al. 1990; Schroeder and Keller 1992). If crucial factors remain unknown, a quasi intact cytoplasmic environment can be maintained by the "slow whole-cell" method, in which low resistance access to the cytoplasm is achieved not by breaking the entire patch but by perforating the membrane patch with ionophores (Schroeder 1988; Korn and Horn 1989; White et al. 1991; Oleson et al. 1993).

A general problem of the patch clamp method on plant membranes is that the necessary contact of the patch pipette with the membrane requires isolation of the membrane from its external structure (tonoplasts from cytoplasm and plasmalemma from cell wall). This usually requires preparation and/or use of protoplasts. Since the physiological state of the protoplast membranes is basically unknown, the biological meaning of the results is problematic without additonal information.

Most patch-clamp experiments on guard cell protoplasts are carried out in the whole-cell configuration i.e., actually *without* a patch in front of the pipette. In this configuration – as in conventional voltage-clamp experiments – only macroscopic currents through an ensemble of transporters are recorded. Documentations of microscopic currents through single channels of guard cells are rare. Usually they represent episodes of certain phenomena rather than the basis of systematic investigations of single channels. In fact, microscopic investigations are very difficult, not only for experimental reasons. Since a channel can change its switching mode upon minor conditional changes or even spontaneously, the question of the identity of a channel type may have no unambiguous answer.

c) Combination of Methods

Guard cells from *Vicia faba*[1] are, like no other plant cell, investigated both as intact cells by conventional voltage-clamp experiments and as protoplasts by patch-clamp techniques. Qualitatively, both preparations show close resemblance regarding the presence of individual transporters as well as in their sensitivity to voltage and other control factors. Nonetheless, in a more detailed scrutiny, subtle differences can be found, for example with respect to pH-dependency (Schroeder 1988; Blatt 1992; Fairley-Grenot and Assmann 1993), the K^+ current activation kinetics, and the dependence on the external K^+ concentration, K^+_o (Schroeder and Fang 1991; Blatt 1992). An extended, systematic comparison could provide valuable information on the role of the cell wall and of turgor pressure in the gating characteristics of individual transporters.

3. Major Transport Elements

In historical terms, the electrophysiology of stomata started with K^+ transport across the plasmalemma of guard cells (Fisher 1968). This historical priority of K^+ is frequently misinterpreted as a causal priority, when transporters for ions other than K^+ are introduced as devices for charge compensation (e.g., Mohr and Schopfer 1992). Now it is generally accepted that a P-type H^+ ATPase (Pedersen and Carafoli 1987) is the dominant primary

---

[1] If not noted otherwise, the data in this chapter refer to measurements on the plasmalemma of guard cells from *Vicia faba*.

active ion transporter in the plasmalemma of vascular plants, and that this ATPase (short the H^+ pump) process eventually energizes transport through channels and cotransporters by an electrochemical gradient.

For economic reasons, several matters are substantially simplified here. For example, only K^+ will be considered for cation relations, and Cl^- for anions, in spite of the important role of organic anions (in particular of malate) in stomatal physiology (Raschke et al. 1988).

With these simplifications, the osmotic and electric relationships in guard cells can be described in terms of five ion transport systems: an electrogenic H^+ pump which serves as motor, an inward rectifying K^+ channel, an outward rectifying K^+ channel, a $(2H - Cl)^+$ symporter for Cl^- uptake, and a Cl^- channel for Cl^- release. Four of these transporters are well characterized from studies with guard cells. The Cl^- uptake by symport with protons has not yet been described in guard cells. However, some anion uptake system must exist in guard cells, and a $(2H - Cl)^+$ symporter such as described for *Chara* (Beilby and Walker 1981; Sanders and Hansen 1981; Sanders 1984) may serve here to complete the model.

The knowledge about the electrical properties of these five elements is briefly summarized in the next paragraphs and listed in condensed form in Table 1. Cross reference between Table 1 and the corresponding, more detailed information in the text is marked with captial superscripts. The electrical relations are illustrated in Fig. 1. They can be expressed in form of an equivalent circuit of the membrane with a parallel arrangement of the five transporters in form of batteries with their inner conductances (left part of Fig. 1). The voltage and time dependencies of these conductances are illustrated in the right columns of Fig. 1; not only the steady-state values are a specific function of the voltage (middle column), also the relaxation kinetics for the conductance changes upon voltage steps (right column) are characteristics for each transporter. These relations represent ex-

Table 1. Direct and/or indirect effects of various ligands on main ion transporters in guard cells

Ligand	H^+ pump	K^+ inw. rectf	K^+ outw. rectf.	Cl^- channel
Inside				
H^+	–	[G]↑	[N]↓	–
$Ca^{2+}{}_c$	–	[D]> 100 nM↓	[O]→	[P]↑
Nucleot.	[A]ATP↑	[F]GTPγS↓ [F]GDPβS↑	–	[Q]A(G)TPγS↑
Na^+	–	[M]→	[M]↓	–
Outside				
H^+	[B]↓	[J]↑	[J]→	–
$Ca^{2+}{}_o$	–	[I]↓	[L]↓	–
K^+	–	[H]↑	[K]↓	–
Auxins	[C]↑	[R]→	–	[R]↑
ABA	–	[E]↓	[E]↑	[E]↑

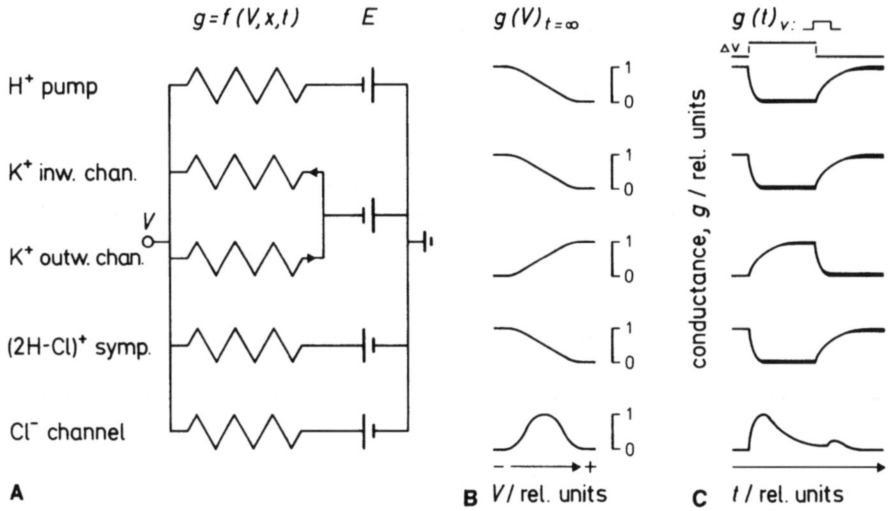

Fig. 1 A–C. Analog circuit of the guard cell plasmalemma. **A** Parallel arrangement of main ion transporters with effects and dependencies on the membrane voltage. **B** Scheme of voltage dependencies of steady-state chord conductances. **C** Scheme of conductance relaxations upon voltage steps of eiter sign. *Table below* reasonable set of system parameters (for explanation see text) according to Gradmann et al. 1993

Transporter		E/mV	k_A^0/S^{-1}	k_I^0/S^{-1}	δ_A-δ_I
Electrogenic H^+ pump		– 400	1	40	–1
K^+ inward-rectifyer		– 100	1	100	–1
K^+ outward-rectifyer		– 100	10	1	1
$(2H – Cl)^+$-symporter		200	0.015	1	–1
	Fast		50	0.5	1
Cl^- channel		100			
	Slow		1	0.1	–1

perimental data – as far as available – and supplementary assumptions. In the following paragraphs, these five major transporters are treated in more detail.

a) Proton Pump

A P-type H^+-ATPase which exports H^+ against its electrochemical gradient, short H^+ pump, is by far the dominating active ion transporter in the plasmalemma of plant cells. It mediates metabolic energy to other ion transporters by creating a proton motive force, with a chemical component, $\Delta pH \cdot F/(RT)$, and an electrical component, the transmembrane voltage V. On a low background conductance of passive transporters (in 0 to 100 μM K^+_o), the pump is able to polarize intact guard cells to $V < -280$ mV (Thiel et al. 1992). In extreme cases, voltage down to –350 mV could be measured (G. Thiel and M.R. Blatt, unpubl. results). Maximum values of –260 mV were reported from protoplasts (Lohse and Hedrich 1992). The few kinetic data from direct investigations of the

H^+ pump in guard cells (Blatt 1987b; Lohse and Hedrich 1992) correspond to the gerneral properties of the H^+ pump in vascular plants, algae, and fungi (Gradmann et al. 1978; Blatt et al. 1990b; Läuger 1991).

By comparison of the membrane current/voltage relationships at different external pH or different energetic conditions, the equilibrium voltage of the pump can be determined ($V_P < -300$ mV, depending, e.g., on pH_o) by kinetic and thermodynamic criteria (Blatt 1987b). The very negative V_P corresponds to a rounded stoichiometry of 1 H^+ transported per ATP hydrolyzed.

The maximum pump currents obtained in intact cells (Blatt 1987b; Thiel et al. 1992) are marginally larger (20–200 mA m^{-2}) than those from protoplasts (30–40 mA m^{-2}: Lohse and Hedrich 1992). The voltage dependence of the pump conductance is equivalent in intact cells (Blatt 1987b; Thiel et al. 1992) and in protoplasts (Lohse and Hedrich 1992): the conductance of the pump has a maximum at V around –200 mV and decreases with positive-going voltage over a wide range of physiologically relevant voltages (Blatt 1987b; Lohse and Hedrich 1992; Thiel et al. 1992). For a simplified, formal evaluation, the nonohmic conductance of the pump can be described with a simple, two-state gating scheme, $A \leftrightarrow I$, with spontaneous but voltage-dependent transitions between an active state, A, and an inactive state, I, where positive voltage pushes the system towards I.

The nature of the electrogenic pump as ATP-driven H^+ translocator provides a crucial role for cytoplasmic ATP[A] and Mg^{2+} as well as pH_c and pH_o[B] for control functions over the pump current (Läuger 1991). In electrical measurements, the thermodynamically expected roles of ATP and pH_o for the electrogenic pump current have been more or less confirmed (Blatt 1987a; Lohse and Hedrich 1992). There is no doubt about (but essentially no experimental evidence for) the crucial role of the cytoplasmic pH and Mg^{2+} also for the ATPase in guard cells.

The presence of these intrinsic factors which potentially affect pump conductance are not indicative per se for a physiological relevance. In fact, relative changes of the high concentration of cytoplasmic ATP[A] for signalling purposes would be energetically very inefficent (Takeuchi and Kishimoto 1983; Hansen 1990).

The physiologically interesting question in which way stimuli that catalyze guard cell swelling, such as blue and red light or auxins[C] (Assmann et al. 1985; Serrano et al. 1988; Lohse and Hedrich 1992) enhance a potential pump conductance in guard cell protoplasts, remains unknown. Apart from modulating one of the substrate factors, additional control mechanisms may be employed. One suggestion is that protein phosphorylation does in some way stimulate pump conductance. Synthetic diacylglycerols (sDG) promoted, possibly through activation of a protein kinase, light-induced guard cell (*Commelina communis*) swelling (Lee and Assmann 1991). The hypothesis of pump modulation through phosphorylation is tempting (Schaller and Sussman 1988) but the electrophysiological identification of an sDG-activated, ATP-dependent current in protoplasts of *Vicia faba* guard cell is as yet tentative at best.

It should be mentioned in this context that the pump activity appears to be reduced during the winter season, both in intact cells (Blatt and Clint 1989; Thiel et al. 1992) and in protoplasts (Lohse and Hedrich 1992) from plants which have been attempted to be grown under a constant temperature and light/dark regime. In the context of photoperiodism, these observations are not sensational. However, such seasonal fluctuations can be very important for planning, performing, and interpreting experimental work.

b) K^+ Inward Rectifier

Inward rectifying currents which reflect an ion-specific influx of K^+ into guard cells have been reported from the plasma membrane of intact guard cells and their protoplasts (e.g., Schroeder 1988; Blatt 1992). The unitary conductance of the principal ion channel which underlies the macroscopic currents is in the range of 4 to 10 pS in {104 K/11 K}[2] (Schroeder and Hagiwara 1989). The available single-channel data demonstrate qualitatively that the inward rectification of the macroscopic current is consequent on the open probability which becomes apparent at negative *V* around −100 mV and rises sigmoidally with progressively negative *V*. This voltage-dependent gating, which makes this channel a K^+ inward rectifier, can be described again by a voltage-dependent equilibrium, $A \leftrightarrow I$, between an active (open) state *A* and an inactive (closed) state *I*, which have the same probability (0.5) at a characteristic gating voltage $V_{1/2}$[3].

Apart from the pronounced voltage-dependency, a number of additional factors have been reported to modulate the characteristics of these channels (Table 1). Among these, the effects of Ca^{2+} and pH from both sides of the membrane are best documented.

Cytoplasmic Ca^{2+} (Ca^{2+}_c) inactivates the K^+ inward rectifier [D] (Schroeder and Hagiwara 1989) over a range from a new nM to some µM Ca^{2+}_c (Schroeder and Hagiwara 1989; Fairley-Grenot and Assmann 1992a) – apparently through an extension of the closed times. Inhibition by Ca^{2+}_c cooperates with the inhibition by positive-going voltages, which cause the inward rectifying characteristics of the macroscopic current-voltage relationship of this channel (Schroeder and Hagiwara 1989).

The K^+ inward rectifier is inhibited in a similar fashion, when intact guard cells are challenged either with 10 µM abscisic acid (ABA)[E] or following evaluation of the concentration of inositol-1,4,5-trisphosphate (IP_3) in the cytoplasm (Blatt et al. 1990a; Thiel et al. 1992). Comparable also is the inhibitory effect of the nonhydrolysable GTP (guanosintrisphosphate) analogue GTPγS[F] when dialyzed at 500 µM into guard cell protoplasts (Fairley-Grenot and Assmann 1991). The impact of these diverse treatments on the K^+ inward current is suppressed by a high cytoplasmic Ca^{2+} buffer capacity (Blatt et al. 1990a; Fairley-Grenot and Assmann 1991; Lemtiri and MacRobbie in prep.). It can, therefore, be anticipated that a rise in Ca^{2+}_c is an intermediate step through which G-protein-linked signal transduction mechanism(s) modulate K^+ inward current in guard cells.

The K^+ inward rectifier is weakly stimulated by cytoplasmic acidification[G] (Blatt 1992; Blatt and Armstrong 1993). Due to the still uncertain relationship between cyto-

---

[2] For an abbreviated presentation of the important information about compsitions of media on the two sides of the membrane, we use the form {cytoplasmic side/outside}. For example, {200 K, 0.1 Ca/50 K} means that the solution on the cytoplasmic side contained 200 mM K^+ plus 0.1 mM Ca^{2+} and the extracellular solution contained 50 mM K^+. Constituents which are not specified (such as Cl^-, H^+ or external Ca^{2+} in our example) are indicated to have small or unknown effects.

[3] For the description of voltage-dependent processes, "threshold potential" or "activation potential" are frequently used for the voltage (range) at which effects become qualitatively apparent. This definition is not precise (Hille 1992). Corresponding to classic enzyme kinetics, we suggst to use a characteristic gating voltage, $V_{1/2}$ (analog to K_D), at which the channel has 50% of its saturation activity.

plasmic pH and Ca^{2+}_c, it cannot be excluded that the effect of pH is secondary and only mediated through a modulation of Ca^{2+}_c (MacRobbie 1992; Blatt and Armstrong 1993).

In intact guard cells, external K^{+}[H] does not seem to alter the activation kinetics of the K^+ inward rectifier upon negative-going V (Blatt 1992). Only in protoplasts do these kinetics appear to be slightly steeper upon a rise in external K^+ (Schroeder and Fang 1991). The insensitivity of gating to K^+_o is typical for inward rectifiers in plant cell membranes, whereas the $V_{1/2}$ of the classic inward rectifying channels from animal cells shifts in parallel with E_K upon changes in K^+_o (Blatt 1992; Kubo et al. 1993).

External Ca^{2+} (Ca^{2+}_o) has multiple effects on the gating of K^+ inward currents[I]. At concentrations between 0.1 and 1 mM Ca^{2+}_o the effect on K^+ channels is dominated by the efficiency of the divalent cation to screen surface charges (Blatt 1992). At ≥ 10 mM Ca^{2+}_o, competition of Ca^{2+} with K^+ for the substrate binding site becomes evident in guard cell protoplasts of *Vicia faba* and of *Zea mays*, resulting in a voltage-dependent block of K^+ inward currents (Busch et al. 1990; Fairley-Grenot and Assmann 1992a, b). However, under these same conditions, an increased net Ca^{2+} influx can be anticipated (Hedrich et al. 1990; Lohse and Hedrich 1992) with its consequent inhibition of K^+ inward currents via elevated Ca^{2+}_o (Blatt 1992).

Gating of K^+ inward currents in intact guard cells is also subject to a direct modulation by extracellular pH[J]. Similar to the K^+ current modulation by elevated Ca^{2+}_c, an alkaline shift of pH_o also cooperates with positive voltage in K^+ current inhibition (Blatt 1992).

c) K^+ Outward Rectifier

Outward rectification of K^+ currents around E_K is well documented by macroscopic currents through the plasma membrane of isolated guard cells (Blatt 1988; Thiel and Blatt 1991) as well as their protoplasts (Schroeder et al. 1987; Schroeder 1988; Fairley-Grenot and Assmann 1992a, 1993). The molecular basis for the voltage- and time dependence of these currents must be sought in the kinetic properties of single K^+ conducting channels.

An elementary device which accounts for these properties is a channel with 10 pS conductance in {175 K/11 K} (Schroeder et al. 1987) or 25 pS in {50 K/50K} (Hosoi et al. 1988). The contribution of a cation conducting channel with only 5 pS in {210 K/35 K} (Schroeder 1988) to the outward rectifying K^+ transport system remains uncertain.

The switching kinetics of the single channels correspond to a model with an open state and two closed states in series ($A \leftrightarrow I_1 \leftrightarrow I_2$), where the equilibration $A \leftrightarrow I_1$ is relatively fast (ms) and voltage-independent, whereas the transition probability for $I_1 \rightarrow I_2$ becomes low ($< 1\ s^{-1}$) for large negative voltages (Hosoi et al. 1988). Correspondingly, the same model is suitable to describe the relaxation kinetics of activation and inactivation of macroscopic K^+ currents upon voltage steps by the sum of two exponentials (Fairley-Grenot and Assmann 1993). Still missing is the explicit analysis (with time constants *and* amplitudes of the exponential component) of these microscopic and macroscopic K^+ currents in terms of the model $A \leftrightarrow I_1 \leftrightarrow I_2$ according to the available formalism (Bertl et al. 1988).

The activation kinetics of the K^+ outward rectifier show a striking K^+-dependence[K]. The basic observation is: if the channel is mostly inactive for K^+ import at a certain $V < E_K$, and E_K is shifted to $E_K' < V$ by reduction of external K^+, the same channel will be activated for K^+ export at the same voltage V (Blatt 1991). In kinetic terms: the characteristic gating voltage, $V_{1/2}$, depends on external K^+ in such a way that $V_{1/2}$ varies in parallel with E_K. Correspondingly, in the activation kinetics (upon a positive voltage step), the time constant of the major exponential component is approximately proportional to the concentration of K^+_o (Blatt 1991). In terms of the gating model, the K^+-dependence is well accounted for by extension of the previous model, $A \leftrightarrow I_1 \leftrightarrow I_2$, by another closed state to $A \leftrightarrow I_1 \leftrightarrow I_2 \leftrightarrow I_3$ via the reaction $I_2 + K^+_o \leftrightarrow I_3$ (Blatt and Gradmann 1994, in prep.). Such K^+-dependent gating kinetics appear to be a feature common to the class of outward rectifiers in plant cells (Blatt 1991; Blatt and Thiel 1993). In the case of guard cells, the physiological benefit of these kinetics is that K^+ efflux through these channels is guaranteed, irrespective of a potentially variable electrochemical equilibrium for K^+ in vivo (Clint and Blatt 1989). For incorporation of this channel into the electrical framework of the entire membrane (Fig. 1, Table 1), a simplified gating scheme ($A \leftrightarrow I$) should be sufficient for a first approach.

Ca^{2+}_o has also been reported to affect the K^+ outward current[L]. In guard cell protoplasts (of *Zea mays*), a rise of Ca^{2+}_o from 1 to about 10 mM reduced the K^+ outward current by about 50%. This inhibitory effect of Ca^{2+}_o on the K^+ channels employs a different mechanism than in the inhibition by K^+_o (Blatt 1991), which was discussed in the preceding paragraph.

The activity of outward rectifying K^+ channels is also affected by ionic conditions at the cytoplasmic side of the membrane[M]. Through electrophoretic injection of Na^+ into intact guard cells, it has been shown that outward K^+ currents are inhibited (in a voltage-dependent manner) by Na^+ (Thiel and Blatt 1991). Considering the high K_D of this Na^+ inhibition (about 80 mM), a physiological role of this process is unlikely; but other cations (e.g., Mg^{2+}) may have a physiological impact on K^+ relations in plants via this inhibition (Blatt 1988; Bertl et al. 1992).

With respect to channel modulation by potential second messengers, the pH-dependency[N] of the outward current is the most interesting feature recognized so far. By simultaneously measuring cytoplasmic pH and membrane currents in guard cells, Blatt and Armstrong (1993) reported that acidification over about half a pH unit effectively eliminated K^+ outward currents. Alkalization of the cytoplasm by about 0.27 pH units in response to the phytohormone ABA enhanced K^+ outward conductance more than twofold, irrespective of the membrane voltage (Blatt and Armstrong 1993). The close correlation between ABA-induced alkalization of the cytoplasm (Irving et al. 1992; Blatt and Armstrong 1993) and the pH sensitivity of K^+ outward currents (Blatt and Armstrong 1993) make this modulation of channel activity a likely mode of action for ABA to promote stomatal closure (Blatt 1990; Thiel et al. 1992).

Also important for the understanding of K^+ channel control in the context of signal transduction mechanisms is the apparent insensitivity of the K^+ outward rectifyer to Ca^{2+}_c[O] in the physiological concentration range > 200 nM. (Hosoi et al. 1988; Schroeder and Hagiwara 1989). The apparent, up to fivefold stimulation of K^+ outward currents by depletion of Ca^{2+}_c from 200 nm down to only few nM (Fairley-Grenot and Assmann 1992a) is unlikely to be physiologically relevant (MacRobbie 1992).

d) $(2H - Cl)^+$ Symporter

It has already been mentioned that a system for anion uptake must exist when anion-exporting transporters (the Cl^- channel) are operative. Unfortunately, such an anion uptake system has not yet been indentified in guard cells. However, in the giant internodal cells of *Chara*, which traditionally serve as model systems for cells of vascular plants, an efficient $(2H - Cl)^+$ symporter has been investigated in great detail (reviewed by Sanders 1984). For an approximate description of its electrical properties, one can calculate the equilibrium voltage of this symporter, and realize that the current-voltage curve becomes steeper for negative-going voltages (Beilby and Walker 1981). This information is the basis for the tentative characterization of the steady-state electrical properties of an equivalent transporter postulated in guard cells (Fig. 1, Table 1). The absolute values of k_A^0 and k_I^0 in Fig. 1, which determine the temporal behavior, are free estimates.

e) Cl^- Channel

In patch-clamp experiments, voltage-dependent anion channels have been found with the selectivity sequence $NO_3^- > I^- > Br^- > Cl^- >$ malate (Marten et al. 1993). The elementary components of this pathway are single channels with a unitary conductance of approximately 30 to 40 pS in {14 K/104 K} (Keller et al. 1989). These channels are mostly closed under steady-state conditions. However, with favorable cytoplasmic preconditioning (see below), these channels can be activated temporarily by positive-going voltage steps, e.g., from –160 to –40 mV, with time constants of some 10 ms for activation and about 10 ms for inactivation (Hedrich and Jeromin 1992). These voltage- and time-dependent activation/inactivation kinetics can be formally described by a gating model with one open state and two parallel closed states, $I_1 \leftrightarrow A \leftrightarrow I_2$, where the channel rests in I_1 at negative V and is pushed by positive voltages through A into I_2. This gating model for Cl^- conductance has already been introduced in the context of action potentials in plants (Mummert and Gradmann 1991). Actually, the described voltage and time dependence of Cl^- channels in plants looks very similar to that of the Na^+ channel in the squid axon (Hille 1992) and seems to play an equivalent role in electrical excitation in plants, including guard cells (see below).

Cl^- channel activity is not solely under the control of membrane voltage, but requires additional cytoplasmic conditions for activation, namely, an elevated concentration of $Ca^{2+}{}_c{}^P$ and the presence of some hundred mM of nucleotides^Q (Hedrich et al. 1990; Hedrich and Jeromin 1992). A detailed study of the concentration-dependency for current activation on $Ca^{2+}{}_c$ has not yet been undertaken. The type of nucleotides required to achive macroscopic anion channel activation was not specific for either an activation of G-proteins or protein phosphorylation through kinase activity. Hence, the detailed gating mechanism is still uncertain (Hedrich et al. 1990).

The Ca^{2+}-dependence of these channels suggests that a rise in $Ca^{2+}{}_c$ precedes their activation to achieve high conductance. In particular, the Ca^{2+} sensitivity is a likely lever through which the open probability of the Cl^- channels is tied in to signal transduction cascades resulting in salt loss from guard cells. One documented example for a possible link between events which elevate Ca^{2+}_c and anion channel activity is the rise in anion channel conductance during the exposure of guard cells to the phytohormone, ABA (Thiel et al. 1992). This treatment is known to favor an elevation of Ca^{2+}_c (McAinsh et al. 1992) and is therefore the likely trigger for anion channel activation.

Worth noting is that spontaneous voltage transitions and voltage oscillations in intact guard cells are correlated with comparable changes of current-voltage relations similar to those produced in the presence of 10 µM of the phytohormone ABA. It might be speculated that fluctuations in Ca^{2+}_c cause excitatory voltage oscillations in guard cells (Thiel et al. 1992).

In contrast to ABA with its indirect effect, 5 to 100 µM auxin[R] (another phytohormone) affects gating of the anion channels directly but only from the extracellular side of the plasma membrane (Marten et al. 1991). The measurements consist of negative maxima of macroscopic Cl^- currents as they occur in the course of continuous voltage changes (ramp) from –200 to +60 mV within 1 s. The apparent effect of auxin is that during the first 45 s of exposure to the drug, the recorded peak currents become about 50% smaller in amplitude and occur at about 35 mV more negative *V*. On the level of single-channel recordings, auxin seems to reduce the lifetime of the open state (Marten et al. 1991). In terms of the gating model $I_1 \leftrightarrow A \leftrightarrow I_2$, auxin might accelerate the transitions $I_1 \leftrightarrow A$.

4. Additional Transporters

The ion transporters listed above are only the better studied of those present in the guard cell plasma membrane, but are by far not a complete list of transporters. Several lines of direct as well as indirect experimental evidence suggest additional transporters for Cl^-, K^+, and Ca^{2+}. Probably, most of these transporters are electrically reflected by the ill-defined, instantaneous activating "leak" currents (Schroeder and Hagiwara 1989; Blatt et al. 1990a; Schroeder and Fang 1991; Fairley-Grenot and Assmann 1992a; Thiel et al. 1992).

a) Ca^{2+} Translocators

The role of Ca^{2+} influx into guard cells has been of major interst since it was realized that it may play a crucial role in signal transduction cascades (DeSilva et al. 1985; McAinsh et al. 1991). Influx of Ca^{2+} through the plasmalemma has been detected by radiotracer fluxes (MacRobbie 1989), a combined electrical and optical measurement (Schroeder and Hagiwara 1990), or indirectly through the activation of Ca^{2+}_c-sensitive anion currents (Hedrich et al. 1990); but so far there is, at least in nonmechanically stressed membranes, not detailed current-voltage description of Ca^{2+}-selective channel or current, either for the guard cell plasma membrane or for the plasma membrane of other higher plant cells (Blatt and Thiel 1993). Interesting, though, is the accumulating evidence that Ca^{2+} may enter the cell through K^+ channels which are to some degree Ca^{2+}-permeable. Evidence in support of a nonselective Ca^{2+} current came from simultaneous recordings of membrane currents and Ca^{2+}_c in ABA challenged guard cells (Schroeder and Hagiwara 1990). This treatment gave rise to fluctuations of Ca^{2+}_c concomitant with the activation of a current. From the reversal potential of this current, these authors deduced that the ABA-evoked current was conducted by a nonselective cation channel passing Ca^{2+} and K^+ equally well. A possible candidate for this channel could be a 10 pS, spontaneously activating cation channel observed by Cosgrove and Hedrich (1991). In intact guard cells, this current could be identical with the largely voltage-insensitive current found to be enhanced by ABA (Thiel et al. 1992).

Another candidate for nonselective Ca^{2+} translocation is the K^+ inward rectifier. Under some yet unknown instances these channels appear to become permeable to the divalent cation (Fairley-Grenot and Assmann 1992a, b). Ca^{2+} permeability through K^+ channels was deduced from current deactivation clamp experiments. Following a deactivating voltage step to the K^+ equilibrium voltage, it was still possible to record a slow deactivating current which could, under these circumstances, not be carried by K^+ but most likely by Ca^{2+}. The ABA-induced Ca^{2+} current (Schroeder and Hagiwara 1990) and Ca^{2+} entry through inward K^+ channels (Fairley-Grenot and Assmann 1992a, b) must reflect two separate translocators, since ABA increases the conductance of the former (Schroeder and Hagiwara 1990) but reduces the conductance of the latter (Blatt 1990; Thiel et al. 1992). Whether the La^{3+}-sensitive Ca^{2+} conductance found in guard cell protoplasts exposed to high Ca^{2+}_o (Lohse and Hedrich 1992) reflects one of these Ca^{2+} pathways or yet an additional one, is unknown.

b) K^+ Translocators

While it is now widely accepted that the inward K^+ channels are the major pathway for K^+ uptake, there is need also to postulate the presence of an additional K^+ uptake system. This hypothesis is based on the observation that unidirectional $^{86}Rb^+$ uptake was detectable in guard cells with free-running V positive of E_K and also in an experimental solution with low external KCl, slightly alkaline pH, and in the presence of the K^+ channel blocker tetraethylammonium (Clint and Blatt 1989). All these conditions, however, are unfavorable for K^+ uptake through a uniporter, because first, it would be against the thermodynamic gradient, and second, because inward K^+ channels have no significant conductance under these conditions (Blatt and Clint 1989; Blatt 1992). The data therefore favor the presence of an additional mechanism for K^+ uptake, possibly a cotransport system.

c) Mechano-Sensitive Channels

The electrophysiological world is divided into those who consider mechano-sensitive gated channels as physiologically relevant transport and signaling entities (Sachs 1989) and those who interpret them as experimental artifacts of the patch-clamp method (Morris and Horn 1991). As for the function of guard cells, it is an attractive idea that mechano-sensitive channels could be turgor sensors (Cosgrove and Hedrich 1991; MacRobbie 1992). Surprising, though, is the great wealth of mechano-sensitive channels reported for guard cells (Cosgrove and Hedrich 1991). The physiological significance of three types of mechano-sensitive channels, which are conductive to Ca^{2+}, Cl^-, and K^+ respectively, remains a matter of speculation (Cosgrove and Hedrich 1991).

Alternatively, the fact that membrane stretch increases Ca^{2+} conductance (Cosgrove and Hedrich 1991) opens the possibility that not all types of channels are directly activated by mechanical stress: stress-induced Ca^{2+} influx may result in an activation of Ca^{2+}-sensitive anion and cation channels. This, in turn, could provide a trigger for membrane excitation (Thiel et al. 1993 and see below).

d) Cl^- Channels

For a while, there was confusion about the current voltage relation of Cl^- channels in the guard cell plasma membrane. While one laboratory reported a largely voltage-independent $Ca^{2+}{}_c$-activated Cl^- current (Schroeder and Hagiwara 1989), others found a Cl^- current also $Ca^{2+}{}_c$-dependent but with a strong voltage-dependent inactivation at negative voltages (Keller et al. 1989; Hedrich et al. 1990). The query has been resolved recently by data indicating the presence of two Cl^- channels or gating modes with the same single channel conductance but with distinctly different kinetic properties: one channel (see above) is characterized by fast (ms) activation and deactivation and the fact that the channel inactivates in a time-dependent manner over seconds (Keller et al. 1989; Hedrich et al. 1990). The second channel (mode) is characterized by a much slower activation (seconds) and the absence of inactivation (Linder and Raschke 1992; Schroeder and Keller 1992). In addition, these channels open well negative of the fairly narrow voltage range in which the fast activating anion channel is active. This feature makes the slow channel/mode a likely conductance to initiate depolarization of extremely polarized membranes (Linder and Raschke 1992; Thiel et al. 1992).

5. Comparative Aspects

The knowledge of the electrical properties of guard cells from *Vicia faba* is outstanding compared to other plant cells. It is tempting but probably dangerous to consider, therefore, the electric properties of guard cells of *Vicia faba* as typical. There are very few studies in which this problem has been explicitly addressed, e.g., with respect to different species or to different tissues. Actually, the few studies which allow such a comparison show only minor or statistically questionable differences between the electrical properties of guard cells from *Vicia faba*, *Zea mays*, *Xanthium strumarium*, and *Commelina communis* (Fairley-Grenot and Assmann 1992a, b, 1993; Linder and Raschke 1992; G. Thiel, unpubl. results).

However, clear differences are reported between the current-voltage relationships of protoplasts from guard cells and from mesophyll cells of *Vicia faba* (Li and Assmann

1993): under experimental conditions which are suitable to evoke the fast activating anion current in guard cell protoplasts, no such currents were detected in protoplasts of mesophyll cells (Marten et al. 1992). In addition, immunological tests with an antibody against an anion channel from kidney membranes were positive only with membrane proteins from guard cells and negative with corresponding mesophyll material. The important conclusion could be that these anion channels are characteristic for excitable membranes.

Differences between tissues were also detected with respect to K^+ channels (Li and Assmann): the inward K^+ rectifier which operates in guard cells appears to be absent in mesophyll cells.

Furthermore, the kinetics of the outward rectifier differed between protoplasts from the two tissues. Differences have also been found with respect to the effect of guanosine-diphosphate (GDPβS): it stimulates K^+ outward currents in mesophyll protoplasts (not in guard cell protoplasts) but enhances K^+ inward currents in guard cell protoplasts (not in mesophyll protoplasts, of course, where the transporter is missing) (Li and Assmann 1993). On the other hand, membrane voltage and K^+ efflux of guard cells and of mesophyll cells respond very similarly to hormonal stimuli, e.g., of ABA (Kasamo 1981; MacRobbie 1992; Thiel et al. 1992). The latter findings jeoparize the simple concept of source and sink tissues for K^+ in the leaf.

6. Integrated Model: Electrocoupling[4]

In this section we give an outline how the ion transporters, which are electrically and osmotically relevant, can be expected to interact in the plasmalemma of guard cells.

The basic idea is the following. From voltage-clamp experiments, the properties of the individual transporters are known in great detail with respect to the voltage- and time-dependence for their conductance. However, under physiological conditions with free-running voltage, these transporters must interact with each other (Fig. 1). A conductance change in one element will cause a voltage change which, in turn, causes every other conductance to alter, and so on. It can be expected that such a system may approach a steady-state or oscillations, depending on the actual set of parameters. This behavior can be simulated by computer calculations.

[4] The model below has been written for processes which are coupled with respect to voltage. Analogue models could be written for processes which are coupled with respect to other factors, such as $Ca^{2+}{}_c$ or $H^+{}_c$. Correspondingly, many processes are known (e.g., in Table 1) from well-buffered conditions, but their interactions via free-running concentrations under physiological conditions are bascially unknown. Nonlinear models of the presented type seem to be important for a realistic view of physiological networks.

a) Description of Algorithm

For a theoretical treatment of the system, we assume (according to experimental results) that each of the five transporters in the membrane has an equilibrium voltage E_i and a maximum conductance (say 1 Sm^{2-}), meaning full activity of all its elements. For the description of the voltage- and time-dependence of the elements, we consider spontaneous transition between the active state and an inactive state, $A \leftrightarrow I$, in the form of an Eyring barrier with the rate constants k_I (for inactivation) and k_A (for activation). These rate constants (corresponding to transition probabilities) depend on the voltage, V, in the form $k = k^0 \exp(\delta u)$, where k^0 is the rate constant at zero voltage, $u = VF/RT$ is the normalized voltage, and δ contains reaction-specific details, such as sign, charge number, stoichiometry, and barrier shape. For an initial approach, $\delta = 0.5$ or $\delta = -0.5$ are useful assumptions to characterize the voltage-dependencies. Values of k^0 determine temporal relations and equilibria between A and I. For the Cl^- channel, the simple scheme $A \leftrightarrow I$ has to be extended by another inactive state: $I_1 \leftrightarrow A \leftrightarrow I_2$.

An explicit description of the algorithm is given by Mummert and Gradmann (1991) for the treatment of the action potential in *Acetabularia* as an interaction between a pump, a K^+ outward rectifier, and a Cl^- channel. This algorithm has simply been extended by a K^+ inward rectifier and a $(2H - Cl)^+$ symporter to account for salt uptake as well (Gradmann et al. 1993). Briefly, for numerical calculations, one can start with steady-state voltage-clamp conditions, when each transporter i has a constant conductance, $g_{i,t<0}$. Upon release of voltage-clamp conditions, the voltage will jump to $V_{t=0} = \Sigma\,(E_i \cdot g_{i,t<0}) / \Sigma\,(g_{i,t<0})$ and the conductances will start to relax from $g_i(V_{t<0})$ towards $g_i(V_{t=0})$. After a small time increment Δt, these conductances will have changed from $g_{i,t<0}$ to $g_{i,t=0} = g_{i,t<0} + \Delta g_i$ and cause a new voltage, $V_{t=0} = \Sigma\,(E_i \cdot g_{i,t=1}) / \Sigma\,(g_{i,t=1})$. Repeating this procedure n times results in the time course of the conductances $g_{i,1..n}$. From here on, the time courses of voltage, currents, fluxes, etc. can be calculated by simple arithmetic.

b) Physiological Voltage Ranges

Scanning through records of free-running voltages from cells of different plants and/or tissues, leads to the general notion of two preferential voltage ranges (Fig. 2, right vertical plane). One range, V_1, which is considerably more negative than the equilibrium voltage of K^+ ($V_1 << E_K$) and another range, V_2, which is some 10 mV more positive than E_K ($V_2 > E_K$). V_1 and V_2 are separated by a range V_0 of low probability. Instability of voltage can be explained by the occurrence of negative slope conductance. In fact, the pump, as well as the K^+ outward rectifier and the Cl^- channel, do exhibit negative slope conductances in the voltage range more negative than E_K. Note that the conductance given in Fig. 1 and used for the outline of the model calculations above are not slope conductances (dI/dV at a given V), but chord conductances [$I/(V - E_i)$].

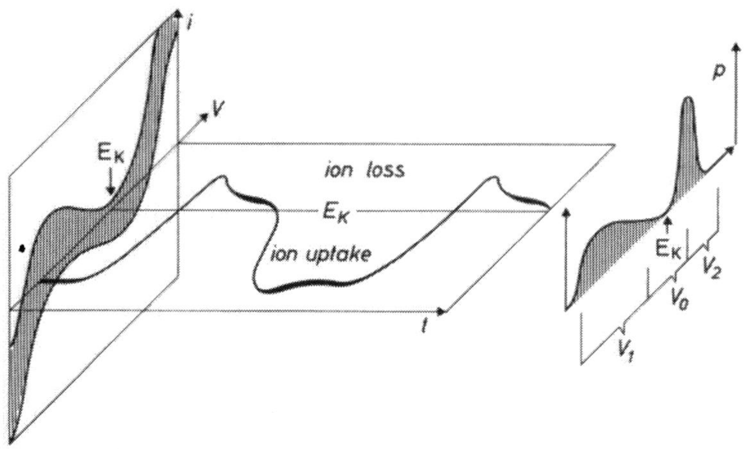

Fig. 2. Global scheme of electric and osmotic relations in guard cells. E_K Equilibrium voltage for K^+ diffusion; *left vertical plane* typical range of current-voltage relationships of the plasmalemma indicating N-shape in central range and saturation phenomena at large voltage displacenment of either sign; *horizontal plane* typical time course of oscillations of free-running voltage (sum of individual membrane currents: zero) and correlated alternation between net uptake and net release of ions; *right vertical plane* typical probability density distribution of free-running voltage in guard cells with two ranges of high probability: a relatively wide range ($V_1 << E_K$), narrower range ($V_2 > E_K$) and a range of V_O, of low probability ($V_1 < V_O < V_2$)

c) Osmotic Problem and Dynamic Solution

Considering only the K^+ relations, it can be seen immediately that there is net salt uptake with osmotic swelling at V_1 and net salt loss with osmotic shrinkage at V_2. An equilibrium with zero K^+ fluxes and osmotic stability (at E_K within V_0) is *very unlikely* to exist. On the other hand, the prime function of guard cells is to maintain certain optimum volume states for intermediate stomatal aperture under various environmental conditions. Apparently, the electrical devices of guard cells are not suited to perform this function in a steady-state mode.

However, the desired long-term (minutes to hours) balance would well be achieved by appropriate switching between periods of osmotic swelling at V_1 and osmotic shrinkage at V_2. Figure 2 illustrates the relations schematically in a three-dimensional plot. The left vertical plane gives a range for possible steady-state current-voltage relationships in plant cells membranes with two-step conductance branches for K^+ inward (negative) currents and for K^+ outward (positive) currents, separated by a range of small or even negative slope in which all transporters may be involved to various extents. In the horizontal plane, an example of nonlinear oscillations of the free-running voltage (total membrane current = sum of individual membrane currents = 0) is drawn according to a model calculation as outlined above and according to experimental recordings in guard cells of *Vicia faba* (Gradmann et al. 1993). Also marked in the horizontal V/t plane is the circumstance that there is net ion uptake at $V < E_K$ and net loss at $V > E_K$. The right vertical plane shows a typical probability density distribution of voltage recordings in plant cells, e.g., of the small section shown in the horizontal plane.

d) Experimental Evidence

In fact, spontaneous transitions between V_1 and V_2 are known from cells of plants (e.g., Antkowiak et al. 1991) and even of fungi (Slayman et al. 1976), but mainly in the context of electrical excitability. Not surprisingly, they have also been reported from guard cells (Thiel et al. 1992; Gradmann et al. 1993).

e) Range of Validity

We consider the bistability of membrane voltages as a general feature of the plasmalemma of plants. Compared to this view, reports of bistable membrane voltages are rare. As judged by the many personal communications received and by our own experience, spontaneous transitions between V_1 and V_2 interfere with the desire to record reproducible steady-state data. Thus, experimental conditions are selected rather to eliminate such transitions. Very stable and very negative resting voltages at V_1 are the pride of the electrophysiologist, although such a state would end in the long run in an osmotic disaster. Voltages in the range of V_2 can also be very meaningful because of their appealing stability and their significance for diffusion-dominated transport through the membrane; but leakage of deleterious electrolytes from micropipettes (Blatt and Slayman 1983; Blatt 1987a) and/or poor electrode seals may frequently result in stable V_2 recordings, whereas the membrane voltage in the intact cell would be in the V_1 range or oscillate between V_1 and V_2.

Compared to the exploding literature about current measurements with patch-clamp techniques, the volume of studies which focus on free-running membrane voltages is small. Nevertheless, the free-running voltage is still a crucial parameter with respect to thermodynamic and physiological issues.

f) Surprises

The used model of nonlinear electrocoupling can produce apparently paradox results. An example: with the parameters given in the legend to Fig. 1 the model yields oscillations with a (small) long-term uptake of Cl^-; stimulation of the $(2H-Cl)^+$ symporter by doubling the rate constant for activation, results actually in a (small) long-term release of Cl^-. Such paradoxes are due to nonlinearities of the oscillating system. Corresponding relations are familiar in electric circuitry: DC signals occur by nonlinear filtering of AC currents.

g) Limitations

The disadvantage of our model is its complexity. By choosing the right parameters, the model can describe about any time course of the membrane voltage within the thermodynamic limits. By the same token, it seems impossible to solve the inverse problem, namely to determine the system parameters from the observed phenomena. This problem will be even greater in reality, because the number of system parameters in our simplified model is unrealistically small.

After all, our approach of electrocoupling of ion transporters under free-running voltage is not an alternative to well-defined voltage-clamp investigations, but an attempt to reconstruct the operation of these transporters within the physiological framework. It appears promising to apply equivalent formalisms to other systems of processes, e.g., those which are coupled by the concentration of free Ca^{2+} and/or H^{+} in the cytoplasm (Table 1). In principle, even the interactions between such entire systems could be treated. However, without knowledge of the model details with mathematical precision, the predictions of such complex models are unreliable.

7. Final Remark

Ten years ago, the cruical role of electrical events in stomatal physiology had been acknowledged as such but were subject to speculations. These were mainly based on analog conclusions drawn from experimental results obtained with other plant cells. This situation has changed fundamentally since concentrated efforts have been undertaken to investigate guard cells by modern methods. Today, electrophysiology of guard cells even sets the standard for the investigation of the electrical properties of cells of vascular plants in general.

Acknowledgments. We thank Anna Amtmann, Ulrike Homann, and Andreas Wolf for critical reading of the manuscript.

References

Antkowiak A, Mayer WE, Engelmann W (1991) J Exp Bot 42:901–910. **Assmann SM**, Simoncini L, Schroeder JI (1985) Nature 318:285–287.

Beilby MJ, Walker NA (1981) J Exp Bot 32:43–54. **Bertl A**, Klieber HG, Gradmann D (1988) J Membr Biol 102:141–152. **Bertl A**, Gradmann D, Slayman CL (1992) Philos Trans R Soc Lond B 338:63–72. **Blatt MR** (1987a) Planta 170:272–287. **Blatt MR** (1987b) J Membr Biol 98:257–274. **Blatt MR** (1988) J Membr Biol 102:235–246. **Blatt MR** (1990) Planta 180:445–455. **Blatt MR** (1991) J Membr Biol 124:95–112. **Blatt MR** (1992) J Gen Physiol 99:615–644. **Blatt MR**, Armstrong F (1993) Planta 191:330–341. **Blatt MR**, Clint GM (1989) Planta 178:509–523. **Blatt MR**, Slayman CL (1983) J Membr Biol 72:223–236. **Blatt MR**, Thiel G (1993) Annu Rev Plant Physiol Mol Biol 44:543–567. **Blatt MR**, Thiel G, Trentham DR (1990a) Nature 346:766–769. **Blatt MR**,

Beilby MJ, Tester M (1990b) J Membr Biol 114:205–223. **Busch H**, Hedrich R, Raschke K (1990) Plant Physiol 93:96A (Abstr)

Clint GM, Blatt MR (1989) Planta 178:495–508. **Cosgrove DJ**, Hedrich R (1991) Planta 186:143–153.

DeSilva DLR, Hetherington AM, Mansfield TA (1985) New Phytol 100:473–482.

Fairley-Grenot KA, Assmann SM (1991) Plant Cell 3:1037–1044. **Fairley-Grenot KA**, Assmann SM (1992a) Planta 186:282–293. **Fairley-Grenot KA**, Assmann SM (1992b) J Membr Biol 128:103–113. **Fairley-Grenot KA**, Assmann SM (1993) Planta 189:410–419. **Fisher RA** (1968) Science 160:784–785.

Gradmann D, Hansen UP, Long WS, Slayman CL, Warncke J (1978) J Membr Biol 39:333–367. **Gradmann D**, Blatt MR, Thiel G (1993) J Memb Biol (in press).

Hamill OP, Marty A, Neher E, Sakmann B, Sigworth FJ (1981) Pflügers Archiv 391:85–100. **Hansen UP** (1990) Bot Acta 103:15–23. **Hedrich R**, Jeromin A (1992) Philos Trans R Soc Lond B 338:31–38. **Hedrich R**, Busch H, Raschke K (1990) EMBO J 9:3889–3892. **Hille B** (1992) Ionic channels of excitable membranes. Sinauer, Sunderland, Mass. **Hosoi S**, Iino M, Shimazaki K (1988) Plant Cell Physiol 9:907–911.

Irving HR, Gehring CA, Parish RW (1992) Proc Natl Acad Sci USA 89:1790–1794.

Kasamo K (1981) Plant Cell Physiol 22:1257–1267. **Keller BU**, Hedrich R, Raschke K (1989) Nature 341:450–453. **Korn SJ**, Horn R (1989) J Gen Physiol 94:789–821. **Kubo Y**, Baldwin TJ, Jan YN, Jan LY (1993) Nature 362:127–133.

Läuger P (1991) Electrogenic pumps. Sinauer, Sunderland, Mass. **Lee Y**, Assmann SM (1991) Proc Natl Acad Sci USA 88:2127–2131. **Li W**, Assmann SM (1993) Proc Natl Acad Sci USA 90:262–266. **Linder B**, Raschke K (1992) FEBS Lett 313:27–30. **Lohse G**, Hedrich R (1992) Planta 188:206–214.

MacRobbie EAC (1988) In: Baker DA, Hull JL (eds) Solute transport in plant cells and tissues. Harlow, Longman, Essex, pp 453–497. **MacRobbie EAC** (1989) Planta 178:231–241. **MacRobbie EAC** (1992) Philos Trans R Soc Lond B 338:5–18. **McAinsh MR**, Brownlee C, Hetherington AM (1991) Proc R Soc Lond B 243:195–201. **McAinsh MR**, Brownlee C, Hetherington AM (1992) Plant Cell 4:1113–1122. **Mansfield TA**, Hetherington AM, Atkinson CJ (1990) Annu Rev Plant Physiol Mol Biol 41:55–75. **Marten I**, Lohse G, Hedrich R (1991) Nature 353:758–762. **Marten I**, Zeilinger C, Redhead C, Landry DW, Al-Awqati Q, Hedrich R (1992) EMBO J 11:3569–3575. **Mohr H**, Schopfer P (1992) Pflanzenphysiologie, 4th edn, Springer, Berlin Heidelberg New York, pp 253. **Moran N**, Ehrenstein G, Iwasa K, Bare C, Mischle C (1984) Science 226:835–838. **Morris CE**, Horn R (1991) Science 251:1246–1249. **Mummert H**, Gradmann D (1991) J Membr Biol 124:265–273.

Oleson DR, DeFelice LJ, Donahoe RM (1993) J Membr Biol 132:229–241.

Pallas JE (1966) Q Rev Biol 41:365–383. **Pedersen PL**, Carafoli E (1987) TIBS 12:146–150. **Pusch M**, Neher E (1988) Pflügers Archiv 411:205–214.

Raschke K (1975) Annu Rev Plant Physiol 26:309–340. **Raschke K**, Hedrich R, Beckmann U, Schroeder JI (1988) 101:283–294.

Sachs F (1989) In: Stein WD, Bronner F (eds) Cell shape: determinants, regulation and regulatory role. Academic Press, San Diego, CA, pp 63–92. **Sanders D** (1984) In: Gerencser GA (ed) Chloride transport coupling in biological membranes and epithelia. Elsevier, Amsterdam, pp 63–120. **Sanders D**, Hansen UP (1981) J Membr Biol 58:139–153. **Schaller GE**, Sussman MR (1988) Planta 173:509–518. **Schroeder JI** (1988) J Gen Physiol 92:667–683. **Schroeder JI** (1992) Philos Trans R Soc Lond B 338:83–90. **Schroeder JI**, Fang HH (1991) Proc Natl Acad Sci USA 88:11583–11587. **Schroeder JI**, Hagiwara S (1989) Nature 338:247–430. **Schroeder JI**, Hagiwara S (1990) Proc Natl Acad Sci USA 87:9305–9309. **Schroeder JI**, Keller BU (1992) Proc Natl Acad Sci USA 89:5025–5029. **Schroeder JI**, Hedrich R, Fernandez JM (1984) Nature 312:361–362. **Schroeder JI**, Raschke K, Neher E (1987) Proc Natl Acad Sci USA 84:4108–4122. **Serrano EE**, Zeiger E, Hagiwara S (1988) Proc Natl Acad Sci USA 85:436–440. **Slayman CL**, Long WS, Gradmann D (1976) Biochim Biophys Acta 426:737–744.

Takeuchi Y, Kishimoto U (1983) Plant Cell Physiol 24:1401–1409. **Thiel G**, Blatt MR (1991) J Plant Physiol 138:326–334. **Thiel G**, MacRobbie EAC, Blatt MR (1992) J Membr Biol 126:1–18. **Thiel G**, Homann U, Gradmann D (1993) J Membr Biol 134:53–66.

White RE, Schonbrunn A, Armstrong DL (1991) Nature 351:570–573. **Wille A**, Lucas WJ (1984) Planta 160:129–142.

Dr. Gerhard Thiel
Professor Dr. Dietrich Gradmann
Pflanzenphysiologisches Institut
der Universität
Untere Karspüle 2
D-37073 Göttingen, FRG

II. Exchange of Energy and Vapor Between Plants and the Atmosphere

By Rainer Lösch

1. Introduction

Biosphere-atmosphere interactions (e.g., Shukla and Mintz 1982; Shugart and Emanuel 1985; Myers 1988; Eamus 1991; Enders et al. 1992) have understably gained in interest with increasing public awareness of the importance of balanced cycles of energy and matter flowing through ecosystem compartments. The geophysical and meteorological processes are investigated by the respective disciplines, while studies about the exchange processes between organisms and their environment belong to fields of interest of biometeorologists, ecosystem researchers, and biologists. The concept of the soil-plant-atmosphere continuum (SPAC) of the water flow through terrestrial ecosystems integrates exchange patterns within and between the various compartments. It quantifies driving forces and fluxes of water through the system, which are controlled by both physically determined concentration gradients and physiological regulations. Water flowing through the SPAC is subjected to an energy-consuming phase change from the liquid to the vapor state, a process which immediately links the water cycle to the system's energy balance. The water flow from the plant to the atmosphere occurs under stomatal control and is, in this way, linked also to the carbon cycle. Quantitative relationships may possibly exist also with the exchange at this leaf-atmosphere interface of gaseous nitrogen and sulfur compounds.

Water flow through SPAC connects the most important exchange processes, whose quantification will be crucial in understanding and predicting function and alterations of terrestrial ecosystems ("global change": e.g., McNaughton and Jarvis 1983; Idso 1989; Bazzaz 1990; Houghton et al. 1990; Kriebitzsch 1991; Woodward and McKee 1991; Neilson et al. 1992). At the integrational level of tissues, organs, and individual plants, knowledge about the control of water flow through SPAC is based on physiological and ecophysiological investigations. Exchange processes between plants and the atmosphere are quantified by ecophysiological and micrometeorological studies. Attempts to quantify energy and matter exchange at the level of whole ecosystems use ground-based measurements, by bottom-up approaches such as analytical studies, by modeling, until now, mostly as top-down integrations (Enders et al. 1992). Remote sensing contributes to our understanding of energy and matter exchange providing area-related data.

About the progress of science at the lower integrational levels reports were given in *Progress in Botany* at regular intervals (Lösch 1989, 1991, 1993; Schmidt 1989, 1992). Progress in the knowledge of exchange processes between plants and the atmospheric

Progress in Botany, Vol. 55

environment which were gained during the last 15 years will be dealt here in greater detail. Two textbooks which give good introductory information on the theme have meanwhile appeared in second editions (Nobel 1991; Jones 1992). A methodical manual dealing also with microclimate, canopy structure and light interception, leaf energy budget studies, and canopy carbon and water balance has been published by Hall et al. (1993).

2. Energy Balance and Radiation Perception of the Vegetation

The general energy balance equation specifies various forms of energy into which the total energy coming from the solar radiation is divided:

$R_n = \lambda E + G + C + S + P$[1].

In practice, G (heat flux by conduction) may be included into C (heat flux by convection), and S (heat storage) is usually small compared with the other terms. The greatest amount of the incoming solar radiation is partitioned to sensible (C + S + G) and latent heat (λE). The most negligible (Thom 1975; Jarvis et al. 1976; Tajchman 1981) is the quantity of energy which is used by metabolic processes (P), and which is supplied by the visible part of the radiation spectrum, more precisely by the photosynthetic active radiation (PAR) which covers the wavelengths between 0.4 and 0.7 μm. PAR was estimated to be 45 ± 10% of the solar radiation at many stations (Pereira et al. 1982; Meek et al. 1984). With data of the atmospheric transmittance, which depends on ozone absorption, Rayleigh scattering, and aerosol extinction, instantaneous values of clear sky solar beam, diffused and total photosynthetic active radiation may be calculated also for remote places whose solar elevation is known (Gueymard 1989). For monthly averages even meteorological standard data will be sufficient (Zaradny and Van der Ploeg 1982; Nikolov and Zeller 1992). Using satellite data, local PAR reception can be estimated on the basis of clear sky radiation data if the attenuation of PAR due to reflectance by clouds is accounted for. The latter is inferred from UV reflectance at 370 nm (Eck and Dye 1991). Algorithms have been developed also for the separation of direct and diffuse components of solar radiation (Bristow and Campbell 1985).

Sensors for appropriate radiation measurements were improved with respect to size (Van der Hage 1984), integration possibilities (Newman 1985; Isoda et al. 1992), and partitioning between sky and ground radiation (Deering and Leone 1986). Salminen et al. (1983) compared different types of sensors when influenced by the within-canopy attenuation of radiation under both clear and overcast sky conditions. Agreement between the calculated and measured penetration coefficients was better under overcast conditions.

---

[1] A list of symbols is given in the Appendix.

a) Absorption and Partitioning of Incoming Radiation

Radiation profiles within canopies are studied particularly in homogeneous stands of crops or forest plantations. For a comprehensive treatment of radiative transfer in non-homogeneous canopies, reference is made to Nilson (1992) and the literature cited there. Radiation exchange and partitioning of soybean fields are described by Andre and Viswanadham (1983) and Baldocchi et al. (1983). Under water stress, energy gain of the soybean plants is directed chiefly into sensible heat. Different crown shapes considerably influence the patterns of light absorption by the individual parts of a canopy (theoretical calculations: Oker-Blom and Kellomäki 1982). Bégué (1992) calculated hemispherical and directional radiative fluxes into an array of opaque cylinders which were taken as a geometrical form representing structures of grain crop fields with regularly clumped canopy elements. The model was validated successfully with data from an African pearl millet field. In the upper part of a wheat field the PAR transmittance (T_{PAR}) and the transmittance of the near-infrared radiation (0.7 –3.0 μm; T_{NIR}) are related by the power function $T_{NIR} = T_{PAR}{}^n$, where n = 0.5 with green plants, increasing in value with maturation yellowing (Rodskjer and Dahlstedt 1977). Torssell and McPherson (1977) developed a mathematical model which allows multiple reflection and transmission in a grass canopy of random leaf distribution. Morris (1989) emphazises that vertically oriented leaves at the top of the canopy and horizontal leaves of lower insertion levels optimally interact in radiation perception (marshes of *Spartina alterniflora*). A radiation interception model developed for *Pinus radiata* plantations was rather insensitive to errors in the determination of crown shape and leaf area when predicting radiant energy interception on a daily basis, but failed in predicting with exactness instantaneous transmittances (Grace et al. 1987). Generally, the prediction of canopy radiation regimes is better for overcast then for clear-sky conditions (Campbell and Marini 1992: PAR regime within an appletree canopy). Diffusive radiation penetration into an individual Norway spruce crown is best simulated if the crown is considered to be a cone with radially uniformly distributed and vertically increasing foliage area density (Koppel and Oja 1984). In a spruce stand, more than 60% of the global radiation is absorbed and converted mostly into latent heat in the upper parts of the canopy (Mayer 1981). Most of the energy absorbed by Himalayan *Alnus* and *Populus* species is lost by re-radiation; transpiration and convection are less important in these trees (Dhyani 1989).

Plant phenological stages play an important role in increasing or decreasing the absorption values (Green and Dawkins 1985: cereal crops). Such a seasonality is particularly obvious in deciduous forests. In the leafy stage of an oak-hickory forest, attenuation of PAR is much higher than that of shortwave and net radiation. During the leafless period, the radiation components are attenuated in the exponentially decreasing order net radiation ≥ shortwave radiation = PAR ≥ global radiation. Due to low solar elevation angles, trunk and twig biomass intercepts the radiation much more in autumn and winter. Still low chlorophyll contents of the expanding leaves are the reason why shortwave attenuation progresses faster than that of PAR during springtime (Baldocchi et al. 1984a, b). The albedo of a tropical dry deciduous forest is higher by one-third in the leafless period than when all trees are leafed (Barradas and Adem 1992). Flowering canopies of oilseed rape reflect radiation in the PAR wavelength range between 0.5 and 0.7 μm and absorb it less than the leafy part of the canopy. During the flowering period, therefore, rape photosynthesis relies mostly on energy gain in the blue part of the PAR spectrum (Yates and Steven 1987). The ratios of absorbance/reflectance relations of different wavelength bands were investigated by Shultis (1991). They depend upon sun angle, diffuse to direct radiation, albedo, and leaf distribution and scattering properties. The latter two components, foliage structure and radiation interaction properties, are treated mathematically by Walthall and Norman (1992) and Knyazikhin et al. (1992). Considering randomly distributed overlappings and gaps of physical leaf models, in both papers probability calculations are developed for the description of radiation profiles through canopies.

Radiation absorption of individual leaves is considerably influenced by structural properties. The outer cuticle surface of the leaf epidermis reflects incident light specularly, as a polarized component (McClendon 1984; Grant et al. 1987). The unpolarized, diffuse part of reflection is influenced by leaf pigmentation and wavelength, which demonstrates that it emanates from the internal leaf structure. Different patterns of epidermal surface structures influence the absorption of solar radiation only to a very small degree (Tanner and Eller 1986). Also, leaf pubescence has little influence on energy input; but reflection of infrared (IR) radiation from the upper leaf surface is higher if the lower one is hair-covered (measurements with *Tussilago farfara*: Eller 1977b; Eller and Willi 1977). Different nitrogen nutrition also changes leaf reflectivities, an aspect which might be used when monitoring crop nitrogen deficits by spectroscopic measurements (Seligman et al. 1983; Kochubei et al. 1984). Temperatures of dust-covered leaves are several degrees higher than those of cleaned ones, due to a higher absorption of IR radiation in the case of the absorption of visible radiation not being much reduced (Eller 1977a).

b) Exchange and Partitioning of Sensible Heat

A great amount of the incoming solar radiation is exchanged in the form of sensible heat. Additionally to the quantities of energy supplied within the PAR range and the highly energy-loaden near-UV-radiation, it is particularly the shortwave-IR range from which energy comes for soil and vegetation temperature gain. Thermal exchange occurs by heat absorption, respective reflection, and convective turbulences. The latter are driven by gradients in space and time of temperature. Sensible heat flux rates can be determined with appreciable exactness by correlation studies of temperature fluctuations along the flux direction (Lloyd et al. 1991). Kawanishi (1986) estimated the heat exchange constant of forests to be in the order of 6–20 W m^{-2} K^{-1}. The aboveground biomass functions as sink or source for heat exchange with the surrounding atmosphere. In a young eucalypt forest, a 1° C h^{-1} air temperature rise is equivalent to approximately 0.4 W m^{-2} heat storage increment for each kilogram of fresh plant material (Aston 1985). Heat transfer parameters of tree leaves were determined by Thorpe and Butler (1977) and Vogel (1984). Knowledge of the parameters of apple bud and blossom heat transfer allows a carefully directed frost protection by appropriate overhead sprinkling (Hamer 1985, 1986a, b). Leuning (1988) developed a detailed theory for predicting leaf-to-air temperature differences as a function of leaf size and temperature of the air, screen, and underlying surface effects, and presented experimental data on the course of leaf temperatures during periods of radiation frost (Leuning and Cremer 1988).

Kondo and Kawanaka (1986) state that (1) bulk transfer coefficients for heat and momentum change with canopy density, (2) maximal heat transfer is ensured by a higher canopy density than is required for maximal momentum transfer, and (3) the two coefficients are determined consistently by peculiarities of the canopy elements. A study by Balding and Cunningham (1976) shows differences in heat transfer characteristics of simple and pinnate leaf forms. Thermal conduction between the two surfaces of a leaf can have significant effects on the leaf energy balance (Dhyani et al. 1986). Heat transfer from leaves in still air relies predominantly upon free convection, which must be considered by theoretical calculations using the appropriate coefficients (Dixon and Grace 1983). Leaf-size variations interact with the canopy temperature regime in a complex manner because of the simultaneous function of the leaf surfaces as exchangers of latent and sensible heat, as screens attenuating the radiation, and as obstacles which influence mass exchange between the various parts of the canopy (Zangerl 1978; Baldocchi et al. 1985). Without spins and ribs which provide an increased surface for convective heat loss, average daytime temperature of a cactus body would be by 5 °C higher (Lewis and Nobel 1977). By contrast, *Lithops* plants, inserted into the soil, do not differ in their temperature regime from their surroundings (Eller and Grobbelaar 1986). The resistance to sensible heat exchange of a turfgrass canopy depends on the volumetric heat capacity of the air divided by the product of the canopy temperature decrease rate upon abruptly diminished radiation, the mass of leaf tissue per unit area of soil surface, and the specific heat capacity of the leaves (Johns et al. 1981). The amount of sensible heat flux over homogeneous grassland canopies can be estimated from ground temperature, windspeed, and air temperature if stable conditions prevail (Schieldge 1978); upwind/downwind effects produce gradients and oasis effects over larger areas (Taha et al. 1991; Hirai 1992).

Thermal stratification above and within canopies influences heat and mass exchange processes (Smolen and Matejka 1982; Jacobs et al. 1992). Just above tall canopies, a roughness sublayer exists where diffusivities are enhanced (Cellier and Brunet 1992). The profile of within-canopy air temperature is governed by the temperature of the soil and plant surfaces, the former playing a particularly important role when deciduous forests are defoliated. Then the highest timeshifts of peak temperatures occur as compared with the situation above the canopy (Van Eimern and Riedinger 1986). During daytime, canopy surface temperatures exceed air temperatures within the stand by several degrees. The foliated canopy is an important heating surface during daytime and a cooling surface at night influencing the air layers above and below accordingly. Energy transfer to the understory and the soil surface occurs preferentially by turbulent fluxes which prevent the buildup of stable layers of cold air near the ground. Temperature layers and gradients are strongly influenced by stand geometry, as is demonstrated by investigations with crops planted in differently spaced rows (Graser et al. 1987). The local heterogeneity of the temperature of the vegetation surfaces can give valuable information about stand structures and plant water status. Holbo and Luvall (1989), analyzing NASA's Thermal Infrared Multispectral Scanner (TIMS) data, were able by use of a model to parallel frequency distribution patterns of temperature fields with surface types of the real canopies of a forested landscape.

c) Heat Storage

The highest amount of storage heat is diverted to the soil heat flux. There is a direct correlation between net radiation and soil heat flux in the case of bare soil, if phase differences in the diurnal variations of radiation gain and soil heat flux are considered (Choudhury et al. 1987). Sensible heat storage in the canopy is of greatest significance during dry, clear weather conditions (McCaughey and Saxton 1988). Then canopy heat storage of a mature mixed forest amounts to 2–3% of the net radiation. It can increase up to 10% if incoming radiation is insufficiently diverted to latent heat consumption (McCaughey 1985a). During diurnal courses, the largest heat gain of biomass occurs in the early morning hours (McCaughey 1985b; McCaughey and Saxton 1988). On cloudy days energy partitioning is different, depending on the various cloud types (O'Rourke and Terjung 1981). Uneven snow cover of taiga regions resulting from various tree forms and snow-drift patterns has consequences for the winter energy balance, because heat flow is twice as high from tree wells (bowl-shaped snow-surface depressions around large spruce stems) than from the undisturbed snow cover (Sturm 1992).

3. Convective and Turbulent Heat and Mass Transfer

a) Wind Effects and Canopy Structure

Heat and mass transfer between different places occurs most effectively by air movements. Principles of vegetation/wind interactions are presented in the textbook of Grace (1977); theories for turbulent transfer processes are outlined by Raupach (1989). Quite a number of original articles have been published since then, trying to adapt the general rules to local situations and to advance knowledge about turbulent air flow within and over inhomogeneous vegetation.

Generally (Grace 1977), a logarithmic relation exists between stand height and wind speed which is modified by several parameters which quantify structural peculiarities of the canopy:

$$u(z) = \frac{u_*}{k} \ln \left(\frac{z-d}{z_0}\right).$$

Here, $u(z)$ is the average wind speed (m s^{-1}) at height z, which is influenced by roughness length z_0 (m), zero plane displacement d (m), and friction velocity u_* (m s^{-1}), latter of which depends upon the shear stress τ: $u_* = \sqrt{\tau/\rho}$; ρ = specific density (kg m^{-3}). The proportionality constant $k = 0.41$ is named von Karman's constant.

The determination of the three stand-specific parameters, z_0, d, and u_*, is necessary in order to describe quantitatively the effects of readily measured wind velocities upon canopies.

De Bruin and Moore (1985), using a mass conservation method (Molion and Moore 1983), estimated a roughness length of 1.3 m and a zero plane displacement height of 12.7 m for an 18-m-tall pine forest. Comparable data from a 10-year measurement period in a pine forest in southern Germany are given by Jäger (1984). Dolman (1986) reports values of 1 and 7.2 m, respectively, for both parameters measured for a 10-m-tall oak canopy. Both parameters were slightly lower in the defoliated state of the oak forest. In quickly growing canopies, like cornfields, the displacement height correlates with the height of the canopy, z_0 is correlated with the difference between canopy height and d (Jacobs and Van Boxel 1988). Typical average values of gramineous canopies are 0.7–0.9 m for d and 0.05–0.2 m for z_0 (maize field: Jacobs and Van Boxel 1988; Sahelian Savanna: Lloyd et al. 1992). The same order of magnitude was found for the aerodynamic resistance of a *Calluna* heathland (Wallace et al. 1984). For a richly structured hilly landscape in Switzerland, Kustas and Brutsaert (1986) determined the two parameters from radioprobe observations. There was good agreement between d/h and z_0/h (h = obstacle height) with corresponding relationships for the roughness of surfaces the obstacles to flow of which were two to four orders of magnitude smaller. This is good evidence that the rules governing airflow over rough surface can be used at different levels of integration, from the single leaf surface to the surface of whole landscapes.

Formation of deserts leads to a reduction of surface roughness as the vegetation perishes and soil erosion ensues. The smoothness of the land surface then causes reduction in rainfall and promotes further desertification (simulations using a general climate model for the summer air circulation of the Sahara region: Sud and Smith 1985). Precipitation is also reduced as a consequence of decreased sensible heat flux from bare desert soil under an altered surface albedo as compared with semidesert areas (Otterman 1989).

Gutman and Berkofsky (1985) provided a set of equations describing vertical turbulent transfer in a well-mixed layer bounded below and above by uneven surfaces, e.g., of a canopy, and of a stable atmospheric layer in higher altitudes. A model simulation of air flow over agricultural fields and plantation forests was developed by Meyers and Paw U (1986). Cionco (1983) defines a coupling parameter which describes the degree of coupling of air flow inside the vegetation to the ambient flow of the surface boundary layer.

Leaf emergence influences within-canopy windspeed profiles and microscale pressure fluctuations of deciduous forests, but has little influence upon windspeed above the canopy (Sigmon et al. 1984), although velocity-pressure interactions exist in the surface layer of canopies (Maitani and Seo 1985a). Downward fluxes of turbulent kinetic energy just at the canopy surface (Maitani 1978) may be altered as a consequence of changed pressure and radiation absorbing surface areas in such a way that air movement above the canopy is not greatly changed during leaf sprouting. In general, windspeed fluctuations are particularly large near canopy surfaces where the foliage has a high density (Desjardins et al. 1978), and radiation absorption and vapor exchange lead to high convective turbulences (Maitani and Seo 1985b). Horizontal wind velocity elements give rise to downward movements of air parcels, eddies, which take place intermittently in downward bursts of kinetic energy transfer (Maitani 1977; similar findings also by Shaw and Zhang 1992). Mesoscale horizontal and vertical windflow patterns above and below forest canopies become complicated if clearings interrupt the otherwise homogeneous stand surfaces (Leahey and Hansen 1987). The other extreme, airflow within and around a single tree, was simulated numerically by Gross (1987). Turbulence spectra over inhomogeneous terrains were measured and statistically evaluated, among others, by Steyn (1982) and Skupniewicz et al. (1989). Surface elements of various roughness play an important role in creating turbulences of airflow (Honjo and Takakura 1986). Realistic exchange models must, therefore, treat energy and mass transfer separately across the single homogeneous surfaces and must take edge effects into account (Klaassen 1992). For practical use, simulations are proposed which take into account terrain structures and standard meteorological data at the level of the atmospheric boundary layer (e.g., Lee and Kau 1984; Detering and Etling 1985). Structural effects of the canopy upon momentum exchange with the atmosphere can be quantified considering the shear stress, τ. Bache (1986) describes profiles of shear-stress and momentum diffusivity within canopies of contrasting vertical structure taking into account a wind-speed-dependent drag coefficient. This latter term is a measure of the "bluffness " (Grace 1977) of the plant body. It characterizes the relation between the actual force of an air stream affecting an obstacle, and the maximal theroretically calculated force which results from air velocity and density. The drag coefficient depends on the roughness of the obstacle which is small in streamlined objects.

b) Heat and Momentum Exchange by Kinetic Energy Transfer

Convective energy and mass exchange due to air movements occurs by transfer of eddies between the leaves and the surrounding atmosphere. Roughness length of the surface plays an important role in the determination of diffusivity values of the eddies (Mehlenbacher and Whitfield 1977). Inside a dense canopy, eddy diffusivity, wind speed, and shear stress can be described by exponential decay functions with depth below the top of the canopy (Brutsaert 1979). A statistical treatment of eddy velocities and momentum concentrations along profiles of sensors within and, particularly, above a canopy during time periods of high-resolution measurement frequencies can give insight into the patterns of exchange processes. This "eddy correlation method" is becoming a reliable tool to describe the variation by time and space of wind gusts accompanied by local changes of air temperature, humidity, CO_2 concentration, and other parameters, since quickly responding sensors of high resolution have now been developed. Eddy fluxes of these quantities over crop fields were studied, among others, by Anderson et al. (1984a, b, 1986) and Ohtaki and Oikawa (1991). Eddy fluxes over forests were studied, e.g., by Leuning and Attiwill (1978), Verma et al. (1986), and Gao et al. (1992). Valentini et al. (1991) analyzed vapor and CO_2 exchange of a Mediterranean macchia by this method. The aerodynamic roughness of this canopy, the structure of which is less homogeneous if compared with dense forests, is intermediate between coniferous and deciduous canopies. CO_2 exchange (A) and transpiration (E) were dependent upon incident PAR, and the relation between both quantities changed with radiation up to a threshold of 400 μmol m^{-2} s^{-1}. At higher radiation values, the ratio E/A was constant. Under ideal conditions, a close correlation exists between radiation densities and eddy transfer rates, but phase lags can occur between sensible and latent heat densities, and these must be considered in the correlation calculations (Jacobs and Palland 1982). Multilayer models of flux densities must take this into account when calculating the exchange processes layer for layer (Chen 1984). Under neutral stability, eddy transfer coefficients of sensible and latent heat will normally be equal. The situation becomes complicated under advective conditions (Lang et al. 1983). Then, fetch parameters will be involved and a suitable fetch-to-height ratio must be given. Heilman et al. (1989) determine it to be at least 20:1 for reliable Bowen ratio calculations, the simplest type of eddy transfer analysis. "Bowen ratio" is the quotient between the fluxes of sensible and latent heat, if the eddy transfer coefficients of both energy fluxes are equal.

4. Evapotranspiration from Canopy Surfaces

Canopy or leaf energy balances are most often used in order to infer amounts of evaporative water loss from radiation and temperature measurements. Whereas transpiration from single leaves can be determined directly by ecophysiological methods, this is quite difficult for plant stands. The sum of sensible and latent heat as the essential terms of the energy balance equation can be written explicitly:

$$(R_n - G) = C + \lambda E = g_H \rho_a c_p (T_s - T_a) + g_W (0.622\rho_a/p)(e^*_{Ts} - e_a).$$

Introducing with Penman (1948) s, the slope of the curve relating saturation vapor pressure to temperature, a transformation becomes possible so that

$$e^*_{Ts} - e_a = (e^*_{Ta} - e_a) - s(T_a - T_s) = \delta e + s(T_s - T_a).$$

Then $(T_s - T_a)$ can be replaced and the energy balance equation is solved to

$$E = \frac{[s(R_n - G) + \rho_a c_p g_H \delta e]}{\lambda[s + (\gamma g_H/g_W)]}.$$

This "Penman-Monteith equation" (Monteith 1965, 1981; Thom 1975) is the most simple linearized form of the application of energy balance calculations upon canopy water loss – more sophisticated ones using nonlinear calculation procedures have been proposed in the literature (e.g., Paw U and Gao 1988; Paw U 1992). In this equation, which calculates E from micrometeorological parameters, short-term variables are only (R_{n-G}), δe and g_W. Only the latter is a physiologically determined parameter and therefore more difficult to measure. g_H can be determined as a parameter which depends on leaf structure. S, ρa, c_p, γ, and λ are physical constants. R_n, G – if needed at all – and δe – determined by air temperature and humidity – are available from micrometeorological measurements.

Canopy conductance for vapor, g_W, may be taken in the simplest way as the product of representative leaf conductance, g_l, (measured by gas exchange analysis or porometry) times leaf area index, LAI (Tan et al. 1978; Dolman 1988; Leuning et al. 1991). This is a sufficient approximation, if time steps of the simulations are not very short, and canopy architecture permits leaf responses to the prevailing mesoclimatic conditions. Combining by such an approach root-surface-related water uptake calculations, pressure-volume-derived plant water storage data, an empirical leaf conductance model the parameters of which are derived from porometrical measurements, and Penman-Monteith calculation of the canopy water exchange, Jensen et al. (1993) successfully simulated and predicted barley water relations under increasing drought stress.

Direct LAI determinations are tedious and, as a rule, impossible with forest canopies. Instead, mostly indirect measurements are used based on radiation penetration through the canopy (Chason et al. 1991; Chen and Black 1991; Lang et al. 1991) or reflectance measurements. Commercially available LAI meters (Bingham and Long 1993) rely on the latter principle. Also normalized difference vegetation index values from Landsat-TM data have been used to determine seasonal dynamics of forest LAI (Curran et al. 1992).

g_l measurements (with single leaves enclosed in cuvettes) as a rule disturb the boundary layer conditions, and therefore might be prone to errors when used for calculations of canopy conductance. Depending on canopy structure, aerodynamic conditions may completely override the stomatal regulation which usually determines the degree of physiologically measured single-leaf conductances. McNaughton and Jarvis 1983 (see also Jarvis 1985; Jarvis and McNaughton 1986; Jones 1990) emphasize this introducing a decoupling coefficient Ω. With values between 0 (total coupling of stomatal responses to the atmospheric situation) and 1 (perfect decoupling), it quantifies the influence of leaf and boundary layer conductance on transpiration rates.

There are feedbacks between transpiration and stomatal responses to the atmospheric environment, particularly to ambient humidity, which must be taken into account when solving the energy balance equation (Bailey and Davies 1981a; Körner 1985; Choudhury and Monteith 1986), and different leaf orientations might considerably change canopy radiation gain, temperature relations, and total transpiration (Forseth and Teramura 1986).

Quite a number of studies therefore prefer measurements of canopy surface temperatures in order to calculate canopy conductances and evapotranspiration from these data and the other meteorological parameters (e.g., Malek et al. 1992; Zhang and Lemeur 1992). Mostly touch-free IR thermometry is used for such measurements (Pinter 1979; Clawson and Blad 1982; Hatfield 1983; Berliner et al. 1984; Epiphanio et al. 1989). Using one-time-of-day IR temperature observations, Choudhury et al. (1986) estimated evapotranspiration from well-watered and water-stressed wheat plots based on energy balance, Penman-Monteith, and root water uptake calculations. The results were checked against measurements with weighing lysimeters, and proved to be correct within 10% deviation limits (similarly: Hatfield et al. 1983). The linear relationship between the canopy-air temperature difference and the saturation deficit of the air (Stark and Wright 1985) is influenced by air turbulences (O'Toole and Hatfield 1983), net radiation (Hipps 1985), and plant physiological stages (Gardner et al. 1981; O'Toole and Real 1986). If the influences of the former two meteorological parameters are known for a certain phenological stage, the water stress situation of a crop can be quantified from the foliage-air temperature differences. Water stress indexes based on such temperature data can be useful for the definition of easily controlled irrigation thresholds for some crops (Jackson et al. 1981; Idso and Reginato 1982; Choudhury et al. 1984; Tubaileh et al. 1986; Peñuelas et al. 1992). Drought influences by changed transpiration rates the leaf energy balance; the plant water content as an energy storage compartment does not play a prominent role (Carlson and Lynn 1991); however, it influences the IR spectrum of a canopy (Shvalaeva et al. 1984).

The reliability of evapotranspiration estimates based on energy balance calculations has been checked by measurements with weighing lysimeters (Choudhury et al. 1986; Campbell 1989; Wendling 1991) and by comparison with water balance calculations

(Papakyriakou and McCaughey 1991). Correspondence also exists between theoretical and measured values if the calculations are applied to the evaporation from a Piché atmometer (Paw U and Gueye 1983), and some proportionality exists also for average data over several-day intervals if comparisons are made with pan A evaporation measurements (Rao et al. 1990; Hsu et al. 1991; Chiew and McMahon 1992). Roth and Günther (1992) emphasize the need for consideration of plant parameters when calculating canopy water loss by the Penman-Monteith equation as well as by other formulas used to estimate evapotranspiration. These other calculation approaches, e.g., calculations according to TURC, Blanley-Criddle, Hargreaves, Priestley-Taylor, Thornthwaite, or Haude (Schrödter 1985; Michalopoulou and Papaioannou 1991; Mohan 1991), are physically and physiologically not as conclusive, but require less input data. National school traditions may also play a role when less concise, more empirical evaporation estimates are preferred (Hennig 1992: neglect of Penman-Monteith-based evapotranspiration estimates in German agronomy and meteorology).

Energy balance and evapotranspiration were determined for various vegetation types under quite different climatic conditions. A New Zealand tussock grassland spends 40% of the received radiation in evapotranspiration, the stomatal regulation of transpiration being considerably influenced by the atmospheric environment ($\Omega = 0.17$; Campbell 1989). A *Digitaria decumbens* pasture used more than two thirds of the energy gain by vaporization of water (Hsu et al. 1991). Data on the long-term energy balance of a moist grassland under a temperate climate can be found in Priban and Ondok (1985). Price (1991) gives information about evaporation from a blanked bog where frequent fog considerably influences radiation and vapor exchange between vegetation and the atmosphere. The depression of evaporation by the foggy conditions plays a key role in keeping appropriate hydrologic conditions for continuous bog development. From papers dealing with tree-dominated natural vegetation units, those of Maitelli and Miranda (1991) and Lafleur (1992) will be cited. The former deals with evapotranspiration and energy fluxes in the Brazilian cerrado. There, in the rainy season, Bowen ratios ranged between 0.3 and 1.2, evapotranspiration consumed 60% of the net radiation under the influence of canopy resistances between 70 and 100 s m^{-1}. In the second paper, the energy balance of a subarctic forest in Canada is analyzed. There, Bowen ratios embraced values between 0.5 and 2.5, less than 50% of R_n was consumed by λE, and r_W attained values between 20 and 66 s m^{-1}, with single-leaf resistances remaining in the order of 200 s m^{-1}. Energy fluxes over dense tropical forests were studied by Pinker et al. (1980), Shuttleworth et al. (1984), and De Abreu et al. (1988). Studies on evapotranspiration and radiation regimes of agricultural crops and fruit trees are comparatively numerous. In addition to the papers already mentioned in the discussion of the various exchange processes, only a few more will be quoted, e.g., Bailey and Davies (1981b: evaporation from soybean), Thorpe (1978: radiation regime and transpiration of apple trees in rows), and Kakayama et al. (1992: evapotranspiration of trellised pear trees in a Japanese orchard).

5. Large-Scale Assessment of Vegetation Structure and Energy and Matter Exchange by Satellite-Based Remote Sensing

The specific absorbance or reflectance of certain wavebands within the electromagnetic spectrum by vegetation structures, associated with energy and matter exchange, allows large-scale monitoring of these processes by airborne (e.g., Clevers 1988) and spaceborne imagery. Several satellites (Landsat-Thematic mapper, NOAA-AVHRR, SPOT, and others) have been launched, which, as a rule, provide pictures and spectral scanning data of the earth's surface in different wavelength ranges of the visible and near-IR part of the spectrum. A good geographical resolution is coupled normally with a broad bandwidth of the spectrum and vice versa. Pixel sizes corresponding to 30 x 30 m ground areas are at present a magnitude of surface area resolution which still provides a good sprectral resolution. Generally, the remote sensing of data on ecosystem structures and processes is still at a level of development where validation by ground-based information is absolutely necessary and most often the actual target of the most detailed study. Identification of signatures, the relations between surface patterns and remotely sensed signals, is a prominent task in many investigations. As a first step, plant reflectance patterns under controlled conditions have to be determined. For instance, Danson et al. (1992) could correlate leaf water contents with the first derivative of the reflectance spectrum at wavelengths corresponding to the edges of IR water absorption bands. Flagella et al. (1992) used the IR reflectance at 1940 nm to evaluate wheat leaf drought resistance; but application of such correlations to spaceborne data often fails. For reasons of atmospheric interactions and view-angle effects, the immediate surface reflectances cannot be used as a rule. Rather, reflectance ratios of at least two wavelength bands have to be used, sometimes only after data conversion procedures and supplemented by ground-based measurements (Ahlrichs and Bauer 1983; Kuusk 1991; Leblon et al. 1991; Caselles et al. 1992a). Jupp and Strahler (1991) emphasize that a peculiar reflectance situation is given if the radiation source and the sensor are aligned ("hot spot"). The brightness change around this situation could be used to infer the particular structure of the reflecting canopy. Boissard et al. (1992) tried to estimate ground cover ratios of wheat fields from reflectance measurements in the red and IR bands of SPOT satellite data; Mulders and Van Logchem (1991) developed equations for the calculation of grass and soil moisture from reflectance values registered by spectral bands of the Landsat-Thematic mapper. Calibration of visible and IR channels of NOAA-AVHRR for identification of vegetation indices and surface temperature estimations was undertaken, among others, by Vidal (1991: correction of atmospheric effects on temperature determinations), Gupta (1992: study of scan angle effects), Kimes and Holben (1992: albedo estimates), Caselles et al. (1992b: calculation of maximal evapotranspiration from AVHRR temperature and albedo data), and Che and Price (1992: general analysis of calibration methods). Radiation in the visible and near-IR bands is heavily disturbed by clouds, and that is why use of satellite data is often limited in tropical and temperate to arctic regions. Microwave and radar sensors may overcome such drawbacks (Haefner and Pampaloni 1992; Hanninen 1992). They have, however, not yet reached an operational level (Holmes 1992; John 1992; Paloscia and Pampaloni 1992). Considerable expectations are placed in remotely sensed, area-related information on vegetation structure, water content, and thermal exchange processes. When knowledge of signatures has passed the present developmental stage, then, in connection with complex energy balance and momentum transfer calculations (Siebert et al. 1992), a real step ahead will have been made in the assessment of biosphere-atmosphere interactions (Moran and Jackson 1991).

Appendix: Symbols, Abbreviations, and Physical Constants

AVHRR = advanced very high resolution radiometer (satellite-based waveband scanner), launched by NOAA (National Oceanic and Atmospheric Administration/USA)

C = heat flux by convection (W m^{-2})
c_p = specific heat capacity (J kg^{-1} K^{-1}) (air: 1010)
γ = psychrometer constant (Pa k^{-1} = p $c_p/0.622\lambda$)
d = zero plane displacement (m)
δe = water vapor pressure deficit of the air (Pa)
e^* = saturation vapor pressure (Pa)
e_a = actual vapor pressure at air temperature (Pa)
λE = latent heat of evaporation (2.454 MJ kg^{-1} at 20 °C) times evaporation (W m^{-2})
G = heat flux by conduction (W m^{-2})
g_H = conductance for heat transfer (mm s^{-1})
g_l = leaf conductance for water vapor transfer (mm s^{-1})
g_W = canopy conductance for water vapor transfer (mm s^{-1})
IR = infra-red (radiation, wavelengths or the like)
K = von Karman's constant (0.41)
LAI = leaf area index (m^2 m^{-2})
P = metabolically used radiant energy (W m^{-2})
PAR = photosynthetic active radiation (µmol m^{-2} s^{-1})
p = pressure (Pa)
R_n = net gain of heat from radiation (W m^{-2})
r_W = canopy resistance to water vapor transfer (s mm^{-1})
ρ = specific density (kg m^{-3})
ρ_a = specific density of dry air (1.204 kg m^{-3} at 20 °C)
S = heat storage (W m^{-2})
s = slope of the curve relating saturation vapor pressure to temperature (Pa K^{-1})
T_s = surface temperature (°C or K)
T_a = air temperature (°C or K)
TM = thematic mapper (satellite-based waveband scanner)
τ = shear stress (kg m^{-1} s^{-2})
u(z) = average wind speed (m s^{-1}) at height z
u_* = friction velocity (m s^{-1})
z = height above earth's surface (m)
z_0 = roughness length (m)
Ω = decoupling coefficient (dimensionless)

References

Ahlrichs JS, Bauer ME (1983) Agron J 75:987–993. **Anderson DE**, Shasi BV, Rosenberg NJ (1984a) Boundary-Layer Meteorol 29:263–272. **Anderson DE**, Verma SB, Rosenberg NJ (1984b) Boundary-Layer Meteorol 29:263–272. **Anderson DE**, Verma SB, Clement RJ, Baldocchi DD, Matt DR (1986) Agric For Meteorol 38:81–100. **Andre RGB**, Viswanadham Y (1983) Agric Meteorol 30:157–174. **Aston AR** (1985) Agric For Meteorol 35:281–298.

Bache DH (1986) Atmos Environ 20:1369–1378. **Bailey WG**, Davies JA (1981a) Boundary-Layer Meteorol 20:187–200. **Bailey WG**, Davies JA (1981b) Boundary-Layer Meteorol 20:417–428. **Balding FR**, Cunningham GL (1976) Bot Gaz 137:65–74. **Baldocchi DD**, Verma SB, Rosenberg NJ, Blad BL, Garay A, Specht JE (1983) Agron J 75:543–548. **Baldocchi DD**, Matt D, Hutchinson B, Mcmillen R (1984a) Agric For Meteorol 32:307–322. **Baldocchi DD**, Hutchinson B, Matt D, Mcmillen R (1984b) Agric For Meteorol 33:177–192. **Baldocchi DD**, Verma SB, Rosenberg NJ, Blad BL, Specht JE (1985) Agric For Meteorol 35:1–20. **Barradas VL**, Adem J (1992) Int J Biometeorol 36:113–117. **Bazzaz FA** (1990) Annu Rev Ecol Syst 21:167–196. **Bégué A** (1992) Remote Sens Environ 40:219–230. **Berliner P**, Oosterhuis DM, Green GC (1984) Agric For Meteorol 31:219–230. **Bingham MJ**, Long SP (1993) Equipment for plant physiology research in a changing environment. In: Hall DO et al. (eds) Photosynthesis and production in a changing environment: a field and laboratory manual. Chapman & Hall, London, pp 357–421. **Boissard P**, Pointel JG, Tranchefort J (1992) Int J Remote Sens 13:1681–1692. **Bristow KL**, Campbell GS (1985) Agric For Meteorol 35:123–132. **Brutsaert W** (1979) Boundary-Layer Meteorol 16:365–388

Campbell DI (1989) Boundary-Layer Meteorol 46:133–152. **Campbell RJ**, Marini RP (1992) J Am Soc Hortic Sci 117:678–684. **Carlson TN**, Lynn B (1991) Agric For Meteorol 57:171–186. **Caselles V**, Sobrino JA, Coll C (1992a) Remote Sens Environ 39:203–211. **Caselles V**, Delegido J, Sobrino JA, Hurtado E (1992b) Int J Remote Sens 13:939–946. **Cellier P**, Brunet Y (1992) Agric For Meteorol 58:93–117. **Chason JW**, Baldocchi DD, Huston MA (1991) Agric For Meteorol 57:107–128. **Che N**, Price JC (1992) Remote Sens Envirom 41:19–27. **Chen J** (1984) Boundary-Layer Meteorol 28:213–226. **Chen JM**, Black TA (1991) Agric For Meteorol 57:1–12. **Chiew FHS**, McMahon TA (1992) Aust J Soil Res 30:101–112. **Choudhury BJ**, Monteith JL (1986) Agric For Meteorol 36:215–226. **Choudhury BJ**, Bhaskar J, Idso SB (1984) Agric For Meteorol 31:69–78. **Choudhury BJ**, Idso SB, Reginato RJ (1986) Remote Sens Environ 19:253–268. **Choudhury BJ**, Idso SB, Reginato RJ (1987) Agric For Meteorol 39:283–298. **Cionco RM** (1983) Boundary-Layer Meteorol 26:325–355. **Clawson KL**, Blad BL (1982) Agron J 74:311–316. **Clevers JGPW** (1988) Neth J Agric Sci 36:75–90. **Curran PJ**, Dungan JL, Gholz HL (1992) Remote Sens Environ 39:3–14.

Danson FM, Steven MD, Malthus TJ, Clark JA (1992) Int J Remote Sens 13:461–470. **De Abreu SLD**, Viswanadham Y, Manzi AO (1988) Theor Appl Climatol 39:1–16. **De Bruin HAR**, Moore CJ (1985) Boundary-Layer Meteorol 31:39–50. **Deering DW**, Leone P (1986) Remote Sens Environ 19:1–24. **Desjardins RL**, Allen LH Jr, Lemon ER (1978) Boundary-Layer Meteorol 14:369–380. **Detering HW**, Etling D (1985) Boundary-Layer Meteorol 33:113–134. **Dhyani PP** (1989) Trop Ecol 30:58–64. **Dhyani PP**, Purohit KD, Purohit AN (1986) Proc Indian Natl Sci Acad Part B Biol Sci 52:665–672. **Dixon M**, Grace J (1983) Plant Cell Environ 6:665–670. **Dolman AJ** (1986) Agric For Meteorol 36:241–248. **Dolman AJ** (1988) J Hydrol 97:225–234.

Eamus D (1991) Plant Cell Environ 14:843–852. **Eck TF**, Dye DG (1991) Remote Sens Environ 38:135–146. **Eller BM** (1977a) Angew Bot 51:9–16. **Eller BM** (1977b) J Exp Bot 28:1054–1059. **Eller BM**, Grobbelaar N (1986) S Afr J Bot 52:403–407. **Eller BM**, Willi P (1977) Oecologia 29:179–188. **Enders G**, Dlugi R, Steinbrecher R, Clement B, Daiber R, Eijk JV, Gaeb S, Haziza M, Helas G (1992) Atmos Environ Part A Gen Top 26:171–190. **Epiphanio JCN**, Carvalho LJCB, Vitorello I, Filho MV (1989) Pesqui Agropecu Bras 24:253–258.

Flagella Z, Pastore D, Campanile RG, Di Fonzo N (1992) J Genet Breed 46:21–27. **Forseth IN**, Teramura AH (1986) Ecology 67:564–571.

Gao W, Shaw RH, Paw U KT (1992) Boundary-Layer Meteorol 59:35–57. **Gardner BR**, Blad BL, Garrity DP, Watts DG (1981) Irrig Sci 2:213–224. **Grace J** (1977) Plant response to wind. Academic Press, London. **Grace JC**, Jarvis PG, Norman JM (1987) N Z J For Sci 17:193–209. **Grant L**, Daughtry CST, Vanderbilt VC (1987) Environ Exp Bot 27:139–146. **Graser EA**, Verma SB, Rosenberg NJ (1987) Agric For Meteorol 41:187–206. **Green CF**, Dawkins TCK (1985) Crop Res 25:13–20. **Gross G** (1987) Boundary-Layer Meteorol 40:311–328. **Gueymard C** (1989) Agric For Meteorol 45:215–230. **Gupta RK** (1992) Int J Remote Sens 13:715–735. **Gutman LN**, Berkofsky L (1985) Boundary-Layer Meteorol 31:287–302.

Haefner H, Pampaloni P (1992) Int J Remote Sens 13:1277–1303. **Hall DO**, Scurlock JMO, Bolhàr-Nordenkampf HR, Leegood RC, Long SP (1993) Photosynthesis and production in a changing environment. Chapman & Hall, London. **Hamer PJC** (1985) Agric For Meteorol 35:339–352. **Hamer PJC** (1986a) Agric For Meteorol 37:159–174. **Hamer PJC** (1986b) Agric For Meteorol 37:175–188. **Hanninen P** (1992) Geol Surv Finl Bull 0:1–71 **Hatfield JL** (1983) Irrig Sci 3:259–268. **Hatfield JL**, Perrier A, Jackson RD (1983) Agric Water Manage 7:341–350. **Heilman JL**, Brittin CL, Neale CMU (1989) Agric For Meteorol 44:261–274. **Hennig A** (1992) Z Pflanzenernaehr Bodenkd 155:169–257. **Hipps LE** (1985) Agric For Meteorol 35:113–122. **Hirai GI**, Nakayama N, Chujo H, Inano T, Hirano T, Shimoda H, Tanaka O (1992) Jpn J Crop Sci 61:41–48. **Holbo HR**, Luvall JC (1989) Remote Sens Environ 27:11–24. **Holmes MG** (1992) Bot J Linn Soc 108:93–109. **Honjo T**, Takakura T (1986) J Agric Meteorol 42:137–143. **Houghton JT**, Jenkins GJ, Ephraums JJ (1990) Climate change: the IPCC scientific assessment. Cambridge University Press, Cambridge. **Hsu SH**, Huang KC, Wang SF (1991) J Agric Assoc China New Ser 0:82–93.

Idso SB (1989) Carbon dioxide and global change: earth in transition. IBR, Tempe. **Idso SB**, Reginato RJ (1982) Water Resour Res 18:1143–1148. **Isoda A**, Yoshimura T, Ishikawa T, Nojima H, Takasaki Y (1992) Jpn J Crop Sci 61:124–130.

Jackson RD, Idso SB, Reginato RJ, Pinter PJ Jr (1981) Water Resour Res 17:1133–1138. **Jacobs AFG**, Palland CL (1982) J Hydrol (Amst) 58:159–166. **Jacobs AFG**, Van Boxel JH (1988) Agric For Meteorol 42:53–62. **Jacobs AFG**, Van Boxel JH, Shaw RH (1992) Agric For Meteorol 58:247–256. **Jäger L** (1984) Arch Meteorol Geophys Bioclimatol Ser B Climatol Environ Meteorol Radiat Res 34:163–180. **Jarvis PG** (1985) Acta Hortic 171:187–205. **Jarvis PG**, NcNaughton KG (1986) Adv Ecol Res 15:1–49. **Jarvis PG**, James GB, Landsberg JJ (1976) Coniferous forest. In: JL Monteith (ed) Vegetation and the atmosphere, vol II. Academic Press, London, pp 171–240. **Jensen CR**, Svendsen H, Andersen MN, Lösch R (1993) Plant Soil 149:1–26. **John B** (1992) Int J Remote Sens 13:481–491. **Johns D**, Van Bavel CHM, Beard JB (1981) Agric Meteorol 25:15–26. **Jones HG** (1990) Hortscience 25:19–26. **Jones HG** (1992) Plants and microclimate, 2nd edn. Cambridge University Press, Cambridge. **Jupp DLB**, Strahler AH (1991) Remote Sens Environ 38:193–210.

Kakayama K, Ichimura T, Kon H, Matsuoka N (1992) Tech Bull Fac Hortic Chiba Univ 0:67–71. **Kawanishi H** (1986) Ecol Model 33:315–328. **Kimes DS**, Holben BN (1992) Int J Remote Sens 13:275–289. **Klaassen W** (1992) Boundary-Layer Meteorol 58:329–354. **Knyazikhin YV**, Marshak AL, Myneni RB (1992) Remote Sens Environ 39:61–74. **Kochubei SM**, Shadchina TM, Kanevskii VA, Shvareva SG (1984) Fiziol Biokhim Kult Rast 16:251–255. **Kondo J**, Kawanaka A (1986) Boundary-Layer Meteorol 37:285–296. **Koppel A**, Oja T (1984) Photosynthetica 18:529–535. **Körner C** (1985) Arch Meteorol Geophys Bioclimatol Ser B 36:83–98. **Kriebitzsch WU** (1991) Forstarchiv 62:179–182. **Kustas WP**, Brutsaert W (1986) Boundary-Layer Meteorol 34:35–54. **Kuusk A** (1991) Remote Sens Environ 37:207–218.

Lafleur PM (1992) Agric For Meteorol 58:163–175. **Lang ARG**, McNaughton KG, Fazu C, Bradley EF, Ohtaki E (1983) Boundary-Layer Meteorol 25:25–42. **Lang ARG**, McMurtrie RE, Benson ML (1991) Agric For Meteorol 57:157–170. **Leahey DM**, Hansen MC (1987) Atmos Environ

21:1227–1230. **Leblon B**, Guerif M, Baret F (1991) Remote Sens Environ 38:147–158. **Lee HN**, Kau WS (1984) Boundary-Layer Meteorol 29:381–396. **Leuning R** (1988) Agric For Meteorol 42:135–155. **Leuning R**, Attiwill PM (1978) Agric Meteorol 19:215–242. **Leuning R**, Cremer KW (1988) Agric For Meteorol 42:123–133. **Leuning R**, Kriedemann PE, McMurtrie RE (1991) Agric Water Manage 19:205–221. **Lewis DA**, Nobel PS (1977) Plant Physiol 60:609–616. **Lloyd CR**, Culf AD, Dolman AJ, Gash JHC (1991) Boundary-Layer Meteorol 57:311–322. **Lloyd CR**, Gash JHC, Sivakumar MVK (1992) Boundary-Layer Meteorol 58:261–272. **Lösch R** (1989) Prog Bot 50:27–50. **Lösch R** (1991) Prog Bot 52:35–60. **Lösch R** (1993) Prog Bot 54:102–133.

Maitani T (1977) Ber Ohara Inst Landwirtsch Biol Okayama Univ 17:29–46. **Maitani T** (1978) Boundary-Layer Meteorol 14:571–584. **Maitani T**, Seo T (1985a) Boundary-Layer Meteorol 33:51–60. **Maitani T**, Seo T (1985b) Boundary-Layer Meteorol 33:197–208. **Maitelli GT**, Miranda AC (1991) An Acad Bras Cienc 63:265–272. **Malek E**, Bingham GE, McCurdy GD (1992) Boundary-Layer Meteorol 59:187–194. **Mayer H** (1981) Arch Meteorol Geophys Bioclimatol Ser B Climatol Environ Meteorol Radiat Res 29:381–392. **McCaughey JH** (1985a) Boundary-Layer Meteorol 31:89–102. **McCaughey JH** (1985b) Boundary-Layer Meteorol 32:1–24. **McCaughey JH**, Saxton WL (1988) Agric For Meteorol 44:1–18. **McClendon JH** (1984) Am J Bot 71:1391–1397. **McNaughton KG**, Jarvis PG (1983) Predicting effects of vegetation changes on transpiration and evaporation. In: Kozlowski TT (ed) Water deficits and plant growth, vol VII, Academic Press, New York, pp 1–47. **Meek DW**, Hatfield JL, Howell TA, Idso SB, Reginato RJ (1984) Agron J 76:939–945. **Mehlenbacher LA**, Whitfield DWA (1977) Boundary-Layer Meteorol 12:153–170. **Meyers T**, Paw U KT (1986) Boundary-Layer Meteorol 37:297–312. **Michalopoulou H**, Papaioannou G (1991) Agric Water Manage 20:209–222. **Mohan S** (1991) Hydrol Sci J 36:447–460. **Molion LCB**, Moore CJ (1983) Boundary-Layer Meteorol 26:115–126. **Monteith JL** (1965) Symp Soc Exp Biol 19:205–234. **Monteith JL** (1981) Coupling of plants to the atmosphere. In: Grace J, Ford ED, Jarvis PG (eds) Plants and their atmospheric environment, Blackwell, Oxford, pp 1–29. **Moran MS**, Jackson RD (1991) J Environ Qual 20:725–737. **Morris JT** (1989) Agric For Meteorol 46:349–362. **Mulders MA**, Van Logchem A (1991) Neth J Agric Sci 39:137–148. **Myers NA** (1988) Environ Conserv 15:293–298.

Neilson RP, King GA, Koerper G (1992) Landscape Ecol 7:27–43. **Newman SM** (1985) Agric For Meteorol 35:243–254. **Nikolov NT**, Zeller KF (1992) Ecol Model 61:3–4. **Nilson T** (1992) Adv Bioclimatol 1:60–88. **Nobel PS** (1991) Physicochemical and environmental plant physiology. Academic Press, New York.

Ohtaki E, Oikawa T (1991) Int J Biometeorol 35:187–194. **Oker-Blom P**, Kelomäki S (1982) Math Biosci 59:291–311. **O'Rourke PA**, Terjung WH (1981) Arch Meteorol Geophys Bioclimatol Ser B Climatol Environ Meteorol Radiat Res 29:365–380. **O'Toole JC**, Hatfield JL (1983) Agron J 75:811–817. **O'Toole JC**, Real JG (1986) Agron J 78:305–310. **Otterman J** (1989) Theor Appl Clim 40:67–80.

Paloscia S, Pampaloni P (1992) Remote Sens Environ 40:15–26. **Papakyriakou TN**, McCaughey JH (1991) Can J For Res 21:1622–1631. **Paw U KT** (1992) Agric For Meteorol 57:297–304. **Paw U KT**, Gao W (1988) Agric For Meteorol 43:121–146. **Paw U KT**, Gueye M (1983) Agric For Meteorol 30:1–12. **Penman HL** (1948) Proc R Soc Lond A 193:120–145. **Peñuelas J**, Save R, Marfa O, Serrano L (1992) Agric For Meteorol 58:63–77. **Pereira AR**, Machado EC, Paes de Camargo MB (1982) Agric Meteorol 26:1–10. **Pinker RT**, Thompson OE, Eck TF (1980) Q J R Meteorol Soc 106:551–558. **Pinter PJ** (1979) Science 205:585–587. **Priban K**, Ondok JP (1985) Folia Geobot Phytotaxon 20:41–56. **Price JS** (1991) Boundary-Layer Meteorol 57:391–406.

Rao, Praveen V, Raikhelkar SV, Sondge VD (1990) J Res Andhra Pradesh Agric Univ 18:131–134. **Raupach MR** (1989) Turbulent transfer in plant canopies. In: Russell G, Marshall B, Jarvis PG (eds) Plant canopies: their growth, form and function. Cambridge Univ Press, Cambridge, pp 41–61. **Roberts G**, Roberts AM (1992) Agric Water Manage 21:155–166. **Roberts G**, France M, Robinson M (1992) Agric Water Manage 21:145–154. **Rodskjer N**, Dahlstedt L (1977) Swed J Agric Res 7:85–88. **Roth D**, Günther R (1992) Z Kulturtech Landentwicklung 33:13–22.

Salminen R, Nilson T, Hari P, Kaipianen L, Ross J (1983) J Appl Ecol 20:897–904. **Schieldge JP** (1978) Agric Meteorol 19:315–328. **Schmidt W** (1989) Prog Bot 51:350–374. **Schmidt W** (1992) Prog Bot 53:408–434. **Schrödter H** (1985) Verdunstung. Springer, Berlin Heidelberg New York. **Seligman NG**, Loomis RS, Burke J, Abshahi A (1983) J Agric Sci 101:691–697. **Shaw RH**, Zhang XJ (1992) Boundary-Layer Meteorol 58:273–288. **Shugart HH**, Emanuel WR (1985) Plant Cell Environ 8:381–386. **Shukla J**, Mintz Y (1982) Science 215:1498–1500. **Shultis JK** (1991) Remote Sens Environ 38:211–228. **Shuttleworth WJ**, Gash JHC, Lloyd CR, Moore CJ, Roberts J et al. (1984) Q J R Meteorol Soc 110:1163–1169. **Shvalaeva AL**, Gusev NA, Cherezov SN (1984) Fiziol Biokhim Kult Rast 16:471–476. **Siebert J**, Sievers U, Zdunkowski W (1992) Boundary-Layer Meteorol 59:1–34. **Sigmon JT**, Knoerr KR, Shaughnessy EJ (1984) Agric For Meteorol 31:329–338. **Skupniewicz CE**, Kamada RF, Schacher GE (1989) Boundary-Layer Meteorol 48:109–128. **Smolen F**, Matejka F (1982) Ekologia-CSSR 1:25–36. **Stark JC**, Wright JL (1985) Am Potato J 62:57–68. **Steyn DG** (1982) Boundary-Layer Meteorol 22:183–192. **Sturm M** (1992) Arct Alp Res 24:145–152. **Sud YC**, Smith WE (1985) Boundary-Layer Meteorol 33:15–50.

Taha H, Akbari H, Rosenfeld A (1991) Theor Appl Clim 44:123–138. **Tajchman SJ** (1981) Bull Am Meteorol Soc 62:1550–1559. **Tan CS**, Black TA, Nnyamah JU (1978) Ecology 59:1221–1229. **Tanner V**, Eller BM (1986) J Plant Physiol 125:285–294. **Thom AS** (1975) Momentum, mass and heat exchange of plant communities. In: Monteith JL (ed) Vegetation and the atmosphere, vol I. Academic Press, London, pp 57–109. **Thorpe MR** (1978) Agric Meteorol 19:41–57. **Thorpe MR**, Butler DR (1977) Boundary-Layer Meteorol 12:61–74. **Torssell BWR**, McPherson HG (1977) Aust J Ecol 2:245–256. **Tubaileh AS**, Sammis TW, Lugg DG (1986) Agric Water Manage 12:75–86.

Valentini R, Mugnozza GES, De Angelis P, Bimbi R (1991) Plant Cell Environ 14:987–994. **Van der Hage JCH** (1984) J Exp Bot 35:762–766. **Van Eimern J**, Riedinger FP (1986) Agric For Meteorol 38:27–46. **Verma SB**, Baldocchi DD, Anderson DE, Matt DR, Clement RJ (1986) Boundary-Layer Meteorol 36:71–92. **Vidal A** (1991) Int J Remote Sens 12:2449–2460. **Vogel S** (1984) Can J Bot 62:741–744.

Wallace JS, Lloyd CR, Roberts J, Shuttleworth WJ (1984) Agric For Meteorol 32:289–306. **Walthall CL**, Norman JM (1992) Agron J 84:101–107. **Wendling U** (1991) Arch Acker-Pflanzenbau Bodenkd 35:251–257. **Woodward FI**, McKee IF (1991) Environ Int 17:535–546.

Yates DJ, Steven MD (1987) J Agric Sci 109:495–502.

Zangerl AR (1978) Oecologia (Berl) 34:107–112. **Zaradny H**, van der Ploeg RR (1982) Z Pflanzenernaehr Bodenkd 145:611–622. **Zhang L**, Lemeur R (1992) Agric For Meteorol 58:209–288.

Professor Dr. Rainer Lösch
Abteilung Geobotanik
Heinrich-Heine-Universität
Universitätsstraße 1/26.13
D-40225 Düsseldorf, FRG

III. Secondary Plant Substances: Ornithine-Derived Alkaloids

By Horst-Robert Schütte

1. Introduction

It is well known that the different alkaloids, a very large group of secondary plant substances, are derived from different amino acids. Ornithine or its derivatives are the source of the pyrrolidine ring system, which is included in the alkaloid skeletons discussed here. Tobacco alkaloids, as well as tropane alkaloids, belong to the plant bases which were first investigated biosynthetically.

2. Tobacco Alkaloids

Figure 1 summarizes the current knowledge on the biosynthesis of the main tobacco alkaloids (Leete 1980, 1983d; Gross 1985). *Nicotiana* species produce principally four main alkaloids, nicotine (XI), nornicotine (XII), anabasine (XVII), and anatabine (XX) (Saitoh et al. 1985). The most common *Nicotiana* alkaloid is (S)-nicotine (XI), which is formed by condensation between nicotinic acid (IX) and the 1-methyl-Δ^1-pyrrolinium salt (X) (Leete 1967). This nicotine synthase reaction has been achieved also in a cell-free system obtained from the roots of *N. glutinosa* as well as from seedlings of *N. tabacum* (Friesen and Leete 1990). Oxygen is necessary for nicotine synthase activity obtained from tobacco seedlings. The nicotine synthase activity was absent in the first 100 h after germination, and reached a maximum after 140 h.

The iminium salt (X) is formed by cyclization of 4-methylaminobutanal (VIII), which is produced from putrescine (IV) via N-methylputrescine (VII) (Mizusaki et al. 1968). The source of putrescine is currently controversial, one route being the direct decarboxylation of ornithine (I), the other being via citrulline (II), arginine (III), agmatine (VI), and N-carbamoylputrescine (V) (Tiburcio and Galston 1986). At this time, the evidence favors the route via arginine, but both routes could be operating.

Some enzymes have been isolated from tobacco roots catalyzing the conversion of ornithine (I) to the 1-methyl-Δ^1-pyrrolinium salt (X) (Mizusaki et al. 1972, 1973; Saunders and Busch 1979; Feth et al. 1985a, b; Wagner et al. 1986; Feth and Wagner 1989). Ornithine decarboxylase catalyzes the formation of putrescine (IV) from ornithine (I). The ornithine as well as the arginine decarboxylase reaction takes place with retention of configuration (Orr and Gould 1982; Richards and Spenser 1982; Robins 1983b). Difluoro-

Progress in Botany, Vol. 55

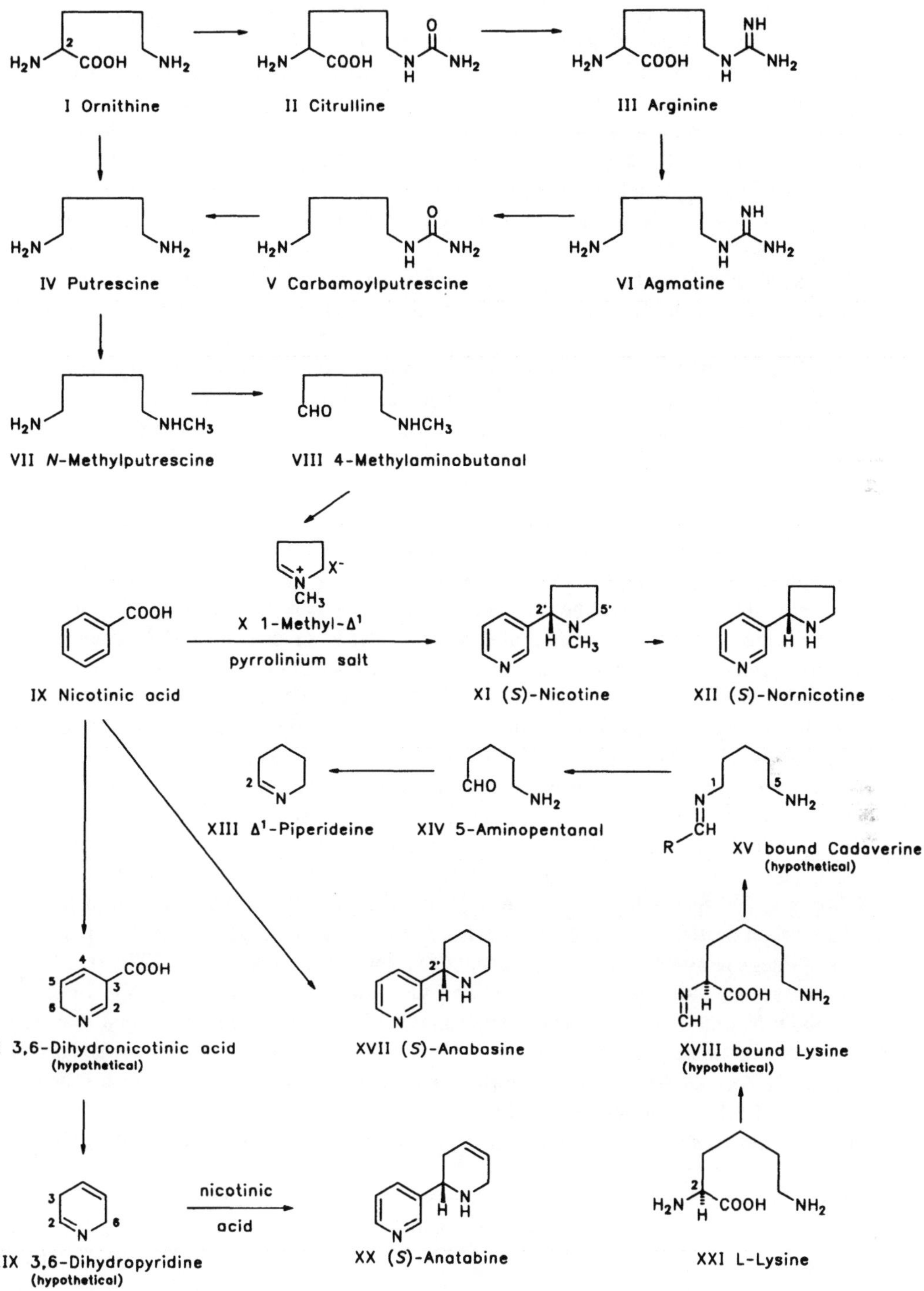

Fig. 1. Biosynthesis of nicotine, nornicotine, anabasine, and anatabine. (Friesen et al. 1992)

methylornithine, as a suicide inhibitor of ornithine decarboxylase, inhibited the formation of nicotine (XI) in a *Nicotiana* callus tissue (Slocum and Galston 1985; Tiburcio et al. 1987; Burtin et al. 1989, 1991). However, the inhibition of nicotine formation was greater with α-difluoromethylarginine, a selective inhibitor of arginine decarboxylase, suggesting that the major pathway to putrescine (IV), at least in this tissue, is from arginine rather than ornithine. Bound putrescine can act as pool for pyrrolidine alkaloid formation in systems where alkaloid biosynthesis is active (Tiburcio et al. 1985). Increased levels of putrescine may enhance the ability of tobacco cells to survive acid stress (Hiatt and Malmberg 1988).

Putrescine N-methyltransferase obtained from tobacco roots catalyzes the reaction between putrescine and S-adenosylmethionine to yield N-methylputrescine (VII). The N-methylputrescine oxidase isolated from *N. tabacum* (Davies et al. 1989; Walton and McLauchlan 1990) affords the 1-methyl-Δ^1-pyrrolinium salt (X), which is suggested to be a quinoprotein. During the oxidation of N-methylputrescine in *N. tabacum*, the pro-S hydrogen is lost from the carbon carrying the primary amino group to yield the N-pyrrolinium ion, which is attached by the pyridine ring precursor at the 1-si,2-re face to yield (S)-nicotine (XI) (Wigle et al. 1982). (S)-Nornicotine (XII) is formed by demethylation of nicotine. Putrescine methyltransferase seems to be the enzyme under stringent control for nicotine biosynthesis, whereas the subsequent methylputrescine oxidase is coregulated, although less stringently (Feth et al. 1986).

As with nicotine, the pyridine ring of anabasine (XVII) is derived from nicotinic acid (IX). The immediate precursor of its piperidine ring is Δ^1-piperideine (XIII) which is derived from lysine (XXI). The metabolic sequence differs from the conversion of ornithine (I) to the iminium salt (X) (Fig. 1). Thus (2-^{14}C)ornithine affords (2',5'-^{14}C) nicotine and is rationalized by the intermediacy of free putrecine. On the other hand, (2-^{14}C)lysine (XXI) affords (2'-^{14}C)anabasine (XVII) with no label at C-6'. The incorporation of the labeled lysine thus cannot proceed via free cadaverine, and it has been proposed that the sequence proceeds via the Schiff base between lysine and pyridoxal phosphate (XVIII) and similarly bound cadaverine (XV). Lysine, but not cadaverine, has been shown to be an intermediate in the biosynthesis of anabasine in hairy root cultures of *Nicotiana* species (Walton and Belshaw 1988; Walton et al. 1988; Watson et al. 1990). A stereospecific oxidation then occures at C-1 of compund XV so that the hydrogen originally present at C-2 of lysine is retained in 5-aminopentanal (XIV) and in Δ^1-piperideine (XIII) at C-2.

Both heterocyclic rings of (S)-anatabine (XX), together with nicotine (XI) the main alkaloids in *N. alata* (Sarychev and Sherstyanykh 1985), are derived from nicotinic acid (IX) (Leete and Slattery 1976; Leete 1978). It is proposed that the Δ^3-piperideine ring of anatabine is derived from 3,6-dihydropyridine (XIX), which is formed by the decarboxylation of 3,6-dihydronicotinic acid (XVI). This dihydronicotinic acid is considered to be the activated form of nicotinic acid, which ultimately becomes the pyridine ring of all the tobacco alkaloids.

When the growth of *N. alata* root cultures was carried out in the presence of increasing concentrations of nicotinic acid (IX), there were profound changes in the amounts and compositions of the alkaloids after 3 days' incubation (Friesen et al. 1992). High concentrations of nicotinic acid inhibited both alkaloids formation and demethylation of nicotine (XI) to nornicotine (XII). Anatabine (XX) production was stimulated with increasing nicotinic acid concentrations. A fixed concentration of nicotinic acid and an increasing amount of 1-methyl-Δ^1-pyrrolinium chloride indicated that the latter compound stimulated nicotine production and also inhibited its demethylation. Increasing concentrations of Δ^1-piperideine (XIII) promoted the production of anabasine (XVII) and also stimulated nornicotine production.

Agrobacterium rhizogenes is responsible for the induction of hairy root desease in many dicotyledonous plants. The molecular basis of this involves the transfer of bacterial DNA from a large root-induced plasmid to the host nuclear DNA (Tepfer 1990). These cultures have been used for the production of secondary metabolites, and this approach should be advantageous for the development of biotechnological processes. Such hairy root cultures of *Nicotiana* species produce nicotine alkaloids and their precursors at levels comparable with those of in vivo roots or even higher (Hamill et al. 1986; Parr and Hamill 1987; Saito et al. 1989; Chriqui et al. 1990; Isogai et al. 1990; Walton and McLauchlan 1990; Mengoli et al. 1992a, b, c). The addition of nicotinic acid (IX) to a hairy root culture of *N. rustica* increased the ratio of anatabine (XX) to nicotine (XI) (Robins et al. 1987), and the addition of cadaverine to this transformed root culture increased the yield of anabasine (XVII) at the expense of nicotine (XI) (Walton and Belshaw 1988; Walton et al. 1988). Transformed root cultures of *N. rustica* have been generated in which the gene from the yeast *Saccharomyces cerevisiae* coding for ornithine decarboxylase has been integrated, with the result of enhanced ornithine decarboxylase activity and enhanced capacity to accumulate both putrescine (IV) and nicotine (XI) (Hamill et al. 1990).

Duboisia leichhardtii and other *Duboisia* species produce both nicotine and tropane alkaloids, while *Nicotiana tabacum* produces only nicotine alkaloids; thus, alkaloid biosynthesis of *Duboisia* and *Nicotiana* are partially overlapping (Luanratana and Griffin 1982; Endo and Yamada 1985; Ishimaru and Shimomura 1989; Gritsanapan and Griffin 1992). Administration of (5-^{14}C)ornithine to a root culture of *D. leichhardtii* afforded radioactive nicotine (XI) and scopolamine (XLIV), in which nicotine was labeled equally at its C-2' and C-5' positions, and the scopolamine was labeled equally at its C-1 and C-5 positions (Leete et al. 1990). These results are consistent with the hypothesis that the pyrrolidine ring of nicotine (XI) and that present in the tropane moiety of scopolamine (XLIV) have a common biosynthetic origin and that ornithine (I) is incorporated via a symmetrical intermediate. Somatic hybrids of *Duboisia* species and *N. tabacum*, obtained by electrofusion followed by individual cloning (Endo et al. 1987) predominantly produce nicotine, while *Duboisia* plantlets produce both tropane and nicotine alkaloids (Endo et al. 1991). The presence in the hybrid plants of activities responsible for the reduction of tropinon (XXXVII), the hydroxylation and epoxidation of hyoscyamine (XLI), and the conversion of nicotine (XI) to nornicotine (XII) demonstrated the presence of the *Duboisia* genes for these enzyme activities.

3. Tropane Alkaloids

Tropane is the 8-methylderivative of the bicyclic heterocyclic compound 8-azabicyclo(3.2.1.)octane. Over 150 alkaloids are known, which contain this bicyclic ring system (Lounasmaa 1988). The majority of the tropane alkaloids are esters of hydroxytropanes with a wide variety of carboxylic acids, some of which are unique to this class of alkaloids, e.g., (S)-tropic acid (XLVI). The tropane alkaloids belong to the first groups of alkaloids subjected to biosynthetic experiments with radioactive precursors. The incorporation of (2-^{14}C)ornithine (I) into hyoscyamine XLI) was reported for the first time as early as 1954 (Leete et al. 1954). Since then, the conversion of this amino acid into the pyrrolidine ring has been confirmed for hyoscyamine (XLI) and hyscine, meteolidine, cochlearine, cocaine (XLIII), and other tropane alkaloids (Liebisch et al. 1965; Baralle and Gros 1969; Ahmad and Leete 1970; Leete 1979, 1980, 1990; Liebisch and Schütte 1985; Yamada et al. 1990). Ornithine (I) is incorporated into the pyrrolidine rings of the different tropane alkaloids unsymmetrically in *Datura innoxia, D. metel*, and *D. stramonium*. In *Nicotiana, Erythoxylum coca, Duboisia leichhardtii, Hyoscyamus albus* and *Nicandra physaloides*, a symmetrical intermediate is involved (Hashimoto et al. 1989a). The unsymmetrical incorporation of (2-^{14}C)ornithine (I) into hyoscyamine (XLI), all the ^{14}C being at the C-1 bridgehead carbon, means the integrity of the C-2 and C-5 carbons of ornithine is illustrated by a mechanism in which ornithine is decarboxylated to a bound form of putrescine (XXII) (Walton et al. 1990), which is then methylated to afford a bound form of N-methylputrescine (XXIII) (Fig. 2). Pyridoxalphosphate is the coenzyme involved, and this could be an integral part of this bound form of putrescine.

On the other hand, hyoscyamine (XLI) formed in root cultures of *Hyoscyamus albus* from (5-^{14}C)ornithine was labeled equally at its C-1 and C-5 position. A considerable amount of free putrescine was isolated from root cultures of *H. albus*, along with the polyamines, spermidine and spermine, which are formed from putrescine and decarboxylated S-adenosylmethionine (Golding and Nassereddin 1985; Smith 1985; McCann et al. 1987; Tiburcio et al. 1990). The formation of putrescine (IV) by decarboxylation of ornithine (I) has been widely investigated and the ornithine decarboxylase has been isolated from *Hyoscyamus albus* (Hashimoto et al. 1989b, c) and several other unrelated plant species (Birecka et al. 1985). δ-N-Methylornithine seems not the be a regular intermediate on the pathway to hyoscyamine, although it is a natural constituent of *Atropa belladonna* (Baralle and Gros 1969; Ahmad and Leete 1970; Hedges and Herbert 1981). Arginine (III) can also serve as a precursor of putrescine (IV). Arginine decarboxylase catalyzes the formation of agmatine (VI), which is converted to putrescine via carbamoylputrescine (V). The activity of the arginine decarboxylase in root cultures of *H. albus* and *D. stramonium* is more important than that of ornithine decarboxylase (Robins et al. 1991a).

All the experimental evidence is consistent with the 1-methyl-Δ^1-pyrrolinium salt being a precursor of the tropane nucleus similarly as in the case of nicotine (XI). The iminium salt is formed by cyclization of 4-methylaminobutanal (XXV) via the carbinolamine, 2-hydroxy-1-methylpyrrolidine (XXVIII). 1-Methyl-2-pyrrolidone (XXVI) has been detected in *Atropa belladonna* (Hartman et al. 1986) and is plausibly formed by oxidation of the carbinolamine (XXVIII). 4-Methylaminobutanal (XXV) is formed by oxidation of N-methylputrescine (VII); the N-methylputrescine oxidase has been isolated from *D. stramonium* (Feth et al. 1985a) and *H. niger* (Hashimoto et al. 1990). N-Methylputrescine has been established as a direct precursor of the iminium salt and the alkaloids derived from it (Liebisch et al. 1966, 1969; Leete and McDonall 1981), and as expected, it is incorporated unsymmetrically into the ultimate alkaloids, e.g., into the tropine moiety of scopolamine (XLIV) in *D. innoxia*. Putrescine N-methyltransferase was partially purified and characterized from cultured roots of *Hyoscyamus albus* (Hibi et al.

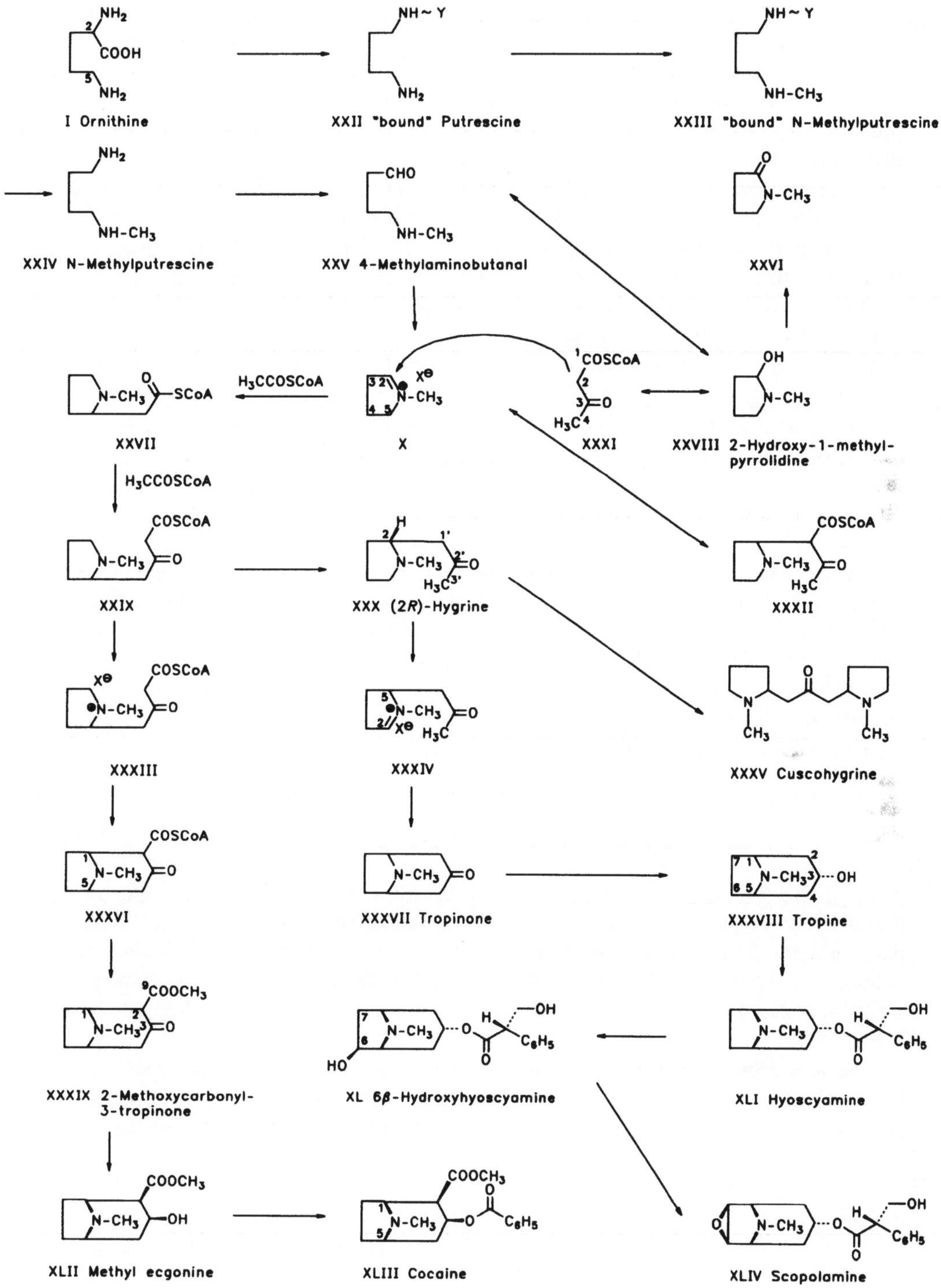

Fig. 2. Biosynthesis of tropane alkaloids

1992). Monoamines, preferably n-butylamine, were good competitive inhibitors of this transferase. From these studies, it is suggested that this transferase is the first committed enzyme specific to alkaloid biosynthesis.

It has been shown that acetic acid is incorporated into carbons 2, 3, and 4 of the tropine moiety of hyoscyamine (XLI) (Kaczkowski et al. 1961; Liebisch et al. 1972a) and these results were interpreted to the effect that two molecules of acetyl coenzyme A yield acetoacetyl coenzyme A (XXXI), which condense at its C-2 position with the 1-methyl-Δ^1-pyrrolinium salt to XXXII. However, recent results have demonstrated that, contrary to conventional wisdom, the non-ornithine-derived moiety of the tropane ring system of 6β-hydroxyhyoscyamine (XL) in *D. stramonium* does not originate from acetoacetate, but by stepwise incorporation of acetate (X → XXVII → XXIX) (Fig. 2; Hemscheidt and Spenser 1992). On the other hand, the formation of the acetate-derived segment of N-methylpelletierine leads via acetoacetate (Hemscheidt and Spenser 1990), providing evidence for unexpected biochemical diversity in biosynthetic processes leading from one and the same substrate to analogous fragments in structurally related natural products.

Hygrine (XXX), which is formed from ornithine (Leete 1985), served as a precursor of the tropine moiety of hyoscyamine and other tropane alkaloids (McGaw and Woolley 1982). In *Datura innoxia*, the (2R)-hygrine (XXX) was a more efficient precursor of hyoscyamine (XLI) than its (2S)-isomer (McGaw and Woolley 1978). In other species such as *Atropa belladonna, Hyoscyamus*, and *Physalis alkekengi*, both isomers of hygrine were equally efficient precursors of tropane skeleton (McGaw and Woolley 1979; Sankawa et al. 1990). Hygrine (XXX) and the pyrrolinium salt (X) are also precursors of cuscohygrine (XXXV) (O'Donovan and Keogh 1969; Leete et al. 1988a). Hygrine was formed nonenzymatically by coupling of 4-methylaminobutanal (XXV) with acetoacetate or acetonedicarboxylic acid (Endo et al. 1988).

The formation of the tropane nucleus from hygrine is considered to proceed via the 5-acetonyl-1-methly-Δ^1-pyrrolinium salt (XXXIV). The final cyclization is a Mannich reaction affording tropinone (XXXVII), which is a key intermediate in the biosynthesis of many of the more complex tropane alkaloids. Reduction of tropinone affords tropine (XXXVIII) or pseudotropine (McGaw and Woolley 1983). The major tropane alkaloid hyoscyamine (XLI) is an ester of tropine with (S)-tropic acid (XLVI). An investigation of the alkaloids of *Datura innoxia* revealed the presence of both tropine (XXXVIII) and pseudotropine along with various esters of both isomers (Witte et al. 1987). Thus, adminstration of (N-methyl-^{14}C)tropinone (XXXVII) to the roots of intact *D. innoxia* plants afforded specifically labeled hyoscyamine (XLI) and scopolamine (XLIV) (Landgrebe and Leete 1990).

The reduction of tropinone has also been studied in *D. stramonium* root cultures (Koelen and Gross 1982). The results regarding the yield of tropine or pseudotropine are contradictory. A tropine-reducing enzyme in root cultures of *Hyoscyamus niger* that formed only pseudotropine was purified 761-fold (Dräger et al. 1988). Another tropinone-reducing activity from roots of intact *D. innoxia* plants formed both tropine (XXXVIII) and pseudotropine in varying ratios depending on NADPH and/or NADH supplied as coenzymes suggesting the presence of two different enzymes forming the isomeric products (Couladis et al. 1991). When NADH was used as coenzyme, the production of tropine was favored over pseudotropine. Two different tropinone reductases were also isolated and completely separated from transformed root cultures of *D. stramonium*

(Portsteffen et al. 1992). Both stereospecifically distinct reductases were purified to near homogeneity from cultured roots of *H. niger* and characterized (Hashimoto et al. 1992). Both enzymes transfer the pro-S hydrogen of NAD(P)H to their substrates.

8-Thiabicyclo(3,2,1)octan-3-one, the sulfur analog of tropinone (XXXVII) was metabolized in *Datura stramonium* transformed root cultures to a large degree including reduction to 8-thiabicyclo(3,2,1)octan-3-ol, analogous to tropine (Parr et al. 1991). In addition, tropane alkaloid synthesis was perturbed. From such inhibitor experiments it is concluded that the relative levels of activity of the two tropinone reductases might play an important role in regulating the balance of tropine to pseudotropine in the spectrum of tropane-alkaloid-producing plants (Dräger et al. 1992).

The biosynthesis of hyoscyamine involves the esterification of tropine (XXXVIII) with tropic acid (XLVI). The reaction by which the ester is formed is yet to be elucidated, but in transformed root cultures of *Datura stramonium* two different acyltransferase activities have been found that can acetylate tropine or pseudotropine respectively using acetyl coenzymes A as cosubstrate (Robins et al. 1991d). Also, in cell cultures of *D. innoxia* and *Duboisia myoporoides*, tropine (XXXVIII) and related compounds were converted into corresponding acetates (Hiraoka and Tabata 1983; Kitamura et al. 1986).

Some of the acids which are present in the esters of the hydroxytropanes are very common in nature. It has been shown that phenylalanine (XLV) is the origin of (S)-tropic acid (XLVI) (Fig. 3; Leete 1990; Schütte and Liebisch 1967; Liebisch et al. 1972b; Kitamura et al. 1992b). During this pathway, the carbonyl group of phenylalanine undergoes a 1,2-migration from C-2 to C-3 with retention of configuration (Leete 1984, 1987a; Platt et al. 1984). At the same time, the displaced hydrogen at C-3 moves to C-2 and ultimately becomes part of the hydroxymethly group of tropic acid (XLVI). The migration of the carboxyl group in the formation of tropic acid is intramolecular (Leete et al. 1975), but it is not known whether the migration of hydrogen from C-3 to C-2 is inter- or intramolecular. Cinnamic acid and 3-hydroxy-3-phenylpropanoic acid are not intermediates between phenylalanine (XLV) and tropic acid (XLVI) (Leete 1983b). Suspensions cultures of *Datura innoxia* as well as *Nicotiana* and *Coffea* are able to convert tropic acid (XLVI) into its glucose esters (Ushiyama and Furuya 1989).

The first step in the conversion of hyoscyamine (XLI) to scopolamine (XLIV) is the 6-hydroxylation. An enzyme preparation has been obtained from cultured roots of *Hyoscyamus niger* which catalyzes the formation of 6β-hydroxyhyoscyamine (XL) from hyoscyamine (XLI) (Hashimoto and Yamada 1986, 1987b; Yamada and Hashimoto 1989). This hydroxylation occurs with retention of configuration and requires 2-ketoglutaric acid and oxygen as cofactors. The next step is not a dehydration to 6,7-dehydrohyoscyamine (Leete and Lucast 1976), but ^{18}O was retained in the scopolanine (XLIV)

XLV [3*R*-^3H]Phenylalanine XLVI (*S*)-Tropic acid

Fig. 3. Formation of tropic acid

isolated from shoot cultures of *Duboisia myoporoides* which had been fed (6-hydroxy-^{18}O)-6β-hydroxyhyoscyamine (Hashimoto et al. 1987). Scopolamine (XLIV) is thus formed by a direct attack of the 6β-hydroxyl group at C-7, displacing the 7β-hydrogen (Hashimoto and Yamada 1989; Hashimoto et al. 1989d).

Using antibodies against hyoscyamine 6β-hydroxylase purified from cultured roots of *Hyoscyamus niger*, it has been shown that this enzyme is abundant in cultured roots, present in plant roots, but absent in leaf, stem, calyx, cultured cells, and cultured shoots, and immunohistochemical studies detected this enzyme only in the pericycle cells of the young roots in several scopolamine-producing plants (Hashimoto et al. 1991). This pericycle-specific localization of scopolamine biosynthesis provides an anatomical explanation for the tissue-specific biosynthesis of tropane alkaloids, and may be important for translocation of tropane alkaloids from the root to the aerial parts. The 6β-hydroxyhyoscyamine epoxidase also requires 2-ketoglutaric acid and oxygen as cofactors; but also meteloidine is derived from tropine (XXXVIII) hydroxylation occurring at C-6 and C-7 (Basey and Woolley 1975; Beresford and Woolley 1975a, b; Leete and Lucast 1975). The biotransformation of hyoscyamine (XLI) to 6-hydroxyhyoscyamine (XL) as well as of 6-hydroxyhyoscyamine to scopolamine (XLIV) has been also demonstrated by suspension cultures of an *Anisodus tanguticus* cell line which produces no alkaloids (Cheng et al. 1987, 1989). For low yields of scopolamine (XLIV) and related alkaloids highly sensitive and specific methods of detection are necessary. Therefore immunological methods such as radioimmunoassays are being developed (Weiler et al. 1981; Lehtola et al. 1982; Oksman-Caldentey and Strauss 1986; Fliniaux and Jacquin-Dubreuil 1987; Savary and Dougall 1990; Kikuchi et al. 1991).

For the biosynthesis of cocaine (XLIII), ornithine and 1-methyl-Δ^1-pyrroliniŭm salt (X) have been established as precursors after administering to intact plants of *Erythroxylum coca* (Fig. 2). (5-^{14}C)Ornithine afforded labeled cocaine (XLIII) in which the ^{14}C was equally distributed between its C-1 and C-5 positions (Leete 1982). This result led to the proposal of putrescine (IV) as an intermediate followed again by methylation yielding N-methylputrescine (VII), oxidation yielding 4-methylaminobutanal (XXV) and cyclization yielding the iminium salt (X). This compound, labeled with ^{15}N and with ^{13}C and ^{14}C, was shown to be a precursor of cocaine (XLIII), which was labeled with ^{15}N and ^{13}C and ^{14}C at its C-5 position (Leete and Kim 1988).

This result was unexpected, since the generally accepted route to cocaine involved the condensation between the iminium salt (X) and acetoacetate (Leete 1983a). Such a condensation would have ultimately yielded cocaine labeled at C-1 and not C-5. Therefore it was proposed that the iminium salt (X) reacts with acetyl coenzyme A to yield the coenzyme A ester of 1-methylpyrrolidine-2-acetic acid (XXVII). Circumstantial evidence favoring the intermediacy of this was the isolation of specific labeled 1-methylpyrrolidine-2-acetic acid from *E. coca* plants after feeding labeled 1-methly-Δ^1-pyrrolinium chloride (Leete 1989). Reaction of the ester with a second molecule of acetyl coenzyme A then affords the coenzyme A ester of 4-(1-methyl-2-pyrrolidinyl)-3-oxobutanoic acid (XXIX). The corresponding methyl ester could be incorporated specifically into cocaine (XLIII) and methylecgonine (XLII) (Leete et al. 1991). The pyrrolidine ring of XXIX is then oxidized to yield the carbinolamine, which will afford the iminium salt XXXIII by elimination of hydroxide. A Mannich reaction then yields 2-methoxycarbonyl-3-tropinone XXXIX. This compound, labeled with ^{13}C and ^{14}C at C-9 and with tritium on its O-methyl group, was incorporated into cocaine with complete retention of tritium relative to ^{14}C (Leete 1983a, c). A stereospecific reduction of this β-ketoester then affords methyl ecgonine, which is a fairly abundant alkaloid in *Erythroxylum coca*. 2,3-Dehydrococaine is not a direct precursor of cocaine (XLIII) in *E. coca* (Leete 1987b).

It has been previously established that phenylalanine (XLV) is a precursor of the benzoyl moiety of cocaine (XLIII) (Gross and Schütte 1963; Leete 1983a). The biosynthetic route leads most probably via benzoyl coenzyme A. Phenylalanine ammonia lyase catalyzes the elimination of ammonia from phenylalanine to yield trans-cinnamic acid and hydration of the cinnamic acid affords 3-hydroxy-3-phenyl-propanoic acid which is then oxidized to benzoylacetic acid in which the carboxyl group is esterified with coenzyme A to inhibit decarboxylation. Benzoyl coenzyme A then arises by cleavage of this β-keto thioester. Cinnamic acid as well as the N-acetylcysteamine thioester of benzoic acid were excellent precursors of the benzoyl moiety of cocaine (XLIII) in E. coca plants. (R)-3-hydroxy-3-phenylpropanoic acid was 11 times more effective than its (S)-isomer, which means that the relative stereochemistry of the hydroxyl group in (R)-3-hydroxyphenyl-propanoic acid is the same as that of the hydroxyl group in 3-hydroxy fatty acid coenzyme A esters which are involved in the β-oxidation of fatty acids (Leete et al. 1988b; Bjorklund and Leete 1992).

Hairy root cultures, mentioned above Section 2, which are formed by infection of dicotyledonous plants with the bacterium *Agrobacter rhizogenes*, but also untransformed root cultures with the ability to accumulate tropane alkaloids, have been obtained from several species of *Datura* (Payne et al. 1987; Jaziri et al. 1988; Knopp et al. 1988; Christen et al. 1989; Parr et al. 1990; Robins et al. 1990, 1991b, c), of *Duboisia* (Yamada and Endo 1984; Deno et al. 1987; Mano et al. 1989; Yoshioka et al. 1989; Yoshimatsu et al. 1990; Kitamura et al. 1991a), of *Hyoscyamus* (Hashimoto and Yamada 1983, 1987a; Hashimoto et al. 1986; Doerk et al. 1991; Sauerwein and Shimomura 1991; Sauerwein et al. 1991, 1992; Shimomura et al. 1991; Christen et al. 1992), from *Atropa belladonna* (Kamada et al. 1986; Jung and Tepfer 1987; Simola et al. 1988, 1989, 1990), from *Nicandra physaloides* (Parr 1992), and *Scopolia japonica* (Mano et al. 1986). The cultures have been established both for the purpose of tropane alkaloid production and for the determination of the biosynthetic pathway which gives rise to these compounds. In most cases, the alkaloid production in these hairy root cultures is the same or even higher than those of normal plants.

Tropane and pyridine alkaloids are biosynthesized in the roots and then transported to and accumulated in the aerial parts. The movement of tropane alkaloids from aerial parts to roots has been suggested on the basis of grafting experiments (Mothes 1955) and the movement from roots to aerial parts is well established (Mothes and Schütte 1969). This is also found for *Duboisia myoporoides*; but no alkaloids were detected in the leaves of 1-month-old regenerated plantlets, even though alkaloids were present in the roots (Kitamura et al. 1985b), which means that these regenerated plantlets did not transport atropine from the roots to the leaves but from the leaves to the roots, whereas the seedlings were able to transport in both directions at all ages examined (Kitamura et al. 1991b). The regenerated plantlets gradually acquired an ability to transport atropine from the roots to the leaves during development and eventually, when they were 12 weeks old, they had a transport capacity comparable to that of the seedlings.

Tropane alkaloids such as atropine are decomposed by atropine esterase to tropine (XXXVIII) and tropic acid (XLVI) in the roots of different tropane alkaloid-producing plants such as *Datura stramonium, D. tatula*, and *Duboisia myoporoides*. In an investigation on alkaloid composition and atropine esterase activity in callus and differentiated tissues from callus and from the original plant of *Duboisia myoporoides* (Kitamura et al. 1985a), it has been found that callus and differentiated shoots from callus contained no tropane alkaloids, though they contained very small amounts of anabasine (XVII) and nicotine (XI). In contrast, differentiated roots from callus contained all the kinds of alkaloids found in the original plant. Neither callus nor differentiated leaves showed atropine esterase activity, but differentiated roots did show such activity. There is no or extremely low atropine esterase activity in the roots of regenerated plants of *D. myoporoides* at the early stage of development, when atropine is not transported from the root to the leaf (Kitamura et al. 1988). After development, the regenerated plants recover the ability of atropine transport from the root to the leaf, and atropine esterase activity is detected in the roots. These results seem to indicate that roots of *D. myoporoides* decompose the alkaloids which are transported back to the roots from the aerial parts. High atropine esterase activity was found in the roots of every tropane alkaloid-producing plant at any stage of development, though much higher activities were detected in the roots of annual and biennal herbs at the fruiting stage (Kitumura et al. 1992a). In contrast to these findings, no cultured roots contained detectable amount of atropine esterase activity in spite of their production of hyoscyamine and scopolamine.

4. Pyrrolizidine Alkaloids

Pyrrolizidine alkaloids constitute a large class of natural products with over 200 known examples. They have a widespread occurrence and are found mainly in *Senecio* spp. (Asteraceae), *Crotolaria* spp. (Leguminosae), and *Heliotropium* and *Cynoglossum* spp. (Boraginaceae) (Robins 1982a, 1984, 1985, 1986, 1987, 1989a, c, 1990, 1991b, 1992). These alkaloids usually consist of a base portion (necine) which is a 1-hydroxymethylpyrrolizidine derivative. A second group is often present at C-7 and the most common necine is retronecine (LII). Pyrrolizidine alkaloids usually occur as monoesters, diesters, or cyclic diesters with necic acids which have unusual structures. Most necic acids are dicarboxylic and contain ten carbon atoms.

Pyrrolizidine alkaloids are important because of the wide range of biological activities which they exhibit. Alkaloids such as monocrotaline (LVII) or retrorsine (LIII) which contain a 1,2-unsaturated necine are hepatotoxic (Mattocks 1986). Many deaths of livestock have resulted from animals grazing on land containing plants which produce pyrrolizidine alkaloids. These plants are also favorites for many herbal remedies known to contribute to human liver disease.

In experiments with *Crotolaria spectabilis, C. retusa, Senecio douglasii, S. vulgaris, S. isatideus*, and *S. magnificus*, ornithine (I), arginine (III), putrescine (IV), spermidine, and spermine have all been demonstrated to be specific precursors of retronecine (LII) (Nowacki and Byerrum 1962; Robins and Sweeney 1981, 1983; Schütte 1982; Robins 1983a, 1989b; Liebisch and Schütte 1985; Hartmann and Toppel 1987; Sander and Hartmann 1989). Difluoromethylarginine, a specific inhibitor of arginine decarboxylase, prevented completely the incorporation of radioactivity from labeled arginine (III) and ornithine (I) into spermidine and the pyrrolizidine alkaloid senecionine (LIV) N-oxid in *S. vulgaris* (Hartmann et al. 1988). In contrast, difluoromethylornithine, a specific ornithine-decarboxylase inhibitor, had no effect on the flow of radioactivity from labeled ornithine and arginine into polyamines and alkaloids. Thus, putrescine is derived in this plant exclusively via the arginine-agmatine route, but this seems to be species-dependent (Birecka et al. 1987, 1988).

Degradations of retronecine (LII) derived from (2-^{14}C)- and (5-^{14}C)ornithine and (1,4-^{14}C)putrescine have shown that in each case ca. 25% of the total radioactivity is located at C-9 and 25% at C-5 to C-7. This was confirmed in experiments with *S. isatideus* and other species using (1,4-^{13}C)-,(1,3-^{13}C)- and stereospecific ^{2}H-labeled putrescine (Grue-Sörensen and Spenser 1981, 1982; Khan and Robins 1981a, 1985a; Robins 1983a). The incorporation pattern into retronecine (LII) and other necines was consistent with the formation of a symmetrical C_4-N-C_4 intermediate, which was shown to be homospermidine (XLVII) (Fig. 4; Khan and Robins 1981b, 1985b; Rana and Robins 1983a; Kelly and Robins 1987a).

An interesting contrast in the biosynthetic pathway is provided by lupinine, which is formed from two molecules of lysine via cadaverine (Schütte 1982; Rana and Robins 1984a, b; Robins and Sheldrake 1987). In this case, however, no later symmetrical intermediate of the type C_5-N-C_5 is involved (Golebieswski and Spenser 1983, 1984; Rana and Robins 1986b). This finding provides another caution against taking anything for granted in biosynthetic work, even when closed structural analogies are involved.

Homospermidine (XLVII) has been detected in *Heliotropium indicum* (Birecka et al. 1984). It has been converted into trachelanthamidine (IL) under physiological conditions using diaminoxidase isolated from pea seedlings followed by reduction involving a coupled dehydrogenase (Robins 1982b). Oxidation of one terminal group of homospermidine (XLVII) leads to an immonium ion (XLVIII), which undergoes further oxidation to an aldehyde, followed by cyclization to give 1-formylpyrrolizidine. A reduction step would then afford trachelanthamidine (IL).

Feeding experiments with ^{2}H-labeled putrescine on *S. isatideus* have shown (Grue-Sörensen and Spenser 1983; Rana and Robins 1983b, 1984c, 1986a, b; Kunec and Robins 1985, 1987; Robins 1991a) that the oxidation of putrescine (IV) to 4-aminobutanal takes place with loss of the pro-S hydrogen. After coupling of the aldehyde and amine, reduction of the imine occurs by attack of a hydride donor on the C-si face of the imine to give homospermidine (XLVII). Of two further oxidation steps, each takes place with loss of pro-S hydrogens to give the dialdehyde. Mannich cyclization leads to the 1-formylpyrrolizidine, and reduction to trachelanthamidine (IL) takes place by addition of a hydride equivalent on the C-re face of the carbonyl group. The hydroxylation at C-7 of the necine does not involve keto or enol intermediates and proceeds with retention of configuration, while formation of the 1,2-double bond on retronecine (LII) involves removal of the pro-S hydrogen and retention of the pro-R hydrogen at C-2.

Alternative cyclization modes of the immonium ion (XLVIII) under enzyme control in the plants could lead to the necines (–)-isoretronecanol (L) and its enantiomer (LI) (Fig. 4). The mode of incorporation of ^{2}H-labeled putrescine into the rosmarinecine (LXI) portion of rosmarinine (LX) in *S. pleistocephalus* plants is again consistent with

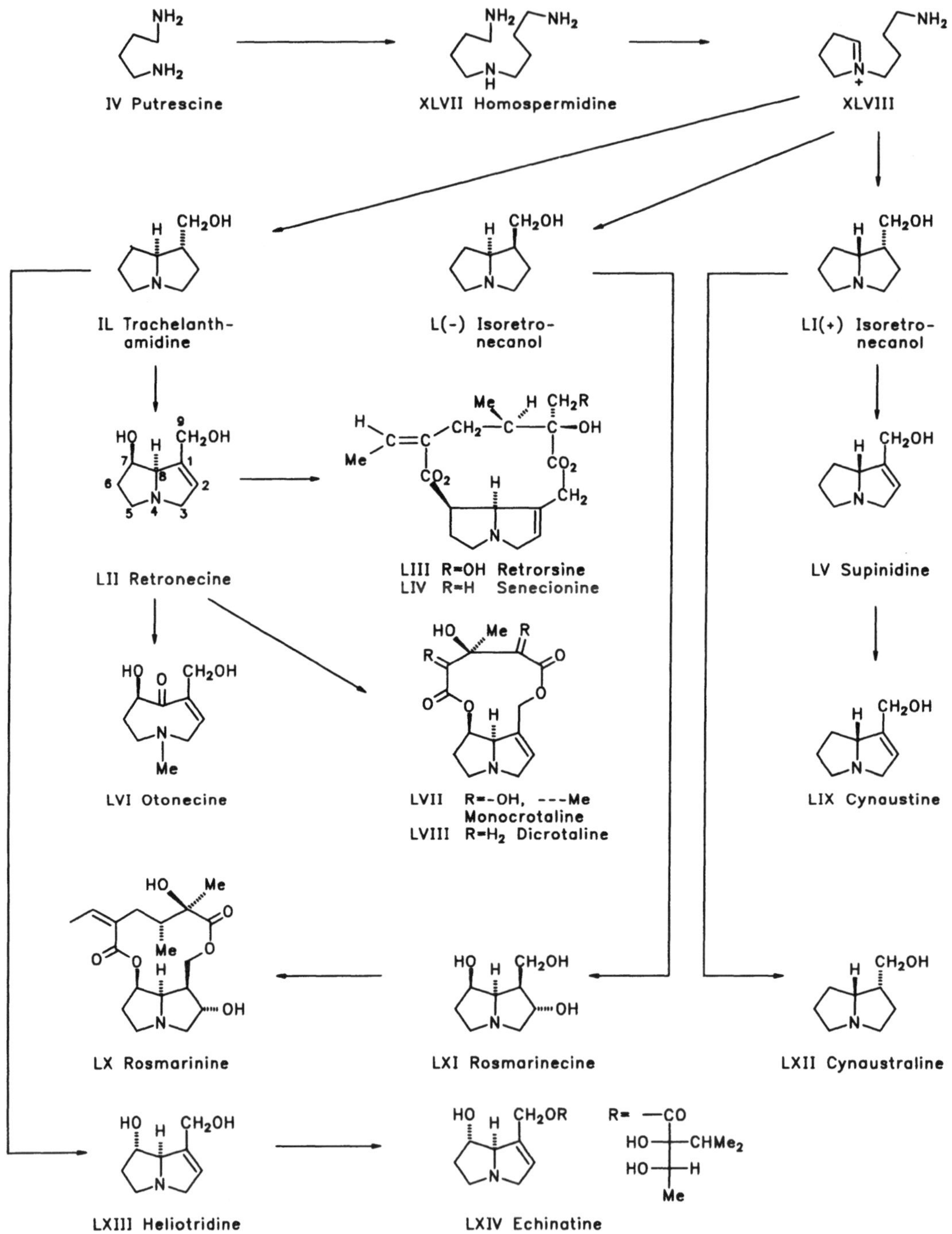

Fig. 4. Biosynthesis of pyrrolizidine alkoloids

oxidation of three of the four amino groups involved in the conversion of two molecules of putrescine (IV) into 1β-formyl-8α-pyrrolizidine with strereospecific loss of the pro-S hydrogens (Kelly and Robins 1987b). Reduction of the 1-formylpyrrolizidine to isoretronecanol (L) takes place by attack of a hydride donor on the C-re face of the aldehyde group. Furthermore formation of the pyrrolizidine ring involves stereospecific removal of the pro-R hydrogen and retention of the pro-S hydrogen on the carbon destined to become C-1 of rosmarinecine (LXI), and hydroxylation at C-2 and C-7 of isoretronecanol (L) therefore proceeds with retention of configuration to form rosmarinecine (LXI).

The facile synthesis of trachelanthamidine (IL) did indicate that the 1-hydroxymethylpyrrolizidines should be tested as intermediates of the more complex necines. Trachelanthamidine (IL) was an efficient precursor for the retronecine (LII) component of retrorsine (LIII) in *S. isatideus* and riddelliine in *S. ridellii*, and the heliotridine (LXIII) part of echinatine (LXIV) in *Cynoglossum officinale* (Rana and Leete 1985; Kunec and Robins 1986, 1989; Leete and Rana 1986). On the other hand racemic isoretronecanol (L, LI) was incorporated well into the rosmarinecine (LXI) portion of rosmarinine (LX) in *S. pleistocephalus* and into the isoretronecanol (LI) part of cynaustraline (LXII) and the supinidine (LV) portion of cynaustine (LIX) in *C. australe* (Hagan and Robins 1990). The immonium ion (XLVIII) has been found to be an efficient precursor for retrorsine (LIII) in *S. isatideus*, for rosmarinine (LX) in *S. pleistocephalus*, for echinatine (LXIV) in *C. officinale* as well as for cynaustraline (LXII) and cynaustine (LIX) in *C. australe* (Kelly and Robins 1988; Denholm et al. 1991). Putrescine (IV), homospermidine (XLVII), N-(4-aminobutyl)1,2-dehydropyrrolidinium salt (XLVIII), trachelanthamidine (IL), and retronecine (LII) are all efficient precursors for the seco pyrrolizidine otonecine (LVI), the base part of the pyrrolizidine alkaloid emiline in *Emilia flammea* (Barbour and Robins 1987; Kelly et al. 1989; Freer et al. 1991). Most probably the N(4)-C(8) bond is broken during the formation of emiline from retronecine. The formation of otonecine (LVI) probably occurs by hydroxylation at C-8 and N-methylation of retronecine (LII) followed by ketone formation and concomitant cleavage of the C-N bond.

Many of the pyrrolizidine alkaloids are macrocyclic diesters of necine and most of the esterifying necic acids are C_{10} diacids. Experiments on the few necic acids that have been studied have shown that they are derived from common amino acids such as isoleucine, its biological precursor threonine, valine, and leucine. Thus senecic acid (in senecionine LIV) and isatinecic acid (in retrorsine LIII) are formed from two molecules of isoleucine with the loss of both carboxy groups (Crout et al. 1966, 1970, 1972; Davies and Crout 1974; Bale et al. 1978; Cahill et al. 1980, 1983). It has been shown in *Senecio* species that the ethyl migration step during the biosynthesis of isoleucine takes place with retention of configuration at the migrating center, and that during conversion into the senecic acid type, both C_5 units are formed with loss of the C-4 pro-S hydrogen atom of isoleucine and retention of the C-4 pro-R hydrogen atom. In a similar manner, trichodesmic acid, the necic acid component of trichodesmine produced by *Crotolaria globifera*, is derived from leucine and isoleucine (Devlin and Robins 1984). Dicrotaline (LVIII) contains the simplest diacid found in these alkaloids, 3-hydroxy-3-methylglutaric acid, but this portion of dicrotaline is formed specifically from isoleucine, label from C-5 of isoleucine is incorporated specifically into the methyl group of this necic acid (Denholm and Robins 1991).

Several lines of evidence indicate that in *Senecio vulgaris* and other *Senecio* species the roots are the site of pyrrolizidine alkaloid biosynthesis and that the N-oxides, for *S. vulgaris* senecionine (LIV) N-oxide, were identified as primary products (Hartmann and Toppel 1987; Toppel et al. 1987; Hartmann et al. 1989). Pyrrolizidine-alkaloid-producing species from other plant families seem to synthesize the alkaloids preferably in the shoot, e.g., *Crotolaria spectabilis* (Fabiaceae) (Nowacki and Byerrum 1962) and *Heliotropium* species (Boraginaceae) (Birecka and Catalfamo 1982; Birecka et al. 1987). The pyrrolizidine alkaloids are not distributed randomly. In *S. vulgaris* the inflorescences, and here particularly the tubular flowers, are the main sites of alkaloid storage (Hartmann and Zimmer 1986), and in the vegetative plant organs the peripheral accumulation in stems is most obvious. The alkaloid N-oxides were found to be selectively taken up and accumulated by cells derived from pyrrolizidine-alkaloid-producing species (von Borstel and Hartmann 1986). A carrier system was identified which catalyzes the specific uptake of senecionine N-oxide into cells and protoplasts and secures its vacuolar accumulation (Ehmke et al. 1988). Dry seeds of *Crotolaria scassellatii* are the only example, so far, where the pyrrolizidine alkaloids exist exclusively as tertiary alkaloids; however, they are rapidly N-oxidized during the first days of seed germination, and only alkaloid N-oxides are detectable in the radicle and emerging shoots, as in the different parts of the mature plant (Toppel et al. 1988).

References

Ahmad A, Leete E (1970) Phytochemistry 9:2345–2347.

Bale NM, Cahill R, Davies NM, Mitchell MB, Smith EH, Crout DHG (1978) J Chem Soc Perkin Trans 1:101–110. **Baralle FE**, Gros EG (1969) Chem Commun 721. **Barbour RH**, Robins DJ (1987) Phytochemistry 26:2430–2431. **Basey K**, Woolley JG (1975) Phytochemistry 14:2201–2203. **Beresford PJ**, Woolley JG (1975a) Phytochemistry 14:2205–2207. **Beresford PJ**, Woolley JG (1975b) Phytochemistry 14:2209–2212. **Birecka H**, Catalfamo JL (1982) Phytochemistry 21:2645–2651. **Birecka H**, DiNolfo TE, Martin WB, Fröhlich MW (1984) Phytochemistry 23:991–997. **Birecka H**, Bitonti AJ, McCann PP (1985) Plant Physiol 79:515–519. **Birecka H**, Birecki M, Fröhlich MW (1987) Plant Physiol 84:42–46. **Birecka H**, Birecki M, Cohen EJ, Bitonti AJ, McCann PP (1988) Plant Physiol 86:224–230. **Bjorklund JA**, Leete E (1992) Phytochemistry 31:3883–3887. **Burtin D**, Martin-Tanguy J, Paynot M, Rossin N (1989) Plant Physiol 89:104–110. **Burtin D**, Martin-Tanguy J, Tepfer D (1991) Plant Physiol 95:461–468.

Cahill R, Crout DHG, Mitchell MB, Müller KS (1980) Chem Commun 419–421. **Cahill R**, Crout DHG, Gregorio MVM, Mitchell MB, Müller US (1983) J Chem Soc Perkin Trans 1:173–180. **Cheng KD**, Zhu WH, Li XL, Meng C, Sun ZM, Yang DH (1987) Planta Med 53:211–213. **Cheng KD**, Fang HJ, Zhu WH, Mang C, Yang DH, Li L (1989) Planta Med 55:391–392. **Chriqui D**, Tepfer M, Prinsen E, van Onckelen H (1990) Bull Soc Bot Fr 137:79–91. **Christen P**, Roberts MF, Phillipson JD, Evans WC (1989) Plant Cell Rep 8:75–77. **Christen P**, Aoki T, Shimomura K (1992) Plant Cell Rep 11:597–600. **Couladis MM**, Friesen JB, Landgrebe ME, Leete E (1991) Phytochemistry 30:801–805. **Crout DHG**, Benn MH, Imaseki H, Geismann TA (1966) Phytochemistry 5:1–21. **Crout DHG**, Davies NM, Smith EH, Whitehouse D (1970) Chem Commun 635–636. **Crout DHG**, Davies NM, Smith EH, Whitehouse D (1972) J Chem Soc Perkin Trans I:671–680.

Davies NM, Crout DHG (1974) J Chem Soc Perkin Trans I:2079–2082. **Davies HM**, Hawkins DJ, Smith LA (1989) Phytochemistry 28:1573–1578. **Denholm AA**, Robins DJ (1991) Chem Commun 19–21. **Denholm AA**, Kelly HA, Robins DJ (1991) J Chem Soc Perkin Trans I:2003–2007. **Deno H**, Yamagata H, Emoto Y, Yoshioka T, Yamada Y, Fujita Y (1987) J Plant Physiol 131:315–323. **Devlin JA**, Robins DJ (1984) J Chem Soc Perkin Trans I:1329–1332. **Doerk K**, Witte L, Alfermann AW (1991) Z Naturforsch 46c:519–521. **Dräger B**, Hashimoto Y, Yamada Y (1988) Agric Biol Chem 52:2663–2667. **Dräger B**, Portsteffen A, Schaal A, McCabe PH, Peerless ACJ, Robins RJ (1992) Planta 188:581–586.

Ehmke A, von Borstel K, Hartmann T (1988) Planta 176:83–90. **Endo T**, Yamada Y (1985) Phytochemistry 24:1233–1236. **Endo T**, Komiya T, Masumitsu Y, Morikawa H, Yamada Y (1987) J Plant Physiol 129:453–459. **Endo T**, Hamaguchi N, Hasimoto T, Yamada Y (1988) FEBS Lett 234:86–90. **Endo T**, Hamaguchi N, Erikson T, Yamada Y (1991) Planta 183:505–510.

Feth F, Wagner KG (1989) Physiol Plant 75:71–74. **Feth F**, Wray V, Wagner KG (1985a) Phytochemistry 24:1653–1655. **Feth F**, Arfmann HA, Wray V, Wagner KG (1985b) Phytochemistry 24:921–923. **Feth F**, Wagner R, Wagner KG (1986) Planta 168:402–207. **Fliniaux MA**, Jacquin-Dubreuil A (1987) Planta Med 53:87–90. **Freer IKA**, Matheson JR, Rodgers M, Robins DT (1991) J Chem Res (S) 46–47. **Friesen JB**, Leete E (1990) Tetrahedron Lett 31:6295–6298. **Friesen JB**, Burkhouse PC, Biesboer DD, Leete E (1992) Phytochemistry 31:3059–3063.

Golding BT, Nassereddin IK (1985) J Chem Soc Perkin Trans I:2017–2024. **Golebiewski WM**, Spenser ID (1983) Chem Commun 1509–1511. **Golebiewski WM**, Spenser ID (1984) J Am Chem Soc 106:1441–1442. **Gritsanapan W**, Griffin WJ (1992) Phytochemistry 31:471–477. **Gross D** (1985) In: Mothes K, Schütte HR, Luckner M (eds) Biochemistry of Alkaloids. VCH, Weinheim, pp 163–187. **Gross D**, Schütte HR (1963) Arch Pharm 296:1–6. **Grue-Sörensen G**, Spenser ID (1981) J Am Chem Soc 103:3208–3210. **Grue-Sörensen G**, Spenser ID (1982) Can J Chem 60:643–662. **Grue-Sörensen G**, Spenser ID (1983) J Am Chem Soc 105:7401–7404.

Hagan DB, Robins DJ (1990) J Chem Res(S) 292–293. **Hamill JD**, Parr AJ, Robins RJ, Rhodes MJC (1986) Plant Cell Rep 5:111–114. **Hamill JD**, Robins RJ, Parr AJ, Evans DM, Furze JM, Rhodes MJC (1990) Plant Mol Biol 15:27–38. **Hartmann T**, Toppel G (1987) Phytochemistry 26:1639–1643. **Hartmann T**, Zimmer M (1986) J Plant Physiol 122:67–80. **Hartmann T**, Witte L, Oprach F, Toppel G (1986) Planta Med:390–395. **Hartmann T**, Sander H, Adolph R, Toppel G (1988) Planta 175:82–90. **Hartmann T**, Ehmke A, Eilert U, von Borstel K, Theuring C (1989) Planta 177:98–107. **Hashimoto T**, Yamada Y (1983) Planta Med 47:195–199. **Hashimoto T**, Yamada Y (1986) Plant Physiol 81:619–625. **Hashimoto T**, Yamada Y (1987a) Agric Biol Chem 51:2769–2774. **Hashimoto T**, Yamada Y (1987b) Eur J Biochem 164:277–285. **Hashimoto T**, Yamada Y (1989) Agric Biol Chem 53:863–864. **Hashimoto T**, Yukimune Y, Yamada Y (1986) J Plant Physiol 124:61–75. **Hashimoto T**, Kohno J, Yamada Y (1987) Plant Physiol 84:144–147. **Hashimoto T**, Yamada Y, Leete E (1989a) J Am Chem Soc 111:1141–1142. **Hashimoto T**, Yukimune Y, Yamada Y (1989b) Planta 178:123–130. **Hashimoto T**, Yukimune Y, Yamada Y (1989c) Planta 178:131–137. **Hashimoto T**, Kohno J, Yamada Y (1989d) Phytochemistry 28:1077–1082. **Hashimoto T**, Mitani A, Yamada Y (1990) Plant Physiol 31:216–221. **Hashimoto T**, Hayashi A, Amano Y, Kohno J, Iwanari H, Usuda S, Yamada Y (1991) J Biol Chem 266:4648–4653. **Hashimoto T**, Nakajima K, Ongena G, Yamada Y (1992) Plant Physiol 100:836–845. **Hedges SH**, Herbert RB (1981) Phytochemistry 20:2064–2065. **Hemscheidt T**, Spenser ID (1990) J Am Chem Soc 112:6360–6363. **Hemscheidt T**, Spenser ID (1992) J Am Chem Soc 114:5472–5473. **Hiatt A**, Malmberg RL (1988) Plant Physiol 86:441–446. **Hibi N**, Fujita T, Hatano M, Hashimoto T, Yamada Y (1992) Plant Physiol 100:826–835. **Hiraoka N**, Tabata M (1983) Phytochemistry 22:409–412.

Ishimaru K, Shimomura K (1989) Phytochemistry 28:3507–3509. **Isogai A**, Fukuchi N, Hayashi M, Kamada H, Harada H, Suzuki A (1990) Phytochemistry 29:3131–3134.

Jaziri M, Legros M, Homes J, Vanhaelen M (1988) Phytochemistry 27:419–420. **Jung G**, Tepfer D (1987) Plant Sci 50:145–151.

Kaczokowski J, Schütte HR, Mothes K (1961) Biochem Biophys Acta 46:588–594. **Kamada H**, Okumura N, Satake M, Harada H, Shimomura K (1986) Plant Cell Rep 5:239–242. **Kelly HA**, Robins DJ (1987a) J Chem Soc Perkin Trans I:177–180. **Kelly HA**, Robins DJ (1987b) J Chem Soc Perkin Trans I:2195–2202. **Kelly HA**, Robins DJ (1988) Chem Commun 329–330. **Kelly HA**, Kunec EK, Rodgers M, Robins DJ (1989) J Chem Res (S) 358–359. **Khan HA**, Robins DJ (1981a) Chem Commun 146–147. **Khan HA**, Robins DJ (1981b) Chem Commun 554–556. **Khan HA**, Robins DJ (1985a) J Chem Soc Perkin Trans I:101–105. **Khan HA**, Robins DJ (1985b) J Chem Soc Perkin Trans I:819–824. **Kikuchi Y**, Irie M, Yoshimatsu K, Ishimaru K, Shimomura K, Satake M, Sueyoshi S, Tanno M, Kamiya S, Sawada J, Terao T (1991) Phytochemistry 30:3273–3276. **Kitamura Y**, Miura H, Sugii M (1985a) Chem Pharm Bull 33:5445–5448. **Kitamura Y**, Miura H, Sugii M (1985b) Planta Med:489–491. **Kitamura Y**, Miura H, Sugii M (1986) Phytochemistry 25:2541–2542. **Kitamura Y**, Miura H, Sugii M (1988) J Plant Physiol 133:316–319. **Kitamura Y**, Sugimoto Y, Samejima T, Hayashida K, Miura H (1991a) Chem Pharm Bull 39:1263–1266. **Kitamura Y**, Matuo N, Takashi T, Miura H (1991b) J Plant Physiol 137:613–618. **Kitamura Y**, Sato M, Miura H (1992a) Phytochemistry 31:1191–1194. **Kitamura Y**, Taura A, Kajiya Y, Miura H (1992b) J Plant Physiol 140:141–146. **Knopp E**, Strauss A, Wehrli W (1988) Plant Cell Rep 7:590–593. **Koelen KJ**, Gross GG (1982) Planta Med 44:227–230. **Kunec EK**, Robins DJ (1985) Chem Commun 1450–1452. **Kunec EK**, Robins DJ (1986) Chem Commun 250–252. **Kunec EK**, Robins DJ (1987) J Chem Soc Perkin Trans I:1089–1093. **Kunec EK**, Robins DJ (1989) J Chem Soc Perkin Trans I:1437–1441.

Landgrebe ME, Leete E (1990) Phytochemistry 29:2521–2524. **Leete E** (1960) J Am Chem Soc 82:612–614. **Leete E** (1967) J Am Chem Soc 89:7081–7084. **Leete E** (1978) Chem Commun 610–611. **Leete E** (1979) Planta Med 36:97–112. **Leete E** (1980) In: Bell EA, Charlwood BV (eds) Secondary plant products, Encyclopedia of plant physiol, New Series, vol 8, Springer, Berlin Heidelberg New York, pp 65–91. **Leete E** (1982) J Am Chem Soc 104:1403–1408. **Leete E** (1983a) Phytochemistry 22:699–704. **Leete E** (1983b) Phytochemistry 22:933–935. **Leete E** (1983c) J Am Chem Soc 105:6727–6728. **Leete E** (1983d) In: Pelletier SW (ed) Alkaloids: chemical and biological perspecitives, vol 1. John Wiley, New York, pp 85–152. **Leete E** (1984) J Am Chem Soc 106:7271–7272. **Leete E** (1985) Phytochemistry 24:953–955. **Leete E** (1987a) Can J Chem 65:226–228. **Leete E** (1987b) J Nat Prod 50:30–35. **Leete E** (1989) Heterocycles 28:481–487. **Leete E** (1990) Planta Med 56:339–352. **Leete E**, Kim SH (1988) J Am Chem Soc 110:2976–2978. **Leete E**, Lucast DH (1975) Phytochemistry 14:2199–2200. **Leete E**, Lucast DH (1976) Tetrahedron Lett 3401–3404. **Leete E**, McDonall JA (1981) J Am Chem Soc 103:658–662. **Leete E**, Rana J (1986) J Nat Prod 49:838–844. **Leete E**, Slattery SA (1976) J Am Chem Soc 98:6326–6330. **Leete E**, Marion L, Spenser ID (1954) Can J Chem 32:1116–1123. **Leete E**, Kowanko N, Newmark RA (1975) J Am Chem Soc 97:6826–6830. **Leete E**, Kim SH, Rana J (1988a) Phytochemistry 27:401–406. **Leete E**, Bjorklund JA, Kim SH (1988b) Phytochemistry 27:2553–2556. **Leete E**, Endo T, Yamada Y (1990) Phytochemistry 29:1847–1851. **Leete E**, Bjorklund JA, Couladis MM, Kim SH (1991) J Am Chem Soc 113:9286–9292. **Lehtola T**, Huhtikangas A, Virtanen R (1982) Planta Med 45:237–239. **Liebisch HW**, Schütte HR (1985) In: Mothes K, Schütte HR, Luckner M (eds) Biochemistry of alkaloids. VCH, Weinheim, pp 106–127. **Liebisch HW**, Ramin H, Schöffinius I, Schütte HR (1965) Z Naturforsch 20b:1183–1185. **Liebisch HW**, Maier W, Schütte HR (1966) Tetrahedron Lett 4079–4082. **Liebisch HW**, Radwan AS, Schütte HR (1969) Liebigs Ann Chem 721:163–167. **Liebisch HW**, Peisker K, Radwan AS, Schütte HR (1972a) Z Pflanzenphysiol 67:1–9. **Liebisch HW**, Bhavsar GC, Schaller (1972b) In: Biochem Physiol Alkaloide, 4th Int Symp, Halle, Saale, 233–237. **Lounasmaa M** (1988) In: Brossi A (ed) The alkaloids, vol 33. Academic Press, New York, pp 1–81. **Luanratana O**, Griffin WJ (1982) Phytochemistry 21:449–451.

Mano Y, Nabeshima S, Matsui C, Ohkawa J (1986) Agric Biol Chem 50:2715–2722. **Mano Y**, Ohkawa H, Yamada Y (1989) Plant Sci 59:191–201. **Mattocks AR** (1986) Chemistry and toxicology of pyrrolizidine alkaloids. Academic Press, London. **McCann PP**, Pegg AE, Sjoerdsma A (1987) Inhibition of polyamine metabolism. Academic Press, Orlando. **McGaw BA**, Woolley JG

(1978) Phytochemistry 17:257–259. **McGaw BA**, Woolley JG (1979) Phytochemistry 18:189–190. **McGaw BA**, Woolley JG (1982) Phytochemistry 21:2653–2657. **McGaw BA**, Woolley JG (1983) Phytochemistry 22:1407–1408. **Mengoli M**, Chriqui D, Bagni N (1992a) J Plant Physiol 139:697–702. **Mengoli M**, Chriqui D, Bagni N (1992b) J Plant Physiol 140:153–155. **Mengoli M**, Ghelli A, Chriqui D, Bagni N (1992c) Physiol Plant 85:697–703. **Mizusaki S**, Kisaki T, Tamaki E (1968) Plant Physiol 43:93–98. **Mizusaki S**, Tanabe Y, Noguchi M, Tamaki E (1972) Phytochemistry 11:2757–2762. **Mizusaki S**, Tanabe Y, Noguchi M, Tamaki E (1973) Plant Cell Physiol 14:103–110. **Mothes K** (1955) Annu Rev Plant Physiol 6:393–432. **Mothes K**, Schütte HR (1969) Biosynthese der Alkaloide. VCH, Weinheim.

Nowacki E, Byerrum RU (1962) Life Sci 5:157–161.

O'Donovan DG, Keogh MF (1969) J Chem Soc (C) 223–226. **Oksman-Caldentey KM**, Strauss A (1986) Planta Med:6–12. **Orr GR**, Gould SJ (1982) Tetrahedron Lett 23:3139–3142.

Parr AJ (1992) Plant Cell Rep 11:270–273. **Parr AJ**, Hamill JD (1987) Phytochemistry 26:3241–3245. **Parr AJ**, Payne J, Eagles J, Chapman BT, Robins RJ, Rhodes MJC (1990) Phytochemistry 29:2545–2550. **Parr AJ**, Walton NJ, Bensalem S, McCabe PH, Routledge W (1991) Phytochemistry 30:2607–2609. **Payne J**, Hamill JD, Robins RJ, Rhodes MJC (1987) Planta Med 53:474–478. **Platt RV**, Opie CT, Haslam E (1984) Phytochemistry 23:2211–2217. **Portsteffen A**, Dräger B, Nahrstedt A (1992) Phytochemistry 31:1135–1138.

Rana J, Robins DJ (1983a) J Chem Res (S) 146–147. **Rana J**, Robins DJ (1983b) Chem Commun 1222–1224. **Rana J**, Robins DJ (1984a) J Chem Res (S) 164–165. **Rana J**, Robins DJ (1984b) Chem Commun 81–82. **Rana J**, Robins DJ (1984c) Chem Commun 517–519. **Rana J**, Leete E (1985) Chem Commun 1742–1743. **Rana J**, Robins DJ (1986a) J Chem Soc Perkin Trans I:983–988. **Rana J**, Robins DJ (1986b) J Chem Soc Perkin Trans I:1133–1137. **Richards JC**, Spenser ID (1982) Can J Chem 60:2810–2820. **Robins DJ** (1982a) Fortschr Chem Org Naturst 41:115–203. **Robins DJ** (1982b) Chem Commun 1298–1290. **Robins DJ** (1983a) J Chem Res (S) 326–327. **Robins DJ** (1983b) Phytochemistry 22:1133–1135. **Robins DJ** (1984) Nat Prod Rep 1:235–243. **Robins DJ** (1985) Nat Prod Rep 2:213–220. **Robins DJ** (1986) Nat Prod Rep 3:297–305. **Robins DJ** (1987) Nat Prod Rep 4:577–590. **Robins DJ** (1989a) Nat Prod Rep 6:221–230. **Robins DJ** (1989b) Chem Soc Rev 18:375–408. **Robins DJ** (1989c) Nat Prod Rep 6:577–589. **Robins DJ** (1990) Nat Prod Rep 7:377–386. **Robins DJ** (1991a) Experientia 47:1118–1122. **Robins DJ** (1991b) Nat Prod Rep 8:213–221. **Robins DJ** (1992) Nat Prod Rep 9:313–321. **Robins DJ**, Sheldrake GN (1987) J Chem Res (S) 256–257. **Robins DJ**, Sweeney JR (1981) J Chem Soc Perkin Trans I:3083–3086. **Robins DJ**, Sweeney JR (1983) Phytochemistry 22:457–459. **Robins DJ**, Bale NM, Crout DHG (1974) J Chem Soc Perkin Trans I:2082–2086. **Robins RJ**, Hamill JD, Parr AJ, Smith K, Walton NJ, Rhodes MJC (1987) Plant Cell Rep 6:122–126. **Robins RJ**, Parr AJ, Payne J, Walton NJ, Rhodes MJC (1990) Planta 181:414–422. **Robins RJ**, Parr AJ, Walton NJ (1991a) Planta 183:196–201. **Robins RJ**, Parr AJ, Bent EG, Rhodes MJC (1991b) Planta 183:185–195. **Robins RJ**, Bent EG, Rhodes MJC (1991c) Planta 185:385–390. **Robins RJ**, Bachmann RJ, Robinson T, Rhodes MJC, Yamada Y (1991d) FEBS Lett 292:293–297.

Saito K, Murakoshi I, Inze D, van Montagu M (1989) Plant Cell Rep 7:607–610. **Saitoh F**, Noma M, Kawashima (1985) Phytochemistry 24:477–480. **Sander H**, Hartmann T (1989) Plant Cell Tissue Organ Cult 18:19–31. **Sankawa R**, Noguchi H, Hashimoto T, Yamada Y (1990) Chem Pharm Bull 38:2066–2068. **Sarychev YF**, Sherstyanykh NA (1985) Tabak (Moscow) 6–12. **Sauerwein M**, Shimomura K (1991) Phytochemistry 30:3277–3280. **Sauerwein M**, Ishimaru K, Shimomura K (1991) Phytochemistry 30:2977–2978. **Sauerwein M**, Wink M, Shimomura K (1992) J Plant Physiol 140:147–152. **Saunders JW**, Bush LP (1979) Plant Physiol 64:236–240. **Savary BJ**, Dougall DK (1990) Phytochemistry 29:1567–1569. **Schütte HR** (1982) Prog Bot 44:165–181. **Schütte HR**, Liebisch HW (1967) Z Pflanzenpyhsiol 57:440–443. **Shimomura K**, Sauerwein M, Ishimaru K (1991) Phytochemistry 30:2275–2278. **Simola LK**, Nieminen S, Huhtikangas A, Ylinen M, Naaranlahti T (1988) J Nat Prod 51:234–242. **Simola LK**, Martinsen A, Huhtikangas A, Jokela R, Lounasmaa M (1989) Acta Chem Scand 43:702–705. **Simola LK**, Parviainen R, Martinsen A, Huhtikangas A, Jokela R, Lounasmaa M (1990) Phytochemistry 29:3517–3523. **Slocum RD**, Galston AW (1985) Plant Cell Physiol 26:1519–1526. **Smith TA** (1985) Annu Rev Plant Physiol 36:117–143.

Tepfer D (1990) Physiol Plant 79:140–146. **Tiburcio AF**, Galston AW (1986) Phytochemistry 25:107–110. **Tiburcio AF**, Kaur-Sawhney R, Ingersoll RB, Galston AW (1985) Plant Physiol 78:323–326. **Tiburcio AF**, Kaur-Sawhney R, Galston AW (1987) Plant Cell Tissue Organ Cult 9:111–120. **Tiburcio AF**, Kaur-Sawhney R, Galston AW (1990) Polyamine Metabolism in The Biochemistry of Plants, vol 16, Academic Press, New York, pp 283–325. **Toppel G**, Witte L, Riebesehl B, von Borstel K, Hartmann T (1987) Plant Cell Rep 6:466–469. **Toppel G**, Witte L, Hartmann T (1988) Phytochemistry 27:3757–3760.

Ushiyama M, Furuya T (1989) Phytochemistry 28:2333–2339.

von Borstel K, Hartmann T (1986) Plant Cell Rep 5:39–42.

Wagner R, Feth F, Wagner KG (1986) Physiol Plant 68:667–672. **Walton NJ**, Belshaw NJ (1988) Plant Cell Rep 7:115–118. **Walton NJ**, McLauchlan WR (1990) Phytochemistry 29:1455–1457. **Walton NJ**, Robins RJ, Rhodes MJC (1988) Plant Sci 54:125–131. **Walton NJ**, Robins RJ, Peerless ACJ (1990) Planta 182:136–141. **Watson AB**, Brown AM, Colquhoun IJ, Walton NJ, Robins DJ (1990) J Chem Soc Perkin Trans I:2607–2610. **Weiler EW**, Stöckigt J, Zenk MH (1981) Phytochemistry 20:2009–2016. **Wigle ID**, Mestichelli LLJ, Spenser ID (1982) Chem Commun 662–664. **Witte L**, Müller K, Arfmann HA (1987) Planta Med 53:192–197.

Yamada Y, Endo T (1984) Plant Cell Rep 3:186–188. **Yamada Y**, Hashimoto T (1989) Proc Jpn Acad Ser B 65:156–159. **Yamada Y**, Hashimoto T, Endo T, Yukimune Y, Kohno J, Hamaguchi N, Dräger B (1990) Proc Phytochem Soc Eur 30:227–242. **Yoshimatsu K**, Hatano T, Katayama M, Marumo S, Kamada H, Shimomura K (1990) Phytochemistry 29:3525–3528. **Yoshioka T**, Yamagata H, Ithoh A, Deno H, Fujita Y, Yamada Y (1989) Planta Med 55:523–524.

Professor Dr. Horst-Robert Schütte
Institut für Pflanzenbiochemie
Weinberg 3
D-06120 Halle/Saale, FRG

IV. Mineral Nutrition: Humid Tropical Forests

By Ernesto Medina and Elvira Cuevas

1. Introduction

Humid tropical forests as considered in this review are those forest ecosystems occurring in areas where rainfall is higher than potential evapotranspiration during most of the year. A convenient separation of the wet and dry realms of tropical environment is attained using the Bailey's index calculated from rainfall and average temperature data (Bailey 1979). According to Bailey's model the lower limit of lowland humid tropics is located in areas with average annual temperature around 26 °C and average annual rainfall above 1500 mm. Distribution of rainfall may vary widely within the humid tropics, determining significant gradients in dynamics of ecosystem processes. As the average temperature decreases with increasing elevation in tropical mountains, the rainfall amount required for humid climatic conditions is reduced.

Tropical humid forests can occur in a wide variety of soils ranging from highly leached, nutrient-poor soils, mainly oxisols, ultisols, and spodosols, as in the Amazon and Rio Negro basins in South America, to montane locations on relative young, fertile soils as in Central America or Papua, New Guinea (Grubb 1977; Vitousek 1984; Jordan 1985). Biomass development (Jordan 1985) or regeneration capacity (Uhl 1987) are not always clearly associated with differences in soil fertility, but they are certainly related to rates and patterns of nutrient cycling (Vitousek and Sanford 1986; Medina and Cuevas 1989).

2. Previous Reviews and Scope

Nutrient relations in tropical forests have been recently reviewed by several authors with emphasis on nutrient cycling and nutrient-use efficiency (Vitousek 1984; Jordan 1985; Vitousek and Sanford 1986; Proctor 1987), as well as specific aspects of utilization and management of forests on nutrient-poor soils (Jordan 1989). The book on mineral nutrients in tropical forests and savannas edited by Proctor (1989) on behalf of the British Ecological Society is particularly useful because of the breadth of topics covered. In this chapter we emphasize the possible mechanisms of nutrient control on forest function and species distribution, as discussed by Grubb (1977). In doing so we will include mostly studies on nutrient inventories, nutrient cycling, and nutrient metabolism, published after the reviews of Vitousek and Sanford (1986) and Proctor (1987).

Progress in Botany, Vol. 55

3. Nutrient Inventories and Nutrient Balance

Estimation of nutrient balance of whole ecosystems has been the goal of several important studies in the humid tropics, but the results are somewhat disappointing due to the large number of uncertainties involved in the measurement of nutrient fluxes through the system. According to Bruijnzell (1991), the output of nutrients in catchment areas can be separated into four groups based on the leaching levels of Ca, Mg, and K, which increase according to site fertility: spodosol + highly depleted oxisols group < oxisol/ultisol group < inceptisols group < mollisol/vertisols group. Most case studies reviewed by Bruijnzeel (1991) indicate accumulation of P, possibly as a result of the low mobility of this element. The data for N balances are not satisfactory, and much research is needed to include input through biological fixation and outputs through denitrification.

a) Nutrient Input Through Precipitation and Throughfall Enrichment by Nutrient Leaching

Input of nutrients in rainfall may be of critical importance for the nutrient balance of forests growing on highly leached soils, such as in oxisols, ultisols, and spodosols in wet lowland areas. Average ion input in tropical areas varies widely both geographically and seasonally, with no discernible correlation between amount of rainfall and amounts of nutrients flowing into the different forests examined. Larger inputs of Mg and Ca compared to K, and a strong tendency towards larger NH_4 than NO_3 fluxes, are the general patterns in these ecosystems (Table 1). Sulfate is comparatively high in the sites where it has been measured, and may explain in part the low pH measured in several remote sites (Table 1).

The annual nutrient accession rates in rainfall were comparatively high in a lower montane rainforest of south Central Java, possibly due to intermediate concentrations and high rainfall (4500 mm) (Bruijnzeel 1989b). Supply of Mg and K in rainfall was consid-

Table 1. Average nutrient concentration and annual inputs in rainfall in several tropical locations throughout the world. (With data collated by Bruinjzeel 1989b; Jordan 1989; Veneklaas 1990)

	Nutrient concentration mmol m^{-3} (range)	n	Nutrient input mmol m^{-2} (range)	n
K	1.0 (0.3–3.1)	15	20 (6–47)	14
Ca	11.1 (tr.–54.4)	15	200 (35–509)	14
Ng	8.2 (tr.–25.1)	14	152 (29-465)	14
NH_4-N	25.9 (3.6–61.4)	11	493 (86–1307)	10
NO_3-N	9.3 (0.7–32.9)	10	196 (36–436)	7
PO_4–P	0.65 (0.10–1.61)	8	15 (3–39)	10
SO_4–S	21.8 (8.1–38.6)	8	424 (100–989)	8
pH	5.3 (4.4–6.4)	12		

ered to be enough to compensate for losses due to wood extraction from *Pinus* and *Agathis* plantations in the area, but not so for N (90 and 50% respectively) and P (60 and 40% respectively).

In the two epiphyte-rich rainforests located at 2550 and 3370 m a.s.l. in the Colombian Andes, nutrient concentration in rainfall was similar, but total flux differed due to higher rainfall in the lower altitude forest (Veneklaas 1990). Throughfall water was enriched in all elements measured (N, P, S, K, Ca, Mg) in the latter. Deposition ratios were smaller for all elements except S in the higher altitude forest, and even a small reduction was observed in N and P. As expected, the largest enrichment factor in throughfall corresponded to K (12.1 and 4.8 in the lower and higher altitude forests, respectively). Similar patterns of throughfall enrichment have been measured in lowland forests in the Amazon basin (Franken et al. 1985). This is contrary to previous reports indicating a certain degree of nutrient scavenging from rainwater in the forest canopy (Jordan et al. 1980).

b) Nutrient Concentration in Leaves

A number of studies report the concentration of nutrients in mature canopy leaves of lowland and montane tropical forests on contrasting soils. In the forests studied, concentration of N is relatively high, while P concentration varies in relation to P availability of the soil. Forests not limited by P availability tend to have higher P/N ratios than those on poor soils, in agreement with previous findings (Table 2).

There appears to be a strong correlation between the biogeochemical properties of river waters and their fringe forests in sites with strong variations in nutrient availability. Furch and Klinge (1989) measured a large number of tree species in forests of contrasting soil fertility in the Amazon basin near Manaus, Brazil (Table 2). Varzea forests with an abundant supply of nutrient-rich sediments have higher concentration of N and P, and P/N ratios larger than 25. Flooded forests in black water river basins ("igapo") and forests on sandy soils have much lower P concentrations and P/N ratios below 20. When nutrient

Table 2. Recent reports on nutrient concentration in canopy leaves of tropical humid forests: (1) Furch and Klinge (1989); (2) Thompson et al. (1992); (3) Bongers and Bopma (1990); (4) Jayasekera (1992)

Site		No. of species	N	P (mmol kg^{-1})	K	P/N (mmol mol^{-1})
(1)	Varzea, Manaus, Brazil	29	1829	57	327	31
	Varzea, Manaus	17	1971	58	435	29
	Varzea, Manaus	13	1779	28	220	16
	Varzea, Manaus	29	1600	43	376	27
	Igapó, Manaus	23	1236	22	168	17
	Tierra Firme, Manaus	183	1286	17	128	14
	Campina, Manaus	14	793	15	168	20
(2)	Maracá, Roraima, Brazil	77	1343	39	263	29
(3)	Los Tuxtlas, Veracruz, México	68	1243	43	269	35
(4)	Montane forest, Sri Lanka	6	1193	31	292	26

concentrations in river water and vegetation in the area of confluence of the Negro and Solimões rivers near Manaus, Brazil, were compared, Klinge and Furch (1991) concluded that concentrations of N, P, Ca, and Mg are considerably larger in the Solimões than in the Río Negro, the dominant cation in the Solimões being Ca.

Thompson et al. (1992) investigated forest structure and leaf chemistry of a high rainfall but strongly seasonal forest in the Maracá island in northern Brazil. They found that in spite of the apparent low soil fertility, mature canopy leaves of dominant trees had average N, P, and K concentrations corresponding more to tropical forests on relatively fertile soils. As expected from their relative high nutrient content, leaves were not sclerophyllous (Medina et al. 1990). The authors remark that the P/N ratio was highly variable and cannot be taken as a reliable indicator of P supply.

Another forest of relatively high P availability, judging from leaf analysis, was studied by Bongers and Popma (1990) in Los Tuxtlas, México. This forest is characterized by moderate concentrations of N but relatively high concentrations of P in their leaf tissues. The P/N ratio is high to very high (35 mmol mol^{-1}). As in the case of the Maracá forest, anatomical leaf characteristics are mesophyll with low average leaf weigth/area ratios (78.8 g m^{-2}). As expected, N and P contents are highly correlated, while leaf weight/area ratio is inversely correlated with both P and N contents.

Jayasekera (1991) also reported comparatively high values of P in the canopy leaves of a montane forest in Sri Lanka. Again, there is a considerable variation both in leaf N and P contents averaging 1.2 mol N kg^{-1}, 31 mmol P kg^{-1}, and a P/N ratio (mmol mol^{-1}) of 26. However, average leaf area/weight ratio was relatively high (106 gm^{-2}).

c) Biomass Accumulation, Species Distribution, and Soil Fertility

Biomass accumulation and tree height are usually associated with nutrient availability. Yamakura and Sahunalu (1990) found that maximum tree height in 32 Southeast Asian rainforests and deciduous forest sites, is negatively correlated with the C/N ratio of the mineral soil. The amount of nitrogen in the mineral soil was positively correlated with tree height in deciduous forests, but negatively correlated in evergreen forests. Gower (1987) observed an inverse relation between total fine root biomass in two Costa Rican forests differing markedly in their P and Ca content per unit area, but with similar values of N to a depth of 50 cm. Gower hypothesizes that investment in fine root biomass is inversely related to P availability.

In the Cameroun, Gartlan et al. (1986) and Newbery et al. (1986) were able to show strong associations between species distribution in gradients of P availability in the uppermost mineral soil layers. In areas of impeded drainage, as in Douala-Edea forest reserve, the effects of P availability on species distribution were somewhat obscured. Besides, in this reserve, P availability was generally too high for a strong selection of caesalpinioid (Newbery et al. 1986). In the Korup forest reserve, with lower average levels of P availability, Gartlan et al. (1986) using direct gradient analysis found significant and different responses of several species towards P availability. Species responses were generally unimodal (basal area increasing or decreasing with P availability, or having a distinct optimum soil P value), but in a few cases basal area was higher at both extremes of the P gradient. These patterns of response seem to be related to the occurrence of ectotrophic mycorrhizal symbiosis (Newbery et al. 1988).

A study relating site characteristics and distribution of tree species in a mixed dipterocarp forest in Sarawak showed that those properties determined by the lithology of the parent material are more important for species distribution than properties such as organic matter and exchangeable cations (Baillie et al. 1987). The analyses, however, showed that P in the top soil is the single most important factor associated with species distribution, while both exchangeable and reserve Mg seem to exert a regulatory factor through their influence on the efficiency of the mycorrhizal root system.

Johnston (1992) used canonical correspondence analysis to investigate the relationships between tree species distribution and soil physicochemical properties in a lower montane rainforest in Puerto Rico. The dominant tree species, *Dacryodes excelsa* and *Manilkara bidentata*, were positively associated with drier soils having low concentrations of Ca and Mg and low pH: *Prestoea montana* was associated with wet soils, high Ca and Mg, and higher pH values (see also Frangi and Lugo 1985). Hydrology patterns determined by geomorphological positions may be more important in separating communities of palms from those of *D. excelsa* and *Sloanea berteriana*.

The use of nutrient content of superficial soil layers to explain species distribution is limited. Organic matter production by the vegetation strongly modifies superficial soil chemistry. The magnitude of variations in superficial contents of nutrients as modified by the predominant tree species was clearly shown in a comparative study of the litter layer and the top soil in tree plantations established in the Luquillo mountains in Puerto Rico (Lugo et al. 1990). Large variations in nutrient content per unit area were found in all nutrients analyzed: N 55–187; P 2.1–8.9; K 15–46; Ca 37–207; Mg 17–43. These differences were also reflected in the mineral soil and could be associated with differences in the amount and quality of litter produced.

4. Epiphytic Communities and the Dynamics of Nutrients in Humid Tropical Forests

Tropical rainforests, particularly montane rainforests, are characterized by a rich epiphytic synusia (Benzing 1990). Ecosystem biomass accounted for by epiphytes is relatively small (1.5–2.2%), but epiphytes can contribute more than 30% of total leaf biomass (Nadkarni 1984). Nadkarni and Matelson (1991), measuring the amount of intercepted tree litter in the canopy of a cloud forest in Costa Rica, established that approximately the same amount of tree litter reaching the soil is intercepted in the canopy. However, it does not accumulate and is frequently blown away. In addition, the rate of litter decomposition in the canopy is lower compared to the rates on the florest floor. Therefore, nutrient supply from intercepted tree litter is not sufficient to cover the requirements of the epiphytic communities in this forest. The contribution of epiphyte litter to litterfall reaching the forest floor is approximately 5–10%. Decomposition of this material was faster than litter derived from terrestrially rooted plants, but nutrient release, except K, was slower. Decomposition rate of litter derived from terrestrial plants had a decay constant of 0.69, while that of epiphytes was 1.3 (Nadkarni and Matelson 1992a, b). Nutrients transferred in epiphyte leaf litter expressed in percent of total nutrient flux in fine litterfall was 7% for N, 8% for P, 4% for Ca, 5% for Mg and 1% for K (Nadkarni and Matelson 1992a). In a montane rainforest in Guadeloupe, with a high cover of epiphytic bryophytes, Coxson (1991) showed that drought-wet cycles of the bryophyte mat on the tree branches can lead to a pulsed release of nutrients amounting to an important fraction of their nutrient capital (28.7 kg ha^{-1} yr^{-1} for K and 0.2 kg ha^{-1} yr^{-1} for P). The dynamics of epiphyte biomass is obviously important to understand nutrient cycling in tropical rain forests.

5. Nutrient Fluxes in Litterfall, Nutrient Availability and Efficiency of Nutrient Utilization

Measurement of litterfall and its associated fluxes of nutrients continues to be one of the more practical ways to evaluate production capacity and availability of nutrients in tropical forests. The within-ecosystem nutrient use efficiency introduced by Vitousek (1984) differentiates tropical forests from nutrient-rich and nutrient-poor soils. Phosphorus/nitrogen ratios may be also used as an index of the relative availability of P. In litterfall, P/N ratios are always lower than in living canopy leaves (on a species basis) because a larger fraction of P, compared to N, is retranslocated before leaf shedding (Medina 1984; Cuevas and Medina 1991).

Analysis of the P/N ratios of the published data of nutrient fluxes in fine litter fall (Cuevas and Medina 1986; Vitousek and Sanford 1986; Proctor 1987; Scott et al. 1992) indicates that they are inversely related to P-use efficiency as measured by the organic matter/P ratio (Fig. 1). Forests growing on relatively fertile soils are clearly separated from forests growing on infertile soils both by the dry mass/P, and the P/N ratios. The set of montane forests included in the figure show a large range of variation from rich to poor soils (Proctor 1987; Bruinjzeel 1989a, b; Veneklaas 1991). Interestingly, there is no relationship between P/N (or N/P) ratios and the efficiency of N use. This may indicate that P/N ratios in humid tropical forests are mostly regulated by the availability of P and not of N, as suggested elsewhere (Grubb 1977).

In northern Brazil, an interdisciplinary study of a seasonal forest on sandy soils (Maracá island, Roraima, Brazil) showed that complex, nonsclerophyllous forests can develop in apparently nutrient-deficient soils (Thompson et al. 1992). High nutrient availability, particularly P, in this forest is indicated by its fine litter production, among the highest reported for tropical forests (8.85–9.52 t ha^{-1} yr^{-1}) (Scott et al. 1992). In addition, the indices of efficiency of nutrient utilization (sensu Vitousek 1984) are medium-low for N and low for P, while the average P/N ratio is medium-high (Fig. 1). The lack of correspondence between leaf chemistry and soil nutrients may be attributed to the strong seasonality of rainfall, which causes peaks in decomposition of soil organic matter and subsequent nutrient release. So the rate of nutrient circulation in the forest floor explains the relatively high production rates, and high fluxes of N and P in litterfall.

The recent data on fine litter fall and nutrient fluxes in litter (Cuevas and Medina 1986, 1991; Puig and Delobelle 1988; Sampaio et al. 1988; Bruijnzeel 1989a; Dantas and

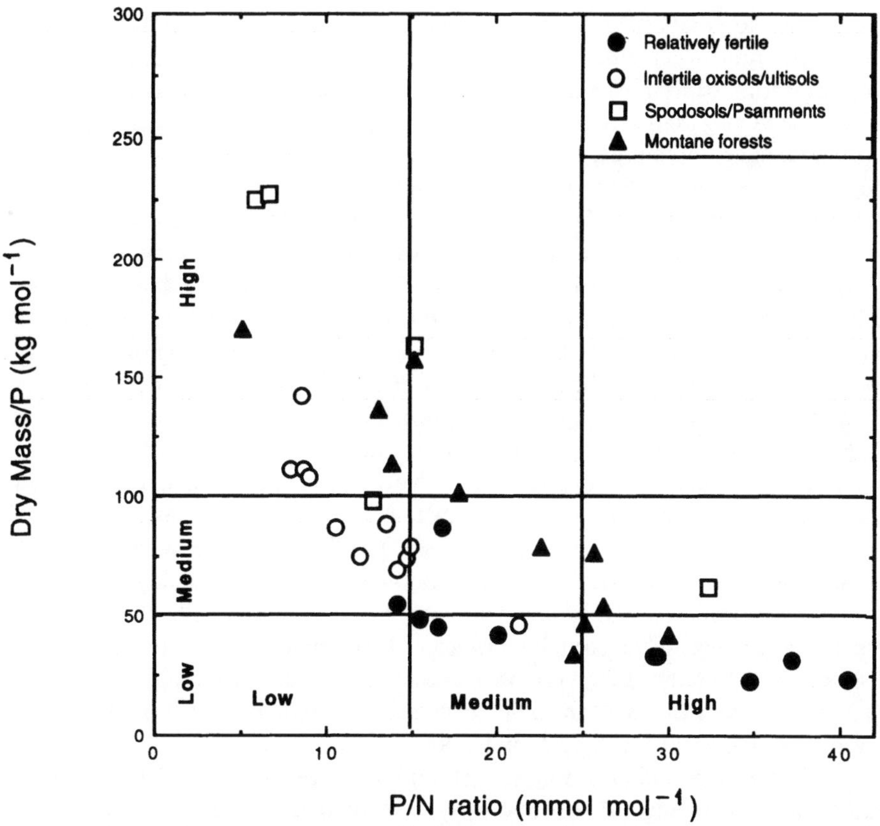

Fig. 1. Distribution of dry mass/P ratios and N/P ratios from tropical forests growing on soils of contrasting fertility. (Data from Tanner 1985; Cuevas and Medina 1986; Vitousek and Sanford 1986; Proctor 1987; Dantas and Phillipson 1989; Furch and Klinge 1989; Veneklaas 1991)

Phillipson 1989; Veneklaas 1991; Wang et al. 1991; Scott et al. 1992) support the generalizations on nutrient use efficiency and site fertility as developed by Vitousek (1984).

6. Evaluation of Soil Fertility: Fertilization Experiments and Bioassays

The indirect evidence that tropical forest soils have a nutrient limitation of some sort (Vitousek 1984; Vitousek and Sanford 1986; Proctor 1987) has led to a number of fertilization assays using either natural vegetation or bioindicators.

Cuevas and Medina (1988) measured the growth of fine roots into plastic-mesh cylinders filled with vermiculite previously soaked individually in solutions of ammonium sulfate, calcium chloride, or potassium phosphate. The experiments were conducted in rainforests of the upper Río Negro basin contrasting in soil chemical and textural characteristics. Results showed that fine root growth was markedly affected by added nutrients, but response varied with forest and soil type. Tierra Firme forests on oxisols responded more significantly to the additions of P and Ca, while Caatinga forests responded more vigorously to N additions. These responses could be correlated with the pattern of nutrient distribution in standing biomass and litterfall (Medina and Cuevas 1989). Long-term fertilization experiments in montane forests in Jamaica showed that responses were species-dependent (Tanner et al. 1990). Leaf N concentration did not increase above control in trees fertilized with urea, but tended to decrease in those fertilized with P after the second year of treatment. *Clethra occidentalis* and *Podocarpus urbanii* had higher P concentration in their leaves after 3 years in the fertilized plots. Responses of leaf production and trunk growth were not homogenous. In a cloud forest in western Venezuela, a similar experiment showed that the rates of trunk growth in N and P fertilized plots were roughly two times higher than those of the controls (Tanner et al. 1992). Litterfall showed a relative increase in fertilized plots only 4 years after the beginning of the experiment. Nitrogen concentrations in litterfall remained unaffected by fertilization with either N or P, while P concentrations increased significantly. Their conclusion was that in these montane forests productivity seems to be limited more by N than P.

Bioassays have also been used to assess nutritional status of rainforest soils. Denslow et al. (1987) used rainforest species with contrasting life history patterns. Growth response was positive in all cases when the soils were fertilized with a complete nutrient solution. Only *Phytolaca rivinoides*, a fast-growing, nonmycorrhizal species, responded positively to the addition of P, N + P, and complete nutrient solutions. The authors attributed the lack of response of the shrubby species to P additions to mycorrhizal associations and slower growth rates of the species. It seems that these species were not limited by P supply, either because they are able to cover their requirements for maximum growth through symbiosis, or that their actual requirements are low.

Healey (1989) fertilized montane forests soils and used *Melinis minutiflora* (an introduced African grass with high P requirements) as a bioindicator. He concluded that growth was stimulated in both forest soils tested by the addition of N and P. Addition of P always increased the concentration of foliar P, while this was not the case with N addition. Results of this bioassay do not coincide with the fertilization experiments in situ, where P did not have a significant effect on any of the parameters measured (Tanner et al. 1990). This emphasizes the shortcomings of the bioassay approach, and the need to document the status of mycorrhizal development in order to evaluate the importance of P supply through the symbiosis.

7. Mineralization of Organic Matter and Nitrogen Availability

Robertson (1984) measured high rates on N mineralization in old soils of La Selva, Costa Rica. Experimental additions of phosphate led to a certain immobilization of N, indicating that microbial activity is limited by P and not by N. The author concluded that P is generally more limiting than N for plant growth in humid tropical forests (Robertson 1989).

Vitousek and Denslow (1987) found that P availability in tropical rainforests soils in La Selva, Costa Rica, vary according to the parent material. Recent alluvium has high level of extractable P (36.6 $\mu g\ g^{-1}$), but extractability is reduced in old alluvium (3.9) and lava flows (2.6 and 1.6). However, nitrogen availability measured as net mineralization was relatively high in all soils, averaging between 50 and 80 µg NO_3-N g^{-1} $month^{-1}$ (Vitousek and Denslow 1986).

Maggs (1991) measured N mineralization in tropical rainforests with contrasting fertility, one on basaltic soils and another on less fertile soils formed on metamorphic rock. Mineralization rates did not differ among forests, and were higher near the soil surface (0–7.5 cm). The values averaged 22 and 10 µg N g^{-1} soil $month^{-1}$, for the upper and lower soil layers, respectively. Relative nitrification (nitrate production/nitrogen mineralized) was higher in the complex forest on basaltic soils (100%) than in the structurally simple rain forest (88%). Therefore NO_3 was the dominant form of N in the complex forest, while both NH_4 and NO_3 had similar importance in the simpler forest.

Marrs et al. (1991) measured both N mineralization and nitrification in a strongly seasonal rainforest in the island of Maracá, Roraima, Brazil. Mineralization rates were not obviously related to soil moisture content. Instead, mineralization peaks were observed in the transition from wet to dry season, possibly indicating saturating conditions during the wet season. Rates measured in this forest are generally lower than those measured in more fertile sites (Costa Rica, Panamá), but were comparable to data from San Carlos in the upper Río Negro basin (Montagnini and Buschbacher 1989; Fig. 2).

Along altitudinal gradients in Costa Rica, mineralization rates decreased with altitude (Marrs et al. 1988). Therefore N availability could be one of the limiting factors causing reduced growth in upper montane forests. Nitrification appears to be partially limited by substrate availability, as shown with ammonium addition experiments. Laboratory exper-

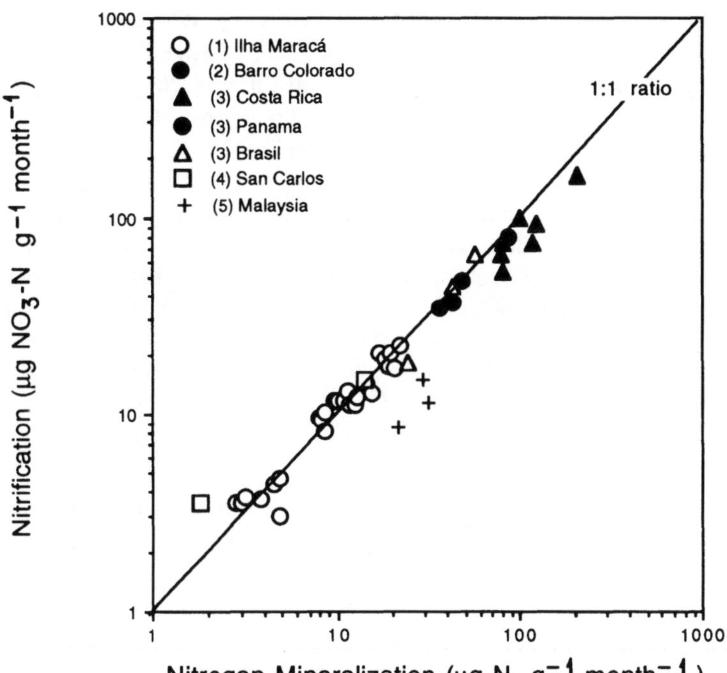

Fig. 2. Mineralization and nitrification rates measured in different tropical forests. (Data from Vitousek and Matson 1988; Chandler 1985; Matson and Vitousek 1987; Marrs et al. 1988, 1991; Yavitt and Wieder 1988; Montagnini and Buschbacher 1989)

iments where both temperature and aeration/moisture were improved showed increases in the mineralization rates of one to three orders of magnitude, and the rates increased with altitude of origin. Possibly saturated soil conditions at the high altitude sites also play a role in inhibiting mineralization.

In relative infertile soils of the Amazon basin, mineralization rates are much larger in the upper soil organic layer (including root mat if present) than in the mineral soil underneath (Vitousek and Matson 1988; Montagnini and Buschbacher 1989). While almost 100% of the N mineralized was also nitrified in the mineral soils, the organic layer nitrification amounted only to 25–39% in Brazil, and 44 to 51% in San Carlos.

Data of potential N mineralization and nitrification from different rainforests indicate that in general all the organic N mineralized is subsequently nitrified, leaving only little or no free NH_4 (Fig. 2). Exceptions were reported in Malaysia (Chandler 1985) and Costa Rica (Vitousek and Matson 1988). Under natural conditions, however, NH_4 could be incorporated into biomass, immobilized in the soil, or taken up by the vegetation, resulting in a reduction of nitrification rates.

8. Denitrification

Tropical forests appear to be an important source of nitrous oxides entering the atmosphere (Keller et al. 1986). Through the pathway of denitrification, significant amounts of N are lost from the system. Rates of nitrous oxide flux in a range of tropical forest ecosystems with contrasting soil fertility are highly correlated with N mineralization rates (Matson and Vitousek 1987). In addition, primary forests and early successional stages loose more N as N_2 and N_2O than intermediate successional stages (Robertson and Tiedje 1988). Denitrification rates averaged 0.11 g N m^{-1} per month in primary forest sites to 0.16 in early-successional sites and less than 0.034 in mid-successional sites. The authors hypothesized that denitrification rates were higher in precolonial times, when a smaller proportion of sites were in mid-successional stages. Further measurements throughout the tropical belt are necessary to clarify the role of tropical forests and deforestation on the N_2O balance of the atmosphere.

9. Assimilation of Inorganic Nitrogen

Nitrogen supply is always critical in determining the level of organic matter production in natural ecosystems. Nitrogen enters ecosystems through nitrogen fixation of both symbiotic and free-living microorganisms, or in rainfall. The proportion of NO_3 and NH_4, inorganic forms of N in the soil, varies considerably among forest types and geographical regions. In climax communities characterized by low rates of nitrification, woody plants appear to take up inorganic N mainly in the form of ammonium (Smirnoff et al. 1984; Smirnoff and Stewart 1985). NH_4 assimilation is associated with lower energy requirements, a fact which might be of importance for understory plants severely limited by light.

Stewart et al. (1988) conducted a detailed study in Australian rainforest communities to establish the form of inorganic nitrogen used by woody species belonging to different ecological classes: pioneers, vines, understory, and canopy species. Species from the closed forests had consistently lower nitrate reductase (NR) activities, in both leaves and roots. Exposure to higher light intensities due to the formation of gaps in the native forests did not significantly increase leaf NR activity of rainforest species. When cultivated with additional supply of NO_3, pioneer species developed consistently higher NR activity, while canopy and emergent species did not show enzyme induction. In pioneer and vine species, enzyme activity was induced mainly in the shoots. In spite of differences in NR activities, leaf N contents were similar, indicating the utilization of NH_4 by the closed forest species. Glutamine synthetase was present in the leaves of all species examined, but in closed forest species activity of chloroplastic glutamine synthetase was consistently lower. In these species the cytosolic isoform of glutamine synthetase accounted for at least 40% of total leaf activity.

Freeden et al. (1991) obtained further confirmation of the differences in NR activities and NR induction in a group of species of *Piper* occurring in rainforests environments in Los Tuxtlas, Mexico. From the six species studied, two are truly shade plants (*P. lapathifolium* and *P. amalago*), two other predominate in gaps or open areas (*P. umbellatum* and *P. auritum*), *P. aequale* requires light for germination but occurs almost only in rainforest shade environments, and *P. hispidum* germinates in gaps but is common in the understory of transitional forest. Gap species had consistently higher in vivo NR activities when grown in high or low light intensities in natural environments. At the other end, the shade species had always low in vivo NR activities under both conditions. In *P. hispidum* NR activity was higher in high-light grown plants.

Another study by Stewart et al. (1992) including species from a coastal rain forest in Brazil showed again that 75% of the species sampled had in vivo activities of NR lower than 50 pkat per g fresh weight. Activity in gaps and forest margins was almost three times higher on average compared to the undisturbed forest. In gallery forest species, the activity of the enzyme increased in average by a factor of 7. Colonizing and early successional species had higher NR activities, which responded markedly to the supply of NO_3. Localization of NR activity was variable. In the group of evergreen forest species analyzed, most of the activity was located in the leaves, with the exception of *Pithecellobium pedicellare*. In the same set of species, additional supply of NO_3 brought about a modest increase of NR activity in both leaves and roots. In a number of species, the leaf NR activity was correlated with the nitrate content in xylem sap, while leaf N content was linearly correlated with the concentration of N (nitrate + organic) in the xylem sap. It appears that in the plant communities analyzed, NO_3 is not the main source of nitrogen. However, most species have the potential to use NO_3 because NR activity can be induced upon addition of the substrate. In fact, annual burning of savanna ecosystems throughout the tropical belt could provide a large source of NO_3 to neighboring forests.

10. Symbiotic Interactions and Nutrient Uptake

Mycorrhizal symbiosis is very widespread in tropical humid forests (Janos 1983; St. John and Uhl 1983; Högberg 1986). Most species develop vesicular-arbuscular (VA) mycorrhizas, but an important group constituted by the Dipterocarpaceae and legumes of the subtribes Amherstieae and Detarieae (Caesalpinioid), and such important genera as *Aldina* and *Swartzia* in the papilionoids, are ectomycorrhizal and can reach dominance over large areas in the humid tropics (Janos 1983; Alexander and Höberg 1986; Alexander 1989). The occurrence of mycorrhiza is of paramount importance to understand the nutrient balance of humid tropical forests. The widespread limitation of P availability over vast tropical areas emphasizes the importance of this biological interaction, considered to increase water and nutrient uptake, particularly P, by higher plants. The seminal paper by Went and Stark (1968) on the occurrence of a "closed" nutrient cycle in tropical forests preventing or reducing nutrient leaching, mediated by mycorrhiza, gave impulse to an exciting period of research on the nutrition of tropical forests (Stark and Jordan 1977; Janos 1983; Jordan 1989). It is now generally accepted that predominance of mycorrhizal symbiosis (both ecto- and endomycorrhizas) in the majority of humid tropical forests is certainly associated with low P availability in the soil (Newbery et al. 1988). It has been shown that frequency and percent of mycorrhizal infection is inversely related to soil pH and available P (van Noordwijk and Hairiah 1986).

In tropical forests with a thick root mat, phosphate solutions sprayed on the soil surface can be efficiently taken up by VA mycorrhiza, thereby preventing nutrient leaching in these forests (Jordan et al. 1979). Salcedo et al. (1991) found in a coastal rainforest in Brazil that a significant fraction of ^{32}P added to the forest floor was retained in the humus layer. The magnitude of the retention in humus was directly related to the microbial biomass, while content in humus roots was directly related to the percent of VA mycorrhizal infection. However, there was a considerable leaching from the organic layer to the mineral soil, which was not affected by root trenching. Vertical P movement in these soils was further evidenced by the increase in P concentration from the upper 5–15 cm of soil (85 mg kg^{-1}) to the layer inmediately below (104 mg kg^{-1}).

There is still a debate if the main role of mycorrhiza is in increasing the surface of nutrient absorption, by penetrating the soils beyond the zone of nutrient depletion around the fine roots, or if mycorrhizal fungi have the capability to utilize organic P sources directly through previous digestion by extracellular phosphatases (Alexander 1989).

11. Final Remarks

In the past 10 years, the study of nutrient relations of tropical forests show qualitative advances, derived from the application of experimental techniques to the analysis of nutrient release, nutrient uptake, and nutrient transformations under natural conditions. In the near future, important advances in the determination of specific nutrient requirements of dominant species, and the operation of particular nutrient uptake mechanisms (e.g., symbiotic interactions) are to be expected. To understand the mechanisms of nutrient recycling and the processes operating in forests regeneration, it is necessary, for instance, to clarify how the nutrient uptake systems work in acid environments markedly depleted in Ca, and which is the role of fine roots, with or without mycorrhizal symbiosis, in the processes or release of nutrients bound into soil organic matter. Moreover, large ecosystem experiments, specifically designed to precisely measure nutrient input and output from humid forest ecosystems, will be required in order to understand the possible changes in forest dynamics to be brought about by global environmental changes.

References

Alexander I (1989) Mycorrhizas in tropical forests. In: Proctor J (ed) Mineral nutrients in tropical forest and savanna ecosystems. Blackwell, Oxford, pp 169–188. **Alexander IJ**, Högberg P (1986) New Phytol 102:541–549.

Bailey HP (1979) Geogr Ann 3–4:196–215. **Baillie IC**, Ashton PS, Court MN, Anderson JAR, Fitzpatrick EA, Tinsley J (1987) J Trop Ecol 3:210–220. **Benzing D** (1990) Vascular epiphytes. Cambridge University Press. Cambridge. **Bongers F**, Popma J (1990) Bot Gaz 151:354–365. **Bruijnzeel LA** (1989a) Malay Nat J 37:199–208. **Bruijnzeel LA** (1989b) J Trop Ecol 5:187–202. **Bruijnzeel LA** (1991) J Trop Ecol 7:1–24.

Chandler G (1985) Soil Biol Biochem 17:347–353. **Coxson DS** (1991) Can J Bot 69:2122–2129. **Cuevas E**, Medina E (1986) Oecologia 68:466–472. **Cuevas E**, Medina E (1988) Oecologia 76:222–235. **Cuevas E**, Medina E (1991) Phosporus/nitrogen interactions in adjacent Amazon forests with contrasting soils and water availibility. In: Tiessen H, López-Hernández D, Salcedo IH (eds) Phosphorus cycles in terrestrial and aquatic ecosystems. SCOPE Regional Workshop 3. South- and Central America. University of Saskatchewan Inst. Pedology, Saskatoon, pp 84–94.

Dantas M, Phillipson J (1989) J Trop Ecol 5:27–36. **Denslow JS**, Vitousek PM, Schultz JC (1987) Oecologia 74:370–376.

Frangi J, Lugo AE (1985) Ecol Monogr 55:351–369. **Franken W**, Leopoldo PR, Bergamin H (1985) Turrialba 35:383–393. **Freeden AL**, Griffin K, Field CB (1991) Oecologia 85:441–446. **Furch K**, Klinge H (1989) Chemical relationships between vegetation, soil and water in contrasting inundation areas of Amazonia. In: Proctor J (ed) Mineral nutrients in tropical forest and savanna ecosystems. Blackwell, Oxford, pp 189–204.

Gartlan JS, Newbery DM, Thomas DW, Waterman PG (1986) Vegetatio 65:131–148. **Gower ST** (1987) Biotropica 19:171–175. **Grubb PJ** (1977) Annu Rev Ecol Syst 8:83–107.

Healey JR (1989) A bioassay study of soils in the Blue Mountains of Jamaica. In: Proctor J (ed) Mineral nutrients in tropical forest and savanna ecosystems. Blackwell, Oxford, pp 273–287. **Heaney A**, Proctor J (1989) Chemical elements in litter in forests on Volcán Barva, Costa Rica. In: Proctor J (ed) Mineral nutrients in tropical forest and savanna ecosystems. Blackwell, Oxford, pp 255–271. **Högberg P** (1986) J Trop Ecol 2:359–372.

Janos DP (1983) Tropica mycorrhizas, nutrient cycles and plant growth. In: Sutton SL, Whitmore TC, Chadwick AC (eds) Tropical rain forest: ecology and management. Blackwell, Oxford, pp 327–345. **Jayasekera R** (1992) Vegetatio 98:73–81. **Johnston MH** (1992) J Trop Ecol 8:253–263. **Jordan CF** (1985) Nutrient cycling in tropical forest ecosystems. John Wiley, Chichester. **Jordan CF** (ed) (1989) An Amazonian rain forest: the structure and function of a nutrient stressed ecosystem and the impact of slash-and-burn agriculture. Man and the Biosphere Series, vol 2. UNESCO

& The Partenon, London. **Jordan C**, Golley F, Hall J, Hall J (1980) Biotropica 12:61–66. **Jordan CF**, Todd RL, Escalante G (1979) Oecologia 39:123–128.

Keller M, Kaplan WA, Wofsy SC (1986) J Geophys Res 91 (D11):11,791–11,802. **Klinge H**, Furch K (1991) Interciencia 16:196–201.

Lugo AE, Cuevas E, Sánchez MJ (1990) Plant Soil 125:263–280.

Maggs J (1991) Aust J Ecol 16:47–51. **Marrs RH**, Proctor J, Heaney A, Mountford MD (1988) J Ecol 76:466–482. **Marrs RH**, Thomson J, Scott D, Proctor J (1991) J Trop Ecol 7:123–137. **Matson PA**, Vitousek PM (1987) Global Biochem Cycles 1:163–170. **Medina E** (1984) Nutrient balance and physiological processes at the leaf level. In: Medina E, Mooney HA, Vázquez-Yanes C (eds) Physiological ecology of plants in the wet tropics. Junk, The Hague, pp 145–154. **Medina E**, Cuevas E (1989) Patterns of nutrient accumulation and release in Amazonian forests of the upper Río Negro basin. In: Proctor J (ed) Mineral nutrients in tropical forest and savanna ecosystems. Blackwell, Oxford, pp 217–240. **Medina E**, García V, Cuevas E (1990) Biotropica 22:51–64. **Montagnini F**, Buschbacher R (1989) Biotropica 21:9–14.

Nadkarni NM (1984) Biotropica 16:249–256. **Nadkarni NM**, Matelson TJ (1991) Ecology 72:2071–2082. **Nadkarni NM**, Matelson TJ (1992a) Biotropica 24:24–30. **Nadkarni NM**, Matelson TJ (1992b) Biotropica 24:113–120. **Newbery DM**, Gartlan JS, McKey DB, Waterman PG (1986) Vegetatio 65:149–162. **Newbery DM**, Alexander IJ, Thomas DW, Gartlan JS (1988) New Phytol 109:433–450.

Proctor J (1987) Appl Geogr 7:135–152. **Proctor J** (ed) (1989) Mineral nutrients in tropical forest and savanna ecosystems. Blackwell, Oxford. **Puig H**, Delobelle J-P (1988) Rev Ecol (Terre et Vie) 43:3–22.

Robertson GP (1984) Oecologia 61:91–104. **Robertson GP** (1989) Nitrification and denitrification in humid tropical ecosystems: potential controls of nitrogen retention. In: Proctor J (ed) Mineral nutrients in tropical forest and savanna ecosystems. Blackwell, Oxford, pp 55–69. **Robertson GP**, Tiedje JM (1988) Nature 336:756–759.

Salcedo IH, Elliot ET, Sampaio EVSB (1991) Mechanisms controlling phosphorus retention in the litter mat in Atlantic coastal forests. In: Tiessen H, López-Hernández D, Salcedo IH (eds) Phosphorus cycles in terrestrial and aquatic ecosystems. SCOPE Regional Workshop 3. South- and Central America. University of Saskatchewan, Inst Pedology, Saskatoon, pp 109–120. **Sampaio EVSB**, Nunes KS, Lemos EEP (1988) Pesq Agropec Brasil 23:1055–1061. **Scott DA**, Proctor J, Thompson J (1992) J Ecol 80:705–717. **Smirnoff N**, Stewart GR (1985) Physiol Plant 64:133–140. **Smirnoff N**, Todd P, Stewart GR (1984) Ann Bot 54:363–374. **Stark N**, Jordan CF (1977) Ecology 59:434–437. **Stewart GR**, Hegarty EE, Specht RL (1988) Physiol Plant 74:26–33. **Stewart GR**, Joly CA, Smirnoff N (1992) Oecologia 91:511–517. **St. John TV**, Uhl C (1983) Acta Cient Venez 34:233–237.

Tanner EJV (1985) J Ecol 73:553–568. **Tanner EJV**, Kapos V, Freskos S, Healey JR, Theobald AM (1990) J Trop Ecol 6:231–238. **Tanner EJV**, Kapos V, Franco W (1992) Ecology 73:78–86. **Thompson J**, Proctor J, Viana V, Milliken W, Ratter JA, Scott DA (1992) J Ecol 80:689–703.

Uhl C (1987) J Ecol 75:377–407.

Van Noordwijk M, Hairiah K (1986) Plant Soil 96:299–302. **Veneklaas EJ** (1990) J Ecol 78:974–992. **Veneklaas EJ** (1991) J Trop Ecol 7:319–336. **Vitousek PM** (1984) Ecology 65:285–298. **Vitousek PM**, Denslow JS (1986) J Ecol 74:1167–1178. **Vitousek PM**, Denslow JS (1987) Biotropica 19:167–170. **Vitousek PM**, Matson PA (1988) Soil Biochem 20:361–367. **Vitousek PM**, Sanford RL (1986) Annu Rev Ecol Syst 17:137–167.

Wang D, Bormann FH, Lugo AE, Bowden RD (1991) For Ecol Manage 46:1–21. **Went FW**, Stark N (1968) Proc Natl Acad Sci 60:497–504.

Yamakura T, Sahunalu P (1990) J Trop Ecol 6:371–378. **Yavitt JB**, Wieder RK (1988) Biotropica 20:2–10.

Professor Dr. Ernesto Medina
Dr. Elvira Cuevas
Centro de Ecología
Instituto Venezolano de
Investigaciones Científicas
Aptdo 21827, Caracas 1020-A,
Venezuela

V. *Geosiphon pyriforme* (Kützing) von Wettstein, a Promising System for Studying Endocyanoses

By Manfred Kluge, Dieter Mollenhauer, and Resi Mollenhauer

1. Introduction

Endosymbiotic consortia are fascinating from several aspects. For instance, they are excellent systems for studying the mechanisms behind cell-to-cell recognition, they open promising possibilities of analyzing metabolic and genetic exchange between the partners, and they give insight into the principles of evolution leading to the organelles of the eukaryotic cell. In context with the evolution of the photoautotrophic eucytes, endosymbioses with cyanobacteria acting as one of the partners (i.e., endocyanoses) are particularly interesting. It is a widely accepted view that the plastids have evolved from endosymbiotic cyanobacteria (Sitte and Eschbach 1992; Sitte et al. 1992). Thus, comparative studies carried out on endocyanoses are not only relevant for research on evolution and on symbiosis per se, they can also contribute substantially to deeper insights into the mechanism of photosynthesis, in particular with respect to the question of how the plastids interact with the rest of the cell. Moreover, the ecological conditions for meeting and coexistence of symbiotic partners are also of interest because a better knowledge in this field could help in understanding the environmental prerequisites which have led to the evolution of different types of eukaryotic cells.

In the past, studies on endocyanoses were carried out mainly with *Cyanophora paradoxa* and various *Glaucocystis* species (for review see, e.g., Geitler 1959; Reisser 1984; Bohnert and Löffelhardt 1992; Kies 1992), and only recently has *Geosiphon pyriforme* been introduced as an experimental system (Mollenhauer 1988; Kluge et al. 1991, 1992). This became possible with the development of reliable methods to cultivate the organism in quantities and qualities sufficient for experimental work (Mollenhauer and Mollenhauer 1988), although cultivation of *Geosiphon* still remains difficult. We are convinced that *Geosiphon* will become a very promising system for future studies on endosymbioses; thus it may be useful to review here the present knowledge on the organism, in particular with respect to the more recent findings on its structure, biochemistry, and physiology. An earlier review on *Geosiphon* by Mollenhauer (1992) dealt mainly with classical phycological aspects.

Progress in Botany, Vol. 55

2. The Organism

Geosiphon represents one of the classic cases of syncyanosis (endocyanosis) as defined by Pascher (1929). The symbiotic nature of the organism was first recognized by the founder of the series *Fortschritte der Botanik* (now *Progress in Botany*), Fritz von Wettstein (1915), who identified *Geosiphon* as a consortium of a fungus with a cyanobacterium. To the present, to our knowledge, *Geosiphon* is the only known example of this kind of endocyanosis.

The taxonomy of the macrosymbiont (host), the fungus, is still under discussion. Presumably the fungus is a phycomycete (Endogonaceae), and there are several reasons for the assumption that it is closely related to the genus *Glomus* (Knapp 1933; Mollenhauer 1988, 1992). However, this assumption remains to be verified.

The endosymbiont belongs to the common genus *Nostoc*. According to Mollenhauer (1988; 1992), it has characters of *N. punctiforme*, but the taxonomic attribution is still under discussion (for details of *Nostoc* taxonomy see Mollenhauer et al. 1993). Furthermore, there is evidence that different *Nostoc* strains occur as endosymbionts in *Geosiphon*. *Nostoc* and the fungus live together on the surface and in the uppermost layers of damp loamy soil, where the latter forms a plagiotropic system of branching and anastomosing nonseptate hyphae (fungal coenocyte). The tips of the growing hyphae incorporate *Nostoc* filament primordia by endocytosis, as will be discussed later in more detail (see Sect. 4. a). Upon endocytosis, the hypha of the fungus forms a typical aboveground, pear-shaped, and multinucleate bladder where the incorporated *Nostoc* filament grows and multiplies (Fig. 1). It should be kept in mind that each *Nostoc*-containing vesicle originates from an individual act of endocytosis. In consequence, different bladders formed by the hyphal systems of the fungus may contain different strains of *Nostoc*. Making use of this peculiarity, we succeeded in producing defined populations of bladders different in the *Nostoc* strains they accommodate. It is obvious that this system opens promising possibilities for cell-biological and metabolic studies.

In nature, *Geosiphon* lives together with a typical, well-defined associaton of mosses, the liverwort *Blasia* and the hornwort *Anthoceros*, as described in detail by Mollenhauer (1992). *Blasia* and *Anthoceros* also accommodate symbiotic *Nostoc* (e.g., Werner 1992). Observations in the field suggest that *Nostoc* filaments from these bryophytes can be incorporated by *Geosiphon*. There is also evidence that *Nostoc* akinetes liberated from senescent, thus decaying *Geosiphon* bladders can continue to live as free soil algae which then can easily invade the slime cavities of the bryophytes. In addition to these interrelations between several hosts with the cyanobacterial symbiont, *Geosiphon* and the associated bryophytes may show interactions. For instance, we have observed that exudates from the bryophytes stimulate the growth of *Geosiphon*. Similar effects on the growth of fungi have been reported by other authors (e.g., Henriksson 1978; Meeks 1990; Schmidt 1991; Grilli Caiola 1992). Further research on the interactions between *Geosiphon* and bryophytes is in progress.

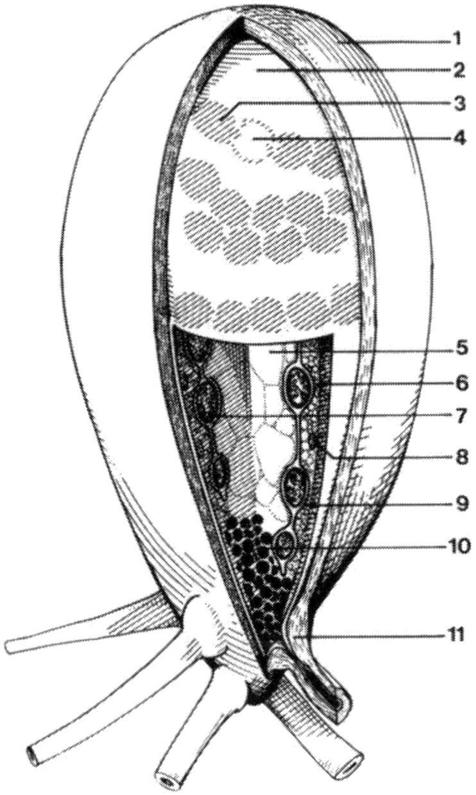

Fig. 1. Organization of a *Geosiphon* bladder (semischematic) *1* Cell wall of the bladder; *2* plasmalemma of the fungus (outside view); *3* endocytotic *Nostoc* seen through the plasmalemma and outer layer of the fungal cytoplasm; *4* heterocyst in a *Nostoc* filament; *5* central fungal cytoplasm with large vacuoles; *6* peripheral fungal cytoplasm with small vacuoles; *7* section of a vegetative *Nostoc* cell; *8* mitochondrion; *9* pericyanobacterial membrane; *10* glycogen granules, lipid droplets, microbodies, and other inclusions; *11* hypha of the mature bladder with narrowed lumen. The function of this constriction is unknown

3. Cytological Structure of the *Geosiphon* Coenocyte

The ultrastructure of the *Nostoc*-accommodating *Geosiphon* vesicles was first investigated by Schnepf (1964). His fundamental study gave rise to Schnepf's theorem explaining the principles of cell compartmentation and providing strong arguments in favor of the endosymbiotic nature of plastids. Recently, new ultrastructural studies by I. Steinmetz and A. Schüßler in the laboratory of E. Schnepf have considerably extended the knowledge on the ultrastructure of *Geosiphon* (data not published), even if it is not yet understood in detail. Figure 1 shows a scheme of the bladder structure summarizing the present state of knowledge.

Towards the environment, the bladder is bordered by an elastic, glossy, and transparent wall, leaving the *Nostoc* filaments inside the bladder clearly visible in the microscope. The wall of the bladder must be quite stable, because it has to sustain a tugor in the bladder of approximately 6 bar (M. Kluge, unpubl.).

The cyanobacteria living in the bladder retain their normal cell wall. They are embedded in a thin sheath of an apparently nonstructured matrix, presumably consisting of *Nostoc*-derived mucilage. Towards the cytoplasm of the fungus, the mucilage sheath is marked off without a gap by a biomembrane most likely produced by the host, and being presumably homologous to its plasmalemma. Thus, the whole structure consists of a long tube bordered by the latter membrane and filled with the *Nostoc* cells and the mucilage surrounding them. Actually, the structure represents nothing else than a huge, sausage-like phagosome. This phagosome is arranged to form a hollow half-sphere casing the central plasma of the bladder. The construction principle of this arrangement is analogous to a half-open cocoon woven by a silkworm. In this model the silk thread stands for the phagosomal tube.

The central cytoplasm of the bladder is kept free of *Nostoc* as it holds true for its basal part. It is important to note that the *Nostoc* cells inside the *Geosiphon* bladder have about the tenfold volume of free-living cells. The mechanism of this enlargement is still unknown, but the phenomenon recalls the morphological changes in *Rhizobium* leading to the formation of bacterioids once the bacteria are incorporated into the cells of the legume nodules (see, e.g., Werner 1992).

Inside a bladder, three structurally different areas of fungal cytoplasm can be distinguished. That is, between the wall of the bladder and the *Nostoc*-containing phagosomal tube, the cytoplasm forms a relatively thin layer with numerous small vacuoles. Also the central cytoplasm inside the cap formed by the *Nostoc*-containing phagosomal structure is highly vacuolated, but here the vacuoles are much larger. In light microscopy the cytoplasm at the basis appears opaque and not translucent as is the case in the upper part of the bladder. The opaque aspect of the much less vacuolated cytoplasm results from its densely packed contents of lipid droplets, glycogen inclusion, and peroxisomes. This part also contains numerous mitochondria. One has the impression that the basal part of the bladder is metabolically particularly active.

4. Establishment of the Symbiosis, Ontogeny, and Senescence of the Coenocyte

a) Life Cycle of *Nostoc*

To explain the process leading to the establishment of the *Geosiphon* symbiosis and controlling its ontogeny, it is necessary to give some information on the endosymbiont itself.

In addition to their role as oxyphotosynthesizers, further special properties (N_2 fixation, heterotrophic growth) enable cyanobacteria to act as productive partners in various symbiotic associations (cf. Stewart 1978; Stewart et al. 1983; Meeks 1990). Moreover, some minor peculiarities of the life cycles of *Nostoc* species are involved in the wide variety of interactions of these cyanobacteria with different macrosymbiotic hosts.

In general, life cycles in the genus *Nostoc* (cf. Komarek and Anagnostidis 1989) comprise three stages clearly differing from each other, namely vegetative filaments, hormogonia, and filament primordia. The vegetative filaments ("trichomes") represent uniserial, immobile chains of cells embedded in a common gelatinous matrix, thus forming a structure called the thallus. Since the thallus represents the active biomass-producing form which grows to maturity, it represents the dominating stage in the life cycle of *Nostoc*.

In the filaments of the thallus the cells grow and divide, and some of them differentiate into intercalary heterocysts lacking photosystem II. Insertion of these special N_2-fixing cells follows a strain-specific basis pattern and is triggered by the level of combined nitrogen in the environment. Another differentiation of the vegetative cell is the so-called akinete, a thick-walled spore which is formed under unfavorable conditions and which eventually gives rise to a new filament. Akinetes, too, are inserted into the filament according to a genus-specific pattern.

If thalli are mature (a status not yet defined physiologically, cf. Campbell and Meeks 1992) they can form hormogonia, i.e., fragments of filaments which are liberated from the thallus and move over the substrate by gliding on hormogonium-derived mucilage. Hormogonia are a means of propagation and have only one type of cells, thus no loger possess heterocysts. The hormogonial cells may divide but do not differentiate further. Cells in hormogonia differ from vegetative cells with respect to the light sensitivity of photosynthesis (Bilger et al. 1993) and possibly also in other physiological properties.

The hormogonia are embedded in a highly fluid slime which is presumably the precondition for the motility of this stage and therefore named "motility slime", whereas the immotile heterocysts possessing vegetative filaments have an envelope of solidified, gelatinous slime ("envelope mucilage").

Hormogonium formation in general and their motility in particular are light-dependent, as holds true for the whole developmental cycle. That is, *Nostoc* hormogonia move only during the light phase. This fact is highly relevant with respect to the incorporation of hormogonia by *Geosiphon*.

Motility of hormogonia requires liquid water and a solid substrate to glide on. Therefore, terrestrial *Nostoc* species form hormogonia only in wet weather. As is true for the development of the fungal hyphae, motion of the hormogonia is stimulated and/or regulated by bioactivators such as IAA and similar substances diffusing from a source (for instance from mosses of the accompanying vegetation) into the aqueous medium. Thus, since both the growth of the fungal hyphae and of the *Nostoc* hormogonia depend on liquid water and bioactivators, the potential partners of the symbiosis can grow together on the substrate and finally establish the symbiotic consortium only under sufficient humidity.

The motility of the *Nostoc* hormogonia is essential to bring fungus and cyanbacterium into contact. It is, however, not the hormogonium itself which is finally incorporated by the host. Rather, cytosis of the cyanobacterium by the fungus is only accomplished with so-called filament primordia, a transient stage on the way from hormogonia to the formation of the vegetative filaments which develops from a hormogonium once the latter becomes sessile. The formation of filament primordia is initiated by transformation of terminal cells of hormogonium into heterocysts. Then the intercalary cells synthesize a new gelatinous sheath and, while doing so, enlarge considerably; however, without dividing.

The filament primordium is transformed into vegetative filaments thus closing the developmental cycle by initiating cytokinesis, which stepwise brings about regular cell shapes and sizes. Intercalary heterocytes are differentiated and finally the cell volume is reduced to "normal". Completion of transformation into vegetative material is indicated by establishing a uniform growth pattern common to all cells: they increase in length up to their "critical cell length" and then divide. This results in a statistic normal distribution of cell length between two extreme values, of which the minor one is exactly half of the larger.

When discussing the special predisposition of cyanobacteria for associations and symbioses, Paerl (1992) stresses the importance of a possibly incomplete enzymatic regulation of metabolic processes which may be the reason that particularly the developmental transient stages of *Nostoc* (hormogonia, filament primordia) release many different metabolites in often substantial amounts into the environment. Such release could be the reason that these states are especially attractive for interaction with prospective symbiotic partners.

If all the pecularities described above are considered in context, it becomes obvious that both the meeting of the prospective symbionts and the process of incorporation of the endosymbiont by the fungus depend strongly on the special features of the life cycle of *Nostoc* which are, among cyanobacteria, unique for the genus

b) The Cystosis of *Nostoc* by the Fungus

As already stressed above, neither hormogonia nor vegetative filaments, but only typical filament primordia can be incorporated by cytosis into the fungal hyphae. Certain observations suggest that enzymes active in formation of the gelatinous matter surrounding the *Nostoc* cells may have some effect on the growing tip of the *Geosiphon* hypha. However, this assumption requires further experimental support.

Once hormogonia have come into contact with a hypha, they tend to assemble at its apical and subapical region. Presumably this area of the hypha is a site of high enzymatic and metabolic activity, and in particular is susceptible to contacts with other cells. It is in this region of the hyphae that cytotic uptake of *Nostoc* by the fungus occurs.

Cyto-cinematographic time-lapse studies enabled us to follow details of cytosis in *Geosiphon*. Incorporation of *Nostoc* is always preceded by arresting a previously motile hormogonium onto the surface of the apical or subapical region of a fungal hypha by the onset of darkness, i.e., under natural photoperiod by the onset of night. This step leads to the transformation of the arrested hormogonium into a filament promordium. Thereafter the following typical sequence of events can be regularly observed, but the mechanisms behind it are not yet understood:

Close to the immobile *Nostoc* primordium the wall of the hypha opens and a portion of the fungal cytoplasm is released. The naked fungal cytoplasm flows round a heterocyst-free part of the *Nostoc* primordium, while in the neighborhood new portions of cytoplasm are released from the hyphae, repeating the whole process until the primordium is incorporated. During this process, the heterocysts are cut off from the primordium by a yet unknown mechanism and left outside the released fungal cytoplasm, so that only heterocyst-free sections of the filament primordium undergo cytosis. Obviously, immediately after release from the hypha and incorporation of the *Nostoc* primordium, the naked fungal cytoplasm begins to regenerate a cell wall. As a consequence of the whole incorporation procedure, the shape of the hyphal tip with the included *Nostoc* filament resembles a small raspberry. While enlarging, the beginning irregular outer contour is smoothed, thus resulting in an at first spherical and finally pear-shaped body, i.e., in the *Geosiphon* bladder shown in Fig. 1.

Shortly after cytosis, the pigmentation of the incorporated *Nostoc* cells bleach reversibly, their habitus shows shrinkage and generally does not look healthy. The bleaching seems to concern mainly the biliproteins, since autofluorescence of the chlorophyll is not changed. Details are under investigation which, however, are made difficult because in very young *Geosiphon* bladders a foamy mass of highly vacuolated cytoplasm covers the cyanobacterial cells, thus hiding them when examined light microscopically. It is conceivable that the initially obvious negative effects of the fungus on the incorporated *Nostoc* are the reminiscence of digestion processes to which particles taken up by a cell via phagocytosis are normally subjected.

c) Maturation of the Conenocyte

After the termination of the endocytotic step, the *Geosiphon* bladder enters a juvenile phase. This is characterized by the reversion of the stress symptoms and the onset of growth and multiplication in the cells of the incorporated *Nostoc* primordium. Simultaneously, a vivid influx of cytoplasm into the young bladder via the nursing hypha begins. The importet plasma derives from older bladders or from other parts of the hyphal network and contains lipid droplets, microbodies, glygogen granules, and mitochondria.

The incorporated cyanobacterial filament first forms a central tangle. This is loosened when *Nostoc* starts to grow. The final arrangement, as shown in Fig. 1, is accomplished by massive elongation of the cyanobacterial filament in numerous, often parallel orientated contours surrounding the central vacuolated host cytoplasm. Presumably by this growth process, much of the lipid matter imported into the bladder is metabolized, since the at first cloudy basal part of the growing *Geosiphon* bladder becomes more translucent.

As already mentioned, only heterocyst-free sections of *Nostoc* primordia are incorporated by the fungus. However, while maturation of the bladder proceeds, the incorporated multiplying endosymbiont develops new heterocysts. As holds true for the vegetative cells of the endosymbiont, also its heterocysts have a larger volume compared with *Nostoc* outside the bladder. Such enlargement of heterocysts is known also from *Nostoc* in other endosymbioses and from those living as cyanobionts in lichen thalli. However, whereas in symbiotic associations with photoautotrophic hosts, the frequency ratio between heterocysts and vegetative cells of *Nostoc* increases considerably upon incorporation, this does not hold true for *Geosiphon*: there the frequency ratio is the same as in *Nostoc* inside and outside the bladder, respectively.

A fully mature *Geosiphon* bladder may live and remain metabolically active for several months, but after several weeks the bladders lose the initially present capability to form new fungal hyphyae spontaneously. However, if growth stimulators are supplied (for instance those from bryophytes accompanying *Geosiphon* naturally), even these senescent bladders can be reactivated, thus regaining the capability to form young, growing hyphae which can incorporate fresh *Nostoc* symbionts (D. and R. Mollenhauer, unpubl.).

d) Specificity of *Nostoc* as Endosymbiont in *Geosiphon*

As far as we can see from current investigations, the cyanobacterial endosymbiont of *Geosiphon* is specific only in a rather wide sense (cf. also Reisser 1983). Preliminary studies in vitro have shown that nearly all the *Nostoc* strains cultivated in our laboratory and tested so far can be incorporated by the fungus, but only some of them are suitable to provoke the formation of a mature bladder and to establish a stable symbiotic consortium. Still much work is required, but we have the impression that *Nostoc* strains isolated from other symbiotic systems (e.g., cycads, *Gunnera*, liverworts, hornworts) are particularly suitable as symbiontic partners in *Geosiphon*.

As already mentioned, in the field, *Geosiphon* regularly coexists with certain bryophytes (*Anthoceros*, *Blasia*) which also accommodate *Nostoc* in their thalli as symbiontic partner (cf. Grilli Caiola 1992; Werner 1992). There is evidence that in the field, *Geosiphon* may exchange *Nostoc* with the mentioned bryophytes of the accompanying cryptogamic flora (Mollenhauer, unpubl.). This phenomenon opens promising perspectives for studies on partner relations in a quite complicated symbiotic network. Presumably this network is even more complex, since it might be true that in the field *Geosiphon* is linked not only with the accompanying bryophytes via the symbiontic *Nostoc* but also via the fungus which penetrates the thallus of *Anthoceros*, thus forming there a kind of mycorrhiza.

5. Culturing of *Geosiphon*

At present, the only known extant population of *Geosiphon* exists in the Spessart Mountains (east of Frankfurt; FRG), near the field station Lochmühle of the Senckenberg Museum Frankfurt am Main. The plant material cultivated and studied by the authors originates from this population. A detailed description of the site and its geographic and geologic characterization is given by Mollenhauer (1992). We have undertaken many excursions to former sampling sites described in the older literature and have forwarded many research requests to field bryologists and mycologists to look for other *Geosiphon* populations. So far, all these attempts have remained unsuccessful, and at present the population in the Spessart is worldwide the only one which we can use.

The techniques to culture *Geosiphon* have been considerably improved during the past 5 years and were already described previously in detail (Mollenhauer 1992). The problem how to obtain axenic cultures of the fungal host remains to be solved because, aiming to understand the biological significance of the symbiosis for the two partners, it is necessary to find out whether or not *Geosiphon* may also exist without symbiotic *Nostoc* and to learn more about the ecological requirements of the fungus. At present, the only criterion to identify *Geosiphon* is the formation of the *Nostoc*-containing bladders. Without these bladders there is no way to distinguish *Geosiphon* from other coenocytic soil fungi.

Fortunately, the developmental cycle of *Geosiphon* includes one stage without *Nostoc*. This is represented by a white, globular, and thick-walled structure functioning as spore. Its formation, and even more its germination, are difficult to manipulate experimentally, but we hope that in future work the white globules might provide a suitable starting point to establish axenic *Geosiphon* cultures.

At present, the most reliable method is to culture *Geosiphon* in the laboratory on soil as follows. Not too old *Geosiphon* bladders obtained from the field or from older cultures are placed on sterilized soil and are allowed to develop new fungal hyphae. These are brought in contact with *Nostoc* which has been precultured to obtain the filament primordia required to establish the symbiosis (see Sect. 4. b). In *Nostoc*, the readiness to produce filament primordia differs strongly from strain, and depends also largely on the conditions under which the cultured *Nostoc* material was previously held. In general, *Nostoc* deriving directly from stock cultures, thus being in a vegetative state, needs some weeks for full activation.

Once the formation of *Geosiphon* bladders has begun, the not incorporated cyanobacteria must be removed from the culture, either mechanically or by addition of lytic enzymes, otherwise they tend to overgrow and thus kill the developing young *Geosiphon* bladders.

Recently, we were able to replace the soil substrate in the cultures by agar-solidified media and even by liquid growth medium and to improve the techniques to obtain larger amounts of bladders.

Considering the quite complicated culture procedures, the differences in the developmental behavior among the various strains of *Nostoc* (see below), and finally the fact that the whole development of *Nostoc* and *Geosiphon* hyphae and bladders are obviously overlaid by strong biological rhythmicity, it is understandable that reproduceable culturing of *Geosiphon* requires much personal experience, is still difficult, and limits extended biochemical and physiological research on the organism.

6. Metabolism of *Geosiphon*

Metabolic studies on *Geosiphon* depend on the availability of sufficient amounts of this organism. In spite of considerable progress in the technique to cultivate it, this is still a problem. Thus, all results reported here were obtained with micromethods requiring only about 10 to 20 bladders per test.

Tracer studies revealed that *Geosiphon* bladders are capable of photoautotrophic carbon acquisition (Kluge et al. 1991). Also dark CO_2 fixation was observed; however, admittedly at much lower rates than in light. The bladders can use free CO_2 as well as bicarbonate as photosynthetic substrate.

$^{14}CO_2$ fixation in light labeled many compunds, mainly products of the reductive pentose phosphate cycle, amino acids, and oligosaccharides, among them raffinose and trehalose. These latter sugars are known to occur as early photosynthetic products, generally in blue-green algae (Feige 1976; Smith 1982).

$^{14}CO_2$ fixation in light, but never in darkness, led quickly to a heavy labeling of insoluble material, presumably of polyglucan. Thus, it is conceivable that the glycogen granula observed in the basal part of the *Geosiphon* bladder derive from photosynthesis by the endosymbiont.

Bilger et al. (1993), by measuring the chlorophyll fluorescence with a PAM fluorometer, found that in both free-living *Nostoc* cells and cells of the same strain enclosed in *Geosiphon*, the rate of photosynthetic electron transport clearly followed different light-saturation curves. The endosymbiotically living *Nostoc* required 200 µmol quanta $m^{-2} s^{-1}$ to reach saturation instead of 100 µmol $m^{-2} s^{-1}$ typical for free-living *Nostoc*. Also the photosynthetic rate at light saturation was nearly twice as high in the endosymbionts compared with the free-living cells. Finally, compared with the free *Nostoc*, the photosynthetic apparatus of the endosymbionts was able to tolerate considerably higher light intensities before photoinhibition occurred. The reason for these differences between the free and symbiotic *Nostoc* in the photosynthetic behavior are not yet clear. A plausible explanation could be that photosynthesis inside the bladder might operate at higher CO_2 concentrations, allowing higher photosynthetic efficiency and better protection against excessive radiation energy. This hypothesis, however, requires further experimental support.

The occurrence of heterocysts in the endosymbiotic *Nostoc* already suggests that *Geosiphon* might be capable of nitrogen fixation. This assumption was further supported by Kluge et al. (1992), who found that *Geosiphon* possesses substantial nitrogenase activity.

A. Schüßler, in the laboratory of M. Kluge, in cooperation with C. Ullrich, found that *Geosiphon* bladders are a nearly ideal system to insert microelectrodes and thus to study the transport of molecules across the plasmalemma by measureing changes of membrane potentials. It was found (unpubl. data) that the *Geosiphon* bladders take up amino acids, in particular glutamate, from outside. Also maltose and glucose are taken up, the latter probably only if Na^+ ions are present. Moreover, uptake of nitrate and chloride was observed, but the transport system involved obviously requires substrate-mediated induction to become active.

Summing up, the results obtained until now with *Geosiphon* allow the conclusion that the organism can harvest carbon (and other nutrients) saprophytically by uptake from outside, as well as via photosynthesis mediated by the endosymbiotic *Nostoc*. It remains an open question whether or not the fungus can exist without its photoautotropic partner.

The modalities of matter exchange between the symbiotic partners inside a *Geosiphon* bladder are totally unknown. Preliminary tracer studies with free-living *Nostoc* isolated from *Geosiphon* show that the cells export large amounts of organic molecules, including amino acids, sugars, and phosphorylated compounds, into the surrounding medium (M. Kluge, unpubl. data). At least the spectrum of the released compounds seems to be influenced by the pH of the medium. Whether or not these results can be directly applied to the *Nostoc* cells in the bladders remains to be clarified.

For studies aiming at better understanding of the metabolic exchange between the host and endosymbiont, it would be very helpful to be able to separate the partners at a given time from each other quickly, for instance after short pulses of $^{14}CO_2$ fixation, in order to measure in the separated fractions the dynamics of changes in the pool sizes of (labeled) metabolites. This method, which has been practised effectively, for instance, with the endocyanelles of *Cyanophora paradoxa* (Bothe and Floener 1978), is for two reasons difficult to apply to *Geosiphon*. First, the method requires relatively large amounts of cells, which our culture methods cannot yet provide. Second, as described earlier, the structure of the phagosome in the bladder is so complex, and the pericyanobacterial membrane sometimes adheres so tightly to the *Nostoc* cells and their mucilage sheath that it would be very difficult to obtain a clean separation of the partners without contamination of the individual fracion by foreign material. Considering these difficulties, the application of electrophysiological methods using microelectrodes to individual *Geosiphon* bladders seems to be a more reliable method to learn about transport and metabolic exchange betwen the symbiotic partners.

7. Conclusion

Most of our recent knowledge on endocyanoses has been derived from studies on *Cyanophora paradoxa* and *Glaucocystis nostochinearum* (Kremer et al. 1979; Reisser 1984; Schlichting et al. 1990; Bohnert and Löffelhardt 1992). We have, however, to keep in mind that these organisms represent a quite advanced form of endosysmbiosis, if not to say the end point in the scale of possible relation between the partners. In *Cyanophora* and *Glaucocystis*, the endosymbionts have given up their individual lives. They are structurally greatly modified, have delivered genetic information to the host, and depend metabolically to such an extent on the host that they are incapable of independent existence.

These are the reasons that the term "cyanelle" has been coined to characterize the type of endosymbiont that exists in *Cyanophora* and *Glaucocystis*. In discussions with researchers involved in the genomic aspects of symbiotic partner relations, we were repeatedly confronted with the opinion that the term endosymbiosis should be restricted to those cases where genomic information has been exchanged between the partners, thus *Geosiphon* should not be included in this category of symbiosis. Indeed, in *Geosiphon*, the endosymbiont can well exist independently of the host, and it is rather unlikely (although not proven) that there is genetic exchange between the partners when living together. However, in our opinion, there is a gradual rather than a principle difference between *Geosiphon* and organisms like *Cyanophora*. Whilst the latter system probably represents an end point in the evolution of endocyanoses, *Geosiphon* marks the initial step. Thus, the comparison of the partner relations in *Geosiphon* with the more advanced forms of endocyanoses is a useful and promising task. It could contribute considerably to the understanding of the evolution culminating the photoautotrophic eukaryotic cell.

References

Bilger W, Büdel B, Mollenhauer D, Mollenhauer R (1993) J Phycol (in press). **Bohnert HJ**, Löffelhardt W (1992) In: Reisser W (ed) Algae and symbioses. Biopress, Bristol, pp 379–397. **Bothe H**, Floener L (1978) Z Naturforsch 33c:981–987.

Campbell EL, Meeks JC (1992) J Gen Microbiol 138:473–480.

Feige GB (1976) Z Pflanzenphysiol 80:386–394.

Geitler L (1959) In: Ruhland W (ed) Handbuch der Pflanzenphysiologie, vol 11. Springer, Berlin Göttingen Heidelberg, pp 530–545. **Grilli Caiola M** (1992) In: Reisser W (ed) Algae and symbioses. Biopress, Bristol, pp 231–254.

Henriksson E (1978) In: Singh SP, Tiwari DN, Kashyap AK, Yadava PK (eds) Advances in cyanophyte research. Varanase, pp 7–13.

Kies L (1992) In: Reisser W (ed) Algae and Symbioses. Biopress, Bristol, pp 353–377. **Kluge M**, Mollenhauer D, Mollenhauer R (1991) Planta 185:311–315. **Kluge M**, Mollenhauer D, Mollenhauer R, Kape R (1992) Bot Acta 105:343–344. **Knapp E** (1933) Ber Dtsch Bot Ges 51:210–216. **Komarek J**, Anagnostidis K (1989) Arch Hydrobiol 82,3 = Algol Stud 56:247–345. **Kremer BP**, Kies L, Rostami-Babet M (1979) Z Pflanzenphysiol 92:303–317.

Meeks JC (1990) In: Rai AN (ed) CRC Handbook of symbiotic Cyanobacteria. CRC, Boca Raton, pp 43–63. **Mollenhauer D** (1988) Natur und Museum, Frankfurt/M. 118:289–309. **Mollenhauer D** (1992) In: Reisser W (ed) Algae and Symbioses. Biopress, Bristol, pp 339–351. **Mollenhauer D**, Mollenhauer R (1988) Endocyt C Res 5:69–73. **Mollenhauer D**, Büdel B, Mollenhauer R (1993) Algol Stud (in press).

Paerl HW (1992) In: Reisser W (ed) Algae and Symbioses. Biopress, Bristol, pp537–565. **Pascher A** (1929) Jahrb Wiss Bot 71:386–462.

Reisser W (1984) In: Linskens HF, Heslop-Harrison J (eds) Encyclopedia of plant physiology, vol 17. Springer, Berlin Heidelberg New York, pp 91–112.

Schlichting R, Zimmer W, Bothe H (1990) Bot Acta 103:392–398. **Schmidt H** (1991) Diss Bot 171:201. **Schnepf E** (1964) Arch Mikrobiol 49:112–131. **Sitte P**, Eschbach S (1992) In: Behnke H-D, Esser K, Kubitzki K, Runge M, Ziegler H (eds) Progress in Botany, vol 53. Springer, Berlin Heidelberg New York, pp 29–43. **Sitte P**, Eschbach S, Maerz M (1992) In: Reisser W (ed) Algae and Symbioses. Biopress, Bristol, pp 711–733. **Smith AJ** (1982) In: Carr NG, Whitton BA (eds) The biology of Cyanobacteria. Botanical Monographs, vol 19. Blackwell, Oxford, pp 47–117. **Stewart WDP** (1978) Endeavour 2:170–179. **Stewart WDP**, Rowell P, Rai AN (1983) Ann Microbiol (Inst Pasteur) 134B:205–228.

von Wettstein F (1915) Österr Bot Z 65:145–156. **Werner D** (1992) Symbiosis of plants and microbes. Chapman & Hall, London.

Professor Dr. Manfred Kluge
Institut für Botanik der Technischen Hochschule
Schnittspannstraße 10
D-64287 Darmstadt

Dr. Dieter Mollenhauer
Resi Mollenhauer
Forschungsinstitut Senckenberg
Außenstelle Lochmühle
D-63599 Biebergemünd

VI. Significance of Glutathione Metabolism in Plants Under Stress

By Heinz Rennenberg and Christian Brunold

1. Introduction

In higher plants the tripeptide glutathione (GSH; γ-glu-cys-gly) and its homologs homoglutathione (hGSH; γ-glu-cys-β-ala) and hydroxymethylglutathione (γ-glu-cys-ser) are generally thought to be the most abundant low molecular weight thiols (Kasai and Larsen 1980; Bergmann and Rennenberg 1993). As products of the plant's primary metabolism, these compounds have received considerable attention during recent years, because they are not only involved in storage and distribution of reduced sulfur within the plant, and hence in the regulation of sulfur nutrition, but are also essential components of the plant's defence system for environmental stress (Fig. 1).

As a consequence, the glutathione concentration of plant cells is determined by nutritional and environmental factors as well. The interaction of these factors is presently not understood, but results in large variations in glutathione levels between organs (e.g., Klapheck 1988), during the day (e.g., Schupp and Rennenberg 1988) or during the year (e.g., Schupp and Rennenberg 1992), and in different environments (e.g., Polle and Rennen-

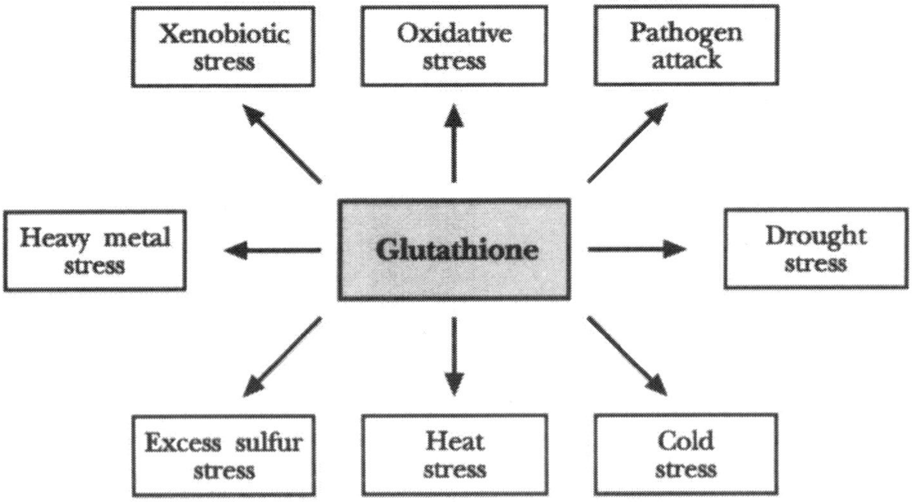

Fig. 1. Stress situations in which glutathione is involved. In this review the contribution of glutathione in the syndroms of reactions during heavy metal stress, xenobiotic stress, oxidative stress, and pathogen attack is discussed

Progress in Botany, Vol. 55

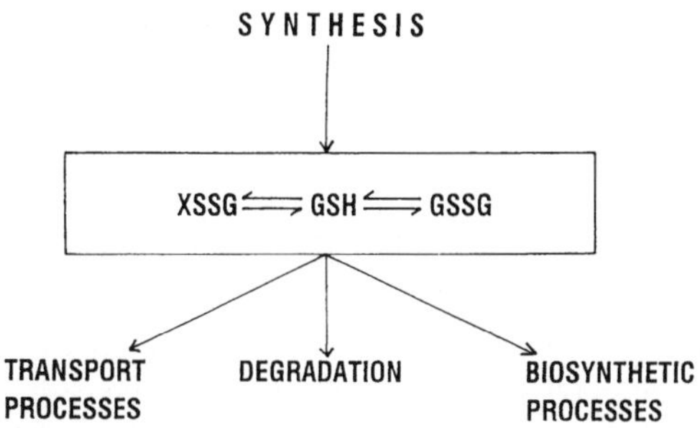

Fig. 2. Processes that modulate the concentration of glutathione. *XSSG* Glutatione mixed disulfides; *GSH* reduced glutathione; *GSSG* oxidized glutathione. (Rennenberg and Lamoureux 1990)

berg 1992). The actual glutathione content found inside plant cells may be the result of glutathione synthesis and degradation, its import and export by membrane and long-distance transport processes, and its use for biosynthesis processes, such as phytochelatin synthesis and conjugation with xenobiotics (Rennenberg and Lamoureux 1990). In addition, an equilibrium is maintained in plant cells between reduced glutathione (GSH), oxidized glutathione (GSSG), and mixed glutathione disulfides (XSSG) by the action of GSSG- and XSSG-reductases (Fig. 2). Although the contribution of GSSG to the overall amount of glutathione in plant cells has only rarely been determined, it appears that more than 95% of the glutathione is usually present in the form of GSH. However, under stressful conditions, GSSG may accumulate (cf. Smith et al. 1990).

Stressful conditions will interact with this complex system of metabolic processes at different levels. Therefore, this review summarizes the present understanding of glutathione synthesis and degradation, before the significance of glutathione metabolism in the compensation of stress from heavy metals, xenobiotics, oxidants, and pathogen attack is discussed. A more detailed view of glutathione synthesis and degradation in plants is given in a recent publication by Bergmann and Rennenberg (1993).

2. Glutathione Synthesis and Degradation

a) Glutathione Synthesis

Synthesis of glutathione proceeds in plant cells in the same two-step procedure previously described for animal cells (Bergmann and Rennenberg 1993; Meister and Anderson 1983). In the first reaction (1), γ-glutamylcysteine is formed from glutamate and cysteine in an ATP-dependent process catalyzed by γ-glutamylcysteine synthetase (EC 6.3.2.2). The enzyme shows low affinity for glutamate, high affinity for cysteine, and feedback inhibition by mM GSH concentrations. Glutathione is formed in the second reaction (2) from γ-glutamylcysteine and glycine. This reaction is also dependent on ATP and is catalyzed by glutathione synthetase (EC 6.3.2.3). From the apparent K_M values and the cellular concentrations of the two substrates it appears that the reaction may be limited in vivo by the availability of γ-glutamylcysteine. Apparently, glutathione synthetase is specific for glycine and does not use β-alanine as a substrate. Homoglutathione seems to be synthetized by a homoglutathione synthetase with high specifity for β-alanine (Bergmann and Rennenberg 1993).

$$\text{glu} + \text{cys} + \text{ATP} \rightarrow \gamma\text{-glu-cys} + \text{ADP} + P_i \quad (1)$$
$$\gamma\text{-glu-cys} + \text{gly} + \text{ATP} \rightarrow \gamma\text{-glu-cys-gly} + \text{ADP} + P_i \quad (2)$$

Both enzymes of glutathione synthesis were found in the chloroplasts and in the cytosol (Hell and Bergmann 1990; Bergmann and Rennenberg 1993). Recently, the enzymes of glutathione synthesis were also demonstrated in root tissue. Whereas γ-glutamylcysteine synthetase was almost equally distributed between proplastids and the cytoplasm, only minor amounts of glutathione synthetase activity were found in the proplastids (Rüegsegger and Brunold 1993).

b) Glutathione Degradation

Degradation of glutathione in plant cells is still a matter of debate (Bergmann and Rennenberg 1993). From a series of experiments with green tobacco cells, it was concluded that degradation of glutathione is initiated by a carboxypeptidase, removing the glycine moiety of glutathione. The resulting γ-glutamylcysteine may either be used for de novo glutathione synthesis or further degraded by γ-glutamylcyclotransferase to cysteine and 5-oxo-proline. Glutamate may be regenerated from 5-oxoproline by 5-oxo-prolinase activity (Fig. 3). Whereas glutathione specific carboxypeptidase activity has so far not been found in plant cells, γ-glutamylcyclotransferase and 5-oxoprolinase were found to be present in green tobacco cells. More recent studies revealed that cysteinylglycine may be an intermediate of glutathione degradation in *Glycine max, Hordeum vulgare* and heterotrophic tobacco cells. From these studies it was proposed that glutathione degradation may be initiated by the hydrolytic activity of γ-glutamyl transpeptidase; the cysteinylglycine produced in this reaction may be cleaved by a dipeptidase (Fig. 3). Both enzymes were found to be present in *G. max* and *H. vulgare*. Further experiments are required to test whether these conflicting results are the consequencee of species- and tissue-specific dif-

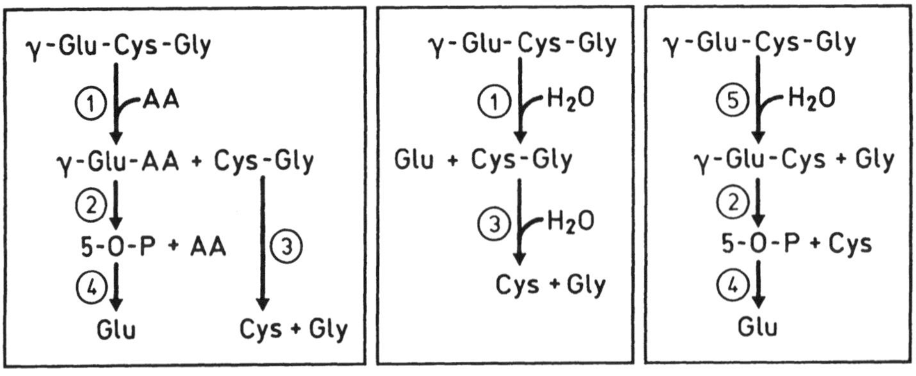

Fig. 3. Pathways of glutathione degradation. *1* γ-Glutamyl transpeptidase; *2* γ-glutamylcyclotransferase; *3* dipeptidase; *4* 5-oxo-prolinase; *5* carboxypeptidase. (Bergmann and Rennenberg 1993)

ferences in glutathione degradation. It must be pointed out that both pathways of glutathione degradation suggested in plant cells are different from the pathway of glutathione degradation established in animal tissue (Fig. 3; Meister 1981).

3. Glutathione and Heavy Metal Stress

a) Heavy Metal Binding Peptides

The reactions of plant cells confronted with excess heavy metals include binding to the cell wall, decreased transport through the plasma membrane, formation of metal complexes in the cytosol, transport of metals or metal complexes into the vacuole, and formation of metal complexes in the vacuole (Tomsett and Thurman 1988; Ernst et al. 1992; Wagner 1993). The reactions are aimed at protecting the plants from the toxic effects of heavy metals and at supplying sufficient amounts of essential heavy metals like Zn, Cu, Ni, and Co. Among the possible reactions between plant cells and heavy metals intracellular chelation involving cysteine-rich polypeptides has been most examined in recent years and several reviews have treated this field in detail (Grill et al. 1990; Rauser 1990, 1991, 1993; Robinson 1990; Steffens 1990; Jackson and Kuske 1993; Wagner 1993).

When exposed to heavy metals, higher plants, algae, and some fungi synthesize poly (γ-glutamylcysteinyl)-glycines. These peptides are also called phytochelatins (PCs), cadystins, γ-glutamylcysteinyl isopeptides or γ-EC isopeptides (Grill et al. 1990; Rauser 1990, 1991, 1993; Steffens 1990; Jackson and Kuske 1993; Wagner 1993). PCs are composed of 2 to 11 γ-glutamylcysteinyl-units and may contain acid-labile S^{2-} (Murasugi et al. 1983; Steffens et al. 1986; Reese and Winge 1988; Verkleij et al. 1990; Reese et al. 1992; Speiser et al. 1992), which may increase the metal binding of the peptides (Reese and Winge 1988).

The functions of PCs have not yet been firmly established. Because of their ability to bind heavy metal ions, PCs are considered to play a role in metal homeostasis and metal detoxification (Grill et al. 1990; Rauser 1990, 1993; Steffens 1990; Jackson et al. 1992). The function of PCs in metal homeostasis was recently substantiated by experiments in which enzymatically inactive, metal-requiring apoforms of diamino oxidase and of carbonic anhydrase were reactivated by Cu- and Zn-PC complexes, respectively (Thumann et al. 1991). Detoxification of Zn and Cd by PCs was shown in vitro with metal-sensitive plant enzymes which tolerated Cd and Zn in the form of a PC complex from 10 to 1000 fold the amount as compared with the free metal ion (Kneer and Zenk 1992). Consistent with the concept of the detoxification function of PCs, *Rauvolfia serpentina* suspension cells cultivated with Cd contained more than 97% of the heavy metal as PC complex and less than 3% associated with the high molecular weight fraction (Kneer and Zenk 1992).

Experiments using *Silene vulgaris* with naturally selected increased Cd or Cu tolerance (de Knecht et al. 1992; Schat and Kalff 1992) show that PCs are not involved in differential metal tolerance in this species. Sensitive plants had a higher Cd uptake than tolerant ones. Exclusion of Cd was not the mechanism of tolerance in tolerant tomato cell cultures obtained by stepwise selection (Gupta and Goldsbrough 1991). In this system the tolerant cell lines accumulated larger amounts of Cd and PCs, and produced higher molecular weight phytochelatins. These results show the important function of PCs for Cd tolerance in these systems. Obviously, plants have evolved several mechanisms for coping with excess heavy metals (Ernst et al. 1992) and different reactions may be used by different systems in dealing with the individual heavy metal. PC formation from glutathione seems to belong to these reactions.

b) Heavy Metals and Glutathione Metabolism

The in vitro formation of PCs from glutathione was first demonstrated using extracts from cell cultures of *Silene cucubalus* (Grill et al. 1989). This reaction is catalyzed by phytochelatin synthase, a γ-glutamylcysteine dipeptidyl transpeptidase and starts by the transfer of a γ-glutamylcysteine moiety of GSH to a second molecule of GSH, forming $(\gamma\text{-glutamylcysteine})_2$-glycine and glycine. Further γ-glutamyl-cysteine moieties can then be added to the growing PC chain. The in vitro reaction was confirmed by Hayashi et al. (1991) with an enzyme preparation from *Schizosaccharomyces pombe*. This preparation also catalyzed the formation of polymeres of γ-glutamylcysteine either by transferring the γ-glutamyl-cysteine moiety of GSH to $(\gamma\text{-glutamylcysteine})_n$ or by polymerizing γ-glutamylcysteine directly. Phytochelatin synthase is a constitutive enzyme which is dependent on Cd or other metals in extracts from *Silene cucubalus* (Grill et al. 1989), but active without Cd in extracts from *Schizosaccharomyces pombe* (Hayashi 1991).

Consistent with the hypothesis that glutathione metabolism plays an important role in PC biosynthesis, the concentration of glutathione declined in plant tissue exposed to Cd (Tuckendorf and Rauser 1990; Meuwly and Rauser 1992; Rüegsegger and Brunold 1992; Bergmann and Rennenberg 1993; Galli et al. 1993). In roots of copper-sensitive *Silene cucubalus* (de Vos et al. 1992), a decrease in glutathione induced by Cu was accompanied by an increase in lipid peroxidation products and a shift from reduced (GSH) to oxidized (GSSG) glutathione, indicating that the depletion of the antioxidant glutathione due to PC formation may induce oxidative stress. Contrary to glutathione, there was a dramatic increase in γ-glutamylcysteine in plant tissue exposed to Cd (Meuwly and Rauser 1992; Rüegsegger and Brunold 1992; Bergmann and Rennenberg 1993). This increase can be explained by the higher level of γ-glutamylcysteine synthetase activity, the higher contents of cysteine, the probably higher sensitivity of glutathione synthetase than γ-glutamylcysteine synthetase to Cd (Rüegsegger and Brunold 1992), and the release of GSH-mediated feedback inhibition of γ-glutamylcysteine synthetase (Bergmann and Rennenberg 1993).

The increase in activity of both enzymes of glutathione synthesis (Rüegsegger et al. 1990; Rüegsegger and Brunold 1992; Bergmann and Rennenberg 1993) during PC production indicates that the increased demand for glutathione is covered by an increased glutathione synthesis. Since higher amounts of cysteine are necessary for the increased glutathione production, changes in the pathway of assimilatory sulfate reduction leading to cysteine (Fig. 4) could be anticipated in plants treated with heavy metals.

An increased level of the extractable activity of both ATP sulfurylase and adenosine 5'-phosphosulfate sulfotransferase was measured in roots and leaves of maize seedlings

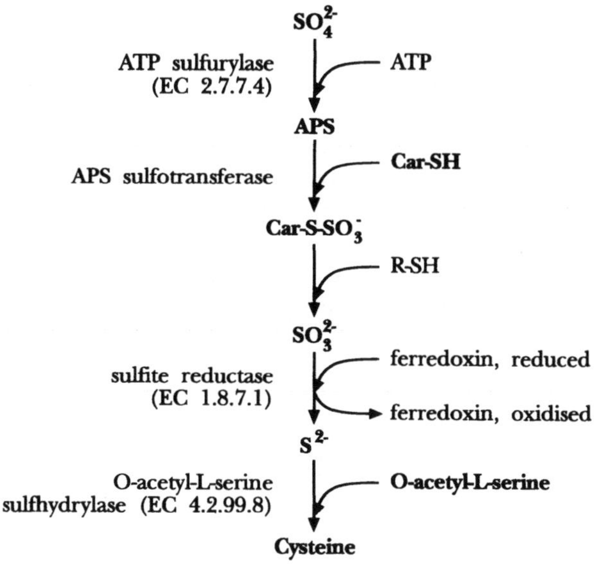

Fig. 4. Formation of cysteine from sulfate in the pathway of assimilatory sulfate reduction. Sulfate is activated to form APS (adenosine 5'-phophosulfate) via ATP sulfurylase. APS is used as sulfonyl donor for APS sulfotransferase which transfers the sulfur to an as yet unidentified carrier molecule (car-SH) to form car-S-SO_3^-. Sulfite (SO_3^{2-}) can then be liberated from the carrier by thiols (R-SH). Sulfite is reduced via sulfite reductase and the formed sulfide (S^{2-}) is incorporated into O-acetyl-L-serine to form cysteine. The reactions presented seem suitable for the discussion in the present review, but other sequences may also be relevant. (Brunold 1990, 1993; Schmidt 1992; Schiff et al. 1993)

cultivated in the presence of Cd. Simultaneously, the incorporation of label from $^{35}SO_4^{2-}$ into higher molecular weight compounds increased (Nussbaum et al. 1988). In pea roots an increase in extractable activity of adenosine 5'-phosphosulfate sulfotransferase (Fig. 4) was detected upon exposure to Cd (Rüegsegger et al. 1990). These effects of heavy metals on assimilatory sulfate reduction show that the increased need for cysteine used for glutathione synthesis is covered by up-regulation of the activities of the key enzymes of the pathway. The molecular basis of this effect is not clear. It seems, however, that glutathione has an important regulatory function (Rüegsegger and Brunold 1992).

Steffens (1990) reported that PC complexes from tomato cell cultures contained labile sulfite in addition to labile sulfide. This finding led Steffens (1990) to the hypothesis that the PC complexes could function as carrier molecules in assimilatory sulfate reduction (Fig. 4). According to the carrier concept of the adenosine 5'-phosphosulfate sulfotransferase reaction (Li and Schiff 1992; Schmidt 1992; Schiff et al. 1993; Fig. 4), almost any thiol will function as an acceptor for the sulfogroup of adenosine 5'-phosphosulfate (APS) in the formation of acid volatile radioactivity in vitro. Therefore, the fact that PCs were active in the in vitro assay of the enzyme (Brunold, unpubl. results) cannot be taken as evidence for the function of PCs as carrier molecules in assimilatory sulfate reduction. Since the adenosine 5'-phosphosulfate sulfotransferase-carrier complex recently described (Li and Schiff 1992) is stable in acid, whereas PC complexes are unstable, they do not seem to be closely related to the carrier molecule involved in assimilatory sulfate reduction (Schiff et al. 1993). This view is supported by the following findings: adenosine 5'-phosphosulfate sulfotransferase activity is localized exclusively or almost exclusively in chloroplasts (Frankhauser and Brunold 1978), whereas PCs were virtually exclusively detected in the vacuoles (Vögeli-Lange and Wagner 1990). PCs were not detected in tomato cell cultures before the addition of Cd (Steffens 1990), even though we can assume that sulfate assimilation was operating. From our present knowledge about the reciprocal regulation of sulfate and nitrate assimilation (Brunold 1993), the initial high levels of both sulfide and sulfite in PC complexes from tomato cells (Steffens 1990) may be explained by assuming that after addition of Cd, assimilatory sulfate reduction was up-regulated, while assimilatory nitrate reduction did not yet produce sufficient O-acetyl-L-serine for incorporating the additionally formed sulfite and sulfide into cysteine.

4. Glutathione-Mediated Conjugation of Xenobiotics

Plants use a series of reactions including oxidation, hydrolysis, conjugation, and reduction for detoxifying xenobiotics (Hatzios 1989; Komives and Dutka 1989; Lamoureux and Rusness 1989, 1993). Most studies on glutathione-mediated conjugation in higher plants used pesticides as substrates. The conjugation reaction involves glutathione and a xenobiotic and leads to a normally nontoxic conjugation product (Fig. 5). The reaction between both substrates may occur nonenzymatically (Lamoureux and Rusness 1989, 1993; Scarponi et al. 1991), but may be enhanced by glutathione-S-transferases (EC 2.5.1.18).

In most organisms studied, glutathione-S-transferases were present in multiple forms. Among other factors, the conjugation rate is dependent on glutathione-S-transferase activity, glutathione level, and type of xenobiotic (Scarponi et al. 1991; Lamoureux and Rusness 1993). The conjugation products may be subject to degradation (Lamoureux and Rusness 1989, 1993; Rennenberg and Lamoureux 1990) and may be actively transported into the vacuole (Martinoia et al. 1993). Glutathione-mediated conjugation has been re-

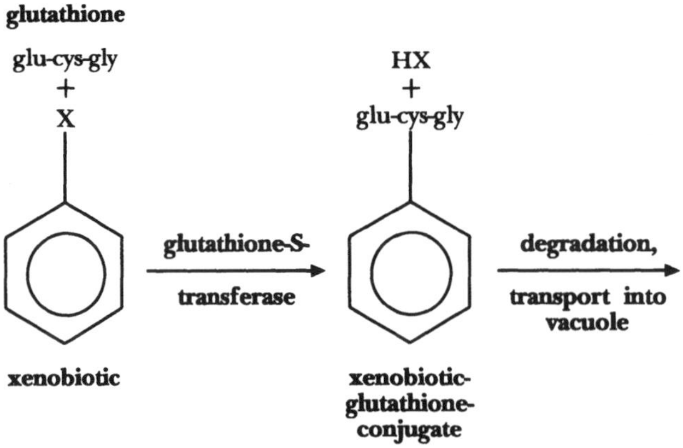

Fig. 5. Scheme of the conjugation of a xenobiotic using glutathione. The nucleophilic displacement of a halogen is presented, but nucleophilic additions have also been reported. The conjugation occurs enzymatically, catalyzed by glutathione-S-transferases, and nonenzymatically

viewed in a comprehensive and perfect way (Lamoureux and Rusness 1993). In this chapter, we focus on recent findings on the significance of herbicide safeners in glutathione conjugation.

Herbicide safeners are xenobiotic compounds which have limited phytotoxicity on their own and selectively protect crop plants against herbicide injury without protecting weeds. It has been proposed that herbicide safeners function by elevating the rate of herbicide detoxification via conjugation with glutathione (Lay and Casida 1976; Hatzios 1989; Komives and Dutka 1989; Lamoureux and Rusness 1989, 1993; Rennenberg and Lamoureux 1990; Dean et al. 1991; Fuerst and Lamoureux 1992). Correspondingly, safeners induce increased levels of glutathione-S-transferases and glutathione (Hatzios 1989; Komives and Dutka 1989; Kreuz et al. 1989; Farago and Brunold 1990; Dean et al. 1991; Kunkel et al. 1991; Viger et al. 1991; Fuerst and Lamoureux 1992), but also affect monooxygenases and glucosylation (Komives and Dutka 1989; Lamoureux and Rusness 1992), glutathione reductase (EC 1.6.4.2) (Komives and Dutka 1989; Han and Hatzios 1991), and enzymes of assimilatory sulfate reduction (Farago and Brunhold 1990). Some of these effects were also obtained with herbicides (Han and Hatzios 1991; Jablonkai and Hatzios 1991). The combined effects of both herbicide and safener may be essential for protection from herbicide action (Han and Hatzios 1991). The effects of a given herbicide safener in a defined crop on the levels of glutathione and glutathione-S-transferases may be dependent on temperature, cultivar, and method of application (Farago and Brunold 1990; Kunkel et al. 1991; Viger et al. 1991). Benoxacor applied to roots of maize seedlings induced an increase in cysteine and glutathione levels as a consequence of increased sulfate assimilation (Farago and Brunold 1990). This increased sulfate assimilation seems to be based on a three- and sevenfold elevated ATP sulfurylase and adenosine 5'-phosphosulfate sulfotransferase activity, respectively. These results indicate that benoxacor may eliminate the previously proposed regulatory mechanism (Brunold 1990), in which increased concentrations of thiols regulate assimilatory sulfate reduction by decreasing the activities of the key enzymes involved (Brunold 1990).

5. Glutathione in the Compensation of Oxidative Stress

The antioxidative system present in plants has the aim to prevent damage by reactive oxygen species. It may either react chemically with these compounds or/and may function as a substrate in the path of enzymatic protection present in the cytoplasm and the chloroplast (Fig. 6; Asada and Takahashi 1987).

In order to prevent toxicity, protection measures have to be available in those cellular compartments where the production of reactive oxygen species takes place. This will be different for stress factors acting from the outside on plants, e.g., air pollutants, and for internally generated stresses, e.g., nutrient deficiences (Polle and Rennenberg 1993a). Oxidative air pollutants, like ozone, sulfur dioxide, or nitrogen oxides, will enter the leaves via the stomata and will first come into contact with the aqueous phase of the cell wall (apoplastic space). Antioxidative systems in this compartment may therefore be of particular significance to prevent damage to the interior of the cells. Although glutathione has been identified as a constituent of the xylem sap (Schupp et al. 1991, 1992), its pre-

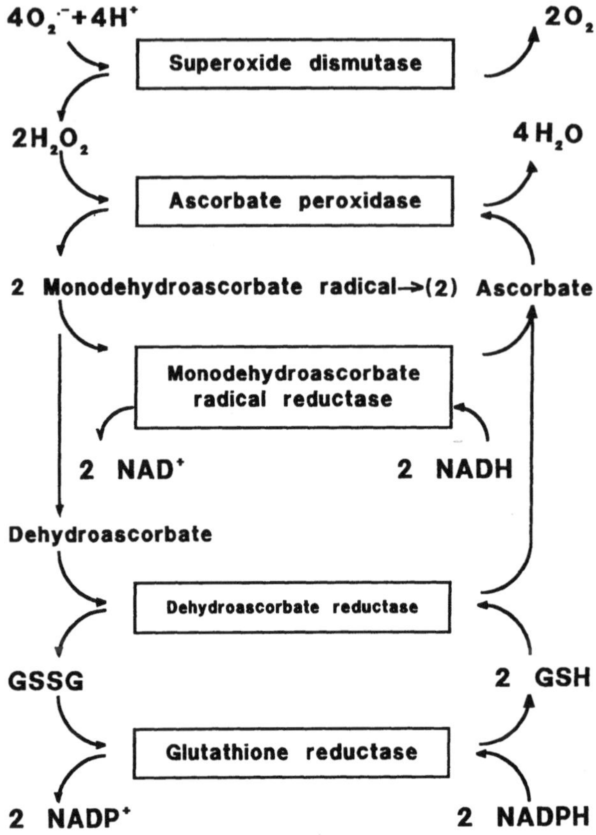

Fig. 6. Pathway of enzymatic detoxification of superoxide radicals and hydrogen peroxide in the cytoplasm and the chloroplast. (Polle and Rennenberg 1992)

sence in the apoplastic space of leaves is still doubtful (Polle et al. 1990). Because of its high apoplastic concentration, ascorbate is thought to be the most important apoplastic component mediating compensation from oxidative stress in this compartment. Also inside the cell, ascorbate may be more suitable as a chemical antioxidant than glutathione. In the chloroplast, e.g., ascorbate seems to be present in 10- to 50-fold higher concentrations than glutathione, but exhibits similar rate constants for singulett oxygen or superoxide radicals (Polle and Rennenberg 1993b). Therefore, ascorbate rather than glutathione is likely to be used in the chemical removal of these toxic oxygen species in the chloroplast. The function of glutathione in the compensation of oxidative stress may be restricted to its role as a substrate of glutathione reductase in the enzymatic detoxification system for reactive oxygen species in chloroplasts and the cytoplasm (Fig. 6).

Numerous studies have shown that cellular glutathione levels change upon exposure of plants to environmental conditions that cause oxidative stress (Smith et al. 1990; Polle and Rennenberg 1993b): thiol levels are higher in sun-exposed than in shaded leaves. Glutathione contents of leaves can increase with increasing light intensity. In conifer needles glutathione accumulates during the winter; its content increases with increasing elevation level. Elevated glutathione levels were observed in leaves exposed to oxidative air pollutants. In many of these studies also enhanced glutathione reductase activity was found. However, in none of these studies it was proven that these increases in glutathione contents or glutathione reductase activity were of adaptive significance. The question whether plants responding to oxidative stress with elevated glutathione and/or glutathione reductase levels are more stress-tolerant than plants with basal levels, was not addressed. Recently, glutathione reductase levels in several species have been elevated by the use of plant molecular biology or classical breeding techniques. Improved protection was observed, when superoxide dismutase levels were also enhanced, whereas an increase in the levels of the individual enzymes had little effect (Polle and Rennenberg 1993a). Whether elevated activities of the enzymes of glutathione synthesis can mediate protection from oxidative stress remains to be elucidated.

6. Glutathione in Plant-Pathogen Interactions

Plants attacked by pathogens start to synthesize phytoalexins and lytic enzymes, and reinforce their cell walls. These defense reactions can also be induced by exogenously applied glutathione (Dron et al. 1988; Wingate et al. 1988; Conrath et al. 1989; Yamada et al. 1989; Choudhary et al. 1990; Lawton et al. 1990; Edwards et al. 1991; Robbins et al. 1991). Treatment of suspension-cultured cells of bean with glutathione induced the formation of hydroxyproline-rich glycoproteins, as well as phenylalanine ammonium lyase (EC 4.3.1.5) and chalcone synthase (EC 2.3.1.74), both enzymes involved in phytoalexin biosynthesis (Wingate et al. 1988). Transcription activation of these enzymes occurred within minutes and was similar to that observed after treatment with an elicitor preparation (Wingate et al. 1988). These results led to the conclusion that glutathione may be involved in signal transduction during defense response (Wingate et al. 1988). If this assumption is correct, glutathione levels should change rapidly in response to fungal elicitor. In bean cells, glutathione did not accumulate after exposure to fungal elicitor, but homoglutathione increased steadily over a 36-h period. However, the changes in homoglutathione levels were slow as compared to the elicitation response. Activation of phenylalanine ammonium lyase transcription occurred within 5 min of exposure to elicitor, whereas changes in homoglutathione levels occurred over hours (Lawton and Lamb 1987). It should be mentioned, however, that the intracellular distribution of glutathione and homoglutathione may change during elicitation and elevated levels may rapidly be obtained at specific sites.

A different hypothesis for the function of glutathione and homoglutathione production in elicited cells is based on the finding that the pathogen-induced wheat gene encodes a protein homologous to glutathione-S-transferases (Dudler et al. 1991). In animal systems, the highly toxic products of membrane lipid peroxidation are substrates of glutathione-S-transferase isoenzymes, and thus may contribute to protection from oxidative damage (Pickett and Lu 1989). As elicitation is known to result in extensive lipid peroxidation (Rogers et al. 1988), it is conceivable that glutathione-S-transferases play a similar protective role in plants. According to this hypothesis, glutathione produced during elicitation would be used for the conjugation of toxic membrane lipid peroxidation products (Dudler et al. 1991).

7. Conclusions

The involvement of glutathione metabolism in the coping of plants with biological as well as abiotic stress is well documented in the recent literature, even though the exact mechanisms involved are mainly unclear. The methods of molecular biology including naturally selected and artificially produced mutants will be essential for the progress of our understanding in this area of research. The potential of molecular biology techniques in this respect has already been demonstrated (Dron et al. 1988; Lawton et al. 1990; Choudhary et al. 1991; Howden and Cobbett 1992; Ortiz et al. 1992; Römer et al. 1992; Yeargan et al. 1992) and future exciting developments are to be expected. Eventually, models for the molecular basis of the reactions of plants in different stress situations will emerge. On the basis of our present knowledge, glutathione is a good candidate for playing a central role in these models.

Acknowledgments. We thank the Deutsche Forschungsgemeinschaft (H.R.), Bayerisches Staatsministerium für Landesentwicklung und Umweltfragen (H.R.), Swiss National Science Foundation (C.B.), Ciba (C.B.), and Körber Foundation (C.B.) for financial support of own work presented here, M. Zeder, R. Hintermann, and T. Brunold for their help in preparing this review, and several colleagues for making preprints available.

References

Asada K, Takahashi M (1987) Production and scavenging of active oxygen in photosynthesis. In: Kyle D, Osmond C, Arntzen C (eds) Photoinhibition. Elsevier, Amsterdam, pp 227–287.

Bergmann L, Rennenberg H (1993) Glutathione metabolism in plants. In: De Kok LJ, Stulen I, Rennenberg H, Brunold C, Rauser W (eds) Sulfur nutrition and sulfur assimilation in higher plants. SPB, The Hague, p 109. **Brunold C** (1990) Reduction of sulfate to sulfide. In: Rennenberg H, Brunold C, De Kok L, Stulen I (eds) Sulfur nutrition and sulfur assimilation in higher plants. SPB, The Hague, p 13. **Brunold C** (1993) Regulatory interactions between sulfate and nitrate assimilation. In: De Kok LJ, Stulen I, Rennenberg H, Brunold C, Rauser W (eds) Sulfur nutrition and sulfur assimilation in higher plants. SPB, The Hague, p 61.

Choudhary AD, Lamb CJ, Dixon RA (1990) Stress responses in alfalfa (*Medicago sativa* L.). Plant Physiol 94:1802–1807. **Conrath U,** Domard A, Kauss H (1989) Chitosan-elicited synthesis of callose and of coumarin derivatives in parsley cell suspension cultures. Plant Cell Rep 8:152–155.

Dean JV, Gronwald JW, Anderson MP (1991) Glutathione-S-transferase activity in nontreated and CGA-154281-treated maize shoots. Z Naturforsch 46c:850–855. **de Knecht JA,** Koevoets PLM, Verkleij JAC, Ernst WHO (1992) Evidence against a role for phytochelatins in naturally selected increased cadmium tolerance in *Silene vulgaris* (Moench) Garcke. New Phytol 122:681–688. **de Vos CHR,** Vonk MJ, Vooijs R, Schat H (1992) Glutathione depletion due to copper-induced phytochelatin synthesis causes oxidative stress in *Silene cucubalus*. Plant Physiol 98:853–858. **Dron M,** Clouse SD, Dixon RA, Lawton MA, Lamb CJ (1988) Glutathione and fungal elicitor regulation of a plant defense gene promoter in electroporated protoplasts. Proc Natl Acad Sci USA 85:6738–6742. **Dudler R,** Hertig C, Rebmann G, Bull J, Mauch F (1991) A pathogen-induced wheat gene encodes a protein homologous to glutathione-S-transferase. Mol Plant-Microbe Inter 4:14–18.

Edwards R, Blount JW, Dixon RA (1991) Glutathione and elicitation of the phytoalexin response in legume cell cultures. Planta 184:403–409. **Ernst WHO,** Verkleij JAC, Schat H (1992) Metal tolerance in plants. Acta Bot Neerl 41:229–248.

Frankhauser H, Brunold C (1978) Localization of adenosine 5'-phosphosulfate sulfotransferase in spinach leaves. Planta 143:285–289. **Farago S,** Brunold C (1990) Regulation of assimilatory sulfate reduciton by herbicide safeners in *Zea mays* L. Plant Physiol 94:1808–1812. **Fuerst EP,** Lamoureux GL (1992) Mode of action of the dichloroacetamide antidote BAS 145-138 in corn. 2. Effects on metabolism, absorption, and mobility of metazachlor. Pestic Biochem Physiol 42:78–87.

Galli U, Meier M, Brunold C (1993) Assimilatory sulfate reduction of cadmium treated non-mycorrhizal and mycorrhizal Norway spruce seedlings (*Picea abies*) and its ectomycorrhizal fungus *Laccaria laccata*. New Phytol (in press). **Grill E,** Löffler S, Winnacker E-L, Zenk MH (1989) Phytochelatins, the heavy-metal-binding peptides of plants, are synthesized from glutathione by a specific γ-glutamylcysteine dipeptidyl transpeptidase (phytochelatin synthase). Proc Natl Acad Sci USA 86:6838–6842. **Grill E,** Winnacker E-L, Zenk MH (1990) Phytochelatins, the heavy metal chelating peptides of the plant kingdom. In: Rennenberg H, Brunold C, De Kok LJ, Stulen I (eds) Sulfur nutrition and sulfur assimilation in higher plants. SPB, The Hague, p 89. **Gupta SC,** Goldsbrough PB (1991) Phytochelatin accumulation and cadmium tolerance in selected tomato cell lines. Plant Physiol 97:306–312.

Han S, Hatzios KK (1991) Effects of the herbicide pretilachlor and the safener fenclorim on glutathione content and glutathione-dependent enzyme activity of rice. Z Naturforsch 46c:861–865. **Hatzios KK** (1989) Development of herbicide safeners: industrial and university perspectives. In: Hatzios KK, Hoagland RE (eds) Crop safeners for herbicides. Academic Press, San Diego, p 3. **Hayashi Y,** Nakagawa CW, Mutoh N, Isobe M, Goto T (1991 Two pathways in the biosynthesis of cadystins $(\gamma EC)_nG$ in the cell-free system of the fission yeast. Biochem Cell Biol 69:115–121. **Hell R,** Bergmann L (1990) γ-Glutamylcysteine synthetase in higher plants: catalytic properties and subcellular localization. Planta 180:603–612. **Howden R,** Cobbett CS (1992) Cadmium-sensitive mutants of *Arabidopsis thaliana*. Plant Physiol 99:100–107.

Jablonkai I, Hatzios KK (1991) Role of glutathione and glutathione S-transferase in the selectivity of acetochlor in maize and wheat. Pestic Biochem Physiol 41:221–231. **Jackson PJ,** Kuske CR (1993) Glutathione-derived metal-binding polypeptides and metallothioneins. In: Iron chelation in plants and soil microorganisms. Academic Press, San Diego, p 157. **Jackson PJ,** Delhaize E, Kuske CR (1992) Biosynthesis and metabolic roles of cadystins $(\gamma\text{-EC})_nG$ and their precursors in *Datura innoxia*. Plant Soil 146:281–289.

Kasai T, Larsen PO (1980) Chemistry and biochemistry of γ-glutamyl derivatives from plants including mushrooms (Basidiomycetes). In: Herz W, Grisebach H, Kirby GW (eds) Progress in the chemistry of organic products. Springer, Berlin Heidelberg New York, pp 173–285. **Klapheck S** (1988) Homoglutathione: isolation, quantification and occurrence in legumes. Physiol Plant 74:727–732. **Kneer R,** Zenk MH (1992) Phytochelatins protect plant enzymes from heavy metal poisoning. Phytochemistry 31:2663–2667. **Komives T,** Dutka F (1989) Effects of herbicide safeners on levels and activity of cytochrome P-450 and other enzymes of corn. In: Hatzios KK, Hoagland RG (eds) Crop safeners for herbicides. Academic Press, San Diego, p 129. **Kreuz K,** Gaudin J, Ebert E (1989) Effects of the safeners CGA 154281, oxabetrinil and fenclorim on uptake and degradation of metolachlor in corn (*Zea mays* L.) seedlings. Weed Res 29:399–405. **Kunkel DL,** Steffens JC, Bellinder RR (1991) Effect of temperature and safeners on glutathione levels and glutathione S-transferase activity in maize. Z Naturforsch 46c:856–860.

Lamoureux GL, Rusness DG (1989) The role of glutathione and glutathione S-transferases in pesticide metabolism, selectivity, and mode of action in plants and insects. In: Dolphin D, Poulson R, Avramovic O (eds) Glutathione: chemical, biochemical, and medical aspects. Part B. John Wiley & Sons, New York, p 153. **Lamoureux GL,** Rusness DG (1992) The mechanism of action of BAS 145 138 as a safener for chlorium ethyl in corn: effect of hydroxylation, glutathione conjugation, glucoside conjugation, and acetolactate synthase. Pestic Biochem Physiol 42:128–139. **Lamoureux GL,** Rusness DG (1993) Glutathione in the metabolism and detoxification of xenobiotics in plants. In: De Kok LJ, Stulen I, Rennenberg H, Brunold C, Rauser W (eds) Sulfur nutrition and sulfur assimilation in higher plants. SPB, The Hague, p 221. **Lawton MA,** Lamb CJ (1987) Transcriptional activation of plant defense genes by fungal elicitor, wounding and infection. Mol Cell Biol 7:335–341. **Lawton AM,** Clouse SD, Lamb CJ (1990) Glutathione-elicited changes in chromatin structure within the promoter of the defense gene chalcone synthase. Plant Cell Rep 8:561–564. **Lay MM,** Casida JE (1976) Dichloracetamide antidotes enhance thiocarbamate sulfoxide content and glutathione S-transferase activity. Pestic Biochem Physiol 6:442. **Li J,** Schiff JA (1992) Adenosine 5'-phosphosulfate sulfotransferase from *Euglena*: enzyme-bound intermediates. Plant Cell Physiol 33:63–72.

Martinoia E, Grill E, Tommasini R, Kreuz K, Amrhein N (1993) An ATP-dependent glutathione S-conjugate "export" pump in the vacuolar membrane of plants. Nature 364:247–249. **Meister A** (1981) On the cycles of glutathione metabolism and transport. In: Estabrook RW, Srere P (eds) Biological cycles. Current topics in cellular regulation, vol 18. Academic Press, New York, pp 21–58. **Meister A,** Anderson ME (1983) Glutathione. Annu Rev Biochem 52:711–760. **Meuwly P,** Rauser E (1992) Alteration of thiol pools in roots and shoots of maize seedlings exposed to cadmium. Plant Physiol 99:8–15. **Murasugi A,** Wada C, Hayashi Y (1983) Occurrence of acid-labile sulfide in cadmium-binding peptide 1 from fission yeast. J Biochem 93:661–664.

Nussbaum S, Schmutz D, Brunold C (1988) Regulation of assimilatory sulfate reduction by cadmium in *Zea mays* L. Plant Physiol 88:1407–1410.

Ortiz DF, Kreppel L, Speiser DM, Scheel G, McDonald G, Ow DW (1992) Heavy metal tolerance in the fission yeast requires transporter. EMBO J 11:3491–3499.

Picket CG, Lu AYH (1989) Glutathione-S-transferases: gene structure, regulation and biological function. Annu Rev Biochem 58:743–764. **Polle A,** Rennenberg H (1992) Field studies on Norway spruce trees at high altitudes. II. Defense systems against oxidative stress in needles. New Phytol 121:635–642. **Polle A,** Rennenberg H (1993a) The significance of antioxidants in plant adaptation ot environmental stress. In: Mansfield T, Fowden L, Stoddard F (eds) Plant adaptation to environmental stress. Chapman and Hall, London, p 263. **Polle A,** Rennenberg H (1993b) Photooxidative stress in trees. In: Foyer C, Mullineaux P (eds) Photooxidative stresses on plants: causes and amelioration. CRC, Boca Raton (in press). **Polle A,** Chakrabarti K, Schürmann W, Rennenberg H (1990) Composition and properties of hydrogen peroxide decomposing systems in extracellular and total extracts from needles of Norway spruce (*Picea abies* L.) Plant Physiol 94:312–319.

Rauser WE (1991) Phytochelatins. Annu Rev Biochem 59:61–86. **Rauser WE** (1991) Cadmium-binding peptides from plants. Methods Enzymol 205:319–333. **Rauser WE** (1993) Metal-binding peptides in plants. In: De Kok LJ, Stulen I, Rennenberg H, Brunold C, Rauser W (eds) Sulfur nutrition and sulfur assimilation in higher plants. SBP, The Hague (in press). **Reese RN,** Winge DR (1988) Sulfide stabilization of cadmium γ-glutamyl peptide complex of *Schizosaccharomyces pombe*. J Biol Chem 263:12832–12835. **Reese RN,** White CA, Winge DR (1992) Cadmium-sulfide crystallites in Cd-$(\gamma EC)_nG$ peptide complexes from tomato. Plant Physiol 98:225–229. **Rennenberg H,** Lamoureux GL (1990) Physiological processes that modulate the concentration of glutathione in plant cells. In: Rennenberg H, Brunold C, De Kok LJ, Stulen I (eds) Sulfur nutrition and sulfur assimilation in higher plants. Fundamental environmental and agricultural aspects. SPB, The Hague, pp 53–65. **Robbins MP,** Hartnoll J, Morris P (1991) Phenylpropanoid defence responses in transgenic *Lotus corniculatus* 1. Glutathione elicitation of isoflavan phytoalexins in transformed root cultures. Plant Cell Rep 10:59–62. **Robinson NJ** (1990) Metal binding polypeptides in plants. In: Shaw AJ (ed) Heavy metal tolerance in plants. CRC, Boca Raton, p 195. **Rogers KR,** Albert F, Anderson AJ (1988) Lipid peroxidation is a consequence of elicitor activity. Plant Physiol 86:547–553. **Römer S,** d'Harlingue A, Camara B, Schantz R, Kuntz M (1992) Cysteine synthase from *Capsicum annuum* chromoplasts. Characterization and cDNA cloning of an up-regulated enzyme during fruit development. J Biol Chem 267:17966–17970. **Rüegsegger A,** Brunold C (1992)

Effect of cadmium in γ-glutamylcysteine synthesis in maize seedlings. Plant Physiol 99:428–433. **Rüegsegger A,** Brunold C (1993) Localization of γ-glutamylcysteine synthetase and glutathione synthetase activity in maize seedlings. Plant Physiol 101:561–566. **Rüegsegger A,** Schmutz D, Brunold C (1990) Regulation of glutathione synthesis by cadmium in *Pisum sativum* L. Plant Physiol 93:1579–1584.

Scarponi L, Perucci P, Martinetti L (1991) Conjugation of 2-chloroacetanilide herbicides with glutathione: Role of molecular structures and of glutathione S-transferase enzymes. J Agric Food Chem 39:2010–2013. **Schat H,** Kalff MMA (1992) Are phytochelatins involved in differential metal tolerance or do they merely reflect metal-imposed strain? Plant Physiol 99:1475–1480. **Schiff JA,** Stern AI, Saida T, Li J (1993) Some molecular aspects of sulfate metabolism in photosynthetic organisms. In: De Kok LJ, Stulen I, Rennenberg H, Brunold C, Rauser W (eds) Sulfur nutrition and sulfur assimilation in higher plants. SBP, The Hague, p 21. **Schmidt A** (1992) Open questions about sulfur metabolism in plants. Annu Rev Plant Physiol Plant Mol Biol 43:325–349. **Schupp R,** Rennenberg H (1988) Diurnal changes in the glutathione content of spruce needles (*Picea abies* L.). Plant Sci 57:113–117. **Schupp R,** Rennenberg (1992) Changes in sulfur metabolism during needle development of Norway spruce. Bot Acta 105:180–189. **Schupp R,** Glavac V, Rennenberg H (1991) Thiol composition of xylem sap of beech trees. Phytochemistry 30:1415–1418. **Schupp R,** Schatten T, Willenbrink J, Rennenberg H (1992) Long-distance transport of reduced sulphur in spruce (*Picea abies* L.). J Exp Bot 41:631–644. **Smith IK,** Polle A, Rennenberg H (1990) Glutathione. In: Alscher R, Cunning I (eds) Stress responses in plants: adaptation and acclimation mechanisms. Wiley-Liss, New York, pp 201–215. **Speiser DM,** Abrahamson SL, Banuelos G, Ow DW (1992) *Brassica juncea* produces a phytochelatin-cadmium-sulfide complex. Plant Physiol 99:817–821. **Steffens JC** (1990) The heavy metal-binding peptides of plants. Annu Rev Plant Physiol Plant Mol Biol 41:553–575. **Steffens JC,** Hunt DF, Williams BG (1986) Accumulation of nonprotein metal-binding polypeptides (γ-glutamyl-cysteinyl)$_n$-glycine in selected cadmium-resistent tomato cells. J Biol Chem 261:13879–13882.

Thumann J, Grill E, Winnacker E-L, Zenk MH (1991) Reactivation of metal-requiring apoenzymes by phytochelatin-metal complexes. FEBS Lett 284:66–69. **Tomsett AB,** Thurman DA (1988) Molecular biology of metal tolerances of plants. Plant Cell Environ 11:383–394. **Tuckendorf A,** Rauser WE (1990) Changes in glutathione and phytochelatins in roots of maize seedlings exposed to cadmium. Plant Sci 70:155–166.

Verkleij JAC, Koevoets P, Van't Riet J, Bank R, Nijdam Y, Ernst WHO (1990) Poly(γ-glutamylcysteinyl) glycines or phytochelatins and their role in cadmium tolerance of *Silene vulgaris*. Plant Cell Environ 13:913–921. **Viger PR,** Eberlein CV, Fuerst EP, Gronwald JW (1991) Effects of GCA-154281 and temperature on metolachlor absorption and metabolism, glutathione content, and glutathione-S-transferase activity in corn (*Zea mays*). Weed Sci 39:324–328. **Vögeli-Lange R,** Wagner GJ (1990) Subcellular localization of cadmium and cadmium-binding peptides in tobacco leaves. Plant Physiol 92:1086–1093.

Wagner GJ (1993) Accumulation of cadmium in crop plants and its consequences to human health. Adv Agron (in press). **Wingate VPM,** Lawton MA, Lamb CJ (1988) Glutathione causes a massive and selective induction of plant defense genes. Plant Physiol 87:206–210.

Yamada T, Hashimoto H, Shiraishi T, Oku H (1989) Suppression of pisatin, phenylalanine ammonia-lyase mRNA and chalcone synthase mRNA accumulation by a putative pathogenicity factor from the fungus *Mycosphaerella pinodes*. Mol Plant Microbe Interact 2:256–261. **Yeargan R,** Maiti IB, Nielsen MT, Hunt AG, Wagner GJ (1992) Tissue partitioning of cadmium in transgenic tobacco seedlings and field grown plants expressing the mouse metallothionein I gene. Transgen Res 1:261–267.

Professor Dr. H. Rennenberg
Institut für Forstbotanik und Baumphysiologie
Professur für Baumphysiologie
Albert-Ludwigs-Universität Freiburg
Am Flughafen 17
D-79085 Freiburg i. Br., FRG

Professor Dr. C. Brunold
Pflanzenphysiologisches Institut
Universität Bern
Altenbergrain 21
CH-3013 Bern, Switzerland

VII. A Stress Physiological Role for Abscisic Acid (ABA) in Lower Plants*

By Wolfram Hartung and Hartmut Gimmler

1. Introduction

The role of the phytohormone abscisic acid (ABA) in cormophytes, especially its action as a stress hormone, and its function in seed germination and stomatal closure, has been well investigated in the last decades. In contrast, the existence of ABA in algae, bryophytes, and fungi has been ambiguous for a long time. Between 1960 and 1970 the detection of ABA in lower plants was limited by proper analytical methods of high sensitivity and lack of sufficient amounts of biological material for the analysis of microorganisms and tiny bryophytes. Nevertheless, by means of simple techniques, the presence of "ABA-like compounds" in at least a few algae appeared likely (Radley 1961; Moss 1965; Jennings 1969; Hussain and Boney 1973). When more precise and sophisticated analytical tools became available between 1970 and 1980, the existence of ABA in algae, bryophytes, and fungi was reinvestigated. First studies failed to demonstrate the presence of ABA in lower plants except mosses. It was concluded that ABA is entirely absent in algae, fungi, and liverworts (Pryce 1972; Bearder 1980; Niemann and Dörffling 1980; Addicott and Carns 1983) and that lunularic acid may replace ABA physiologically in protophytes and thallophytes (Pryce 1972). However, when analytical methods became more and more sensitive, and especially with the introduction of RIA – and ELISA – tests for ABA, the picomole range of sensitivity was reached and this hypothesis collapsed. It could be demonstrated convincingly that ABA does occur in algae (Kingsham and Moore 1982; Tietz and Kasprik 1986; Sabbatini et al. 1987; Boyer and Dougherty 1988; Tietz et al. 1989), fungi (Assante et al. 1977; Dahyia et al. 1988), and liverworts (Hartung et al. 1987). This breakthrough initiated studies of the physiological role of ABA in lower plants, of the evolution of this compound from a secondary product of plant metabolism to a phytohormone, and, last but not least, of its involvement in stress resistance in lower plants (Bopp 1990a). In this chapter, data on the biology of ABA, scattered throughout the literature, are summarized and analyzed with special emphasis on the potential role of ABA as stress hormone.

* Dedicated to Prof. Dr. O.H. Volk on the occasion of his 90th birthday.

Progress in Botany, Vol. 55

2. Abscisic Acid in Algae

a) Distribution of Abscisic Acid Within the Algal Kingdom

Almost 100 different algal species have been tested so far for the presence of ABA, about 96% of which contained significant amounts (Kentzer et al. 1975a, b; Hirsch et al. 1989 and the references therein; Tietz et al. 1989; Zahradnickova et al. 1990, 1991; Cowan and Rose 1991; Kentzer and Mazur 1991; Marsalek et al. 1992, in press). Studies included the following algal divisions: *Cyanobacteria*, *Rhodophyta*, *Heterokontophyta*, *Haptophyta*, *Eustigmatophyta*, *Cryptophyta*, *Dinophyta*, *Euglenophyta*, and *Chlorophyta*. We assume that the rare cases in which ABA could not be detected (*Enteromorpha compressa*, Niemann and Dörffling 1980; *Fritschiella tuberosa*, Tietz and Kasprik 1986; *Caulerpa paspaloides*, Jacobs 1985) were due to insufficiencies in the analytical procedures applied. This assumption is confirmed by reinvestigations with improved techniques (e.g., in the case of *Draparnalida*, Tietz and Kasprik 1986; Tietz et al. 1989) and the observation that related genera or even other species of the same genus do contain ABA. It has to be concluded that, in contrast to former assumptions (Pryce 1972), ABA is universally distributed within the algal kingdom. From the evolutionary point of view, it is of twofold interest that ABA synthesis already occurs in the prokaryotic *Cyanobacteria*, but is apparently absent in bacteria. The latter was concluded from the failure to detect ABA in 13 species from both *Archebacteria* and *Eubacteria* (Hirsch et al. 1989; Müller et al. 1989). Thus, the ability to synthetize ABA was developed already at the prokaryotic level, but its exact origin within *Bacteria* or *Cyanophyta* remains obscure. However, in the light of the endosymbiontic descent of chloroplasts of higher plants from *Cyanobacteria*, it is puzzling that only the latter, but not the former, are able to synthetize ABA (Hartung et al. 1981; Quarrie and Lister 1984; Cowan and Railton 1986; Hirsch et al. 1989).

b) Abscisic Acid Content and Permeability Coefficients

In nonstressed algal cells, ABA contents have been reported to vary between 7 and 34 nmol ABA kg^{-1} fresh weight, 6 and 200 nmol ABA g^{-1} chlorophyll, or 10 and 140 nmol ABA g^{-1} protein, respectively (Hirsch et al. 1989). The ABA content of algae is comparable to the reported range for submerged living bryophytes (Table 2), but is significantly lower than the ABA content of terrestrial bryophytes. In nonvacuolated algal cells, the measured ABA contents can be converted into cytoplasmic ABA, concentrations between 0.1 and 40 μM. This range corresponds approximately to internal ABA concentrations measured in higher plants.

Higher plants regulate their ABA levels by the control of ABA synthesis, on the one hand, and tuning of the ABA metabolism (= ABA degradation) on the other. Since algae live in an aquatic environment, they have an additional option to remove excess ABA by excretion into the surrounding medium. In fact, algae excrete measurable amounts of ABA into the external medium (Hirsch et al. 1989; Zahradnickova et al. 1990; Cowan and Rose 1991; Marsalek et al. 1992), and excretion was stimulated by salt, acid, and drought stress (Zahradnickova et al. 1990, 1991; Cowan and Rose 1991; Marsalek et al. 1992, in press). Studies of the ABA distribution between cells of *Dunaliella, Eremosphaera, Ankistrodemus,* and *Anacystis* and the culture media of these species demonstrated that in batch cultures 60 and 98% of the total ABA is localized within the cells and that the ABA gradient across the plasma membranes of these algae varies between 1600 and 55 000 (Hirsch, unpubl.). The ABA distribution was calculated to be far from the equilibrium predicted by the Henderson-Hasselbalch equation (Hirsch et al. 1989). This indicates that rates of ABA synthesis in algal cells are always higher than rates of ABA efflux from the cell into the medium. The permeability coefficients of the plasma membrane of algal cells for ABA seem to be lower than corresponding values for fungi and higher plants. For *Dunaliella acidophila* and *Eremosphaera viridis*, Gimmler (unpubl.) found values in the range of 0.03 and 0.3 x 10^{-9} m s^{-1}. At present, it is not known whether ABA excretion from the algal cell has additional functions apart from the regulation of the internal ABA pool, as is discussed for fungal ABA exrection (see below).

c) Biosynthesis and Metabolism of Abscisic Acid

Incubation of *Dunaliella* cells with ^{14}C-labeled CO_2, acetate, and mevalonic acid results in the formation of ^{14}C-ABA (Bopp-Buhler et al. 1991; Cowan and Rose 1991). Mevalonic acid and isopentylpyrophosphate could be identified as precursors of ABA in algae. ABA synthesis from prelabeled precursors is stimulated by salt stress (Cowan and Rose 1991). In principle, similar results were obtained using cell-free extracts of algae. Hypersalinity induction of ABA synthesis was found to be inhibited by actinomycin D, cycloheximide, and chloramphenicol. These observations have been interpreted to indicate that both transcription and de novo protein synthesis are required for the increased production of ABA upon salt shock. Results indicate that ABA synthesis in algae follows the terpenoid pathway, as has been suggested also for higher plants.

Since for higher plants also a photooxidative or enzymatic hydrolyzation of violaxanthin via xanthoxin has been suggested as a possibility of biosynthesis of ABA, the effect of the inhibitors of carotene synthesis and terpenoid synthesis, norflurazone and AMO 1618, on the formation of ABA in algae has been investigated (Bopp-Buhler et al. 1991). Both inhibitors produced a strong decrease in the carotenoid level and parallel a strong increase in the phytoene pool. Simultaneously, a two- to sevenfold increase in the endogenous ABA pool was observed. This clearly demonstrates that at least in algae ABA cannot be derived from violaxanthin, because in such a case the ABA pool of algal cells would be expected to be decreased rather than increased by norflurazone and AMO 1618, as has been shown for fungi by Okamoto et al. (1988c). Thus, in algae, ABA and carotenoids compete for the same precursors of the terpenoid pathways. Also, the following observations cannot be explained by an indirect biosynthesis of ABA from carotenoids in algae: (1) colorless algae such as *Astasia* and *Beggiatoa* contain considerable amounts of ABA (Hirsch et al. 1989). (2) In several algal divisions, such as the *Cyanobacteria*, *Dinophyta*, or *Euglenophyta*, a lack of the ability to synthesize violaxanthin is observed (Rowan 1989), whereas the ability to synthesize ABA is well expressed. Similar arguments may be applied also to the divisions of *Rhodophyta* and *Euglenophyta*: within these divisions, the formation of violaxanthin is observed in only very few species.

Incubation of *Dunaliella parva* and *Dunaliella salina* with ^{14}C-mevalonic acid or ^{14}C-ABA resulted in the synthesis of metabolic products of ABA, putatively identified as phaseic acid (PA), dihydrophaseic acid (DPA), and glucose esters of ABA (ABA-GE) (Hirsch et al. 1989; Cowan and Rose 1991). A suprisingly high percentage of labeled ABA metabolites was excreted into the medium (Hirsch et al. 1989). Tetcyclacis, an inhibitor of ABA metabolism, blocking the hydroxylation of ABA (Daeter and Hartung 1990), caused an increase in the internal ABA level in *Dunaliella acidophila* (not shown). All results imply that there is no significant difference between ABA catabolism in algae and higher plants.

d) Effect of Stress on the Endogenous ABA Level

Higher plants respond to various kinds of stress with an accumulation of ABA, which is believed to be a signal for regulation. For the physiological significance of a stress-induced signal, it is important to ensure that the supposed stress is not lethal and that the kind of stress is encountered by the test organism also in nature. These prerequisites have been taken into account in the following studies. Salt stress, a stress factor common for many algae in nature, increases the internal ABA pool and excretion of ABA into the medium in *Dunaliella* (Hirsch et al. 1989; Tietz et al. 1989; Cowan and Rose 1991), *Draparnaldia* (Tietz et al. 1989), and *Chlorella, Anabaena, Nostoc, Trichormus,* and *Synechococcus* (Zahradnickova et al. 1990, 1991; Marsalek et al. 1992, in press). Similar responses apply to drought stress in green algae (Marsalek et al. in press). Also, pH changes in the medium affect the endogenous ABA pool. In alga adapted to neutral pH values, a transfer to acidic pH values causes an increase in the ABA pool (Zahradnickova et al. 1990, 1991; Marsalek et al. in press), whereas in algae adapted to extremely high acidities, a transfer to higher pH values causes a rise in the ABA pool (Hirsch et al. 1989). The latter two effects are a nice demonstration of the fact that a phylogenetic adaptation of an organism to extreme conditions is not identical with an actual physiological stress. The latter occurs only when a sudden change in the external conditions to which the organism is adapted takes place. Finally, also strong irradiation of *Dunaliella* cells, known to cause photoinhibition, increases the internal ABA pool (Bopp-Buhler et al. 1991). In none of the cases of stress-induced increases in the internal ABA pool investigated so far, could it be demonstrated (as is the case in higher plants) that ABA serves as a chemical signal for regulatory processes involved in the overall process of stress resistance which is essential for the survival under stress conditions.

e) The Physiological Role of Abscisic Acid

Several attempts were made to elucidate the physiological role of ABA in algae by treatment of algal cells with external ABA. Many of these attempts were carried out in a situation of extremely poor knowledge of the biology of ABA in algae (existence as native compound was uncertain, the pool size, the mode of uptake, the permeability coefficients, etc were all unknown). However, an effect of external ABA on physiological reactions of algal cells can be expected only if during the applied range of investigation sufficient external ABA is able to enter the cell thereby causing a significant increase in the internal pool. Considering our knowledge of the interaction of ABA in higher plants with stomatal closure and seed germination, the first experiments in which external ABA was applied to microscopic or macroscopic algae which possess neither stomata, buds, or seeds, had to be initially arbitrary and naive. Therefore, it is not too surprising that the observed effects of externally applied ABA do not as yet yield a clear and consistent picture. In some cases, results even contradict each other. Table 1 summarizes some of the observed ABA effects in algae. In charophytes, the hydraulic conductivity of the plasma membrane was stimulated (Wanless et al. 1973; Ord et al. 1977), whereas in *Dunaliella*, the water permeation was either slightly inhibited or not affected (Hirsch et al. 1989). The ion uptake was found to be stimulated (Wanless et al. 1973; Ullrich and Kunz 1984), whereas other authors reported a reduced ion permeability (Huang 1991). The K^+ content of *Dunaliella* was not affected at all (Hirsch et al. 1989); the same applies to K^+ channels in the plasma membrane of *Chara* (Tester and MacRobbie 1990). In regard to energy metabolism, it is consistently reported that photosynthesis of algal cells is not affected by ABA, whereas with respect to respiration, results contradict each other: stimulations (Ullrich and Kunz 1984), no effects (Hirsch et al. 1989), and inhibitions (Huang 1991) by ABA have been reported. In contrast to CO_2 fixation in the light (via ribulose bisphosphate carboxylase), the CO_2 fixation in darkness (via PEP carboxylase) is strongly suppressed in *Dunaliella* by ABA (Bopp-Buhler and Gimmler, unpubl.), with half-maximal inhibition at 10^{-6} M ABA. This result matches an inhibition of the PEP-carboxylase in crude extracts obtained from *Dunaliella* cells observed by the same authors. The latter two results may be of particular interest, since in higher plants ABA seems to stimulate gene expression of the PEP carboxylase (McElwain et al. 1992).

Of future interest will be observations that several processes of algal differentiation are reported to be affected by external ABA. Developmental processes in algae stimulated by ABA are the formation of heterocysts in *Cyanobacteria* (Ahmad et al. 1978; Huddart et al. 1986; Smith et al. 1987), the gametogenesis in *Chlamydomonas* (Ishiura 1976), the abscission of oospores in *Chara* (Sabbatini et al. 1987), the senescence of *Stigeoclonium* (Tietz et al. 1989), and the growth of *Euglena, Anacystis, Nostoc,* and *Stigeoclonium* (Bralczyk et al. 1978; Ahmad et al. 1978; Tietz et al. 1989). However, growth reduction by ABA has also occasionally been observed (*Coscinodiscus*, Kentzer and Mazur 1991). Hirsch et al. (1989) speculated that a positive correlation may exist between the ABA content of various algal species and the organizational level of algae. If this is true, effects of ABA on developmental processes may not be surprising. However, some care is necessary if one compares the effect of ABA on membrane reactions, bioenergetic processes, and enzymes on one hand, and on developmental processes of algae on the other: the former are short-term reactions and the effects of ABA are usually

Table 1. Effects of external ABA on physiological reactions in algae

Classification of ABA effect	Specification of ABA effect	Algal genus	Reference
Membrane effect	Stimulation of hydraulic conductivity	*Nitella*	Ord et al (1977)
	Stimulation of transcellular water flow	*Nitella*	Wanless et al. (1973)
	Inhibition (or no effect) of the permeability of the plasma membrane for water	*Dunaliella*	Hirsch et al. (1989)
	Stimulation of ion uptake	*Nitella*	Wanless et al. (1973)
	Stimulation of nitrate uptake	*Chlorella*	Ullrich and Kunz (1984)
	No effect on phosphate uptake	*Scenedesmus*	Tillberg (1970)
	No effect on the K^+ pool	*Dunaliella*	Hirsch et al. (1989)
	No effect on K^+ channels	*Chara*	Tester and MacRobbie (1990)
Enzyme effect	Stimulation of sucrose phosphate synthetase	*Chlorella*	Salerno (1985)
	Inhibition of PEP carboxylase	*Dunaliella*	Bopp-Buhler (unpubl.)
	Inhibition of dark CO_2 fixation		
Energy metabolism	No effect on photosynthesis	*Scenedesmus*	Tillberg (1970)
		Ankistrodesmus	Ullrich and Kunz (1984)
		Dunaliella	Hirsch et al. (1989)
Differentiation and growth	Stimulation of growth	*Euglena*	Bralczyk et al. (1978)
		Anacystis	Ahmad et al. (1978)
		Nostoc	
		Stigeoclonium	Tietz et al. (1989)
	Inhibition of growth	*Coscinodiscus*	Kentzer and Mazur (1991)
	Stimulation of heterocyst formation	*Nostoc*	Smith et al,. 1987
			Huddart et al. (1986)
			Ahmad et al. (1978)
	Stimulation of gametogenesis	*Chlamydomonas*	Ishiura (1976)
	Stimulation of abscission of oospores	*Chara*	Sabbatini et al. (1987)
	Promotion of senescence	*Stigeoclonium*	Tietz et al. (1989)
	Delay of cap formation	*Acetabularia*	Saidova et al. (1988)

tested within the range of less than 1 h, whereas the latter are long-term processes, which require testing the possible effect of ABA in a range of days or at least several hours. The probability that externally applied ABA causes a significant change of the internal ABA pool (and this is a prerequisite for any ABA effect) is much higher in long-term than during short-term incubation experiments. Future studies on the effect of ABA on the physiology of algae have to take this consideration much more into account than has been done until now.

3. Abscisic Acid in Bryophytes

Abscisic acid has been detected in *Hepaticae*, especially in the *Marchantiales*, in a few species of *Anthocerotae* (Hartung et al. 1987), and in protonemata of *Funaria hygrometrica* (Werner et al. 1991). In *Marchantiales*, a large variability in ABA content could be observed. It ranges from 30 nmol g^{-1} in the extremely xerophilic *Exormotheca* species

Table 2. ABA in different species of *Marchantiales*. The species are listed in the order of decreasing drought resistance. Some of the data after Hartung et al. (1987)

Species	ABA (pmol g^{-1}FW)
Exormotheca bullosa (Link) K.Müller	30 755
Exormotheca holstii St.	1 700
Plagiochasma rupestris (Forster) St.	1 131
Riccia okahandjana Arnell	516
Riccia duthieae Volk & Perod	404
Riccia glauca L.	366
Targionia hypophylla L.	302
Riccia gougetiana Mont.	300
Riccia angolensis St.	259
Riccia nigrella D.C.	251
Oxymitra paleacea Bisch.	173
Riccia albovestita Volk	162
Riccia discolor L. & L.	103
Marchantia polymorpha L.	101
Conocephalum conicum (L.) Lindb.	96
Riccia cavernosa Hoffm. emend. Raddi	83
Riccia atropurpurea Sim	81
Reboulia hemisphaerica (L.) Raddi	30
Riccia ampullacea Perold	27
Riccia simii Perold	26
Riccia elongata Perold	22
Lunularia cruciata (L.) Dum	20
Ricciocarpus natans (L.) Corda	17
Corsinia coriandrina (Spreng.) Lindb.	13
Marchantia berteroana L. et L.	10
Riccia fluitans L.	10
Dumortiera hirsuta (SW) R.Bl.Nees	4
Cyathodium africanum Mitten	2

down to 1–10 pmol g^{-1} in submerged freshwater species such as *Riccia fluitans* (Table 2).

a) Biosynthesis and Metabolism

No experimental data are available as to whether ABA is synthesized in bryophytes directly via farnesylpyrophosphate or indirectly via violaxanthin and xanthoxin. Violaxanthin, the precursor for ABA in the indirect pathway, is present in bryophytes. On the other hand, one could argue that because of the close relationship of the *Hepaticae* to algae, ABA could also be formed directly (see also Sect. 2. c).

Almost no information is available about metabolism and degradation of ABA in bryophytes. For *Riccia fluitans* Hellwege (unpubl.) observed that ABA is metabolized to phaseic and dihydrophaseic acid, similarly known for higher plants and algae. This metabolic step can be inhibited by tetcyclacis, an inhibitor of cytochrome P-450 monooxygenases, similarly described by Daeter and Hartung (1990) for leaf tissues. After alkaline hydrolysis of aqueous fractions ABA is released from conjugates, indicating that most probably the ABA glucose ester is present in *Riccia fluitans*.

b) The Physiological Role of ABA

α) Developmental Processes – The Conversion of Submerged Thalli into Terrestrial Forms

Some members of the *Marchantiales* (*Riccia fluitans, Ricciocarpus natans*) live under or on the surface of water and undergo characteristic developmental changes when occupying the land, as can occur when a pond is drying out. During the early stages of this process, ABA is increased in these thalli, most probably in the meristems, by a factor of 20–30. Treatment of the waterform with ABA also induces the conversion of the water- into the landform (Hellwege et al. 1992). This process is accompanied by characteristic changes of the pattern of polypeptides, which allow the conclusion that during the occupation of the land, a specific set of proteins is formed under the influence of ABA (Hellwege et al. 1994a). The function of these proteins is as yet unknown. They seem to be different from ABA-induced dehydrins of the resurrection plant *Craterostigma*. Antibodies which were raised against those by Bartels et al. (1992) and Schneider et al. (1993) do not recognize the ABA-induced proteins of *Riccia fluitans*. On the other hand, however, in both the ABA-treated waterform and the landform, antibodies raised against a specific 31-kDa tonoplast protein (Betz and Dietz 1991) recognize one of the ABA-inducible proteins. The formation of the landform seems to be accompanied by the synthesis of a specific tonoplast protein. It should be emphasized that the stress to which the *Riccia* thalli are exposed is very mild. We obviously have an extremely sensitive stress response leading to an ABA accumulation. This is different from many tissues of higher plants, where threshold levels of stress up to –2.0 MPa have to be reached to induce ABA accumulation (Hartung and Heilmeier 1993).

Table 3. Physiological responses of bryophytes after treatment with exogenous ABA

Species	Physiological process	Response	Reference
Funaria hygrometrica	Bud formation	Inhibition	Valadon and Mummery 1971
Physcomitrium pyriforme	Growth and differentiation	Inhibition	Menon and Lal 1974
Funaria hygrometrica	Growth of protonema Formation of caulonema	Inhition Inhibition	Lehnert and Bopp 1983
Timmiella anomala	Formation of gemmae	Promotion	Chopra and Kapur 1989
Hyophila involuta	Formation of gemmae	Inhibition	Mehta 1990
Hepaticae, musci	Gametangial development	Inhibition	Chopra and Sarla 1990
Marchantia palmata callus culture	Callus growth Callus differentiation	Inhibition Promotion	Dhingra-Babbar and Chopra 1986
Mannia dichotoma callus culture	Callus growth Rhizoid formation	Inhibition Inhibition	Kapur and Chopra 1991

Other developmental changes which are induced by exogenous ABA are listed in Table 3. In most cases, exogenous ABA seems to inhibit growth processes, except the formation of gemmae in *Timmiella*. Most of these data are difficult to discuss. Detailed dose-response curves are lacking, as well as attempts to correlate the courses of endogenous ABA with the physiological phenomena observed. The physiological significance of these observations remains obscure.

β) Stomatal Reactions

The stoma-bearing sporophytes of *Anthocerotae* contain particularly high ABA amounts which can be further increased under stress conditions (Hartung et al. 1987). The stomata also react to exogenous ABA by closing, whereas fusicoccin (FC) causes opening. The ABA dose response strongly resembles that of higher plants, such as *Valerianella locusta* (Hartung 1983). Data on ABA concentrations of the stoma-bearing sporophytes of *Musci* are not available. However, Garner and Paolillo (1973) have demonstrated that stomata of sporophytes of *Funaria* respond to ABA in a very similar manner. The physiological significance of the ABA effect on guard cells of *Anthoceros* is unclear. The guard cells of *Anthocerotae* are not connected to a water-transporting system. Some of them do not even have a significant substomatal cave. Their role in gas exchange is negligible. Obviously, a stress physiological system for ABA was developed at a very early stage of evolution, although no ecophysiological need for such a system existed.

γ) Induction of Desiccation Tolerance

Table 2 shows that the great variability of endogenous ABA content in *Marchantiales* can be attributed to a different desiccation tolerance. The extremely desiccation-tolerant species of the genus *Exormotheca* have the highest, the extremely hydrophilic species such as *Riccia fluitans* or *Cyathodium africanum* exhibit the lowest ABA amounts. Similarly, the desiccation-tolerant tubers of *Anthoceros dichotomus*, which survive longer periods of dryness, are particularly rich in ABA, when compared with the thalli and sporophytes of this species.

Werner et al. (1991) and Bopp and Atzorn (1992) have shown that protonemata of *Funaria hygrometrica* can survive desiccation when drying out slowly, a condition which allows a significant increase in endogenous ABA. This is not possible during rapid drying. When treated with ABA, nonhardened *Funaria* protonemata can also survive desiccation treatment. The same is true also for the xerophilic species *Exormotheca holstii*. ABA content fluctuated, depending on the water status, by 10- to 25-fold during cycles of dehydration and hydration. In *Marchantia polymorpha*, ABA increases only by two- to three-fold, which is not sufficient for induction of desiccation tolerance (Hellwege et al. 1994b). Bopp and Werner (1993) and Werner and Bopp (1993) have shown that in desiccated or ABA-treated protonemata, several proteins were synthesized which are comparable to the dehydrins of barley and corn described by Close et al. (1989). ABA stimulated the phosphorylation of these proteins in *Funaria* (Werner and Bopp 1993).

Bopp and Werner (1993) have probed the ABA-induced polypeptides of *Funaria hygrometrica* protonemata with antibodies raised by Schneider et al. (1993) against some of the *Craterostigma* dehydrins. They were able to recognize the dehydrins of *Funaria*, which indicates that those of both systems are very similar.

Hellwege et al. (1994b) also detected a group of proteins in ABA-treated and desiccated thalli of *Exormotheca*. One of these is recognized by an antibody of a cytosolic dehydrin of *Craterostigma* (Bartels et al. 1992) but not by the antibody raised against a plastidal dehydrin (Schneider et al. 1993). Drought- and ABA-dependent polypeptides are absent in nonhardened thalli which were cultivated in the glasshouse for several weeks under moist, unstressed conditions, as long as they are not incubated with ABA, a treatment which revives their desiccation resistance. These studies indicate that mechanism of desiccation tolerance occured very early in the evolution.

4. Abscisic Acid in Fungi

In 1977, Assante et al. showed that *Cercospora rosicola* produces ABA in large amounts. Since then, ABA has been detected in many species of *Ascomycetes*, *Fungi imperfecti*, and some *Basidiomycetes*. Recent reports were published by Dörffling et al. (1984), Dahiya et al. (1988); Filimonova and Talieva (1989), Crocoll et al. (1991) and Talieva et al. (1991). Baker's yeast was shown to produce optically active cyclohexanone analogs of ABA (Lamb and Abrams 1990).

a) Biosynthesis and Metabolism

The biosynthetic pathway of ABA has been investigated in detail in different *Cercospora* species (*C. rosicola:* Norman et al. 1988; Al-Nimri and Coolbaugh 1990a, b, 1991; Bennett et al. 1990. *C. cruenta*, Oritani et al. 1988, 1990; Yamamoto et al. 1988, 1990. *C. pini densiflori*, Okamoto et al. 1988a, b, c). All these investigations provided evidence that ABA was synthesized directly via farnesyl-diphosphate, different ionylidene derivatives, and deoxy-ABA. This is different from higher plants, where nowadays the indirect pathway via violaxanthin and xanthoxin seems to be well accepted. Norman (1991), however, concludes from her experiments that β-carotene, which was detected in *Cercospora rosicola*, could also serve as an ABA precursor. She concludes this also from inhibitor studies with fluridone and decylimidazole. These compounds inhibit both β-carotene and ABA formation. Takahama et al. (1988) showed that addition of plant extracts to the culture medium stimulated fungal ABA production. Hartung and Eggeling (1984), Kettner and Dörffling (1987), and Kettner (1990) were not able to find conversion of radioactive ABA to phaseic acid (PA) and dihydrophaseic acid (DPA), as seems to be the case in green organisms, including algae and liverworts. One may conclude that internal ABA levels are regulated by release to the extracellular space rather than metabolism. On the other hand, Arnone et al. (1990) observed biotransformation of (±)-ABA with cultures of *Aspergillus niger* to (-)2'.3'-dihydro-ABA.

b) Transport of ABA Across Fungal Membranes

Hartung and Eggeling (1984) have studied the uptake of radioactive ABA into mycelia of *Cercospora cruenta*. They observed that, other than in cells of higher plants, ABA is taken up under alkaline conditions (pH 8) by the mycelia at only slightly reduced rates, compared to acid conditions. This indicates a high permeability of the fungal plasma membrane for the ABA-anion (ABA^-) (Fig. 1). Consequently, ABA cannot be distributed within the mycelium compartments according to the anion trap concept (Hartung and Slovik 1991). ABA which is synthesized in the cytosol of the fungus should therefore be released very quickly and effectively to the extracellular space. When we compare fluxes across the plasma membrane of different cells with *Cercospora cruenta* (Behl and Hartung 1987; Table 4), those of *Cercospora* prove to be the highest. This fits well to the fact that in liquid cultures large amounts of ABA, up to the milligram range, can be detected within a relatively short time of culture in the medium. It also explains the absence of metabolizing systems in the cytosol, and it also could play an important role in the possible functions of ABA in fungus-plant interactions.

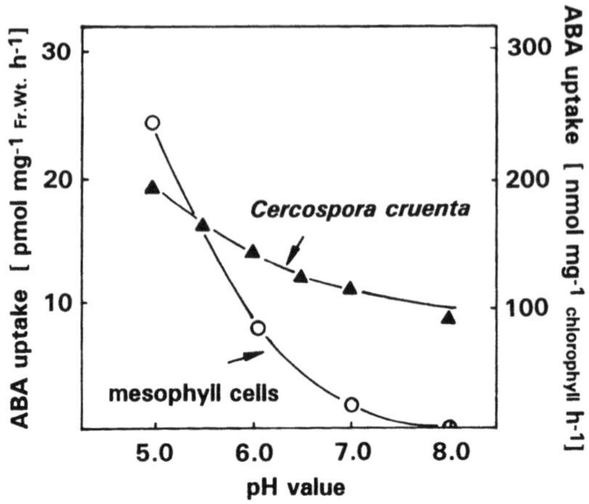

Fig. 1. Uptake of radioactive ABA in leaf mesophyll cells and in mycelium of *Cercospora cruenta*, dependent on the external pH

Table 4. ABA fluxes (Φ_{CO}) and rate constants (k_{CO}) across the plasma membrane. External pH: 5.8, external ABA concentration: 1 μM. MC = mesophyll cell; GC = guard cell. (After Behl and Hartung 1987)

	Valerianella locusta		*Hordeum distichon*	*Cercospora cruenta*
Cell type	MC	GC	Root cortex	Mycelia
Φ_{CO} (nmol g^{-1}FW h^{-1})	–	–	0.35	2.94
k_{CO} (h^{-1})	0.7	0.9	0.97	2.60

c) Functions of Fungal ABA

There are only few published data on the effects of exogenous ABA on growth and developmental processes in fungi. Stopinska and Michniewicz (1988) found stimulation of mycelium growth and spore germination and inhibition of production of perithetia and cylindrial spores by ABA in *Ceratocystis fimbriata.* However, no correlation between growth of mycelium and endogenous ABA could be observed. Gogala (1989) observed a weak stimulation of growth of the mycorrhizal fungus *Suillus variegatus*. We could find no ABA effects on the growth of *Cercospora cruenta* and *Cercospora rosicola* mycelia.

α) ABA of Pathogenic Fungi

It has been shown that fungus-infected plants increase (Chigrin 1986; Bousquet et al. 1990, 1991) or decrease (Cahill and Ward 1989a, b) in ABA level. Peltonen et al. (1991) could observe no ABA changes in spring wheat infected with *Septoria nodorum.* Careful investigations by Kettner (1990) with fungus-infected ABA-deficient tomatoes showed that fungal-produced ABA participated strongly in the increase in ABA level after infection and that ABA of fungal origin is metabolized by the host to PA and DPA. Due to the high permeability of fungal membranes for ABA and the high biosynthetic rates of ABA in fungi, this finding is not surprising. Bertrand et al. (1989) performed an immunocytochemical investigation using the immunogold technique with healthy and infected tomato root tips (*Fusarium oxysporum* f.sp. *radicis-lycopersici*). In their system, ABA accumulated in root cap cell walls and in the cells surrounding the columella. In diseased plants the labeling pattern was more irregular. Cell walls and cytoplasm were labeled by only a few scattered gold particles.

It also has been reported that ABA, either exogenous or endogenous, may be involved in the resistance or susceptibility of plants to fungal diseases. Bostock et al. (1990); Li and Heath (1990) Hammerschmitt (1984), Salt et al. (1986), and Michniewicz et al. (1990) have described increased susceptibility of hosts infected with *Peronospora, Phytophtora*, rust fungi, and different *Fungi imperfecti*. There are also a few examples of beneficial effects of ABA on plants after fungal infections (Dunn et al. 1990; Ersek et al. 1991). For more information and references, see the review article of Fraser (1991).

β) ABA in Mycorrhizal and Rhizospheric Fungi

In higher plants, ABA plays a very important ecophysiological role as a stress signal which is sent from roots growing in a drying soil to the leaves to affect stomatal reactions and leaf growth (Davies and Zhang 1991). It has been pointed out repeatedly that ABA does not originate exclusively from the root system, but may also be released from soil microoorganisms such as fungi and algae to the rhizospheric water (Hartung et al. 1990). Müller et al. (1989) speculated that plant hormones of the rhizosphere affect roots and may be taken up by the root system. Preliminary computer modeling of stress-dependent ABA redistribution in roots (Daeter et al. 1993) postulates the existence of at least 0.1–1.0 nM ABA in the rhizosphere. Otherwise a significant portion of root-synthesized ABA would be wasted by release into the surrounding soil. Indeed, Hartung et al. (1990) have found water-extractable ABA in significant amounts in desert soil taken from the direct vicinity of root systems. Lower ABA concentrations were found in compost soils. This ABA may also be of fungal origin and released to the soil water because of the high permeability of the fungal plasma membrane for ABA. Also mycorrhizal fungi seem to be able to synthesize ABA and to release it to the root systems. Dannenberg et al. (1993) have shown that in roots of VAM-infected maize plants, of all hormones investigated only the ABA level was increased significantly. Similar observations were made by Murakami-Mizukami et al. (1991) with nodules of soybean plants bearing VA mycorrhizas. On the other hand, Coleman et al. (1990) analyzed ectomycorrhizal Douglas fir seedlings and could find no altered ABA contents. More work is required in this important field, especially the effect of stress on fungal production of ABA.

5. ABA in Lichens

No ABA studies with lichens have been published until now. We have analyzed ABA in *Ramalina maciformis* and found that the ABA level was increased transiently when dry thallus was moistened, and decreased afterwards significantly. We would like to propose an important putative stress-physiological role of ABA for lichens. The fungal partner may be a potent site of ABA biosynthesis, especially under stress conditions. This ABA could be released easily to the algal partner because of the special permeability properties of the fungal plasma membranes. The synthesis and release may be increased significantly during the course of desiccation. Strongly increased ABA concentrations could induce dehydrins, as already found and discussed for resurrection plants, *Cyanobacteria* (Close and Lammers 1993), xerophilic liverworts, and moss protonemata. Thus, a biochemical basis for the impressive desiccation tolerance of lichens could be explained, as well as a beneficial role of the fungus as a potent and effective ABA supplier for the algal partner and the whole lichen system. Experimental work on that interesting problem is in progress.

6. Conclusions

At present, we are far from understanding the function of ABA in algae. ABA is universally distributed within the algal kingdom. As in higher plants, ABA accumulates under stress conditions and its metabolism seems to be similar to that in higher plants. However, the indirect biosynthetic pathway via violaxanthin and xanthoxin seems to be unlikely. At the present stage, it cannot be excluded that in many *Cyanobacteria* and algae ABA is still a nonessential, secondary product of algal metabolism, which appeared during the progress of algal evolution to gain the functions of a phytohormone later at the bryophyte level. The situation seems to be much clearer in bryophytes. Here, ABA acts as a stress hormone by inducing desiccation resistance in several species and developmental processes. It would be worth investigating aerial algae such as *Trentepohlia* or algae on temporarily moistened and drying habitats, e.g., on rocky seashores, for their ABA relations. We cannot exclude that ABA may induce desiccation resistance also in such species. ABA of fungal origin may be released into the surrounding environment in order to affect stress physiological processes in plant partners, such as algae in the lichen systems, or plant roots which are in contact with rhizopheric and mycorrhizal fungi.

Acknowledgment. This study was supported by the Deutsche Forschungsgemeinschaft (SFB 251).

References

Addicott FT, Carns HR (1983) In: Addicott FT (ed) Abscisic acid. Praeger, New York, pp 1–21. **Ahmad MR,** Saxena PN, Amla DV et al. (1978) Beitr Biol Pflanz 54:33–40. **Al-Nimri LF,** Coolbaugh RC (1990a) J Plant Growth Regul 9:221–226. **Al-Nimri LF,** Coolbaugh RC (1990b) J Plant

Growth Regul 10:63–66. **Al-Nimri IF,** Coolbaugh RC (1991) Plant Physiol (Suppl) 93:71. **Arnone A,** Cardillo R, Nasini G, DePava OV (1990) J Chem Soc Perkin Trans I, 1990:3061–3064. **Assante G,** Merlini C, Nasini G (1977) Experientia 33:1556–1557.

Bartels D, Hanke C, Schneider K, Michel D, Salmini F (1992) EMBO J 11:2772–2778. **Bearder JR** (1980) In: Pirson A, Zimmermann MH (eds) Encyclopedia of plant physiology, Springer, Berlin Heidelberg New York, vol 9, pp 9–112. **Behl R,** Hartung W (1987) Bull Growth Regul Soc Am 15:1–13. **Bennett RD,** Norman SM, Maier VP (1990) Phytochemistry 29:3473–3478. **Bertrand S,** Benhamou N, Nadeau P, Dostaler D, Gosselin A (1989) Can J Plant Pathol 11:186. **Betz M,** Dietz KJ (1991) Plant Physiol 97:1294–1301. **Bopp M** (1990a) In: Pharis RP, Rood SB (eds) Plant growth substances. Springer, Berlin Heidelberg New York, pp 1–10. **Bopp M** (1990b) In: Chopra RN, Bhatla SC (eds) Bryophyte development: physiology and biochemistry. CRC, Boca Raton, pp 55–78. **Bopp M,** Atzorn R (1992) Naturwissenschaften 79:337–346. **Bopp M,** Bhatla SC (1990) Cr Rev Plant Sci 9:317–328. **Bopp M,** Werner O (1993) Bot Acta, 106:103–106. **Bopp-Buhler ML,** Wabra P, Hartung W, Gimmler H (1991) Cryptogam Bot 2:192–200. **Bostock RM,** MacDonald JD, Duniway JM, Stiltes J (1990) Phytopathology 80:1023. **Bousquet JF,** Touraud G, Piollat MT, Bosch U, Trottet M (1990), J Agron Crop Sci 165:297–300. **Bousquet JF,** Touraud G, Piollat MT, Bosch U, Trottet M (1991) Agronomie 11:777–786. **Boyer GL,** Dougherty SS (1988) Phytochemistry 27:1521–1522. **Bralczyk J,** Wielgat B, Wsielewskajy-Dabrowska LD, Kleczkowski K (1978) Plant Sci Lett 12:265–272.

Cahill DM, Ward EWB (1989a) Phytophathology 79:1238–1242. **Cahill DM,** Ward EWB (1989b) Physiol Mol Plant Pathol 35:483–494. **Chigrin VV** (1986) Zh Obshch Biol 47:310–327. **Chopra RN,** Kapur A (1989) Plant Sci 61:203–206. **Chopra RN,** Sarla (1990) In: Chopra RN, Bhatla SC (eds) Bryophyte development: physiology and biochemistry. CRC, Boca Raton, pp 167–190. **Close TJ,** Lammers PJ (1993) Plant Physiol 101:773–779. **Close TJ,** Kortt AA, Chandler PM (1989) Plant Mol Biol 13:95–108. **Coleman MD,** Bledsoe CS, Smit BA (1990) New Phytol 115:275–284. **Cowan AK,** Railton ID (1986) J Plant Growth Regul 4:211–224. **Cowan AK,** Rose PD (1991) Plant Physiol 97:798–803. **Crocoll C,** Kettner J, Dörffling K (1991) Phytochemistry 30:1059–1060.

Daeter W, Hartung W (1990) J Plant Physiol 136:306–312. **Daeter W,** Slovik S, Hartung W (1993) Philas Trans R Soc Lond (in press). **Danneberg G,** Latus C, Zimmer W, Hundshagen B, Schneider-Poetsch JH, Bothe H (1993) J Plant Physiol 141:33–39. **Dahiya JS,** Tewari JP, Woods DL (1988) Phytochemistry 27:2983–2984. **Davies WJ,** Zhang J (1991) Annu Rev Plant Physiol 42:55–76. **Dhingra-Babbar S,** Chopra RN (1986) Beitr Biol Pflanz 61:467–476. **Dörffling K,** Petersen W, Sprecher E, Urbasch I, Hansen HP (1984) Z Naturforsch 39:683–684. **Dunn RM,** Hedden P, Bailey JA (1990) Physiol Mol Plant Pathol 36:339–350.

Ersek T, Ladyzhenskaya EP, Korableva NP, Adam A (1991) J Phytopathol 132:146–152.

Filimonova MV, Talieva MN (1989) Izv Akad Nauk SSSR, Ser Biol 1989:540–546. **Filimonova MV,** Talieva MN, Andreev LN (1989) Dokl Akad Nauk SSSR 307:1019–1024. **Fraser RSS** (1991) In: Davies WJ, Jones HG (eds) (1991) Environmental plant biology series: abscisic acid: physiology and biochemistry. Bios, Oxford, pp 189–200.

Garner LB, Paolillo DJ (1973) Bryologist 76:423–427. **Gimmler H,** Hartung W (1988) J Plant Physiol 133:165–172. **Gogala N** (1989) Agric Ecosys Environ 28:151–154.

Hammerschmidt R (1984) Physiol Plant Pathol 24:33–42. **Hartung W** (1983) Plant Cell Environ 6:427–428. **Hartung W,** Eggeling S (1984) Tagung Deutsch Bot Ges, Wien, Abst, 1984:43. **Hartung W,** Heilmeier H (1993) In: Jackson MB (ed) Interaction stresses on plants in a changing climate. Springer, Berlin Heidelberg New York (in press). **Hartung W,** Slovik S (1991) New Phytol 119:361–382. **Hartung W,** Heilmann B, Gimmler H (1981) Plant Sci Lett 22:235–242. **Hartung W,** Weiler EW, Volk OH (1987) Bryologist 90:393–400. **Hartung W,** Heilmeier H, Wartinger A, Kettemann I, Schulze ED (1990) Isr J Bot 39:373–382. **Hellwege E,** Volk OH, Hartung W (1992) J

Plant Physiol 140:553–556. **Hellwege E,** Dietz KJ, Hartung W (1994a) (in preparation). **Hellwege E,** Dietz KJ, Volk OH, Hartung W (1994b) Planta (submitted). **Hetherington AM,** Quatrano RS (1991) New Phytol 119:9–32. **Hirsch R,** Hartung W, Gimmler H (1989) Bot Acta 102:326–334. **Huang CY** (1991) Bot Bull Acad Sin (Taipeh) 32:265–270. **Huddart H,** Smith RJ, Langton PD et al. (1986) New Phytol 104:161–174. **Hussain H,** Boney AD (1973) New Phytol 72:403–410.

Ishiura M (1976) Plant Cell Physiol 17:1141–1150.

Jacobs WP (1985) In: Bopp M (ed) Plant growth substances 1985. Springer, Berlin Heidelberg New York, pp 249–265. **Jennings RC** (1969) New Phytol 68:683–688.

Kapur A, Chopra RN (1991) J Hattori Bot Lab 1991:107–116. **Kentzer T,** Mazur H (1991) Acta Physiol Plant 13:153–157. **Kentzer T,** Borowaczak E, Szczepkowska E (1975a) Pol Arch Hydrobiol 22:413–427. **Kentzer T,** Borowczak E, Szczepkowska E (1975b) Pol Arch Hydrobiol 22:413–427. **Kettner J** (1990) Diss Univ Hamburg 1990. **Kettner J,** Dörffling K (1987) Physiol Plant 69:278–282. **Kingsham AR,** Moore J (1982) Bot Mar 25:149–154.

Lamb N, Abrams SR (1990) Can J Chem 68:1151–1162. **Lehnert B,** Bopp M (1983) Z Pflanzenphysiol 110:379–391. **Li A,** Heath MC (1990) Physiol Mol Plant Pathol 37:245–254.

Malliga P, Subramanian G (1989) Curr Sci (Bangalore) 58:634–635. **Marsalek B,** Simek M, Lukesova A (1991) Dev Plant Soil Sci 1991:529–530. **Marsalek B,** Zahradnickova H, Hronkova M (1992) J Plant Physiol 139:506–508. **Marsalek B,** Zahradnickova H, Polisenska The extracellular production of abscisic acid by green algae under salt, acid, drought stress and senescence. J Planar Chromatogr (in press). **McElwain EF,** Bohnert HJ, Thomas JC (1992) Plant Physiol 99:1261–1264. **Mehta P** (1990) Phytomorphology 40:119–124. **Menon MKC,** Lal M (1974) Planta 115:319–328. **Michniewicz W,** Czerwinska E, Rozej B (1990) Acta Physiol Plant 12:41–48. **Moss B** (1965) New Phytol 64:387–399. **Müller M,** Deigele C, Ziegler H (1989) Z Pflanzenernähr Bodenkd 152:247–254. **Murakami-Mizukami Y,** Yamamoto Y, Yamaki S (1991) Soil Sci Plant Nutr 37:291–298.

Niemann DI, Dörffling K (1980) J Phycol 16:383–389. **Norman SM** (1991) Plant Growth Regul 10:103–108. **Norman SM,** Poling VP, Pon DL (1988) Agr Biol Chem 52:1309–1310.

Okamoto M, Hirai N, Koshimizu K (1988a) Phytochemistry 27:2099–2104. **Okamoto M,** Hirai N, Koshimizu K (1988b) Phytochemistry 27:3465–3470. **Okamoto M,** Hirai N, Koshimizu K (1988c) Mem Coll Agric Kyoto Univ 132:76–115. **Ord GNSG,** Cameron UF, Fensom DS (1977) Can J Bot 55:1–4. **Oritani T,** Sasaki K, Yamashita K (1988) Agric Biol Chem 52:2119–2122. **Oritani T,** Yamamoto H, Yamashita K (1990) Agric Biol Chem 54:125–130.

Peltonen J, Kittilä S, Peltonen-Sainio P, Karjalinen R (1991) Crop Protec 10:260–264. (1991). **Pryce RJ** (1972) Phytochemistry 11:1759–1761.

Quarrie SA, Lister PG (1984) Z Pflanzenphysiol 114:309–314.

Radley M (1961) Nature 191:684–685. **Rowan KS** (1989) Photosynthetic pigments in algae. Cambridge University Press, Cambridge.

Sabbatini MR, Arguelo JA, Fernandez OA, Bottini RA (1987) Aquat Bot 28:189–194. **Saidova NS,** Kof EM Kefeli VI (1988) Ontogenez 19:200–208. **Salerno GL** (1985) Physiol Plant 64:259–264. **Salt SD,** Tuzun S, Kuc J (1986) Physiol Mol Plant Pathol 28:287–297. **Schneider K,** Wells B, Schmelzer E, Salamini F, Bartels D (1993) Planta 189:120–131. **Smith RJ,** Hobson S, Ellis IR (1987) New Phytol 105:543–550. **Stopinska J,** Michniewicz M (1988) Bull Pol Acad Sci Biol 36:253–258.

Takahama T, Yoshida H, Araki K, Nakayama K (1988) Biotechnol Lett 5:55–58. **Talieva MN,** Filimononva MV, Andreev LN (1991) Izv Akad Nauk SSSR Ser Biol 1991:194–200. **Tester M,** MacRobbie EAC (1990) Planta 180:569–581. **Tietz A,** Kasprik W (1986) Biochem Physiol Pflanz 181:269–274. **Tietz A,** Ruttkowski U, Koehler R, Kasprik W (1989) Biochem Physiol Pflanz 184:259–266. **Tillberg JE** (1970) Physiol Plant 23:647–653.

Ullrich WR, Kunz G (1984) Plant Sci Lett 37:9–14.

Valadon LRG, Mummery R (1971) Physiol Plant 24:232–234.

Wanless IR, Bryniak N, Fenson DS (1973) Can J Bot 51:1055–1070. **Wattal RK,** Sachar RC (1991) In: Verna R, Nishizuka Y (eds) Serono Symp Ser Adv Exp Med. Raven New York, vol 4, pp 323–331. **Weiler EW** (1979) Planta 144:255–264. **Werner O,** Bopp M (1993) J Plant Physiol 141:93–97. **Werner O,** Maria Ros Espin R, Bopp M, Atzorn R (1991) Planta 186:99–103.

Yamamoto H, Oritani T, Yamashita K (1988) Agric Biol Chem 52:2683–2684. **Yamamoto H,** Oritani T, Yamashita K (1990) Agric Biol Chem 54:1923–1930.

Zahradnickova H, Marsalek B, Polisenska M (1990) J Planar Chromatogr 3:243–246. **Zahradnickova H,** Marsalek B, Polisenska M (1991) J Chromatogr 555:239–246.

Prof. Dr. Wolfram Hartung
Prof. Dr. Hartmut Gimmler
Lehrstuhl Botanik I
Julius-von-Sachs-Institut für Biowissenschaften
Mittlerer Dallenbergweg 64
D-97082 Würzburg, FRG

VIII. UV-B Effects on Terrestrial Plants and Aquatic Organisms

By Manfred Tevini

1. General Introduction

UV-B effects on plants have been studied more frequently since local and global ozone depletion scenarios have given rise to the expectancy of increasing UV-B radiation (280-320 nm) on the earth's surface. This increase has already been observed in Antarctica during the ozone hole development (Lubin et al. 1989; Frederick and Alberts 1991) as well as in the Northern hemisphere (Blumthaler and Ambach 1990; Blumthaler 1993). The danger of increased UV-B to the biosphere has been documented in two UNEP (United Nations Environmental Programme) reports of 1989 and 1991, by Scope (1992) and van der Leun and Tevini (1992). Further reviews of UV-B effects in terrestrial or aquatic plants have been published during the past 5 years by Caldwell et al. (1989), Bornman (1989), Smith (1989), Tevini and Teramura (1989), Krupa and Kickert (1989), ElSayed et al. (1990), Voytek (1990), Karentz (1991), Raven (1991), Stapleton (1992), Tevini (1993), Häder (1993), Bornman and Teramura (1993).

UV-B radiation can be applied by artificial UV-B sources to biological material either in reaction vessels, plastic bags, quartz bottles, growth chambers, greenhouses, or in the field supplementing solar radiation. In all cases, the biological effectiveness of the UV-B radiation has to be determined in order to compare the various simulated ozone depletion scenarios and to assess the biological consequences of solar UV-B radiation changes due to ozone changes. UV-B can have damaging effects on DNA and/or physiological processes of the organisms as various mono- and polychromatic action spectra demonstrate (Caldwell 1971; Setlow 1974; Caldwell et al. 1986; Coohill 1989, 1991; Häder et al. 1991; Cullen et al. 1992).

Figure 1 shows action spectra in the UV range, all of which indicate increases in biological effectiveness with shorter wavelengths. Ozone depletion shifts the solar cutoff to shorter wavelengths and thus increases the irradiance at a given wavelength due to ozone absorption (Fig. 1). Depending upon the shape of the action spectrum which is used as a weighting function for the irradiation source, the ozone simulation scenarios vary from drastic cases where a DNA action spectrum (Setlow 1974) is used, through moderately damaging cases from the use of the generalized plant action spectrum (Caldwell 1971) or lower effects when a polychromatic action spectrum for seedling growth are used (Tevini and Steinmüller 1988). Polychromatic action spectra (Rundel 1983; Caldwell et al. 1986; Steinmüller 1986) include the beneficial effects of background white light on photorepair of DNA damage, or on accumulation of target-shielding pigments and are thus closer to

Progress in Botany, Vol. 55

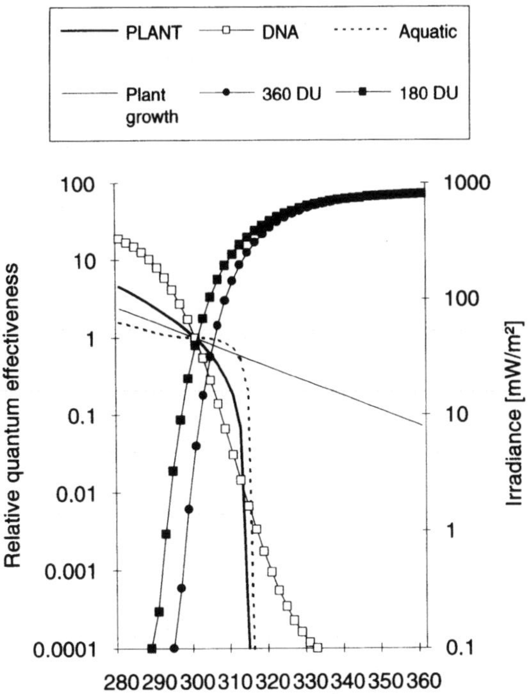

Fig. 1. UV-action spectra for PLANT-damage (Caldwell 1971), DNA-damage (Setlow 1974), Inhibition of motility in *Euglena* (Aquatic, Häder et al. 1991) and plant growth reduction (Plant growth, Tevini and Steinmüller 1988) and computed UV irradiance due to an ozone depletion from 360 Dobson units (DU) to 180 DU over Germany in summer on cloudless days. (Green et al. 1980)

the natural radiation conditions. A new absolute action spectrum for pyrimidine dimer induction in intact alfalfa leaves also indicates that the slope in the short wavelength band is less steep when DNA is shielded (Quaite et al. 1992a), which consequently lowers the predicted impact of ozone depletion when only DNA-related responses are considered. This impact could even be absent where DNA dimers are repaired by visible light, a process called photoreactivation. This emphasizes the importance of maintaining a realistic balance between UV-B, UV-A, and longer wavelengths, which is largely not the case under artificial irradiation conditions. Furthermore, it is well known that physiological processes like photosynthesis are also negatively affected by UV-B. However, the damage is much lower in the presence of high white light (Mirecki and Teramura 1984; Cen and Borman 1990; Adamse and Britz 1992), indicating a modulating contribution of polychromatic radiation to the net effect of the UV-induced responses. Thus, the use of only one weighting function for serious predictions is both very problematic and dangerous, since action spectra are very complex and strongly related to the individual plant and to its different physiological responses which change with the environment. In addition, due to the absorption of many interacting compounds (e.g., UV-absorbing proteins and pigments), a photoreceptor molecule cannot be identified which clearly matches the response curve.

There are only a few action spectra available for specific UV-B effects on phytoplankton (Häder et al. 1991; Cullen et al. 1992; Helbling et al. 1992). Recently, one was presented for the diatom *Phaeodactylum* based on photosynthetic measurements (Cullen et al. 1992); this is similar to an action spectrum for photoinhibition of photosynthesis in *Rumex patientia* (Caldwell et al. 1986). Applying this action spectrum to weight biological effectiveness of increased UV-B in antarctic phytoplankton, a reduction in near-surface photosynthesis by 12–15% was calculated for the antarctic ozone hole. Since the action spectra for inhibitory effects differ from the DNA absorption spectrum in some organisms (Häder et al. 1991), and since UV radiation impaired motility can be measured after 10 min – too short a time to account for protein resynthesis – other mechanisms must mediate the UV-B response (Häberlein and Häder 1993).

2. Effects on Terrestrial Plants

The biological effects and consequences of enhanced UV-B radiation for terrestrial plants are numerous but nevertheless still insufficiently known. Most knowledge has come from crop plants mainly of temperate regions, whereas tropical plants and natural ecosystems have been more or less neglected. Studies of more than 300 plant species and cultivars (out of about 300 000 seed plant species) have been carried out, and about 50% of these plants have been considered as UV-sensitive, where sensitivity is defined as any morphological, physiological, or biochemical change induced by UV-B.

Tree species, especially conifers in the seedling stage, are also susceptible to enhanced UV-B radiation (Sullivan and Teramura 1991). Of ten conifer species, four were susceptible to UV-B in terms of reductions in height or biomass (Sullivan and Teramura 1988).

a) Growth

In many sensitive plant species reduced leaf area and/or stem growth has been found as, for example, in wheat, rice, maize, rye, soybean, sunflower, and cucumber (Tevini and Teramura 1989). Recently, two comparable studies on rice cultivars of different geographical regions have been performed in greenhouses using 19 kJ m^{-2} d^{-1} (plant weighted, normalization at 300 nm) as an enhanced daily dose (Dai et al. 1992) and 15.7 kJ m^{-2} d^{-1} (Teramura et al. 1991) respectively, compared to about 11 kJ m^{-2} d^{-1} typically found in rice-growing areas near the equator (Bachelet et al. 1991). From 16 rice cultivars native to the Philippines, India, Thailand, China, Vietnam, Nepal, and Sri Lanka, about one-third showed statistically significant decreases in total biomass and leaf area. Tiller number, highly correlated with yield (Paulsen 1987), was reduced in 6 of these 16 cultivars. The Sri Lanka cultivar Kurkaruppan, however, showed increases in both total biomass and tiller number, indicating that selective breeding might be a succesful tool for obtaining UV-B tolerant cultivars (Teramura et al. 1991). In the second study on four cultivars from the Philippines based on total biomass changes, different cultivar responses were reported, with IR74 the most sensitive and IR64 the least sensitive (Dai et al. 1992). A further study on four Japanese rice cultivars revealed similar results, but unfortunately did not state the UV-radiation used in the phytotron in biological effective units (Kumagai and Sato 1992), realizing that also UV-C radiation was emitted, which is not present in solar radiation. As stated before, proper irradiation conditions are the most crucial factors in evaluating and assessing effects of solar UV-B changes. One of the newest techniques to employ solar UV-B only was developed by Tevini et al. (1990). Two identical growth chambers were located at a southern latitude with naturally high UV-B levels. These levels were attenuated by passing ozone through the Plexiglas cuvette on top of one growth chamber. Plants beneath this cuvette received less than ambient UV-B which may be considered as a control UV-B radiation received at more northern latitudes. Plants in the second growth chamber were covered by a cuvette containing ambient air and thus received relatively enhanced UV-B compared to the control. With this ozone technique a difference in solar UV-B radiation of 20% was simulated and growth, function, and composition of sunflower and corn seedlings have been intensively studied over several summer seasons (Tevini et al. 1990, 1991a, b; Mark 1992). In both sunflower and corn seedlings, growth paramaters like plant size and leaf area were significantly reduced, as well as maximum photosynthesis based on chlorophyll content which remained unchanged under high UV-B radiation. These observations confirm results of artificial UV-B applications in the field or in growth chambers, although not in all cases were growth reductions correlated with lower biomass accumulation or photosynthesis. Changes in morphological responses, such as increased branching and number of leaves, can lead to changes in competition for light (Barnes et al. 1990). In a competition study on wheat and wild oat, it was shown that under enhanced UV-B levels wheat had a competitive advantage over wild oat when they were grown together. By using a canopy microclimate and radiation interception model, these competition differences can be quantitatively explained by shifts solely in the growth form of these species (Ryel et al. 1990). Growth reductions were also found in wild-type and stable-phytochrome-deficient mutants of cucumber, indicating that growth in the latter case is regulated by an unknown UV-B receptor, not by phytochrome (Ballaré et al. 1991a).

The molecular reasons for growth reductions can be attributed to changes in DNA and/or phytohormones. Phytohormones can affect growth by altering their concentration in the growth sensitive tissue and by changing phytohormone dependent processes. Elongation growth is related to indole acetic acid (IAA), which absorbs in the UV-B range and is readily photodestroyed by UV-B in vitro and in vivo as shown in sunflower seedlings under low white light conditions. Furthermore, the plastic epidermal cell wall extensibility, which is enhanced in auxin-induced elongation growth, was also reduced (Ros 1990). Peroxidase activity, which can reduce cell elongation at high activity by different mechanisms, was enhanced in UV-B irradiated sunflower (Ros 1990) and sugar beet plants (Panagopoulos et al. 1990). Another phytohormone, ethylene, which changes elongation to radial growth, is produced to a greater extent in UV-B irradiated sunflower seedlings (Ros 1990). In UV-B exposed cucumber and bean seedlings growth could be stimulated by gibberellins (Ballaré et al. 1991b; Saile-Mark 1993).

b) Photosynthesis and Yield

Photosynthesis is one of the most studied physiological processes under UV-B accompanied mainly by growth experiments. Multiple sites of inhibition have been demonstrated throughout the years with the most sensitive site around photosystem II (Iwanzik et al. 1983; Bornman 1989; Renger et al. 1989; Strid et al. 1990). The reason for different action even in photosystem II may be again the insufficient irradiation sources often emitting UV-C in addition to UV-B. From previous experiments it has been assumed that UV-C acts primarily on the water-splitting system or on the oxidizing site of PSII (Yamashita and Butler 1968; Kulandaivelu and Noorudeen 1983; Renger et al. 1989). However, UV-B acts on either the reaction center itself, producing dissipative sinks which quench the variable fluorescence (Iwanzik et al. 1983), and/or on the reducing site of PSII (Melis et al. 1992). Evidence for the first action comes from the additional appearances of polypeptide fragments from the PSII reaction center under UV-B irradiation (Greenberg et al. 1989a; Nedunchezhian and Kulandaivelu 1991; Melis et al. 1992). However, recently it was also shown that modification of Q_A by UV-B can prevent electron transport from reduced pheophytin to Q_A, thus by charge recombination reactions between P-680$^+$ and Pheo$^-$ quenching the variable fluorescence (Melis et al. 1992). Further targets seem to be the quinone Q_B and/or the quinone pool inducing the degradation of the D1 protein (formerly called herbicide or Q_B-binding protein) as seen in photoinhibition (Greenberg et al. 1989a). However, UV-B damage is different to damage by visible radiation because the D1-degradation rate is much slower (5–7 h) than that during photoinhibition (Greenberg 1989b). Trebst and Depka (1990) also found the degradation of the D1 protein subunit of photosystem II, but they indicated the use of UV-C (254 nm) radiation. UV-C destroys plastoquinone (Trebst and Pistorius 1965) and plastohydroquinone (Lichtenthaler and Tevini 1969), whereas UV-B does not, except when photosynthetic electron flow is already low as the result of prolonged irradiation periods with high UV-B irradiances, as demostrated in radish leaves (Tevini and Iwanzik 1983). In the field, when longer UV-B irradiation periods occur, photoinhibition and UV-B damage may be additive and thus increase impact on net photosynthesis. This was also deduced from the inhibition of violaxanthin deepoxidase by UV-B, which is thought to have a protective role against photoinhibition. This enzyme synthesizes zeaxanthin (via antheraxanthin) which dissipates some of the absorbed excessive light energy (Pfündel et al. 1992). Photosystem I seems much more resistant to UV-B radiation (Iwanzik et al. 1983; Renger et al. 1989; Strid et al. 1990; Prasil et al. 1992).

In contrast, ribulose 1,5-bisphosphate carboxylase (rubisco), the key enzyme of the Calvin cycle, has recently been shown, at least in pea leaves, to be very UV-sensitive (Jordan et al. 1992). For example, one day of UV-B exposure declined the enzyme activity by 40%, the large (LSU) and small subunits (SSU) of Rubisco by 10–15% (Jordan 1993). Consistent with this observation are recent results on the gene expression of pea chloroplast proteins (Jordan et al. 1991, 1992). Both the nuclear-encoded SSU and chloroplast-encoded LSU-mRNA transcripts are severely reduced with a time delay for chloroplast transcripts. Chlorophyll a/b binding protein transcripts were also quantitatively reduced in pea and *Arabidopsis* mutants (Jordan et al. 1993). The down regulation of gene expression for chloroplast protein can be recovered to 60% after 3 days which correlates with a similar pattern in the quantum yield of photosynthesis (Chow et al. 1992).

Since photosynthesis is essential for plant productivity, UV-induced damage to photosynthetic function will often manifest itself in less biomass or lower yield. In the soybean cultivar Essex, a 25% reduction of photosynthesis was coupled with the same reduction in soybean yield when the plants were irradiated in a realistic arrangement with supplemental UV-B in the field (Teramura and Sullivan 1988). A 20% reduction in yield was further established in a 6-year field study for Essex when simulating a 25% ozone reduction (5.1 kJ m^{-2} d^{-1} supplemental UV-B), whereas in the cultivar Williams the yield increased under the same conditions (Teramura et al. 1990a). This again indicates sensitivity differences within the same species. Variation in UV-B sensitivity in plants from natural ecosystems along a 3000-m elevation gradient has been shown in 33 plant species collected in Hawaii and then grown in greenhouses under enhanced UV-B-levels of 15.5 kJ m^{-2} d^{-1} and 23.1 kJ m^{-2} d^{-1} (Sullivan et al. 1992). Among them, 14 species had reduced growth and 8 species reduced biomass, with the general rule that sensitivity to artificial UV-B radiation is reduced with increasing elevations where seeds are collected. Similar results have been obtained with two *Aquilegia* species from alpine and nonalpine environments in the columbines (Larson et al. 1990). Altogether, these results confirm earlier findings along an arctic-alpine life zone, showing that sensitivity to UV damage is lower in species from low latitudes and high elevations where daily doses are highest (Caldwell et al. 1982).

c) UV-Protection and Adaptation

During evolution, plants have adapted to enhanced UV-B radiation, mainly by developing two protection mechanisms: photoreactivation and accumulation of UV-absorbing pigments such as phenylpropanoids, flavonoids, and anthocyanines. Photoreactivation is directed towards the repair of thymine dimers of the cyclobutane type which may be responsible for most UV-induced damage of the DNA in plants. Recently, it was shown that the UV-B fluence response for the dimer formation is linear up to 690 J m^{-2} as applied to alfalfa spout leaves (Quaite et al. 1992b). The photolyase responsible for this repair uses visible light and is UV-B inducible, as shown in *Arabidopsis thaliana* (Pang and Hays 1991) but phytochrome-regulated in bean hypocotyls (Langer and Wellmann 1990). Several studies have confirmed the importance of high white light levels in the reduction of UV-B induced damage normally found under low white light levels (Mirecki and Teramura 1984; Cen and Bornman 1990; Panagopoulos et al. 1990). Therefore, the repair capacity as far as photolyase is involved must be high in sunlight, since all active light qualities are present in high amounts. However, enhanced UV-B may have damaging effects on the enzyme activity itself (by destruction of UV-absorbing amino acids), and thus reduces the repair capacity, especially when the photolyase is located in the epidermal layer of the leaves.

UV-resistant species may be well protected by shielding pigments synthesized via the phenylpropanoid pathway in the epidermal layer or in leaf hairs (Jahnen and Hahlbrock 1988; Hahlbrock and Scheel 1989; Karabourniotis et al. 1992). Penetration of UV-B radiation into the mesophyll passing through the epidermal layer has been measured by using a fiber-optic microprobe (Bornman and Vogelmann 1991). Two subalpine conifer species with increased growth under supplemental UV-B radiation (Sullivan and Teramura 1988) together with 11 species from the Rocky Mountains, had no measurable 300-nm radiation penetration into the mature needles (DeLucia et al. 1991; Day et al. 1992). In contrast, in postemerging and elongating leaves, the photosynthetic mesophyll received small but measurable UV-B fluxes, indicating that there are smaller amounts of screening properties in early needle stages (DeLucia et al. 1992).

The protective function of flavonoids in the mechanism of photosynthesis has been clearly demonstrated in rye seedlings, which accumulate isovitexins and other phenylpropanoids exclusively in the epidermal layer, and thus, damage to photosynthesis is prevented (Tevini et al. 1991a).

The biosynthesis of these compounds is regulated by *cis-trans* isomerization of cinnamic acids (Braun and Tevini 1993). The equilibrium between the *cis* and *trans* form is shifted to the *cis* form by shorter wavelengths in the UV-B, thus lowering the feedback inhibition of the key enzyme PAL, and possibly also of the subsequent enzymes, normally performed by *trans* cinnamic acids. The UV regulation at the genetic level was intensively studied in cultured cells and leaves of *Petroselinum* by Hahlbrock's group, demonstrating that the increase in mRNA transcripts for PAL, 4-coumarate: CoA ligase (4CL) and CHS is caused by UV-induced changes in transcription (Chappell and Hahlbrock 1984; Kuhn et al. 1984; Schmelzer et al. 1988). Schulze-Lefert et al. (1989) defined DNA sequences necessary for UV activation of the parsley chalcone synthase (CHS) gene, and Block et al. (1990) further characterized a core of essential nucleotides containing ACGTGGC. This *cis*-acting DNA sequence lies in the unit 1 of the CHS promoter with two separate boxes necessary for UV inducibility. Three binding factors to the box II core sequence have been isolated and are called common plant regulatory factors (CPRF), indicating that several other environmental factors apart from UV radiation can also induce CHS responses. This was shown for pathogen challenge in soybean and snapdragon (Wingender et al. 1990; Fritze et al. 1991). CPRF-1 accumulation induced by UV radiation is consistent over time with CHS mRNA accumulation, supporting the suggestion that this factor is involved in UV regulation (Weisshaar et al. 1991a, b). The enzyme 4CL is also UV-inducible and requires an additional *cis*-acting element which is, however, distinct from that controlling development and being located in the exogenic region of the gene (Douglas et al. 1991).

In addition to UV radiation, blue and red/far-red light are also involved in the regulation of gene expression as shown for CHS in parsley cells, petunia and mustard seedlings (Bruns et al. 1986; Koes et al. 1989; Batschauer et al. 1991). In parsley cells blue light abolishes the 2 h-lag phase after UV induction, whereas phytochrome alters the extent of the CHS induction. In petunia, tissue-specific light regulation occurs since CHS expression is UV dependent in seedlings and controlled by red light in petals. In mustard, the UV-B and blue light influence on CHS expression were found in primary leaves but not in cotelydons demonstrating that the yet unidentified UV-B photoreceptor either appears later in seedling development or that responsiveness for UV-B in the gene expression changes. Recently, it has been shown that flavin is an endogenous photoreceptive chromophore in parsley cells by feeding the cells with riboflavin, thus increasing the levels of CHS and flavonoids when irradiated for 6 h with UV-B plus white light compared to unfed controls. Blue light had no effect in this respect (Ensminger and Schäfer 1992).

In addition to flavonoids, polyamines which stabilize the membrane structure and inhibit lipid peroxidation (Smith 1985; Tadolini 1988), may also play a role in protecting plants from UV-B stress (Kramer et al. 1992).

d) Combination Effects

CO_2-concentration and temperature increases due to the greenhouse effect are also anticipated in future. Model calculations predict a temperature increase of 1 to 5 °C and a doubling of the ambient CO_2 content by the middle of the next century. It is therefore of interest to investigate whether the joint stress factors show simply cumulative enhancing or compensative effects. Over the last 3 years, progress has been made in answering this question.

Several studies describe the joint effects of enhanced UV-B and CO_2. Wheat, rice, and soybean were grown in greenhouses at 350 or 650 ppm CO_2 at 8.8 and 15.7 kJ m^{-2} d^{-1}UV-B. As expected for C_3 plants, an increase in CO_2 increases light-saturated photosynthesis, apparent quantum efficiency, water use efficiency, biomass, and seed yield in all three species. However, increased UV-B radiation reduced the CO_2-induced increase in biomass and seed yield in wheat and rice, whereas in soybean biomass increased further and seed yield remained high under both environmental stresses (Teramura et al. 1990b). As pointed out in the previous chapters, the reasons for UV-B-induced reductions may be both reduced PSII activity and the biochemistry of CO_2 fixation. At least in rice, this damage of the photosynthetic apparatus cannot be ameliorated by increased CO_2. Van de Staaij et al. (1992) reported that in *Elymus athericus*, a C_3 grass, the UV-induced dry weight reductions could not be significantly ameliorated by doubling of CO_2. The agricultural species pea and tomato also showed the same negative effect (Rozema et al. 1990). In both studies this lack of response may be due to unrealistically high UV-B levels used in these experiments. In sunflower and maize seedlings grown in growth chambers using the ozone filter technique, doubling of CO_2 could not further increase biomass (over soil) normally found under increased solar UV-B (Mark 1992; Tevini and Mark 1993). An increase in temperature of 4 °C over the normal daily temperature course can also compensate for biomass reductions normally found at lower temperatures and enhanced UV-B as also shown for sunflower and maize (Tevini 1993).

Ozone in combination with UV-B reduces pollen tube growth more than either stress alone. A UV exposure of 3 W m^{-2} for 30 min followed by O_3 at 120 ppb for 3 h reduced pollen growth of *Nicotiana tabacum* by 79% and *Petunia hybrida* by 75%. The effect appeared to be additive, indicating that different targets may be affected by the two environmental factors (Feder and Shrier 1990).

Heavy metals are in most cases toxic to plants. Only one study describing the joint effects of 5 mM cadmium and UV-B radiation of only 6.17 kJ m^{-2} d^{-1} to *Picea abies* L. has appeared so far (Dubé and Bornman 1992). In general, the combined treatment accentuates the reductions of growth and photosynthesis normally found under separate treatments, showing that certain environmental factors reveal additive responses in plants, especially in the field where plants are usually subjected to more than one stress factor.

Leaf diseases are often found in agricultural crops like the *Cercospora* leaf spot disease, induced by the fungus *Cercospora beticola* Sacc. resulting in substantial losses of sugar beet storage roots. UV-B, in combination with infection by this fungus, increases the losses in storage root weight compared to separately given stresses (Panagopoulus et al. 1992). In one cucumber cultivar, susceptible to *Colletotrichum lagenarium* or *Cladosporium cucumerinum*, pre-infection treatment with UV-B (11.6 kJ m^{-2} d^{-1}) led to a greater disease development, especially on the cotyledons and not on true leaves, whereas post-infection UV-B irradiation did not have this effect (Orth et al. 1990). Considering all the results of combined studies with abiotic and biotic factors, too little is known to make any general prediction of the effects of possible changes in and interactions with the future environment.

3. Effects on Aquatic Organisms

Estimates show that about 2 x 10^{11} tons of carbon annually are photosynthetically converted into biological matter on our planet. More than half of this carbon is incorporated by the primary producers, mainly phytoplankton, the base of the aquatic food web (Houghton and Woodwell 1989). In turn, these organisms serve as food for primary consumers: the larvae of fish and shrimps. Hence any reduction in the productivity of phytoplanktonic biomass production could have drastic consequences for the marine ecosystem (Häder et al. 1989).

The primary producers of antarctic waters in particular, have been subjected to environmental fluctuations, especially to enhanced UV-B radiation due to the ozone hole during Austral spring (Lubin et al. 1989; Frederick and Alberts 1991). A few measurements in Australia and New Zealand have also confirmed the higher UV-B levels during the ozone hole (Roy et al. 1990; McKenzie et al. 1991) which in 1991 reached total ozone values as low as 108 DU (WMO 1991).

Other external factors such as white light, gravity, chemical gradient, and temperature, control the orientation and vertical migration of the organisms in the water column. Recent experiments were carried out in order to measure the biologically effective UV-B radiation using a bacterial cell dosimeter (Karentz and Lutze 1990) or for full spectral irradiance a submersible spectroradiometer (Gieskes and Kraay 1990; Smith et al. 1992). The results showed that UV-B radiation was still present at a depth of 60–70 m in clear Antarctic waters, with highest biological effectiveness up to 10 m. The attenuation of Photosynthetic Active Radiation (PAR) to 1% of the incident light, which in blue light reaches depths greater than 200 m, gives an indication of the depth of the euphotic zone, largely dependent upon the transparency of the water to solar radiation (Häder et al. 1989). The transparency of the water determines the extent of the euphotic zone, where most planktonic organisms are found (Nultsch and Häder 1988).

Total Ozone Mapping System (TOMS) estimates, as well as surface data from the Palmer National Science Foundation Monitoring Station show that large concentrations of primary producers are present in stable water columns with low surface mixture. Primary producers are found in the Marginal Ice Zone (MIZ) near the edge of the pack ice, and also in waters with high turbidity such as the upwelling areas on continental shelves. This indicates the importance of both the quality and quantity of light penetrating the water column.

The dependence of phytoplankton upon light implies that orientation mechanisms which react to light sources must be present, and accounts for the abundance of these organisms in the top layer of the water column.

The orientation and motility of phytoplankton are impaired by UV-B radiation (Häder and Worrest 1991). Consequently, general metabolism, photosynthetic production, and nitrogen fixation and assimilation are adversely affected, reducing growth and endangering survival.

a) Orientation and Motility

In order to cope with a changing environment, planktonic organisms need to be able to adjust their position in the water column. In a number of ecologically significant phytoplankton groups, orientation mechanisms depend on light and gravity (Häder and Liu 1991). In 1980, Foster and Smyth reported that most motile microorganisms use various bands in the visible and longwave UV range for photoorientation. UV-B radiation impairs the motility of motile microorganisms such as flagellates or cyanobacteria by reducing their ability to move towards or away from a light source and thus disordering their orientation in the water column. Even after only a few hours of exposure to unfiltered solar radiation, the percentage of motile flagellates of *Cryptomonas maculata* and the freshwater *Cryptomonas* was shown to decrease experimentally; as well as indicating reduced speed of movement. Escape from harmful radiation is thus prevented (Häder and Häder 1989, 1991). Figure 2 illustrates the positive phototaxis of *Gymmodinium* (Fig. 2 a) which can be easily impaired by enhanced UV-B radiation (Fig. 2 b).

Studies of freshwater *Euglena* demonstrate the responsiveness of such organisms to light: negative phototaxis at high fluence rates and positive phototaxis at lower fluence rates. However, negative phototaxis seems to be impaired more easily, resulting in a movement towards the surface under elevated UV-B radiation. The action spectrum for the inhibition of motility in this organisms shows a considerable peak at about 270 nm, a smaller one at 305 nm and a shoulder at 290 nm (Häder et al. 1991).

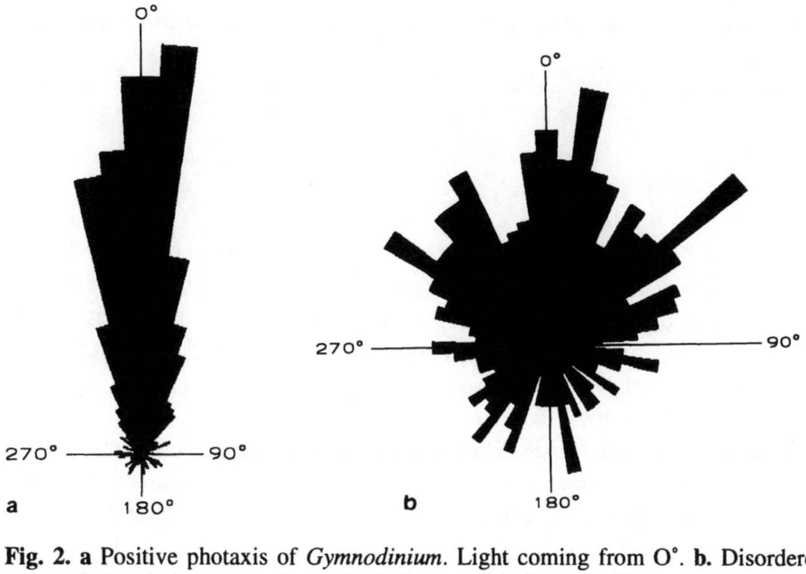

Fig. 2. a Positive photaxis of *Gymnodinium*. Light coming from O°. **b.** Disordered phototaxis of *Gynmodinium* induced by enhanced UV-B. (**a**, **b** Schäfer et al. 1993)

Generally, organisms tolerated solar radiation significantly longer when the short wavelength component was absorbed by cutoff filters, or on inserting an artificial layer of ozone using the ozone filter technique (Häder et al. 1990; Tevini et al. 1990). UV-A and unfiltered visible radiation can also affect motility in a number of groups of phytoplankton when subjected to it unfiltered, over a prolonged period of time.

An inverse relationship between motility and UV-B radiation has been clearly demonstrated, and this may be one factor for decreased algal growth during the summer.

As the action spectrum for inhibition of motility in *Euglena gracilis* differs from the DNA absorption spectrum (Häder et al. 1991), specific photoreceptor systems (pigments) must be responsible for the orientation of motile phytoplankton (Häder et al. 1991).

In *Euglena* the paraflagellar body (PFB), a bulge at the lower end of the emerging flagellum, is thought to be the location of the photoreceptor. Both pterins and flavins are present in the PFB, and experimental evidence shows that the molecular mechanisms of orientation operates in a way similar to that in the photosynthetic apparatus, with antenna pigments absorbing at shorter wavelengths, and a reaction center (Galland et al. 1990).

Many flagellates concentrate in horizontal bands, as a result of phototaxis and/or gravitaxis; gravity being another external factor influencing the orientation of many organisms (Kessler 1985; Fenchel and Finlay 1986; Häder 1993). In some of them, solar and UV radiation also have a negative effect on gravitaxis over a period of a few minutes. Vertical diurnal fluctuations of these bands occur when conditions such as light change (Häder 1991).

Prorocentrum micans, a marine dinoflagellate, uses photo- and gravitaxis to migrate vertically, high irradiances resulting in negative phototaxis and low irradiances in positive phototaxis (Eggersdorfer and Häder 1991). Populations of phytoplankton are adversely affected when these important regulatory mechanisms are impaired. Since some species appear to be more tolerant than others to high doses of UV-B radiation, a change in the composition of species in phytoplanktonic communities may occur, consequently affecting the entire aquatic food web (Worrest 1982; Häder et al. 1989). Toxic algal blooms (such as dinoflagellates or cyanobacteria) could develop, which would poison fish and other organisms (Häder 1993).

b) Photosynthesis

Excessively intense irradiation bleaches the photosynthetic pigments of many phytoplankton species, and damages cellular structures: some of the accessory pigments are bleached within 15 min of exposure to sunlight. In *Cryptomonas maculata* the accessory biliprotein pigments are bleached first, followed by the carotenoids and finally the chlorophylls. By using an artificial UV radiation source, it can be shown that these effects are largely due to the UV-B component (Häder et al. 1988).

After shorter periods of exposure to UV-B, an increase in chlorophyll fluorescence can be detected, indicative of a loss of excitation energy for photochemical reactions. After prolonged exposure, fluorescence decreases as the absorbing pigments are destroyed. These detrimental processes consequently decrease photosynthetic oxygen production (Häder et al. 1989; Zündorf and Häder 1991).

The recent dispute over the use of polyethylene bags instead of quartz glassware as sample containers has focused attention upon the problems with calculating percent UV inhibition when using only chlorophyll-specific data. A significant difference in photosynthetic assimilation rates was found between the samples in glassware compared with those in bags, from which was inferred that a certain level of toxicity was produced resulting in artifacts. However, it seems to be the case that the inhibition of photosynthetic rates can be masked in chlorophyll-related data, as response rates can differ under different light treatments (Holm-Hansen and Helbig 1993; Prézelin and Smith 1993).

c) Nitrogen Assimilation

The rate of protein synthesis is determined by nitrogen assimilation. Nitrogenase is the enzyme essentially responsible for nitrogen assimilation, activated by light, and inactivated by UV-B radiation (Döhler 1985). Further incorporation of nitrogen into amino acids and proteins is brought about by other enzymes, also adversely affected by UV-B in many phytoplanktonic organisms (Döhler 1990).

Prokaryotic organisms such as cyanobacteria are able to utilize atmospheric nitrogen dissolved in water. Particularly prevalent in rice paddies, they play an important role in providing nitrogen for higher plants, since about 35 million tons of the element are assimilated by this group of organisms annually (Kumar and Kumar 1988). These microorganisms are highly sensitive to UV-B radiation (Häder and Häder 1990). Therefore, a loss of natural fertilizers must be compensated for by artificial supply, production of which contributes to the general energy problem and thus, in turn, to the greenhouse effect.

d) UV-B Protection

Phytoplankton organisms contain a number of UV-B-absorbing compounds which may function as protective pigments although this function has not yet clearly been proven (Bidigare 1989; Vernet et al. 1989). Most of these compounds are mycosporine-like amino acids (MAAs) with an absorption maximum in the UV-A range and only lower absorption peaks in the UV-B (Carretto et al. 1990; Karentz et al. 1991). Recently, MAAs have been found in *Phaeocystis pouchetii*, one of the most abundant arctic species. Furthermore, enhanced concentrations of carotenoids were detected in macroalgae, phytoplankton, and invertebrates (Smith et al. 1992).

However, to date, there is no evidence to suggest that production of these mycrosporine-like amino acids increases with enhanced levels of UV-B; only then would compensation be possible (Raven 1991).

Some cyanobacteria can actively prevent UV-B damage by producing scytonemine, which absorbs light in the UV-B region, and the production of which is induced by UV-B radiation (Garcia-Pichel and Castenholz 1991). In the cyanobacterium *Nostoc commune*, a similar UV-B protecting pigment was found in high amounts (up to 10% of dry weight) when grown under solar UV radiation (Scherer et al. 1988).

Preliminary studies on UV-induced damage to the DNA in phytoplankton indicate repair mechanisms ensuring the survival of these organisms (Karentz 1988). However, photorepair mechanisms mediated by the enzyme photolyase as in higher plants have not been demonstrated (Häder et al. 1986).

e) Atmospheric Carbon Dioxide Increase

The emission per year of about 5 Gt of carbon by fossil fuel consumption and 2 Gt by the deforestation of tropical rainforests has disturbed the balance of the natural carbon cycle. Phytoplankton is an important biological sink for atmospheric carbon dioxide, and any decrease in the prevalence of these organisms will thus contribute further to the greenhouse effect, and to global climactic change. The most recent experiments performed in Antarctica in and outside the ozone hole area have shown, that the biomass production of the phytoplankton has declined by 6–12% compared to that outside this area (Smith et al. 1992). This loss in biomass is equivalent to losses in CO_2 fixation. It has been calculated that global losses of the order of 10% of total phytoplankton productivity are equal to all the anthropogenic output to the atmosphere (Häder et al. 1991).

In summary, UV-B radiation has adverse effects upon the ability of organisms such as phytoplankton both to adapt to changes in their environment, and to continue the normal photosynthetic process.

However, our knowledge of the stress capacity of the marine ecosystem is limited to cope with the stress: studies of the long-term effects of UV-B upon ecosystem responses are sparse, and at an early stage. In addition, there is little historical data on oceanic productivity for comparisons; it is thus not possible to make any accurate predictions. If any meaningful results are to be obtained, more detailed analysis is required in two main areas: on action spectra for phytoplankton, based upon dose-response data; and on the molecular mechanisms involved in UV-B damage.

A predictive model is currently under development, its aim being to illuminate these uncertain areas: if results show the effects of UV-B upon these organisms to be as damaging as early studies seem to imply, the long-term implications could be both far reaching and devastating.

Acknowledgments. Thanks are extended to Prof. Dr. D-P. Häder and Thomas Hietzker for valuable comments on the manuscript. Work in the author's laboratory has been supported by the BMFT.

References

Adamse P, Britz SJ (1992) Photochem Photobiol 56:645–650.

Bachelet D, Barnes PW, Brown D, Brown M (1991) Photochem Photobiol 54:411–422. **Ballaré CL,** Barnes PW, Kendrick RE (1991a) Physiol Plant 83:652–658. **Ballaré CL,** Casal JJ, Kendrick

RE (1991b) Photochem Photobiol 54:819–826. **Barnes PW,** Flint SD, Caldwell MM (1990) Am J Bot 77:1354–1360. **Batschauer A,** Bruno E, Schäfer E (1991) Plant Mol Biol 16:175–185. **Bidigare RR** (1989) Photochem Photobiol 50:469–477. **Block A,** Dangl JL, Hahlbrock K, Schulze-Lefert P (1990) Proc Natl Acad Sci USA 87:5387–5391. **Blumthaler M** (1993) In: Tevini M (ed) Ultraviolet-B radiation and ozone depletion. Effects on humans, animals, plants, microorganisms, and materials. Lewis, Boca Raton, pp 71 ff. **Blumthaler M,** Ambach W (1990) Science 248:206–208. **Bornmann JF** (1989) J Photochem Photobiol B 4:145–158. **Bornman JF,** Teramura AH (1993) In: Björn LO (ed) Environmental UV photobiology (in press). **Bornman JF,** Vogelmann TC (1991) J Exp Bot 42:547–554. **Braun J**, Tevini M (1993) Photochem Photobiol 57:318–323. **Bruns B,** Hahlbrock K, Schäfer E (1986) Planta 169:393–398.

Caldwell MM (1971) In: Giese AC (ed) Photophysiology. Academic Press, New York, pp 131–177. **Caldwell MM,** Robberecht R, Nowak RS, Billing WD (1982) Arct Alp Res 14:195–202. **Caldwell MM,** Camp LB, Warner CW, Flint SD (1986) In: Stratopheric ozone reduction, solar ultraviolet radiation and plant life. In: Worrest RC, Caldwell MM (eds) NATO ASI Ser 8:87–111. **Caldwell MM,** Teramura AH, Tevini M (1989) Trees 4:363–367. **Carretto JJ,** Cariguana MO, Daleo G, deMarco SG (1990) Plankt Res 12:909–912. **Cen YP,** Bornman JF (1990) J Exp Bot 41:1489–1495. **Chappel J,** Halbrock K (1984) Nature 311:76–78. **Chow WS,** Strid A, Anderson JM (1992) Plant Cell 4:129–134. **Coohill TP** (1989) Photochem Photobiol 50:451–457. **Coohill TP** (1991) Photochem Photobiol 54:859–870. **Cullen JJ,** Neale PJ, Lesser MP (1992) Science 258:646–650.

Dai Q, Coronel VP, Vergara BS, Barnes PW, Quintos AT (1992) Crop Sci 32:1269–1274. **Day TA,** Vogelmann TC, DeLucia EH (1992) Oecologia 92:513–519. **DeLucia EH,** Day TA, Vogelmann TC (1991) In: Current topics in plant biochemistry and physiology, vol 10, pp 32–48. Publ Univ Missouri. **DeLucia EH,** Day TA, Vogelmann TC (1992) Plant, cell and environ 15:921–929. **Döhler G** (1985) J Plant Physiol 118:391–400. **Döhler G** (1990) In: Ulrich WR, Rigano C, Fuggi A, Aparicio JP (eds) Springer, Berlin Heidelberg New York, pp 354–359. **Douglas CJ,** Hauffe KD, Ites-Morales M-E, Ellard M, Paszowskii U, Hahlbrock K, Dangl JL (1991) EMBO J 10:1767–1775. **Dubé LS,** Bornman JF (1992) Plant Physiol Biochem 30:761–767.

Eggersdorfer B, Häder D-P (1991) Eur J Biophys 85:319–326. **El-Sayed SZ,** Stephens FC, Bidigare BB, Ondrusek ME (1990) In: Kerry KR, Hempel G (eds) Antarctic ecosystems. Ecological change and conservation. Springer, Berlin Heidelberg New York, pp 379–385. **Ensminger PA**, Schäfer E (1992) Photochem Photobiol 55:437–447.

Feder WA, Shrier R (1990) Environ Exp Bot 30:451–454. **Fenchel T,** Finlay BJ (1986) J Protozool 33:139–145. **Foster KW,** Smyth RD (1980) Microbiol Rev 44:572–630. **Frederick JE,** Alberts AD (1991) Geophys Res Lett 18:1869–1871. **Fritze K,** Staiger S, Czaja I, Walden R, Schell J, Wing D (1991) Plant Cell 3:893–905.

Galland P, Keiner P, Dörnemann D, Senger H, Brodhun B, Häder D-P (1990) Photochem Photobiol 51:675–680. **Garcia-Pichel F,** Carstenholz RW (1991) J Phycol 27:395–409. **Gieskes WC,** Kraay GW (1990) Biomass Newsl 12:12–14. **Green AES,** Cross KR, Smith LA (1980) Photochem Photobiol 31:59–65. **Greenberg BM,** Gaba V, Canaani O, Malkin S, Mattoo AK, Edelman M (1989a) Proc Natl Acad Sci USA 86:6617–6620. **Greenberg BM,** Gaba V, Mattoo AK, Edelman M (1989b) Z Naturforsch 44c:450–452.

Häberlein A, Häder D-P (1993) Acta Protozool (in press). **Häder D-P** (1991) In: Riklis E (ed) Photobiology. Plenum, London, pp 497–510. **Häder D-P** (1993) In: Tevini M (ed) UV-B radiation and ozone depletion Effects on humans, animals, plants, microorganisms, and materials. Lewis, Boca Raton, pp 155–192. **Häder D-P,** Häder M (1989) Bot Acta 102:236–240. **Häder D-P,** Häder M (1990) Acta Protozool 29:291–303. **Häder D-P,** Häder M (1991) Environ Exp Bot 31:33–41. **Häder D-P,** Liu SM (1991) Curr Microbiol 21:161–168. **Häder D-P,** Worrest RC (1991) Photochem Photobiol 53:717–725. **Häder D-P,** Watanabe M, Furuya M (1986) Plant Cell Physiol 27:887–894. **Häder D-P,** Rhiel E, Wehrmeyer W (1988) Fems Microbiol Ecol 53:9–18. **Häder D-**

P, Worrest RC, Kumar HD (1989) In: UNEP (ed) Environmental effects panel report. Aquatic ecosystems. United Nations Environment Programme, Nairobi, Kenya, pp 39–48. **Häder D-P,** Häder M, Liu S-M, Ullrich W (1990) Biosystems 23:335–343. **Häder D-P,** Worrest RC, Kumar HD (1991) In: UNEP (ed) Environmental effects of ozone depletion: 1991 update. United Nations Environment Programme, Nairobi, Kenya, pp 33–40. **Hahlbrock K** Scheel D (1989) Annu Rev Plant Physiol. Plant Mol Biol 40:347–369. **Helbling EW,** Villafane V, Ferrario M, Holm-Hansen O (1992) Mar Ecol Prog Ser 80:89 ff. **Holm-Hansen O,** Helbling EW (1993) Science 259:534. **Houghton RA,** Woodwell GM (1989) Sci Am 18–26.

Iwanzik W, Tevini M, Dohnt G, Voss M, Weiss W, Gräber P, Renger G (1983) Physiol Plant 58:401–407.

Jahnen W, Hahlbrock K (1988) Planta 173:453–458. **Jordan BR** (1993) In: Jackson MB, Black CR (eds) Interacting stresses on plants in a changing climate. NATO ASI Subseries Global environmental change. Springer, Berlin Heidelberg New York (in press). **Jordan BR**, Chow WS, Strid A, Anderson JM (1991) FEBS 284:5–8. **Jordan BR,** He J, Chow WS, Anderson JM (1992) Plant Cell Environ 15:91–98. **Jordan BR,** James PE, Anthony RG (1993) J Exp Bot 44 Suppl, pp 17.

Karabourniotis G, Papadopoulos K, Papamarkou M, Manetas Y (1992) Physiol Plant 86:414–418. **Karentz D** (1988) Antarct J US 23:114–115. **Karentz D** (1991) Antarct Sci 3:3–11. **Karentz D,** Lutze LH (1990) Limnol Oceanogr 35:549–561. **Karentz D,** McEnen FS, Land MC, Dunlop WC (1991) Mar Biol 108:157–166. **Kessler JO** (1985) Nature 313:218–220. **Koes RE,** Spelt CE, Mol JNM (1989) Plant Mol Biol 12:213–225. **Kramer GF,** Krizek DT, Mirecki RM (1992) Phytochemistry 31:1119–1125. **Krupa SV,** Kickert RN (1989) Environ Pollu 61:263–393. **Kuhn DN,** Chappell J, Boudet A, Hahlbrock K (1984) Proc Natl Acad Sci 81:1102–1106. **Kulandaivelu G,** Noorudeen AM (1983) Physiol Plant 58:389–394. **Kumagai T,** Sato T (1992) Jpn J Breed 42:545–552. **Kumar A,** Kumar HD (1988) In: Seu SP (ed) Plant Physiology Research, New Dehli, Society for Plant Physiology and Biochemistry, 1st Int. Congr Plant Physiology.

Langer B, Wellmann E (1990) Photochem Photobiol 52:861–863. **Larson RA,** Garrison WJ, Carlson RW (1990) Plant Cell Environ 13:983–987. **Lichtenthaler HK,** Tevini M (1969) Z Naturforsch 24b:764–769. **Lubin D,** Frederick JE, Booth CR, Lucas T, Neuschuler D (1989) Geophys Res Lett 16:783–785.

Mark U (1992) Karls Beitr Entw Ökophysiol 11:1–220. **McKenzie RL,** Matthews WA, Johnston PV (1991) Geophys Res Lett 18:2269–2272. **Melis A,** Nemson JA, Harrison MA (1992) Biochim Biophys Acta 1100:312–320. **Mirecki RM,** Teramura AH (1984) Plant Physiol 74:475–480.

Nedunchezhian N, Kulandaivelu G (1991) Physiol Plant 81:558–562. **Nultsch W,** Häder D-P (1988) Photochem Photobiol 47:837–869.

Orth AB, Teramura AH, Sisler HD (1990) Am J Bot 77:1188–1192.

Panagopoulos I, Bornman JF, Björn LO (1990) J Photochem Photobiol B: Biol 8:73–87. **Panagopoulos I,** Bornman JF, Björn LO (1992) Physiol Plant 84:140–145. **Pang Q,** Hays JB (1991) Plant Physiol 95:536–543. **Paulsen GM** (1987) In: Heyne EG (ed) Wheat and wheat improvement. Am Soc Agron, vol 13. Madison, pp 384–391. **Pfündel EE,** Pan RS, Dilley RA (1992) Plant Physiol 98:1372–1380. **Prasil O,** Adir N, Ohad (1992) In: Barber J (ed) The photosystems. Elsevier, Amsterdam, pp 295–348. **Prézelin BB,** Smith RC (1993) Science 259:535.

Quaite FE, Sutherland BM, Sutherland JC (1992a) Nature 358:576–578. **Quaite FE,** Sutherland BM, Sutherland JC (1992b) Appl Theor Electrophor 2:171–175.

Raven JA (1991) J Photochem Photobiol B 9:239–244. **Renger G,** Völker M, Eckert HJ, Fromme R, Hohm-Veit S, Gräber P (1989) Photochem Photobiol 49:97–105. **Ros J** (1990) Karls Beitr Entw

Ökophysiol 8:1–157. **Roy CT,** Gies HP, Grahne (1990) Science 347:235–236. **Rozema J,** Lensen GM, van de Staaij JVM (1990) In: Goudriaan J, van Keulen H, van Laar HH (eds) The greenhouse effect and primary productivity in European agro-ecosystems. Pudoc, Wageningen, pp 68–71. **Rundel RD** (1983) Physiol Plant 58:360–366. **Ryel RJ,** Barnes PW, Beyschlag W, Caldwell MM, Flint SD (1990) Oecologia 82:304–310.

Saile-Mark M (1993) Karlsr Beitr Entw Ökophysiol 14 (in press). **Schäfer J,** Sebastian C, Häder D-P (1993) Arch Microbiol (submitted). **Scherer S,** Chen TW, Böger P (1988) Plant Physiol 88:1055–1057. **Schmelzer E,** Jahnen W, Halbrock K (1988) Proc Natl Acad Sci 85:2989–2993. **Schulze-Lefert P,** Dangl JL, Becker-Andre M, Hahlbrock K, Schulz W (1989) EMBO J 8:651–656. **SCOPE** (1992) Effects of increased ultraviolet radiation on biological systems, Paris. **Setlow RB** (1974) Proc Natl Acad Sci 104:471–474. **Smith RC** (1989) Photochem Photobiol 50:459–468. **Smith RC,** Prézelin BB, Baker KS, Bigidare RR, Boucher NP, Coley T, Karentz D, MacIntyre S, Matlick HA, Menzies D, Ondrusek M, Wan Z, Waters KJ (1992) Science 255:952–959. **Smith TA** (1985) Annu Rev Plant Physiol 36:117 ff.

Stapleton AE (1992) Plant Cell 4:1353–1358. **Steinmüller D** (1986) Karlsr Beitr Entw Ökophysiol 6. **Strid A,** Chow WS, Anderson JM (1990) Biochim Biophys Acta 1020:260–268. **Sullivan JH,** Teramura AH (1988) Am J Bot 75:225–230. **Sullivan JH,** Teramura AH (1991) Trees 6:115–120. **Sullivan JH,** Teramura AH, Ziska LH (1992) Am J Bot 79:737–743.

Tadolini B (1988) Biochem J 249:33 ff. **Teramura AH,** Sullivan JW (1988) Environ Pollut 93:416–469. **Teramura AH,** Sullivan JW, Lydon J (1990a) Physiol Plant 80:5–11. **Teramura AH,** Sullivan JH, Ziska LH (1990b) Plant Physiol 94:470–475. **Teramura AH,** Ziska LH, Sztein AE (1991) Physiol Plant 83:373–380. **Tevini M** (1993) In: Tevini M (ed) UV-B radiation and ozone depletion. Effects on humans, animals, plants, microorganisms, and materials. Lewis, Boca Raton, pp 125–154. **Tevini M,** Iwanzik W (1983) Physiol Plant 58:395–400. **Tevini M,** Mark U (1993) In: Shima A, Ichahashi M, Fujiwara Y, Takebe H (eds) Frontiers of photobiology. Excerpta Medica, Amsterdam, pp 541–546. **Tevini M,** Steinmüller D (1988) In: Coohill P Photochem Photobiol 50:451–457. **Tevini M,** Teramura AH (1989) Photochem Photobiol 50:479–487. **Tevini M,** Mark U, Saile M (1990) In: Payer HD, Pfirrmann T, Mathy P (eds) Environmental research with plants in closed chambers. air pollution research report 26, CEC, Brussels. **Tevini M,** Braun J, Fieser G (1991a) Photochem Photobiol 53:329–333. **Tevini M,** Mark U, Fieser G, Saile M (1991b) In: Riklis E (ed) Photobiology. Plenum, New York, pp 635–649. **Tevinin M,** Ros J, Mark U, Hietzker T (1993) In: Johnson JB (ed) Cambridge (in press). **Trebst A,** Depka B (1990) Z Naturforsch 45c:765–771. **Trebst A,** Pistorius E (1965) Z Naturforsch 20b:143–147.

UNEP (1989) Environmental effects panel report. Nairobi, Kenya, pp 1–64. **UNEP** (1991) Environmental effects of ozone depletion: 1991 update. Nairobi, Kenya, pp 1–51.

Van der Leun JC, Tevini M (1992) Photochem Photobiol B: Biol 13:187–190. **Van de Staaij JWM,** Lensen GM, Strotenga M, Rozema J (1992) Vegetatio 0:1–8. **Vernet M,** Mitchel BG, Holm-Hansen O (1989) Antarct J US 24:12–13. **Voyteck MA** (1990) Ambio 19:52–61.

Weisshaar B, Armstrong GA, Block A, Da Costa E, Silva O, Hahlbrock K (1991a) EMBO J 10:1777–1786. **Weisshaar B,** Block A, Armstrong GA, Herrmann A, Schulze-Lefert P, Hahlbrock K (1991b) In: Jenkins GI, Schuch W (eds) Soc Exp Biol SEB Symp Ser 45, pp 191–210. **Wingender R,** Röhring H, Höricke C, Schell J (1990) Plant Cell 2:1019–1026. **WMO** (1991) Scientific assessment of ozone depletion. World meteorological organization, global ozone research and monitoring project report No. 25, Geneva. **Worrest RC** (1982) In: Calkins J (ed) The role of ultraviolet radiation in marine ecosystems. Plenum, New York, pp 429–457.

Yamashita T, Butler WL (1968) Plant Physiol 43:2037–2040.

Zündorf J, Häder D-P (1991) Arch Microbiol 156:405–411.

Professor Dr. Manfred Tevini
Botanisches Institut II
Universität Karlsruhe
Kaiserstraße 12
D-76128 Karlsruhe, FRG

IX. Molecular Aspects of Natural and Induced Incompatibility in Flowering Plants

By Christine Gietl

1. Introduction

Evolutionary success of the angiosperms is based to a considerable extent on the development of a great diversity of flowers, which are adapted to ensure efficient reproduction. Thus, carpels protect the ovules and the stigma receives the pollen grains, which after germination form a tube growing through the stylar tissue. The pollen tube delivers two generative nuclei to the embryo sac for a double fertilization. In order to increase cross-pollination leading to genetic recombination, a number of mechanisms have evolved. Besides architectural features of the flower and adaptations to insect pollination, cell biological systems contribute to the promotion of cross-fertilization.

This chapter reviews two such systems: self-incompatibility and male sterility. Only cases are considered in which molecular or biochemical insight has been obtained. In the last 2 years, a number of pertinent reviews have appeared: Ebert et al. 1989, Nasrallah and Nasrallah 1989, Levings CS III 1990, Hanson 1991, Mau et al. 1991, Nasrallah et al. 1991.

2. Gametophytic Self-Incompatibility

In the gametophytic self-incompatibility system the phenotype of the pollen is determined by its own haploid genotype. In a majority of cases, the incompatibility is effected by a single gene (S-gene) occurring in multiple alleles in the population of a species. A pollen tube carrying nuclei with a particular allele is inhibited in its growth if the same allele is present in the pistil. Fertilization succeeds if the S-allele of the pollen differs from either of the two alleles residing in the nuclei of the diploid stylar tissue. In young flowers the recognition between self and nonself has not yet developed, and in old flowers it collapses. The successful molecular analysis of this incompatibility system in *Nicotiana alata* depended on this possibility of obtaining self-pollination of immature flowers with mature pollen of the same plant (Anderson et al. 1986). This results in plants homozygous for an S-allele. The first cDNA for the S_2-allele product was obtained by identifying a 32-kDa glycoprotein from stylar tissue which cosegregated with the S_2-allele in progenies of crosses, obtaining the N-terminal amino acid sequence of the protein, synthesizing corresponding oligonucleotides, and using these to screen cDNA libraries prepared

Progress in Botany, Vol. 55

from mRNA of appropriate stylar tissue (Anderson et al. 1986). Subsequently, cDNAs for the S_1-, S_3-, and S_6- alleles were obtained by differential screening procedures (Anderson et al. 1989; Haring et al. 1990). These cDNA clones were used to study the S-locus in Southern hybridizations of genomic DNA from plants homozygous for the S_1-, S_2-, S_3-, S_6-, and S_7- alleles. Generally, single endonuclease restriction fragments hybridized with all these probes, consistent with the expectation that all these cDNAs are products of alleles from the same gene (Bernatzky et al. 1988).

Alignment of the deduced amino acid sequences indicated an amino terminal signal peptide of 15 amino acids in agreement with the fact that the glycoprotein is secreted into the style mucilage. Within the deduced mature protein (204 amino acids), two conserved regions are interspersed with hypervariable regions (Haring et al. 1990). The conserved amino acid domains showed homology to secreted fungal RNases from *Aspergillus* and *Rhizopus* (McClure et al. 1989), including two histidine residues in the catalytic sites of the fungal RNases and nine conserved cysteine residues. It was indeed shown that RNase activity co-eluted with individual S-allele glycoproteins during cation exchange chromatography (Jahnen et al. 1989a; McClure et al. 1989). The isolated S-glycoproteins inhibited the pollen tube growth of *Nicotiana alata* in vitro to a limited extent in an S-allele-specific manner (Jahnen et al. 1989b). Isolated S-RNase was supplied to in vitro-grown pollen tubes from S_2-, S_3-, and S_6-plants (Gray et al. 1991). Similar uptake of the S_2-RNase was found by immuno-electron microscopy in compatible and incompatible combinations. The uptake resulted in inhibition of protein synthesis in experiments with exogenously supplied labeled amino acids. Also this inhibition did not differ in compatible and incompatible combinations. The question whether the S-RNase is indeed involved in the incompatibility reaction has been analyzed with the following experiment. S_2- as well as S_1- and S_3-pollen RNA was labeled by growing plants in the presence of ^{32}P. The "hot" pollen was placed on stigmas of unlabeled flowers in compatible and incompatible combinations. After 48 h, autoradiographs of the styles revealed inhibition of the growth of incompatible pollen tubes. Extraction of rRNA from the stylar tissue revealed the labeled 28S and 18S rRNA from the pollen to be degraded in incompatible combinations, but not in compatible combinations (McClure et al. 1990).

Expression of the S_2-mRNA was studied by in situ hybridization with the corresponding cDNA clone, and found to be restricted to the transmitting tissue cells of the style and to the inner epidermis of the placenta (Cornish et al. 1987). The expression is also developmentally regulated and most pronounced in mature flowers, but weak in immature styles (Cornish et al. 1987). This pattern of expression parallels the presence of the S_2-RNase analyzed by electrophoresis of style extracts (Anderson et al. 1986) and by immunocytochemistry with an antibody raised against a synthetic 22 amino acid peptide of the major variable region of the S_2-RNase (Anderson et al. 1989). The S_2-RNase was present in the intercellular matrix of the transmitting tract of the style and in the inner wall of the placental epidermal cells (McFadden et al. 1992).

Self-compatible *Nicotiana* species such as *Nicotiana tabacum* have a low level of RNase activity within the style mucilage. It is notable that crosses between *N. alata* and *N. tabacum* succeed only if *N. tabacum* is the female parent. Possibly, the style of *N. alata* rejects the *N. tabacum* pollen because of a high content of RNase in its style and a missing means of defense in the *N. tabacum* pollen tube.

In order to investigate whether *N. tabacum* could be changed to incompatibility, transgenic plants expressing the S_2-cDNA or a 5' truncated S_2-gene under the control of the ^{35}S promoter of cauliflower mosaic virus were produced (Murfett et al. 1992). The appropriately glycosylated S_2-RNase was expressed preferentially in mature floral organs, but the amount of S_2-RNase in styles of the transgenic plants was about 100-fold lower than that found in styles of *Nicotiana alata*. The transgenic plants self-pollinated spontaneously and were as successfully pollinated with *N. alata* pollen as untransformed *N. tabacum*. It remains to be seen whether overexpression of S_2-RNase in the style of transgenic plants can inhibit the self-compatibility of *N. tabacum*.

With the aid of the S-cDNAs isolated from *Nicotiana alata*, cDNAs encoding homologous S-glycoproteins have been isolated from other Solanaceae such as *Petunia inflata* (Ai et al. 1990), *Petunia hybrida* (Clark et al. 1990), *Solanum cacoense* (Xu et al. 1990), and *Solanum tuberosum* (Haring et al. 1990). Ribonuclease activity of the encoded S-glycoproteins has been found in *P. hybrida, P, inflata*, and *S. tuberosum*. Correlation between the presence of specific glycoproteins with the inheritance of S-alleles has been established in *S. tuberosum* (Kirch et al. 1989) and *Lycopersicum peruvianum* (Mau et al. 1986). In *Papaver rhoeas*, isolated S-glycoproteins elicit allele-specific inhibition of pollen tube growth in vitro (Franklin-Tong et al. 1989).

Future work is directed to the question of whether the pollen expresses the S-allele, and how the incompatibility reaction is mediated between the S-RNase glycoproteins and the as yet unknown receptor component of the pollen tube. The recognition must involve the hypervariable domain in the S-glycoproteins. Possibly, a cell wall protein in the pollen tube permits the entrance of the S-protein only if the hypervariable region is identical.

3. Sporophytic Self-Incompatibility

In the sporophytic self-incompatibility system, the genetic control also frequently operates through a single S-locus with multiple alleles. However, the incompatibility response is determined by the alleles carried by the pollen-producing plant, rather than by the S-allele of the individual pollen grains. Pollen germination is inhibited when an allele in the pollen parent is matched by one of the pistil, even if the haploid pollen grain does not contain the particular allele. In the absence of a match, the pollen can germinate and grow through the style to the embryo sac. Two hypotheses have been put forward to explain sporophytic control: either the S-locus is active premeiotically in the pollen mother cell (mRNA in the cytoplasm) or S-gene action is in the tapetum, a sporophytically derived cell layer that surrounds the developing microspores; in this case, subsequent transfer of the S-gene products occurs onto the exine surface of microspores during degradation of the tapetum. The S-alleles may express dominance relationships or may act independently in the pollen and pistill.

In contrast to the gametophytic system, where arrest of pollen tube growth after self-pollination usually occurs within the style, the arrest in sporophytic systems is characteristically at the stigma surface, or at a very early stage of germination. The papillar cells of the stigma are the major barrier, but once this barrier has been successfully passed, the pollen tubes can continue growth into the ovary.

The sporophytic system of self-incompatibility has been studied in the diploid species *Brassica oleracea* and *Brassica campestris* (Brassicaceae). Allotetraploid *Brassica napus* (2n = 38) containing the *B. campestris* (2n = 20) and *B. oleracea* (2n = 18) genomes generally occurs as a self-compatible plant. The molecular genetics of the self-incompatibility system was built on the key observation that specific glycoproteins of stigma extracts were correlated with individual S-alleles (Hinata and Nishio 1978; Nasrallah and Nasrallah 1984; Kandasami et al. 1989). These S-locus-specific glycoproteins (SLSG) appear on SDS-PAGE gels as a group of bands in the M_r range of 57×10^3 to 65×10^3 (Nasrallah and Nasrallah 1984). The first cDNA clone encoding part of an SLSG from *Brassica oleracea* was identified in two ways (Nasrallah et al. 1985). A β-galactosidase: S-gene product fusion protein reacted in Western blots with an antibody to SLSG, and Southern blot analysis showed restriction fragment length polymorphisms that segregated with particular S-alleles. Subsequently, cDNAs from *B. oleracea* (var. *acephala*, var. *capitata*, var. *alboglabra*, var. *gemmifera*), representing the S_2-, S_5-, S_6-, S_7-, S_{13}-, S_{14}-, S_{22}-, and S_{29}- alleles, were obtained and sequenced (Nasrallah et al. 1987; Lalonde et al. 1989; Trick and Flavell 1989; Chen and Nasrallah 1990; Scutt et al. 1990). SLSGs from *B. campestris* have been identified by amino acid sequencing of peptide fragments (Isogai et al. 1987; Takayama et al. 1987). S-alleles of *B. oleracea* have been arranged in a complex dominance series based on their genetic behavior relative to other alleles in stigma and pollen (Thompson and Taylor 1966). High activity (class I) alleles exhibit a strong incompatibility phenotype, in which an average of 0–10 pollen tubes develop per self-pollinated stigma, and are placed relatively high on the dominance scale of S-alleles. Low-activity (class II) alleles have a weak incompatibility phenotype in which 10–30 pollen tubes develop per self-pollinated stigma, and exhibit recessive and competitive interactions in pollen. Class I is exemplified by the S_6-, S_{13}-, S_{14}-, S_{22}-, S_{29}-, and S_{63}-alleles, and class II includes the S_2-, S_5-, and S_{15}-alleles (Nasrallah et al. 1991). Among class I alleles, comparison of the amino acid sequences derived from the S-locus-specific genes showed 79–85% identity, when pairs of sequences are compared (Nasrallah et al. 1987; Trick and Flavell 1989). The S_2-allele shares less than 70% overall DNA sequence similarity with class I alleles, and is 32% divergent at the amino acid level (Chen and Nasrallah 1990).

The nucleotide sequence of SLSG cDNA derived from the S_6 homozygote of *B. oleracea*, predicts a polypeptide of 436 amino acids. Amino acids hydropathy profiles reveal a 31-amino acid hydrophobic signal peptide at the N-terminus of the polypeptide, in keeping with the localization of SLSG in the papillar cell wall. The mature SLSG polypeptides (405 amino acids) predicted from the S_6-, S_{13}-, S_{14}-, and S_{22}-alleles differ along their length by short gaps and amino acid substitutions. The S-locus-specific glycoproteins can be divided into three regions: the amino terminal region (residue 1–181) shows 80% conserved positions; the middle part (residue 182-268) is the most variable with only 52% conserved positions; the carboxy terminal region with 12 invariant cysteins contains 78% conserved positions. Asparagine residues of a potential N-linked glycosyl chain attachment site (Asn-X-Thr/Ser, where X is any amino acid) are frequently conserved. Since no differences in carbohydrate compositions among S-genotypes could be discerned, it was concluded that the specificity of pollen recognition is not a function of the structure of glucan moieties (Takayama et al. 1986, 1989).

In addition to the S-locus genes described above, a second S-locus gene has been cloned from *B. oleracea*. This gene, the S-locus receptor kinase (SRK) gene, shows sequence homologies at the N-terminal end to SLSG genes and at the C-terminal end to serine/threonine kinases (Stein et al. 1991).

Goring and Rothstein (1992) isolated and characterized an S-receptor kinase cDNA, SRK-910, from the W1 self-incompatible *B. napus* ssp. *oleifera* line, which was produced by introgressing a *B. campestris* S-locus into the self-compatible Westar canola cultivar. The predicted protein sequence of 858 amino acids shows a putative signal peptide at the N-terminal end and a putative transmembrane domain separating the SLSG homologous N-terminus from the rest of the peptide chain. The SRK-910 allele is most closely related to its SLSG counterpart at the same locus, the SLSG-910 allele: there is 90% homology at the DNA level and 84% similarity at the amino acid level, including the 12 cysteine residues conserved in all SLSG sequences and 7 potential glycosylation sites. The predicted amino acid sequence at the 3'end of the gene (after the transmembrane domain) contains conserved amino acids found in serine/threonine kinases (Hanks et al. 1988), which are also highly conserved in the other receptor kinases reported for plants: the SRK-6 and SRK-2 from *B. oleracea* (Stein et al. 1991), the ARK-1 from *Arabidopsis* (Tobias et al. 1992), and ZMPK-1 from corn (Walter and Zhang 1990). By expressing the SRK-910 protein in *Escherichia coli*, it could be demonstrated that SRK-910 is a functional serine/threonine kinase (Goring and Rothstein 1992). It is possible that self-recognition is mediated by direct associations between the SLSG and SRK proteins. Alternatively, an additional gene at the S-locus encoding for a protein ligand recognized by the receptor kinase might be involved in signal transduction.

Also active in the female and male reproductive organs are a number of other members of the S-multigene family, the S-locus-related (SLR) genes, which are unlinked to the S-locus. The best characterized of these SLR-genes are the SLR1 (Lalonde et al. 1989; Trick and Flavell 1989; Isogai et al. 1991) and SLR2 (Boyes et al. 1991; Scutt and Croy 1992) genes, which encode secreted glycoproteins. SLSG and SLR genes have a sequence similarity between 70 and 90%, which is reflected in the similar primary structures of their protein products. SLR proteins possess a hydrophobic signal peptide at the amino terminus and are localized in the papillar cell wall (Umbach et al. 1990). At the carboxy terminus 12 cysteine residues are arranged in the same precise order. The SLR genes, like SLSG, contain a number of potential N-glycosylation sites. The main difference between the SLSG gene and the two SLR genes, in addition to the differences in their linkage relationships to the S-locus, is in the extent of sequence variability exhibited by allelic variants in each gene. SLR-1 is expressed in self-incompatible as well as in self-compatible strains of *Brassica*; in the self-fertile Brassicaceae *Arabidopsis thaliana* the S-related gene AtSt1 was characterized (Dwyer et al. 1992). The observation that these genes are functional in self-compatible Brassicaceae and are preferentially expressed in reproductive tissues suggests that they may be involved in aspects of the pollination process other than the inhibition of self pollen.

Nasrallah et al. (1992) identified a new locus, SCF1, which regulates S-locus function in *B. campestris*. In the stigma of plants homozygous for a mutation in this locus (scf1), the S-locus-linked SLSG gene and two other unlinked members of the S-multigene family, the SLR1 and SLR2, all of which encode secreted glycoproteins, are coordinately down-regulated at the RNA level, while the expression of the S-locus-derived SRK gene, which encodes a putative receptor protein kinase, is unaffected. As a consequence, the stigma is receptive to all pollen, including self pollen, but the pollination phenotype of the male gametophyte is not affected. It is proposed that the SCF1 locus encodes a positive transacting factor required for the enhanced transcription of several genes in the stigmatic papillar cells during pistil maturation.

In contrast to the extensive polymorphisms at the S-locus and the associated variability observed among alleles of the SLSG gene, SLR1 and SLR2 genes isolated from different *Brassica* strains are highly conserved (Lalonde et al. 1989). A comparison of two SLR1 sequences derived from *B. oleracea* homozygous for the S_{29} and S_{63} self-incompatibility alleles showed 99% conservation at the DNA level (Trick 1990). A similar degree of conservation (99%) was observed at the SLR2 locus (Boyes et al. 1991). Based on the sequence conservation in different *Brassica* strains and the fact that they are unlinked to the S-locus, the SLR1 and the SLR2 genes are not directly implicated in the determination of S-allelic specificity. The extreme sequence conservation, however, and their expression to high levels even in self-compatible strains in which the SLG gene is either not functional or not expressed (Lalonde et al. 1989) suggests an important role for the SLR proteins in pollination events in *Brassica*.

The analysis of the *Brassica* S-multigene family has demonstrated that the S-locus, although behaving as a single Mendelian locus, has a complex organization. Southern blot analysis of genomic DNA with certain SLSG-cDNAs shows multiple bands. These bands reflect the presence of a family of at least 11 sequences related to the SLSG structural gene of which only one was demonstrated to be active per haploid genome (Nasrallah et al. 1988). A probe isolated from the 3' untranslated region of the S_{13} cDNA clone hybridizes specifically to the active gene copy in S_{13} plants. Sequence data confirm that inactive members of the gene family are related but divergent. The active gene contains no introns.

In Brassicaceae, the inhibition of incompatible pollen at the stigma surface predicts that the S-locus genes are expressed in the papillar cells of the stigma surface. Sporophytic control of pollen self-incompatibility in this family predicts that the S-locus is expressed in sporophytic cells of the anther. In situ hybridization experiments have shown that the SLSG genes are expressed in the papillar cells of the stigma: no transcripts are evident in very young buds; as the stigma matures, and while the differentiation and elongation of the papillar cells occurs, a hybridization signal of increasing intensitiy is obtained. No hybridization signal above background could be detected in the underlying cells of the stigma, style, and ovary, or in anthers (Nasrallah et al. 1988). Immunogold labeling of thin sections has further demonstrated that the S-locus-specific glycoproteins are secreted into the papillar cell walls, where they accumulate (Kandasamy et al. 1989).

In transgenic *B. oleracea* (Sato et al. 1991), the SLSG genes and the SLSG-glucuronidase (SLSG-GUS) fusions were similarly expressed at high levels in the papillar cells of the stigma. The sensitivity of assays for GUS activity allowed the demonstration of SLSG-promoter activity in the stigma, in the transmitting tissue of the style and ovary, as well as in the tapetal cells of the anther and in microspores. Staining was detected simultaneously in the cells of the tapetum and in the microspores (late uninucleate mircospore stage). After degeneration of the tapetum, GUS staining of the microspores persisted into the mature trinucleate pollen stage. GUS activity levels were approximately 40 000-fold higher than control levels in the papillar cells, 100-fold higher in the tapetum/microspores and 40-fold higher in the transmitting tissue of the styles and ovaries. The observation that the SLG gene was expressed during pollen maturation in the cells of the tapetum and in haploid microspores is consistent with the view that sporophytic control of pollen self-incompatibility phenotype includes both S-gene action in the tapetum, with subsequent transfer of S gene products onto the surface of microspores after the degeneration of the tapetum, and activity in the premeiotic mother cells.

The same SLSG-GUS constructs were transformed into *Arabidopsis thaliana*, a self-compatible Brassicaceae (Toriyama et al. 1991a) and into *Nicotiana tabacum*, a self-compatible Solanaceae (Thorsness et al. 1991).

The overall conclusion from the analysis in the three genera, *Arabidopsis*, *Nicotiana*, and *Brassica*, is that the SLSG promoter activity is limited to the pistil and anther tissues.

Interestingly, significant differences in the temporal and spatial patterns of transgene expression were observed. In transgenic *Arabidopsis*, SLSG promoter activity was detected in only two cell types, the papillar cells of the stigma and the tapetal cells of the anther. In transgenic tobacco, on the other hand, the SLSG promoter was maximally active throughout the transmitting tissue of the stigma and style, with lower activity observed in the placental epidermis of the ovary, but it was also active gametophytically in the pollen grains. No SLSG promoter activity was detected in the tapetum of transgenic tobacco anthers. This pattern is consistent with the known features of gametophytic control of pollen selfincompatibility in Solanaceae and would reflect the distribution of endogenous transacting factors available for interaction with the SLSG promoter elements.

The analysis of transgenic plants and self-compatible *Brassica* lines allows further interesting conclusions. Introduction of the SLG-gene from a self-compatible *B. campestris* line into the self-incompatible *B. oleracea* resulted in self-compatibility of the transformed plants. In these cases the self-compatible phenotype was associated with a drastic reduction of S-locus-specific glycoproteins in the stigma, but without change in the pollen phenotype (Toriyama et al. 1991b). This self-compatible transgenic phenotype mimics the generation of the self-compatible amphidiploid genus *B. napus* following naturally occurring combinations of the genomes of the self-incompatible species *B. oleracea* and *B. campestris*.

On the other hand, introduction of class I SLG genes into self-compatible *B. napus* did not generate self-incompatible plants (Nishio et al. 1992). Realization of self-incompatibility requires at a minimum the expression of the two S-locus-linked genes, SLSG and SRK, suggesting that pollen recognition is based on a cascade of events involving cell to cell signaling between pollen and papillar cells.

4. Cytoplasmic and Nuclear Male Sterility

Self-incompatible plants produce viable pollen, but the pollen tubes cannot grow within stigma or styles of an ovary containing the same incompatibility allele. Another very efficient possibility to prevent self-fertilization and to ensure cross-fertilization is male sterility.

When plants fail to produce functional pollen grains, they are called male sterile. Male sterility may be conditioned by either nuclear or cytoplasmic genes. If the sterility trait is inherited in a non-Mendelian fashion, it is designated cytoplasmic male sterility (CMS). Cytoplasmic genes of plants are most often maternally inherited. CMS is found in many crop plants as *Zea mays, Sorghum, Petunia, Helianthus,* or *Phaseolus*. Much of the interest in CMS plants stems from their use, or potential use, in producing hybrid seeds. Presently, hybrid seed of some crops is produced with CMS inbred lines, while production of others depends upon hand emasculation. Crop plants grown from hybrid seed derived from CMS/fertile crosses would be predicted to be male sterile. However, the economically useful portion of hybrid crops results from self-pollination (e.g., corn kernels, sunflower fruit), as male fertility can be restored by pollination with plants carrying dominant nuclear restorer genes.

Both the chloroplast and the mitochondrial genomes can be maternally inherited in higher plants; however, to date, only mitochondrial genome mutations have been correlated with cytoplasmic male sterility.

The most striking aspect of the higher plant mitochondrial genome is its large size variation among plant species ranging from 200–2400 kb (Newton 1988). The plant mitochondrial genome displays great recombinational capacity and flexibility. Repeats of any size can play a key role in the evolution of mtDNA. Numerous chloroplast sequences have been imported to mtDNA. It is quite conceivable that nuclear DNA has been integrated into the mtDNA, but no detailed screening has ever been carried out (Fauron et al. 1991).

Rearrangements of mtDNA result in chimeric genes. Besides creating new open reading frames, rearrangements can result in genes with new promoter activities. Many chimeric genes with portions of identified and unidentified reading frames are known in mtDNA of fertile lines; others are present in CMS lines but not correlated with sterility (Hanson 1991). The first CMS-correlated loci were identified in maize T-CMS and *Petunia* CMS lines, and have provided models for locating mitochondrial loci encoding sterility in other species.

Three major types of maize male sterile cytoplasms, cms-T, cms-C, and cms-S, are recognized. They are distinguished by the specific nuclear genes that restore pollen fertility.

The mitochondrial DNA sequence responsible for cms-T of maize carries the gene T-urf13 (Dewey et al. 1986) within a 3547-nucleotide sequence combining portions of the flanking and/or coding regions of the maize mitochondrial 26S ribosomal gene, the ATPase subunit 6 gene, and the chloroplast tRNA-Arg gene. It contains two open reading frames – T-urf13 and orf221 – encoding polypeptides of 12 961 M_r and 24 675 M_r. The larger open reading frame hybridizes to transcript present in all maize cytoplasms, the smaller only in T-cytoplasm. T-urf13 is constitutively expressed in cms-T mitochondria and its transcripts are present in coleoptiles, ear shoots, leaves, roots, and tassels. Antibodies raised to a chemically synthesized oligopeptide corresponding to ORF13 readily identified the 13 kdaltons polypeptide in cms-T mitochondria from all organs (Wise et al. 1987; Dewey et al. 1987). The dominant nuclear alleles of the Rf1 and Rf2 loci act jointly to suppress pollen abortion in cms-T maize (Laughnan and Gabay-Laughnan 1983). Presence of nuclear restorer gene Rf1 in cms-T plants alters the transcription profile of T-urf13 and decreases the abundance of URF13 by aproximately 80% (Dewey et al. 1987). The abnormal *Petunia* gene termed pcf for "*Petunia* CMS-associated fused", shows a comparable organization. It contains part of the coding region of atp9, part of each exon of coxII, and an unidentified reading frame termed urf-S (Young and Hanson 1987). Complete, functional atp9 and coxII genes are located elsewhere in the mitochondrial genome, as is the case with functional copies of the chimeric parts in T-urf13 of maize. Immediately downstream of pcf are two cotranscribed essential mitochondrial genes coding for subunit 3 of NADH dehydrogenase (nad3) and for ribosomal protein S12 (rps12) (Gualberto et al. 1988; Rasmussen and Hanson 1989). Although its protein product is unknown, urf 25 in maize also appears to be a functional single copy mitochondrial gene. In maize the 5' transcribed region is identical to that of the normal atp6 gene, while the *Petunia* 5'transcribed region corresponds to the sequence of a normal atp9 gene (Dewey et al. 1987; Young and Hanson 1987). The pcf gene has been associated with CMS by its cosegregation with male sterility in somatic hybrids and by the demonstration that expression of the pcf-encoded polypeptide is affected by the nuclear restorer gene conferring fertility (Nivison and Hanson 1989).

For both cms-T maiza and cms *Petunia*, secondary phenotypes are described: susceptibility to fungal pathogens for maize and lower alternative oxidase activity in *Petunia*.

The Texas cytoplasm of maize carries two cytoplasmically, inseparably inherited traits, male sterility and sensitivity to the fungal toxins of *Bipolaris maydis* race T (BmT; formerly known as *Helminthosporium maydis*) and *Phyllosticta maydis* (Pm); both traits are associated with the unusual mitochondrial gene T-urf13, which encodes a 13-kDa polypeptide (URF13). Frame shift mutations within T-urf13 resulting in a premature stop codon (Wise et al. 1987b) or homologous recombination resulting in loss of the 13-kDa polypeptide (Rottmann et al. 1987) were associated with mutation to male fertility and toxin insensitivity. The coding region of T-urf13 has been cloned into inducible expression vectors and transformed into *E. coli*. When induced, *E. coli* expressed the 13-kDa polypeptide encoded by T-urf13 as shown by an antiserum prepared to the protein. After addition of BmT- or Pm-toxin, inhibition of whole cell respiration, swelling of spheroplasts, and massive ion leakage were observed in *E. coli*; these effects are similar to those observed with isolated cms-T mitochondria (Dewey et al. 1988; Braun et al. 1989). Incubation of toxin with mitochondria isolated from cms-T maize produces rapid leakage of Ca^{2+} and NAD^{+} (Holden and Sze 1987, 1989). With ^{3}H-labeled Pm toxin it could be shown that the fungal toxin binds to the URF13 protein in a positive cooperative fashion in maize mitochondria and *E. coli* (Braun et al. 1990). URF13 is a membrane protein in both organisms; it is located in the plasma membrane of *E. coli* and in the inner mitochondrial membrane of maize, forming a channel in the presence of toxin. From the size of NAD^{+} passing through pores generated by the interaction of T-toxin with URF13, three putative membrane spanning helical domains and an oligomeric structure within the membrane are suggested (Hack et al. 1991; Korth et al. 1991).

In *Petunia* a lower alternative oxidase activity has been correlated with CMS (Connett and Hanson 1990). In higher plants, reducing equivalents may be oxidized either through the cyanide-sensitive cytochrome oxidase pathway that is also found in animals or through the cyanide-resistant alternative pathway. When oxygen consumption was measured in the presence of appropriate inhibitors in cells or isolated mitochondria of isonuclear CMS and fertile pairs of *Petunia* lines, a consistent reduction in alternative pathway activity was noted. Similar results were found when immature anthers of fertile and CMS *Petunia* were compared or suspension cultures derived from stable CMS somatic hybrids and fertile hybrids. A culture from a fertility restored line was also restored in alternative pathway activity. The *Petunia* Rf gene thus specifies three characteristics: restoration of pollen development, increase in alternative oxidase activity, and decrease in abundance of pcf gene products.

Despite the impressive correlation between CMS and disease susceptibility in maize or alteration in electron partitioning in *Petunia*, respectively, there are no conclusive experimental data indicating the mechanism by which the additional proteins found in CMS-mitochondria cause pollen abortion. It is perplexing that a mitochondrial gene mutation specifically interferes with pollen formation and does not noticeably affect other plant developmental processes. In several types of male sterility such as cms-T maize (Warmke and Lee 1977), *Petunia* (Bino 1985a, b), *Capsicum annuum* (Horner and Rogers 1974), and *Helianthus annuus* (Horner 1977), ultrastructural and histological examinations have revealed that the first signs of abnormality are found not in the microsporocyte but in the tapetum. Rapid division of mitochondria occurs in tapetal and sporogenous cells of both fertile and sterile anthers during early development, resulting in a 20- to 40-fold increase in the number of mitochondria per cell. Female development is normal in cms-T maize, and rapid mitochondrial replication does not occur in the female reproductive structures. The increased requirement for mitochondria in pollen formation might be the reason that a mitochondrial gene mutation, such as T-urf13 or pcf, could seriously impair mitochondrial biogenesis, whereas in other less demanding developmental processes the mutation may not be limiting. It remains to be examined if an anther-specific substance exists that affects mitochondria in a fashion similar to the fungal toxins. The *Phyllosticta maydis* and *Bipolaris (Helminthosporium) maydis* toxins (PM- and HmT-toxin) have been characterized, and each consists of a family of related compounds. HmT-toxin consists of compounds with a linear β-oxo-dioxo polyketol structure and chain lengths which vary from C_{35} to C_{45}. The PM-toxin family of compounds have shorter chain lengths, C_{33} to C_{35}, and a linear oxo-oxo polyketol structure. Methomyl, an insecticide, bears little structural resemblance to those toxins, but demonstrates selective toxicity with a much lower potency (Frantzen et al. 1987).

Structure of a fungal pathoxin and methomyl:

PM-toxin C (C_{35})

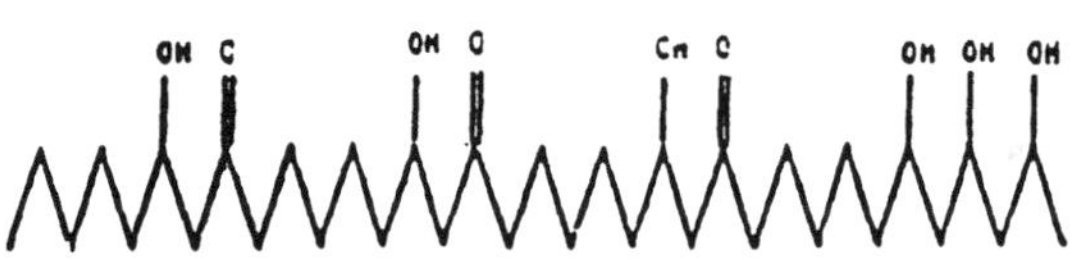

Methomyl

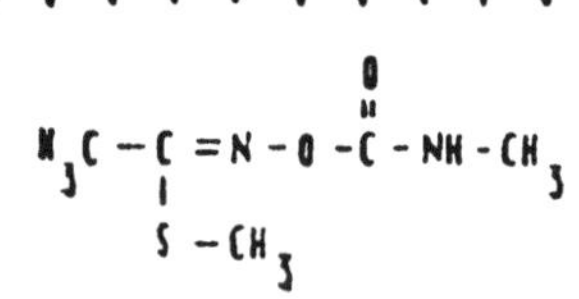

A greater understanding of the control of electron flux through the respiratory pathways may be necessary to identify the molecular basis of CMS in *Petunia*. It should be noted that alternative oxidase is also a protein in the inner mitochondrial membrane.

Molecular data are also available for CMS in *Helianthus annuus* (Köhler et al. 1991; Laver et al. 1991). A rearrangement involving an inversion and an insertion has occurred immediately downstream of the gene encoding the α-subunit of F_1ATPase (atpA) in the mitochondrial DNA from sterile sunflower. This rearrangement has resulted in the creation of a new open reading frame which is cotranscribed with atpA. Translation within the mitochrondrion in the presence of radioactivity shows that a new 15-16-kDa protein is synthesized (Horn et al. 1991). The new ORF could encode the protein causally related to the CMS phenotype.

In common bean (*Phaseolus vulgaris*), cytoplasmic male sterility is associated with the presence of a 3-kb unique mitochondrial sequence (pvs) approximately 1 kb downstream (3') to atpA. The pvs sequence encodes at least two open reading frames (297 and 720 bp in length) with portions derived from the chloroplast genome. Fertility restoration by the nuclear restorer gene F_r results in the loss of this transcriptionally active region (Johns et al. 1992). The complete loss of this 3-kb region upon restoration, and its absence from other common bean mitochondrial genomes indicate that its presence may cause the CMS phenotype.

In nearly all cases of CMS, with the exception of CMS-S in maize (Lee et al. 1980), microspore abortion is preceded by abnormal tapetal development or premature tapetal breakdown (Horner and Rogers 1974; Horner 1977; Warmke and Lee 1977; Bino 1985b). In common bean, no differences in tapetal development were observed at the light microscope level between fertile and male sterile lines; the tapetal cells remained uninucleate throughout their development. In the fertile line, the young microspores were released from their tetrad configuration as the callose broke down. In the CMS line, however, following meiosis the callose wall dissolved, but cytokinesis was often incomplete and the microspores retained their tetrahedral configuration (Johns et al. 1992). Fertility restoration by restorer gene F_r results in a semisterile phenotype in F_1 progeny (Mackenzie and Basset 1987), which is characterized by the presence of fully fertile, sterile, and intermediate buds on a single plant. The observation of both viable and aborted pollen within a single tetrad suggests a gametophytic, or pollen-specific factor in pollen abortion or fertility restoration. In this way, CMS in common bean is similar to S-CMS in maize, which is known to be gametophytic sterile, whereas T-CMS and C-CMS in maize are sporophytic-type steriles (Palmer et al. 1992).

A very elegant method to generate cytoplasmic male sterility in one step was described by Melchers and coworkers (1992): mesophyll protoplasts of *Lycopersicum esculentum* were treated with iodoacetamide to inactivate mitochondria, and protoplasts of *Solanum acaule* and *Solanum tuberosum* were irradiated with X-rays to inactivate nuclei; protoplasts thus modified were fused. Among the fusion products were some tomato plants that were indistinguishable from the original cultivars, but exhibited various degrees of male sterility, which was inherited maternally over several generations. Analysis of mitochondrial DNA revealed that the mitochondrial genome of the CMS hybrids does not contain all elements of the mitochondrial DNA of either parent but includes sequences of a recombinational nature not present in either parent. The advantages of this method over others for generating male sterility are: only one step is required, because backcrossing and selection are not necessary, and the nuclear genotype of the cultivar is unaffected. A limitation, however, might be that fusion partners for the production of male sterility should not be too closely or too remotely related.

Besides cytoplasmic mutations, about 50 nuclear mutations have been described that prevent normal pollen development and result in male sterility. Many male sterile mutations interfere with tapetal cell differentiation and/or function, indicating that this tissue is essential for the production of functional pollen grains (Kaul 1988). Male sterile lines of both crop and horticultural species are an essential part of programs to produce hybrid seed; and several approaches to engineer nuclear male sterility are possible: with the use of an anther-specific promoter tapetum abnormalities can be created, specific functions in microsporogenesis can be altered, anther genes can be inhibited by antisense RNA, or the hormone balance in anthers can be altered. C. Mariani and coworkers (1990) transformed tobacco and oilseed rape plants *(Brassica napus)* with constructs containing a cDNA encoding a RNase from *Bacillus amyloliquifaciens* (barnase) driven by a tapetum-specific promoter (TA29). The expression of the TA29-barnase gene leads to the production of male sterile plants, presumably by hydrolyzing tapetal cell RNAs. An analogous process may occur naturally in the reproductive structures of self-incompatible plants (McClure et al. 1989). In crop plants, where fruit is not the harvested product (for example, carrot, cabbage), male sterile plants can be crossed with any pollinator line to produce hybrid seed. In other crops such as tomato, wheat, rice, and corn, it will be necessary to restore full male fertility in the offspring. It was therefore demonstrated with tobacco and rape that fertility can be restored by introducing a ribonuclease-inhibitor gene under the control of the tapetum-specific promoter: male sterile plants expressing the chimeric ribonuclease gene in the anther tapetal cell layer were crossed with transgenic male fertile plants carrying a chimeric tapetal-cell-specific ribonuclease-inhibitor gene. F_1 progeny expressing both genes are restored to male fertility by the formation of the cell-specific RNase/RNase inhibitor complexes (Mariani et al. 1992).

Another approach to creating nuclear male sterility is based on earlier observations of the mistiming of β-1,3-glucan hydrolase activity in developing anthers of male sterile *Petunia* lines (Izhar and Frankel 1971). Normally, in angiosperms microsporocytes synthesize before meiosis between the cellulose cell wall and plasma membrane a special cell wall consisting of callose. After completion of meiosis and initiation of microspore exine wall formation, the callose wall is broken down by callase, a tapetally secreted β-1,3-glucanase activity, releasing free microspores into the locular space (Steiglitz 1977). The early appearance of active callase in male sterile *Petunia* lines was mimicked in transgenic tobacco by engineering the secretion of a modified pathogenesis related vacuolar β-1,3-glucanase (with the C-terminal vacuolar targeting sequence removed) from the tapetum prior to the appearance of callase activity in the locule (Worall et al. 1992). Plants expressing the modified glucanase from the tapetum-specific promoter exhibited reduced male fertility, ranging from complete to partial male sterility. The resultant microspores have an abnormally thin cell wall that lacks sculpturing. The tapetum shows hypertrophy. Premature callose degradation is sufficient to cause male sterility, and callose is essential for the formation of a normal microspore cell wall.

A third successfull approach to nuclear male sterility resulted from antisense inhibition of flavonoid biosynthesis in *Petunia* anthers (Van der Meer et al. 1992). Flavonoids are phenylpropanoid-based secondary metabolites that, besides their role in the pigmentation of flowers and fruit, display many other functions such as protection against UV light. Chalcone synthase (CHS) is the key enzyme of flavonoid biosynthesis. CHS-deficient mutants of maize and *Petunia* are known to produce white pollen that does not germinate and is sterile in self crosses. The CHS gene was placed in reverse "antisense" orientation under the control of the CaMV 35S promoter with an "anther box" inserted at the 3'end of the 35S promoter. The "anther box" is a conserved sequence found in genes expressed in anthers, and the chimeric promoter resulted in expression of the antisense CHS gene in the tapetum of transgenic plants. This led to inhibition of pigment synthesis in the anthers, arrest of gametophyte development, and male sterility. This finding indicates that flavonoids play an essential role in male gametophyte development.

It will be of interest to see, if the procedures to engineer pollen sterility and fertility restoration as described above will be useful additions to heterosis breeding and hybrid seed formation of crop and horticultural plants.

References

Ai Y, Singh A, Coleman CE, Ioerger TR, Kheyr-Pour A, Kao TH (1990) Sex Pl Reprod 3:130–138. **Anderson MA,** Cornish EC, Mau SL, Williams EG, Hoggart R, Atkinson A, Bonig I, Grego B, Simpson R, Roche PJ, Haley JD, Penschow JD, Niall HD, Tregar GW, Coghlan JP, Crawford RJ, Clarke AE (1986) Nature 321:38–44. **Anderson MA,** McFadden GI, Bernatzky R, Atkinson A, Oprin T, Dedman H, Tregar G, Fernley R, Clarke AE (1989) Plant Cell 1:483–491.

Bernatzky R, Anderson MA, Clarke AE (1988) Dev Genet 9:1–12. **Bino RJ** (1985a) Protoplasma 127:230–240. **Bino RJ** (1985b) Theor Appl Genet 69:425–428. **Boyes DC,** Chen CH, Tantikanjana T, Esch JJ, Nasrallah JB (1991) Genetics 127:221–228. **Braun CJ,** Siedow JN, Williams ME, Levings CS III (1989) Proc Natl Acad Sci USA 86:4435–4439. **Braun CJ,** Siedow JN, Levings CS III (1990) Plant Cell 2:153–161.

Chen CH, Nasrallah JB (1990) Mol Gen Genet 222:241–248. **Clark K,** Okuley JJ, Collins PD, Sims TL (1990) Plant Cell 2:815–826. **Connett MB,** Hanson MR (1990) Plant Physiol 93:1634–1640. **Cornish EC,** Pettitt JM, Bonig I, Clarke AE (1987) Nature 326:99–102.

Dewey RE, Levings CS III, Timothy DH (1986) Cell 44:439–449. **Dewey RE,** Timothey DH, Levings CS III (1987) Proc Natl Acad Sci USA 84:5374–5378. **Dewey RE,** Siedow JN, Timothy DH, Levings CS III (1988) Science 239:293–295. **Dwyer KG,** Lalonde BA, Nasrallah JB, Nasrallah ME (1992) Mol Gen Genet 231:442–448.

Ebert PR, Anderson MA, Bernatzky R, Altschuler M, Clarke AE (1989) Cell 56:255–262.

Fauron CMR, Havlik M, Casper M (1991) In: Herrmann RG, Larkins BA (eds) Plant molecular biology. Plenum, London, pp 345–363. **Franklin-Tong VE,** Ruth E, Marmey P, Lawrence MJ, Franklin FCH (1989) New Phytol 112:307–315. **Frantzen KA,** Doly JM, Knoche HW (1987) Plant Physiol 83:863–868.

Goring DR, Rothstein SJ (1992) Plant Cell 4:1273–1281. **Gray JE,** McClure BA, Bönig I, Anderson MA, Clarke AE (1991) Plant Cell 3:271–283. **Gualberto JM,** Wintz H, Weil JH, Grienenberger JM (1988) Mol Gen Genet 215:118–127.

Hack E, Lin C, Yang H, Horner HT (1991) Plant Physiol 95:861–870. **Hanks S,** Quinn AM, Hunter T (1988) Science 241:42–52. **Hanson MR** (1991) Annu Rev Genet 25:461–486. **Haring V,** Gray JE, McClure BA, Anderson MA, Clarke AE (1990) Science 250:937–941. **Hinata K,** Nishio T (1978) Heredity 41:93–100. **Holden MJ,** Sze H (1987) Plant Physiol 84:670–676. **Holden MJ,** Sze H (1989) Plant Physiol 91:1296–1302. **Horn R,** Köhler HR, Zetsche K (1991) Plant Mol Biol 17:29–36. **Horner HT Jr** (1977) Am J Bot 64:745–759. **Horner HT Jr,** Rogers MA (1974) Can J Bot 52:435–449.

Isogai A, Takayama S, Tsukamoto C, Ueda Y, Shiozawa H, Hinata K, Okazaki K, Suzuki A (1987) Plant Cell Physiol 28:1279–1291. **Isogai A,** Yamakawa S, Shiozawa H, Takayama S, Tanaka H,

Kono T, Watanabe M, Hinata K, Suzuki A (1991) Plant Mol Biol 17:269–271. **Izhar S,** Frankel R (1971) Theor Appl Genet 41:104–108.

Jahnen W, Batterham MP, Clarke AE, Moritz RL, Simpson RJ (1989a) Plant Cell 1:493–499. **Jahnen W,** Lush WM, Clarke AE (1989b) Plant Cell 1:501–510. **Johns C,** Lu M, Lyznik A, Mackenzie S (1992) Plant Cell 4:435–449.

Kandasamy MK, Paolillo DJ, Faraday CD, Nasrallah JB, Nasrallah ME (1989) Dev Biol 134:462–472. **Kaul MLH** (1988) Monographs on theoretical and applied genetics, 10 Male sterility in higher plants. Springer, Berlin Heidelberg New York. **Kirch HH,** Uhrig H, Lottspeich F, Salamini F, Thompson RD (1989) Theor Appl Genet 78:581–588. **Köhler RH,** Horn R, Lossl A, Zetsche K (1991) Mol Gen Genet 227:369–376. **Korth KL,** Struck F, Kaspi CJ, Siedow JN, Levings CS III (1991) In: Herrmann RG, Larkins BA (eds) Plant Molecular Biology. Plenum, London, pp 375–381.

Lalonde B, Nasrallah ME, Dwyer KG, Chen CH, Barlow B, Nasrallah JB (1989) Plant Cell 1:249–258. **Laughnan JR,** Gabay-Laughnan S (1983) Annu Rev Genet 17:27–48. **Laver HK,** Reynolds SJ, Moneger F, Leaver CJ (1991) Plant J 1:185–193. **Lee SLJ,** Earle ED, Gracen VE (1980) Am J Bot 67:237–245. **Levings CS III** (1990) Science 250:942–947.

Mackenzie S, Basset M (1987) Theor Appl Genet 74:642–645. **Mariani C,** De Beuckeleer M, Truettner J, Leemans J, Goldberg RB (1990) Nature 347:737–741. **Mariani C,** Gossele V, De Beuckeleer M, De Block M, Goldberg RB, De Greef W, Leemans J (1992) Nature 357:384–387. **Mau SL,** Williams EG, Atkinson A, Anderson MA, Cornish EC, Grego B, Simpson RJ, Kheyr-Pour A, Clarke AE (1986) Planta 169:184–191. **Mau SL,** Anderson MA, Heisler M, Haring V, McClure BA, Clarke AE (1991) Society for Experimental Biology, Great Britain, pp 245–269. **McClure BA,** Haring V, Ebert PR, Anderson MA, Simpson RJ, Sakiyama F, Clarke AE (1989) Nature 342:955–957. **McClure BA**, Gray JE, Anderson MA, Clarke AE (1990) Nature 347:757–760. **McFadden GJ,** Anderson MA, Bönig J, Gray JE, Clarke AE (1992) J Microsc 166:137–148. **Melchers G,** Mohri Y, Watanabe K, Wakabayashi S, Harada K (1992) Proc Natl Acad Sci USA 89:6832–6836. **Murfett J,** Cornish EC, Ebert PR, Bönig I, McClure BA, Clarke AE (1992) Plant Cell 4:1063–1074.

Nasrallah JB, Nasrallah ME (1984) Experentia 440:279–281. **Nasrallah JB,** Nasrallah ME (1989) Annu Rev Genet 23:121–139. **Nasrallah JB,** Kao TH, Goldberg ML, Nasrallah ME (1985) Nature 318:263–267. **Nasrallah JB,** Kao TH, Chen CH, Goldberg ML, Nasrallah ME (1987) Nature 326:617–619. **Nasrallah JB,** Yu SM, Nasrallah ME (1988) Proc Natl Acad Sci USA 85:5551–5555. **Nasrallah JB**, Nishio T, Nasrallah ME (1991) Annu Rev Plant Physiol Plant Mol Biol 42:393–422. **Nasrallah ME,** Kandasami MK, Nasrallah JB (1992) Plant J 2:497–506. **Newton KJ** (1988) Annu Rev Plant Physiol Plant Mol Biol 39:503–532. **Nishio T,** Toriyama K, Sato T, Kandasamy MK, Paolillo DJ, Nasrallah JB, Nasrallah ME (1992) Sex Plant Reprod 5:101–109. **Nivision HT,** Hanson MR (1989) Plant Cell 1:1121–1130.

Palmer RG, Albertsen MC, Horner HT, Skorupska H (1992) Nucleus 35:1–18.

Rasmusson J, Hanson MR (1989) Mol Gen Genet 215:332–336. **Rottmann WH,** Brears T, Hodge TP, Lonsdale DM (1987) EMBO J 6:1541–1546.

Sato T, Thorsness MK, Kandasamy MK, Nishio T, Hirai M, Nasrallah JB, Nasrallah ME (1991) Plant Cell 3:867–876. **Scutt CP,** Croy RRD (1992) Mol Gen Genet 232:240–246. **Scutt CP,** Gates PJ, Gatahouse JA, Boulter D, Croy RRD (1990) Mol Gen Genet 220:409–413. **Steiglitz H** (1977) Dev Biol 57:87–97. **Stein JC,** Howlett B, Boyes DC, Nasrallah ME, Nasrallah JB (1991) Proc Natl Acad Sci USA 88:8816–8820.

Takayama S, Isogai A, Tsukamoto C, Ueda Y, Hinata K, Okazaki K, Koseki K, Suzuki A (1986) Agric Biol Chem 50:1673–1676. **Takayama S,** Isogai A, Tsukamoto C, Ueda Y, Hinata K, Okazaki K, Suzuki A (1987) Nature 326:102–105. **Takayama S,** Isogai A, Tsukamoto C, Shiozawa H, Ueda Y, Hanata K, Okazaki K, Koseki K, Suzuki A (1989) Agric Biol Chem 53:713–722. **Thompson KF,** Taylor JP (1966) Heredity 21:345–362. **Thorsness MK,** Kandasamy MK, Nasrallah ME, Nasrallah JB (1991) Dev Biol 143:173–184. **Tobias CM,** Howlett B, Nasrallah JB (1992) Plant Physiol 99:284–290. **Toriyama K,** Thorsness MK, Nasrallah JB, Nasrallah ME (1991a) Dev Biol 143:427–431. **Toriyama K,** Stein JC, Nasrallah ME, Nasrallah JB (1991b) Theor Appl Genet 81:769–776. **Trick M** (1990) Mol Gen Genet 218:112–117. **Trick M,** Flavell RB (1989) Mol Gen Genet 218:112–117.

Umbach AL, Lalonde BA, Kandasamy MK, Nasrallah JB, Nasrallah ME (1990) Plant Physiol 93:739–747.

Van der Meer JM, Stam ME, Van Tunen AJ, Mol JNM (1992) Plant Cell 4:253–262.

Walker JC, Zhang R (1990) Nature 345:743–746. **Warmke HE,** Lee SLJ (1977) J Hered 68:213–222. **Wise RP,** Fliss AE, Pring DR, Gengenbach BG (1987a) Plant Mol Biol 9:121–126. **Wise RP,** Pring DR, Gengenbach BG (1987b) Proc Natl Acad Sci USA 84:2858–2862. **Worall D,** Hird DL, Hodge R, Paul W, Draper J, Scott R (1992) Plant Cell 4:759–771.

Xu B, Mu J, Nevins DL, Grun P, Kao TH (1990) Mol Gen Genet 224:341–346.

Young EG, Hanson MR (1987) Cell 50:41–49.

PD Dr. Christine Gietl
Institut für Botanik und Mikrobiologie
der Technischen Universität München
Lehrstuhl für Botanik
Arcisstraße 21
D-80290 München, FRG

C. Genetics

I. Replication and Cell Cycle Control

By Walter Nagl

1. Introduction

The main topics of this chapter are associated with each other. Initiation of a replication origin depends on its chromatin connection. DNA replication in eukaryotes is indeed chromatin replication. This occurs in a specific spatial and temporal pattern, and hence is part of the cell cycle. The rapidly increased understanding of the regulation of the cell cycle will constitute the central section of this essay. The plasticity of cell cycle control, and with it of genome replication, is shown in terms of new aspects of DNA and gene amplification. The latter leads to questions of evolution of both the genome and DNA replication.

2. Origins: to Be or not to Be Localized

For a long time, models of DNA replication were derived from studies of prokaryotes, and were generalized to higher organisms due to high similarities found in in vitro studies of eukaroytic DNA. Recently, however, evidence accumulated for the view that chromosomal origins of DNA replication in higher eukaryotes differ significantly from those of *E. coli (oriC)* and the tumor virus, SV40 (*ori* sequences). Initiation events appear to occur throughout broad zones surrounding the OBR (origin of bidirectional replication) rather than at specific origins.

Several models exist that account for this behavior. Benbow et al. (1992) suggest that unwinding of duplex DNA may be uncoupled, both temporally and spatially, from DNA replication, resulting in transient single-stranded intermediates that function in lieu of conventional replication forks during chromosomal DNA replication. DNA synthesis begins subsequently at multiple sites within the unwind regions rather than at specific origin sequences. DePamphilis (1993) suggests that nuclear organization (chromatin structure) establishes an OBR, within which replication can begin at many sites, but actually starts at a few only at a given time. He called this view "Jesuit model" according to the Bible text "many are called, but few are chosen". Accordingly, a site-specific cellular origin may function only in the context of native chromosomal nucleoprotein, not in the context of a plasmid introduced by standard transfection methods. Heintz et al. (1992) envisage eukaryotic replication origins as *promoters* of bidirectional DNA synthesis, this means that origins are multipartite regulatory elements resembling promoters of transcription. The participation of transcription factors in the regulation of DNA replication may facilitate temporal control of transcription and replication during the cell cycle, as well as providing a mechanism for integrating origin

Progress in Botany, Vol. 55

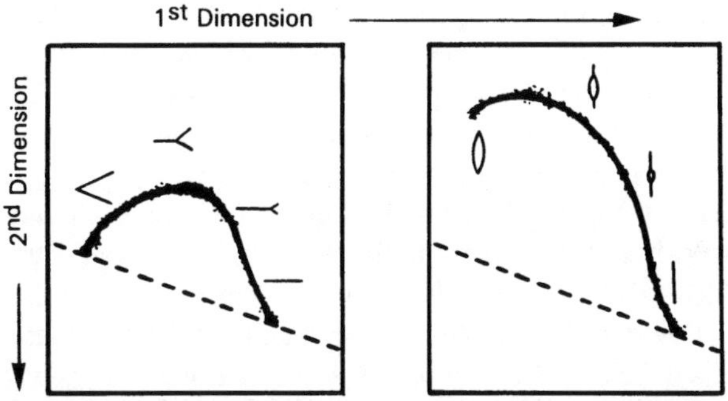

Fig. 1. Interpretation of typical patterns of replication intermediates separated by the two-dimensional gel electrophoretic method of Brewer and Fangman (1987). Panels show idealized autoradiographic images that would be obtained when a restriction digest of replicating DNA is hybridized with probes for fragments that contain different intermediates. *Solid lines* Gel patterns corresponding to the represented replication forks and bubbles (with a centered origin), respectively; *dashed lines* show the position of unreplicated total DNA fragmented with restriction enzymes. Actually, different curves normally occur within one and the same gel

selection with the cellular transcriptional program (see also Biamonti et al. 1992). It must, however, be remembered that most results in eukrayotes are deduced from the most intensively studied example, a methotrexate-resistant Chinese hamster ovary cell line (CHOC 400) that has amplified one allele of the dihydrofolate reductase (DHFR) gene and flanking sequences more than 1000 times. The multiple amplicons occur as tandem repeats in homogeneously staining regions (HSRs) as shown earlier (Nagl 1990, 1992). In plants (*Pisum sativum*), Van't Hof and Lamm (1992) studied the site of initiation of the ribosomal genes.

Vassilev and DePamphilis (1992) published an excellent "guide" to indentification of origins of DNA replication in eukaryotic chromosomes. In this guide, and in other reviews (e.g., Hamlin 1992), also the analysis of replication by two-dimensional gel electropheresis (Brewer and Fangman 1987; Nawotka and Huberman 1988) is described. This method has drastically increased our understanding of replication in eukaryotes. The principles of interpretation of the now frequently seen autoradiograms are, therefore, shown in Fig. 1.

Although our knowledge of DNA replication is far from complete and many elusive regulatory mechanisms still remain beyond our reach, many enzymes and a multiplicity of biochemical details have been discovered. The recent knowledge and controversies on the regulatory mechanism of DNA replication have been collected by Hughes et al. (1992).

3. Nucleosomes: the Problem Solved?

The question of how nucleosomes are distributed during replication was already discussed in this series and several models have been shown (Nagl 1981). The problem is apparently solved now. Convincing data support the model of distributive segregation, i.e., of random distribution of the parental nucleosomes to the daughter strands (e.g., Krude and

Knippers 1991). The assembly of histones on the newly synthesized naked DNA strands occurs by a two-step mechanism, starting with the deposition of H3/H4 tetramers onto the DNA, followed by H2A/H2B dimers. Several assembly factors, such as nucleoplasmin and others, have been identified (reviewed by Gruss and Sogo 1992). The replicative and nonreplicative chromatin assembly pathways are evidently controlled by different factors (Lässle et al. 1992). Upon deacetylation and histone H1 addition, chromatin maturation and compaction takes place (Wolffe 1991). Using multiparameter flow cytometry, Bruno et al. (1991) recorded several alterations of expression of nuclear proteins during the S period and reached the conclusion that the structure and composition of chromatin associated with nonreplicated DNA is different from chromatin associated with the newly replicated DNA. This aspect leads to questions of the role of chromatin structure in the pattern of replication, and the context of replication with cell cycle regulation (see below).

4. Replication Pattern and Chromatin "Flavors"

Early studies have shown that DNA synthesis follows a spatial and temporal pattern in the interphase (S phase) nucleus, concentrated in discrete foci. New techniques (e.g., immunofluorescence double staining and confocal microscopy) allow the visualization of two subsequent replication patterns in the same nucleus (e.g., Manders et al. 1992). At the molecular level DNA synthesis is coordinated in groups of adjacent replication units, named replicon clusters. Within a period of 1 h, replication of DNA in one cluster is nearly complete. It is now well established that nuclear matrix organization, chromatin structure, and gene activities are involved in regulation of initiation of cellular DNA replication in a coordinate and temporal manner.

For a long time, early replicating euchromatin and late replicating heterochromatin have been discerned, but the differentiation can be much more refined. An interesting step in that direction was made by Holmquist (1992), who detected a correlation between the time of replication and the chromatin "flavor". He discernes, in connection with other aspects of human chromosome structure and function, five flavors of euchromatic bands. These are G bands and, based upon combinations of extreme *Alu* richness and GC richness, four different R-band flavors. The two flavors with extremely GC-rich components represent only 15% of all bands. However, they contain 65% of mapped genes, 19 of 25 mapped oncogenes, most cancer-associated rearrangements, evolutionary rearrangements, meiotic chiasmata, and X-ray-induced breaks. The bands of a given flavor replicate predominately during a certain interval of the S phase. These findings corroborate earlier correlations between local chromatin organization and local genome organization (e.g., Chen and Manuelidis 1989). Plant chromosomes easily exhibit C bands, but normally do not allow a sophisticated banding like those of vertebrates. Nevertheless, replication bands could be visualized to some extent in plants, also (Taniguchi and Tanaka 1991).

Structural and functional organization of the nuclei in a given cell line and tissue must be maintained through DNA replication and mitosis. As that organization is mainly brought about by the pattern of DNA methylation, hence chromatin condensation, and hence gene recognizability, the state of DNA methylation must be preserved during DNA replication. Actually, a number of studies showed that tissue-specific patterns of methylated cytosine residues in the genome are transmitted by clonal inheritance through the action of a DNA(cytosine-5)methyltransferase. Faithful clonal transmission of methylation pattern requires the recognition and methylation of all hemimethylated CpG sites prior to the subsequent S phase and involves close coordination of replication and methylation. Although this concept is not new, recent techniques allow the localization of the methylase in the replication foci exactly (Leonhardt et al. 1992).

5. Cdc2 and the New Cell Cycle Philosophy

During the last decade, two hypotheses for the regulation of the mitotic cell cycle evolved independently. Yeast geneticists, studying cell cycle mutants, realized that a number of successive gene activities are necessary to pass critical checkpoints and to complete the mitotic cell cycle. They developed the "domino model" (Fig. 2). Another group of scientists studying egg cleavage in *Xenopus* found an oscillating, or cycling, cytoplasmatic factor, that promoted egg maturation and mitosis (MPF) and suggested a "clock model" (Fig. 2). Recently, both models fused to a "universal model" for the cell cycle of eukaryotes. In brief, on the one hand, there is a rather stable protein, p34, of the cdk family, encoded by the cell division cycle gene, *cdc2* (for explanation and homologes, see Table 1), that requires successive gene activities for its respectively phosphorylation and dephosphorylation on certain of its amino acid residues. On the other hand, there is an oscillating family of proteins, the cyclins, that transiently bind to $p34^{cdc2}$, thereby forming complexes which act as protein serine kinases, and phosphorylate transcription factors and replication proteins during the S period, and chromosomal histones (particularly H1), nuclear lamins, spindle factors and proto-oncogene products during mitosis (Fig. 3). For this activity, p34 has to be dephosphorylated at Tyr15

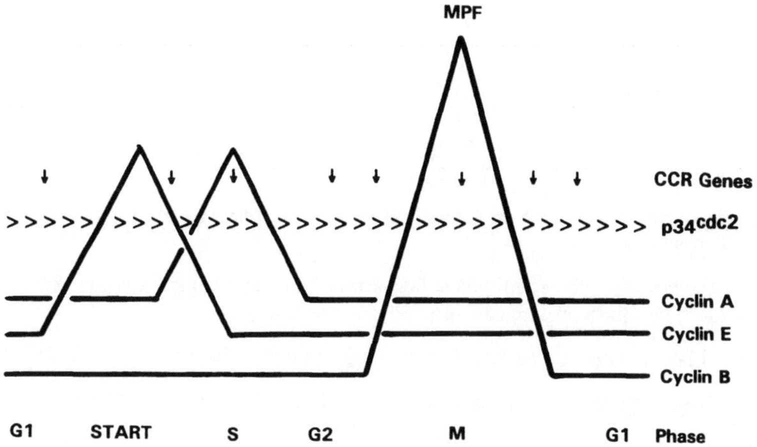

Fig. 2. Diagram to illustrate the coincidence of both aspects, the domino model and the oscillation model, of the mitotic cell cycle. The $p34^{cdc2}$ kinase is rather stable throughout the cell cycle (although it occurs in various forms and complexes) and requires the successive function of a number of cell cycle regulating (CCR) genes (*arrows*), which have been detected in mutants of yeasts. The proteins of the cyclin family (here are shown some of the human forms) are oscillating, or cycling, indicating a cell cycle clock. Actually, the checkpoints at G_0/G_1 (*START*), G_1/S (and consequently DNA replication) and G_2/M (and consequently mitosis) require an active complex of both protein types. The best-known complex is the MPF (maturation, or mitosis, promoting factor). Actually, more checkpoints, more cyclins, more cdks of the p34 family, and different complexes formed by them, are involved in the regulation of the cell cycle

Table 1. Nomenclature of some gene products of importance in the regulation of the cell cycle. (For further details, see Pelech et al. 1990; Jacobs 1992; Xiong e al. 1992; for oncoprotein phosphorylation involved in the activation of some of the cell cycle-specific kinases, see Meek and Street 1992; Wang 1992)

Abbreviation	Comments
C_1	G_1-specific and START-specific cyclins (e.g., cyclins C, E, D in mammals, cyclins 1, 2, 3 in yeast)
C_S	S phase-specific cyclins
C_M	Mitosis-specific cyclins (related to cyclin B; at least five in *S. cerevisiae*)
CAK	Cycle-activating kinase, phosphorylating $p34^{cdc2}$ at Thr161 (165, 167)[a] at G_1/S, S, and G_2
cdk1	Cyclin-dependent kinase 1 (= $p34^{cdc2}$; other members of the cdk family are cdk2, cdk3, PSSALRE, PLSTIRE, PCTAIRE-1, -2, 3. PSK-J3 and others)
cdk2	S phase-specific cyclin-dependet kinase ($p33^{cdk2}$)
INH	Inhibiting protein phosphatase (PP2A), dephosphorylates $p34^{cdc2}$ at Thr161 (165, 167)[a] at the end of mitosis (metaphase/anaphase)
MAP	1. Mitogen-activated protein 2. Microtubuli-associated protein
MPF	Maturation (or mitosis) promoting factor, complex consisting of active $p34^{cdc2}$, cyclin B and other subunits, operative as kinase towards histone H1, nuclear lamin, and other proteins, and acting as regulator of other protein kinases
p33	p34-Related protein in S phase = cdk2
$p34^{cdc2}$	Serine-specific protein kinase (34 kDa), product of cdc2 gene in *Schizosaccharomyces pombe* (fission yeast); $p34^{cdc28}$ is the homolog in *Saccharomyces cerevisiae* (budding, or bakers, or brewers yeast); at least 11 forms are known from human cells
p53	Tumor suppressor protein, in general: growth suppressor protein, inhibits the G_0/G_1 transition
PCNA	Proliferating cell nuclear antigen (= a cyclin, or part of a complex with cyclin and cdk2; auxiliary protein of DNA polymerase delta)
RB (pRb)	Retinoblastoma nuclear antigen (tumor suppressor protein at G_1)
$p80^{cdc25}$	Phosphatase, dephosphorylates $p34^{cdc2}$ at Thr14 and Tyr15 at onset of mitosis
$P107^{wee1}$	Kinase, phosphorylates $p34^{cdc2}$ at Thr14 and Tyr15 in G_2, probably also in G_1 (relative of RB)
RPA	Replication protein A, to be phosphorylated at the onset of S phase by cdc2 kinase; the complex $p34^{cdc2}$/RPA may function as an origin recognition factor and stimulate unwinding of origins

[a] Terms in parenthese indicate equivalent residues in different organisms.

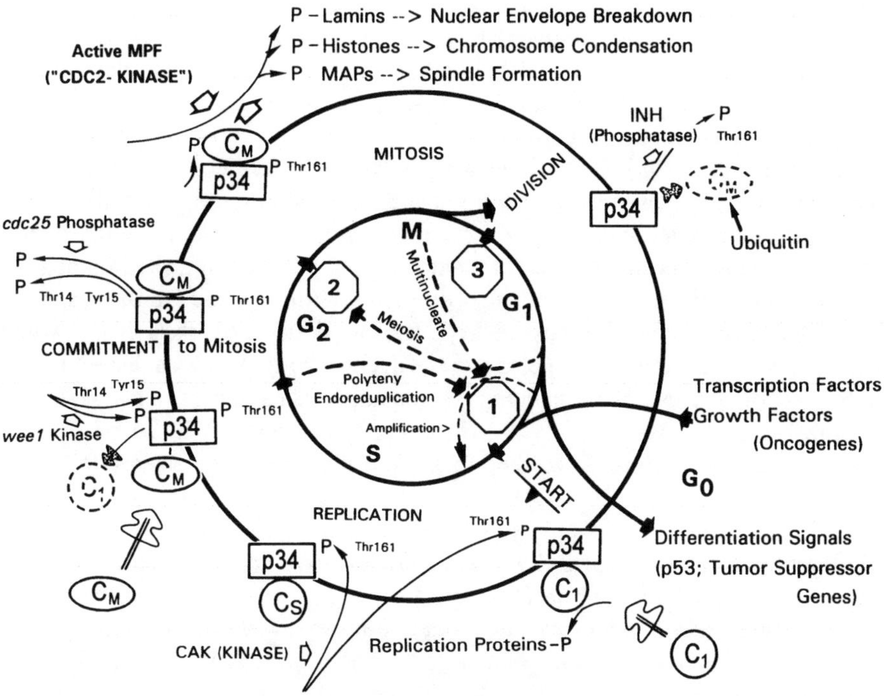

Fig. 3. Principles of the universal eukaryotic cell cycle. *Inner circle* Stages and checkpoints; *outer circle* important regulatory events. Signs: *Octagons* checkpoints; *rectangles* $p34^{cdc2}$ kinase forms; *circles* and *ellipses* cyclin forms; *dotted symbols* degradation; *P* phosphate groups (with indication of which amino acid residue becomes respectively phosphorylated and dephosphorylated). The factors leading to G_0 and differentiation and those leading to START and proliferation are best known from cancer research, but also share some universal components. For explanation of abbreviations see Table 1. Overriding a checkpoint leads to various altered cell cycles (*dotted lines*), so that polyploidy, polyteny, amplification, multinucleate cells, meiosis, etc. are the result

and, in some organisms, at Thr14 as well, and has itself to phosphorylate its cyclin subunit autocatalytically. At the end of mitosis, cyclin is also dephosphorylated, and as a consequence, the complex dissociates and cyclin becomes degraded.

Under appropriate conditions, G1-specific cyclin is synthesized and binds to $p34^{cdc2}$, which leads to the start of a new cycle. Upon phosphorylation and exchange of G1-specific cyclins with S phase-specific cyclins, DNA replication is inititated and DNA synthesized and repaired. The progression through S phase apparently requires the cyclin A/$p33^{cdk2}$ complex, which carries out phosphorylation of the replication initiation factor RPA. At the end of the S phase, the cyclins are exchanged by mitosis-specific cyclin, but due to the phosphorylation of Thr14 and Tyr15 the complex is still inactive. Actually, the system is much more complicated and different forms of $p34^{cdc2}$ ("cdc2-kinase") and cyclin are active in various organisms, in different developmental stages and tissues, and at different cell cycle stages, but the principle is apparently the same in all eukaryotes, from yeast to higher plants and humans. It should be kept in mind that, on the one hand, the primary targets of the cdc2-kinase are in many cases protein kinases themselves, which in a cascade-like manner are regulating the downstream factors. On the other hand, the cdc2-kinase is activated by either phosphatases and kinases. In the M phase, a phosphorylation cascade downstream the mitogen-activated protein (MAP) of *Xenopus*, a MAP kinase kinase (MAPKK) was found, wich is either phosphorylated by its downstream target kinase, MAPK, in a feedback way, or by an upstream MAP kinase kinase kinase (MAPKKK). Therefore, we find a very complex phosphorylation network (e.g., Pelech et al. 1990; Matsuda et al. 1993). In addition, $p34^{cdc2}$/cyclin is further complexed with a number of regulatory peptides, and last but not least, the small G proteins are also involved in some way. Due to the independent work of several laboratories, the homologous genes and gene products received different names, so that some complicated nomenclature was the result (Table 1). For convenience, all members of the gene familiy responsible for the p34-related proteins are now called cdc2-kinase (or cdc2-like kinase), while the oscillating proteins are thought to belong to the cyclin family. Reviews dissecting the cell cycle for specific organisms, or in comparison among different organisms, were, for example, published by Brooks et al. (1989), Nurse (1991), Jacobs (1992), Marsh (1992), and stressing the transition from G_1 to the DNA replication phase, by Reed (1992) and Stillman et al. (1992).

In plants, the cdc2-model of the cell cycle holds, too (e.g., John et al. 1989; Feiler and Jacobs 1990; for a review see Francis 1992). The participation of $p34^{cdc2}$ in the control of both onset of DNA replication and onset of mitosis was shown. Differential expression has, however, been found in various tissues (Bergounioux et al. 1992), and also different isoforms may be involved (Miao et al. 1993).

It is worthwile to note that the transition from a differentiated cell to a replicating and proliferating *tumor* cell in mammals and humans is characterized by oncogene-mediated activation of *cdc2* and the gene for cyclin D1, a G_1/S-specific cyclin (Sala and Calabretta 1992). In general, oncogenes and tumor viruses inducing cell transformation and immortality apparently stimulate directly, or via a complex network, transcription of genes involved in the G_0/G_1 transition and/or the G_1/S progression (Middleton et al. 1991; Takai et al. 1992). Moreover, also many proto-oncogene products are phosphorylated by $p34^{cdc2}$ during mitosis, and, at G_1/S, the retinoblastoma protein (RB) and the p53 growth (tumor) suppressor protein, which are negative regulators of entry into S phase. As cellular oncogenes control normal cell proliferation in mammals, and as oncogene homologes evidently are present also in plants (e.g., Zabulionis 1988), oncoproteins may be involved in the regulation cascade also in this kingdom. Overexpression and amplification of cyclin genes were identified in breast cancerogenesis (Keyomarsi and Pardee 1993). The p53 tumor suppressor gene seems to be the most important gene controlling cell cycle checkpoints, particularly the start of DNA replication and amplification. Therefore, defects in the cell cycle regulatory machinery may be envisaged as a key to the cancer problem (Hartwell 1992).

There are several feedback controls in the cell cycle, normally guaranteeing the completion of one stage before onset of the next step is allowed. $p34^{cdc2}$ is evidently involved in establishing the dependency of S phase on mitosis, and of mitosis on the completion of DNA replication (e.g., Broek et al. 1991; Enoch et al. 1992; Murray 1992; Smythe and Newport 1992). The multiplicity of cyclins and cdks and the number of complexes that can be formed still raises the question of exactly how many checkpoints there are in the cell cycle.

Among others, mutations and metabolic states have been found that lead to uncoupling from each other of the succession of serial cell cycle stages, or of events occurring in parallel in the nucleus and the cytoplasm, by overriding a checkpoint (cf. Fig. 3). For instance, successive S phases may occur in the absence of mitosis (e.g., Handeli and

Weintraub 1992). Smith and Orr-Weaver (1991) reported that the product of the *string* gene (the *Drosophila* homologe of cdc25) is not required for the endo-cycles during polytenization, i.e., the MPF (or cdc2-kinase) is not activated. A similar effect can be obtained by certain inhibitors of protein kinases, such as staurosporine and its analog, K-252a (Usui et al. 1991). Nagl (1993a) induced high levels of polyploidy in cell suspension cultures of *Phaseolus* by K-252a, and he concluded that the widespread phenomenon of somatic polyploidy is another aspect of the regulative versatility of the cdc2 kinase complex (Nagl 1993a; also Francis 1992). It might turn out that the different derivative cell cycles leading to polyploidy, polyteny, nuclear restitution, meiosis, and others are in fact controlled by this system. Moreover, it is possible that higher organisms have developed additional regulatory systems to those present in protozoans and yeasts. In this context, one should note that the nimA protein kinase in *Aspergillus* has been shown to act on regulatory pathways parallel to cyclin/Cdk.

Senescence and programmed cell death (apoptosis) marks the end of the life in most cells. Stein et al. (1991) found that senescent cells fail to express cdc2 and cyclins in response to mitogen stimulation. Hence, evidence is increasing that the cell cycle-regulation genes and gene products play the key role not only in cancerogenesis, but also in normal life from cleavage to senescence (Berges and Isaacs 1993). Thus, many questions have been anwered in recent years, but in turn the answers have engendered more questions, as is characterized for scientific progress.

6. Amplification: from Complexity to Perplextiy

Gene (DNA) amplification can be seen as an uncoupling process of the completion of S phase and mitosis, or as loss of the G_1/S checkpoint control (Hartwell 1992). Although no specific gene or protein responsible for this switch is known from plants, there is a good deal of evidence that a mutant p53 (cf. Fig. 3, Table 1) is involved in gene amplification in cell transformation and tumor progression (Livingstone et al. 1992; Yin et al. 1992). As oncogenes themselves, or genes activating them, are often amplified, carcinogenesis is further stimulated by a positive feedback of cell cycle regulation. In addition, genome and chromosome instability is caused leading to irreversible cytogenetic changes in the cells. This instability itself may lead to amplification events again. Gaudray et al. (1992), therefore, described the situation as changing "from complexity to perplexity". We may, in addition, be also perplexed to see the genome more and more as a bubbling soup rather than as a stable bone.

Gene amplification is now studied with a variety of techniques, like pulse field gel electrophoresis, Southern blotting, etc. Excellent visualization is also possible by fluorescence in situ hybridization (FISH), particularly in human chromosomes, and recently by PCR in situ hybridization (Nuovo 1992). The huge number of publications cannot be cited here, but the reader will find an example in every issue of the relevant journals. As a result, a number of models for the possible mechanism of DNA amplification have been elaborated (cf. also Nagl 1989, 1990). Among them are a branching process model (Kimmel et al. 1992), a U-turn replication model (Aladjem and Lavi 1992), a hairpin formation or Hoogsteen pairing model (Bianchi et al. 1991), loop and circular DNA models (Schwab 1992; Stahl et al. 1992), and a retrotransposon model (Yoshioka et al. 1992).

Underreplication of DNA in polyploid and polytenic nuclei is another example of uncoupling of completion of DNA replication from the cell cycle, although that event normally occurs only in endo-cycles. However, as a recent model interprets underrepresentation of sequences in *Drosophila* rather as the consequence of elimination of unconnected DNA fragments than as consequence of nonreplication or underreplication (Glaser et al. 1992), the same event could occur at the diploid level and during a mitotic cycle. Loss (and gain) of DNA (evidently repetitive, noncoding DNA) during the in vitro cultivation and regeneration of plants (e.g., Cecchini et al. 1992), during in vivo selection (e.g., McMurphy and Rayburn 1992), and during differentiation (Altamura et al. 1991) are well established. Arnholdt-Schmitt et al. (1991) and Arnholdt-Schmitt (1993) reported the occurrence of rapid DNA variation due to amplification and loss of sequences, and in addition, due to methylation, in carrot root explants. Mellerowicz et al. (1989, 1992) observed genomic variability in the vascular cambium of *Abies balsamina* during different seasons of the year. Shang and Wang (1991) recognized DNA amplification, the formation of double minutes and polytenization in differentiating cells of *Triticum aestivum* in vitro and in roots of regenerated plants in vivo. Natali et al. (1993) reported that the 2C nuclear DNA content varied within one and the same progeny of *Helianthus*, due to variation of repetitive DNA fractions. The genome and the phenotype were dependent on the location of the seeds in the flowering heads. These examples of DNA elimination and DNA amplification again prove the concept of high plasticity of the plant genome (Nagl 1978, 1990; Bassi 1991). Cycles of gene (UMP synthase) amplification and deamplification (loss) were also found in animal cells (Qumsiyeh et al. 1993). New findings on amplification of protein-coding genes in plant cells were published by Harms et al. (1992). These authors found herbicide resistance due to amplification of a mutant acetohydroxyacid synthase gene in tobacco cells. Amplification of the genes encoding 5-enolpyruvylshikimate-3-phosphate synthase (EPSPS) in glyphosate-tolerant cells (as reported earlier, cf. Nagl 1990, 1992) was further studied by Wang et al. (1991) and Shyr et al. (1993).

Further reviews on amplification were published by Lara et al. (1991) for sciarid flies, by Beverey (1991) for the diploid protozoan parasite *Leishmania*, and by Hamlin et al. (1991) for mammalian cells.

There is also an amplification-chromatin connection. Dijkwel (1992) detected by fluorescence in situ hybridization of CHOC400 interphase nuclei with a FITC labeled DHFR-specific probe that the DHFR amplicons remained clustered. Moreover, these regions invariably appeared to bulge out of the nuclear lamina, possibly indicating that the nuclear architecture is disturbed in these regions.

7. Evolution and Replication

The role of DNA amplification in genome and species evolution was reviewed earlier (Nagl 1978, 1990). New evidence and models for the amplification and evolution of repetitive sequences have been put forward for plants (Kumar et al. 1992; Leeton and Smyth 1993) and mammals (Deininger et al. 1992). The evolution of the amplified chromosome segments in tobacco doublet haploids was studied by Reed et al. (1992), that of histone genes in *Vicia* by Rogers and Bendich (1992) and that of tRNA genes in *Arabidopsis* by Beier et al. (1991).

In this essay, a brief excursion will be made to the origin of life and the present-day hypotheses on the evolution of prebiotic genome replication. As biologists and chemists began to trace the evolution of self-replicating molecules, they found "molecular fossils" (Weiner and Maizels 1987) and detected the RNA world. Today, many researchers are convinced that the first molecules able to store information and catalyze their own reproduction were made of RNA, probably some tRNA-like structure (Mans et al. 1991; Piccirilli et al. 1992). In particular, group-I self-splicing introns ("ribozymes") catalyze phosphodiester exchange reactions similar to those catalyzed by DNA and RNA polymerase (e.g., Dounda et al. 1991). Volume 7, No. 1 of the FASEB Journal (1993) was entirely devoted to these aspects. The recent experimental studies on the evolution of the informative genetic code was updated by Lacey et al. (1992). One of the consequences of ribozymes might have been the evolution of the exon universe, and hence of the evolution of modern genes. The findings on the genetic conservation of protein structure and function across very broad taxonomic boundaries, the mosaic structure of genomes and genetic loci, and the molecular mechanisms of genetic change all point to a view of biological evolution as involving the rearrangement of basic ancient genetic motifs (Nagl 1986; Shapiro 1992).

Synthetic structures capable of self-replication, cooperation, and mutation (and hence competition, selection, and inheritance) have been synthesized by Hong et al. (1992) by coupling the imide ester with amines bearing adenine nuclei in $CHCl_3$, which yields self-complementary amides, which dimerize through hydrogen bounding and become replicating. One of the replicators bears a photochemically active function that is cleaved on irradiation. A different approach was also successful. Autocatalytic self-replicating micelles as models for prebiotic structure form from polar molecules of sodium octanoate, filling them with the alcohol 1-octanol and using permanganate as an oxidizing agent to form the polar acid. The system was further improved by alkaline hydrolysis of ethyl caprylate (Bachmann et al. 1992). Interestingly, the micelles can be converted into more robust vesicles by a pH change induced by dissolved CO_2. On a mathematical level, Reggia et al. (1993) demonstrated that self-replication is not an inherently complex phenomenon but rather an emergent property arising from local interactions in systems that can be simpler than is generally believed. The basic forces and constraints of evolution might be seen in some simple physical laws related to thermodynamics and self-organization, and the genes of today as the result of fractal unfolding of the ancient motifs (cf. Nagl 1983, 1986, 1993b).

Maddox (1991) interpreted the progress in chemical construction of self-replicating molecules and their meaning for the evolution of life as a goal in the reductionist agenda. This is without doubt true for the evolution of living matter; but should we reduce humans to self-replicating living matter? What is the evolutionary basis of human ethics, if there is any? (Nagl 1993b). Pessimistically seen, the devastation of our globe and the slaughter of humans in many wars all over the world seem to justify an absolute reductionism; but we should then be honest enough rename our species "*Homo insapiens*".

References

Aladjem MI, Lavi S (1992) Mutat Res 276:339–344. **Altamura MM**, Cavallini A, Cionini G, Monacelli B et al. (1991) Cytobios 67:85–93. **Arnholdt-Schmitt B** (1993) Theor Appl Genet 85:793–800. **Arnholdt-Schmitt B**, Holzapfel B, Schillinger A, Neumann K-H (1991) Theor Appl Genet 82:283–288. **Bachmann PA**, Luisi PL, Lang J (1992) Nature 357:57–59. **Bassi P** (1991) Biol Zentralbl 110:1–13. **Beier D**, Stange N, Gross HJ, Beier H (1991) Mol Gen Genet 225:72–80. **Benbow RM**, Zhao J, Larson DD (1992) BioEssays 14:661–670. **Berges R**, Isaacs JT (1993) Clin Chem 39:356–361. **Bergounioux C**, Perennes C, Hemerly AS, Qin LX, Sarda C, Inze D, Gadal P (1992) Plant Mol Biol 20:1121–1130. **Beverley SM** (1991) Annu Rev Microbiol 45:417–444. **Biamonti G**, Perini G, Weighardt F, Riva S et al. (1992) Chromosoma 102:S24–S31. **Bianchi NO**, Bianchi MS, Kere J (1991) Genes Chrom Cancer 3:136–141. **Brewer BJ**, Fangman WL (1987) Cell 51:463–471. **Broek D**, Bartlett R, Crawford K, Nurse P (1991) Nature 349:388–393. **Brooks R**, Fantes P, Hunt T, Wheatley D (eds) (1989) The Cell Cycle. J Cell Sci, Suppl 12. **Bruno S**, Crissman HA, Bauer KD, Drzynkiewicz Z (1991) Exp Cell Res 196:99–106.

Cecchini E, Natali L, Cavallini A, Durante M (1992) Theor Appl Genet 84:874–897. **Chen TL**, Manuelidis L (1989) Chromosoma 98:309–316.

Deininger PL, Batzer MA, Hutchison CA III, Edgell MH (1992) Trends Genet 8:307–311. **DePamphilis ML** (1993) J Biol Chem 268:1–4. **Dijkwel PA** (1992) Cell Biol Int Rep 16:725–737. **Dounda JA**, Couture S, Szostak JW (1991) Science 251:1605–1608.

Enoch T, Carr AM, Nurse P (1992) Genes Dev 6:2035–2046.

Feiler HS, Jacobs TW (1990) Proc Natl Acad Sci USA 87:5397–5401. **Francis D** (1992) New Phytol 122:1–20.

Gaudray P, Szepetowski P, Escot C, Birnbaum D, Theillet C (1992) Mutat Res 276:205–212. **Glaser RL**, Karpen GH, Spradling AC (1992) Chromosoma 102:15–19. **Gruss C**, Sogo JM (1992) BioEssays 14:1–8.

Hamlin JL (1992) BioEssays 14:651–659. **Hamlin JL**, Leu T-H, Vaughn JP, Ma C, Dijkwel PA (1991) Progr Nucleic Acid Res Mol Biol 41:203–239. **Handeli S**, Weintraub H (1992) Cell 71:599–611. **Harms CT**, Armour SL, DiMaio JJ, Middlestaedt LA et al. (1992) Mol Gen Genet 233:427–435. **Hartwell L** (1992) Cell 71:543–546. **Heintz NH**, Dailey L, Held P, Heintz N (1992) Trends Genet 8:376–381. **Holmquist GP** (1992) Am J Hum Genet 51:17–37. **Hong J-I**, Feng Q, Rotello V, Rebek J Jr (1992) Science 255:848–850. **Hughes P**, Fanning E, Kohiyama M (eds) (1992) DNA replication: the regulatory mechanisms. Springer, Berlin Heidelberg New York.

Jacobs T (1992) Dev Biol 153:1–15. **John PCL**, Sek FJ, Lee MG (1989) Plant Cell 1:1185–1193.

Keyomarsi K, Pardee AB (1993) Proc Natl Acad Sci USA 90:1112–1116. **Kimmel M**, Axelrod DE, Wahl GM (1992) Mutat Res 276:225–239. **Krude T**, Knippers R (1991) Mol Cell Biol 11:6257–6267. **Kumar LS**, Sivaraman L, Ranjekar PK (1992) Plant Syst Evol 179:155–165.

Lacey JC Jr, Wickramasunghe NMD, Cook GW (1992) Origins Life Evol Biosphere 22:243–275. **Lara FJS**, Stocker AJ, Amabis JM (1991) Braz J Med Biol Res 24:233–248. **Lässle M**, Richter A, Knippers R (1992) Biochim Biophys Acta 1132:1–10. **Leeton PRJ**, Smyth DR (1993) Mol Gen Genet 237:97–104. **Leonhardt H**, Page AW, Weier H-U, Bestor TH (1992) Cell 71:865–873. **Livingstone LR**, White A, Sprouse J, Livanos E et al. (1992) Cell 70:923–935.

Maddox J (1991) Nature 354:351. **Manders EMM**, Stap J, Brakenhoff GJ, Van Driel R, Aten JA (1992) J Cell Sci 103:857–862. **Mans RMW**, Pleij CWA, Bosch L (1991) Eur J Biochem

201:303–324. **Marsh J** (ed) (1992) Regulation of the eukaryotic cell cycle. Ciba Foundation Symp 170, Wiley, Chichester. **Matsuda S**, Gotoh Y, Nishida E (1993) J Biol Chem 268:3277–3281. **McMurphy LM**, Rayburn AL (1992) Theor Appl Genet 84:798–802. **Meek DW**, Street AJ (1992) Biochem J 287:1–15. **Mellerowicz EJ**, Riding RT, Little CHA (1989) Can J Bot 67:990–996. **Mellerowicz EJ**, Riding RT, Little CHA (1992) Physiol Plant 85:526–530. **Miao G-H**, Hong Z, Verma DPS (1992) Proc Natl Acad Sci USA 90:943–947. **Middleton T**, Gahn TA, Martin JM, Sugden B (1991) Adv Virus Res 40:19–55. **Murray AW** (1992) Nature 359:599–604.

Nagl W (1981) Progr Bot 43:119–131. **Nagl W** (1983) Biol Zentralbl 102:257–269. **Nagl W** (1986) In: Raup DM, Jablonski D (eds) Patterns and process in the history of life. Dahlem Conferences, Springer, Berlin Heidelberg New York, pp 223–232. **Nagl W** (1989) Prog Bot 51:173–180. **Nagl W** (1990) In: Bajaj YPS (ed) Biotechnology in agriculture and forestry vol 11. Springer, Berlin Heidelberg New York, pp 153–201. **Nagl W** (1992) Progr Bot 53:166–180. **Nagl W** (1993a) Plant Cell Rep 12:170–174. **Nagl W** (1993b) Ethik Sozialwiss 4:3–109. **Natali L**, Cavallini A, Cionini G, Sassoli O et al. (1993) Theor Appl Genet 85:506–512. **Nawotka KA**, Huberman JA (1988) Mol Cell Biol 8:1408–1413. **Nuovo GJ** (1992) PCR in situ hybridization. Raven, New York. **Nurse P** (1991) Philos Trans R Soc Lond 33B:271–276.

Pelech SL, Sanghera JS, Daya-Makin M (1990) Cell Biol 68:1297–1330. **Piccirilli JA**, McConnell TS, Zaug AJ, Noller HF, Cech T (1992) Science 256:1420–1424.

Qumsiyeh MB, Goorha S, Suttle DP (1993) Cytogenet Cell Genet 62:162–168.

Reed SI (1992) Annu Rev Cell Biol 8:529–561. **Reed SM**, Burns JA, Wernsman EA (1992) Crop Sci 32:362–366. **Reggia JA**, Armentrout SL, Chou H-H, Peng Y (1993) Science 259:1282–1287. **Rogers SO**, Bendich AJ (1992) Theor Appl Genet 84:617–623.

Sala A, Calabretta B (1992) Proc Natl Acad Sci USA 89:10415–10419. **Schwab M** (1992) Bichim Biophys Acta 1114:43–50. **Shang XM**, Wang WC (1991) Genome 34:799–809. **Shapiro JA** (1992) Genetica 86:99–111. **Shyr Y-YJ**, Caretto S, Widholm JM (1993) Plant Sci 88:219–228. **Smith AV**, Orr-Weaver TL (1991) J Cell Sci 112:997–1008. **Smythe C**, Newport JW (1992) Cell 68:787–797. **Stahl F**, Wettergren Y, Levan G (1992) Mol Cell Biol 12:1179–1187. **Stein GH**, Drullinger LF, Robetorye RS (1991) Proc Natl Acad Sci USA 88:11012–11016. **Stillman B**, Bell SP, Dutta A, Marahrens Y (1992) In: March J (ed) Regulation of the eukaryotic cell cycle. Wiley, Chichester, pp 147–160.

Takai Y, Kaibuchi K, Kikuchi A, Kawata M (1992) Int Rev Cytol 133:187–230. **Taniguchi K**, Tanaka R (1991) Jpn J Genet 66:485–489.

Usui T, Yoshida M, Abe K, Osada H et al. (1991) J Cell Biol 115:1275–1282.

Van't Hof J, Lamm SS (1992) Plant Mol Biol 20:377–382. **Vassilev LT**, DePamphilis ML (1992) Crit Rev Biochem Mol Biol 27:445–472.

Wang JYJ (1992) Biochim Biophys Acta 1114:179–192. **Wang Y**, Jones JD, Weller SC, Goldsbrough PB (1991) Plant Mol Biol 17:1127–1138. **Weiner AM**, Maizels N (1987) Proc Natl Acad Sci USA 84:7383–7387. **Wolffe AP** (1991) Trends Cell Biol 1:61–66.

Xiong Y, Zhang H, Beach D (1992) Cell 71:505–514.

Yin Y, Tainsky MA, Bischoff FZ, Strong LC, Wahl GM (1992) Cell 70:937–948. **Yoshioka K**, Kanda H, Takamatsu N, Togashi S et al. (1992) Gene 120:191–196.

Zabulionis RB (1988) Genome 30:820–824.

Professor Dr. Walter Nagl
Abteilung Zellbiologie
Fachbereich Biologie
Universität Kaiserslautern
D-67653 Kaiserslautern, FRG

II. Recombination: Effects on Structure and Function of the Mitochondrial Genome

By Renate Horn, Volker Hahn, and Wolfgang Friedt

1. Introduction

Three decades ago, it was shown for the first time that mitochondria contain DNA (Luck and Reich 1964, cited in Lonsdale and Grienenberger 1992). In the meantime, the mitochondrial DNA (mtDNA) of eukaryotes has been studied in detail and some of the mitochondrial genomes have already been sequenced. In many cases, mtDNAs are small (approx. 16 kb) and circular. Furthermore, it can now be generalized for all vertebrates and invertebrates that the genetic information of mitochondria is compactly organized and predominantly on one strand of the mtDNA (for ref. cf. review of Lonsdale and Grienenberger 1992).

In contrast, the size of the mitochondrial genome of plant species differs in a wide range. For fungi, sizes of predominantly circular genomes reported to date vary between 18.9 and 176 kb. Mitochondrial genomes of higher plants can be even larger; the largest one described so far is that of muskmelon, with a size of 2400 kb (cf. Lonsdale and Grienenberger 1992). Most of the plant species studied up to now possess multi-circular mitochondrial genomes, where the circular organization is determined by repeated sequences (Palmer 1992). This multipartite structure of the mitochondrial DNA is a consequence of homologous recombinations between large sequence duplications and inversions. Recombination events within and between circular structures can possibly lead to multimeric circles and to highly complex forms of the mitochondrial genomes (cf. Lonsdale and Grienenberger 1992).

The evolution of plant mitochondrial genomes has obviously been driven by rearrangements and by the loss of sequences as a result of recombination events. The present chapter summarizes recent findings in this field.

2. Organization of the Mitochondrial Genome

a) Recombination Repeats and Rare Recombination Events

The high frequency of recombination via repeated sequences is partly responsible for the complexity of plant mitochondrial DNA. Repeated sequences which are present in two

Progress in Botany, Vol. 55

copies relative to other sequences in the genome, but occur in four genomic environments because they are involved in frequent recombination events, are called recombination repeats (Stern and Palmer 1984).

Depending on the orientation of the repeats, either direct or indirect, intramolecular recombination in the master chromosome leads to subgenomic circles or to isomeric forms (Fauron et al. 1991). Recombination repeats have been found in all mitochondrial genomes mapped to date, with the exception of *Brassica hirta* (Palmer and Herbon 1987) and *Marchantia polymorpha* (Oda et al. 1992). However, the lack of recombination repeats in these two genomes shows that recombination repeats are not an essential feature of plants mitochondrial DNAs. The size of most known recombination repeats range from 1 to 12 kb (Palmer and Herborn 1986; Stern and Palmer 1986; Siculella and Palmer 1988); the smallest known to date is 266 bp (Schuster and Brennicke 1986) but larger recombination repeats have also been reported (Fauron et al. 1990; Folkerts and Hanson 1991). A summary of the known recombination repeats, the master circle sizes, and the subgenomic circles that can be predicted is provided by Hanson and Folkerts (1992).

In addition to (and distinct from) the frequently occurring recombination across defined recombination repeats, rare recombinations between very short repeats also occur in plant mitochondrial DNA. The result are DNA molecules present in substoichiometric amounts relative to the main mitochondrial genome which are termed "sublimons" (Small et al. 1987). To explain structural changes observed in plant mitochondrial genomes, models involving frequent recombination events as well as rare recombinations have been developed.

Three copies of a recombination repeat, two in direct and a third in indirect orientation, are present on the 443 kb master chromosome in *Petunia* line 3704 (Folkerts and Hanson 1989). In *Petunia* CMS line 3688, one of these three 6.6-kb repeats is truncated (Folkerts and Hanson 1991). Sequence analysis revealed that the normal and the truncated repeats are nearly identical over 1.42 kb (Conklin and Hanson 1993). The deletion in the truncated repeat has probably arisen from a rare recombination event between two copies of an 18 bp repeat which are localized in direct orientation at the position of the deletion breakpoints in the truncated repeat. The lack of sequence homology shared by different recombination repeats suggests that recombination may occur by a homologous mechanism rather than being strictly site-specific, although there seems to be also some site specificity, as the mere presence of repeats is not sufficient for apparent high-frequent recombination (Hanson and Folkerts 1992).

When a number of recombination repeats in direct and indirect orientation are present in a master chromosome, the situation becomes very complex, as for example, in maize (Fauron et al. 1991).

A recombination model involving two steps can explain the different rearrangements in the mitochondrial genome observed in fertile revertants in maize CMS-T after tissue culture (Fauron et al. 1992). At first, intramolecular recombination across two different sets of direct repeats gives rise to subgenomic circles. A second intermolecular recombination between defined subgenomes creates a newly rearranged master chromosome containing deletions and duplications. In the revertant V3, intramolecular recombination between two sets of direct repeats R2 (0.127 kb) and R1 (4.6 kb) is followed by intramolecular recombination either through the R1 repeat, the R2 repeat, or other homologous sequences resulting in a new master chromosome of 735 kb (Fauron et al. 1990).

The new genome organization is characterized by a 423 bp deletion between the repeats R1 and R2 including the *urf*13 gene, and a duplication of 165 kb of the sequence localized between R'1 and R'2 repeats.

Evidence for homologous recombination can be obtained by recovery of clones containing the repeat in the different genomic environments, i.e., four genomic environments for a two-copy repeat, nine genomic environments for a three-copy repeat (Fauron et al. 1989; Conklin and Hanson 1991). The presence of a recombination repeat in four genomic environments can be demonstrated by hybridizing a probe internal to the repeat against mtDNA which has been digested with restriction enzymes that do not cut within the repeat. Four signals reflect the four genomic environments of the repeat which are called repeat "versions" (Hanson and Folkerts 1992).

The various subgenomic circles generated by recombination have, however, not been directly visualized for any plant mitochondrial DNA so far. As no in vitro recombination system is established for plant mtDNA, the theories concerning the mechanism of recombination across the repeats have not yet been experimentally verified. No function has been assigned to any of the characterized repeats other than the functions of the genes present in the repeated sequences like *rps*19, *rps*3, and *rpl*16 in the 6.6 kb recombination repeat in *Petunia* line 3704 (Hanson and Folkerts 1992; Conklin and Hanson 1993; Sutton et al. 1993).

b) Implications of Nuclear and Chloroplast Sequences

Homology to chloroplast DNA could be detected in the mitochondrial genome of higher plants (Stern and Palmer 1984). Most of the sequences homologous to the chloroplast seem to be nonfunctional in mitochondria, but some tRNA genes of chloroplast origin are transcribed and processed, although others are nontranscribed (Stern and Lonsdale 1982; Wintz et al. 1988; Joyce and Gray 1989; Hanic-Joyce and Gray 1990; Izuchi et al. 1990; Sangaré et al. 1990). Mitochondrial loci associated with CMS in *Petunia*, maize, and common bean also contain chloroplast sequences (Dewey et al. 1986; Hanson and Folkerts 1992; Johns et al. 1992). In *Petunia* the 5'-flanking region of the S-*pcf* locus which contains three genes (*pcf*, *nad*3, and *rps*12) shows homology to a part of the chloroplast *atp*B gene. In maize, the locus T-*urf*13 consists of *urf*13 and *orf*25, which contains part of a *tRNA* gene homologous to the chloroplast DNA at its 3'-end. Also parts of the *pvs* sequences which are associated with CMS in common bean are derived from the chloroplast genome. The transcriptionally active region of the *pvs* sequence encodes at least two open reading frames (297 and 720 bp in length) and is lost upon fertility restoration.

Estimation of the content of chloroplast homologous sequences in the mitochondrial genome of some crucifer species revealed that the same 12 to 14 kb chloroplast-like sequences are detectable in *Brassica napus, B. hirta*, and *R. sativa* (Nugent and Palmer 1988).

Southern hybridizations were also used to estimate the extent of chloroplast DNA present in the mitochondrial genome of cowpea (Orellano et al. 1992). Plastid genes showing homology with mtDNA include those encoding ribosomal proteins, RNA polymerase, subunits of photosynthetic complexes, and the two major rRNAs.

In sunflower, eight regions dispersed throughout the mtDNA show homology to the plastid DNA (Siculella and Palmer 1988). In addition, a homology to the nuclear DNA could be shown for part of the 5 kb insertion which is involved in causing the CMS phenotype in the PET1 cytoplasm (Zetsche and Horn 1993). There also exists some homology to the nuclear DNA in the plasmids P1 and P2 in sunflower, but no homology to the main mitochondrial DNA (Crouzillat et al. 1989; De la Canal et al. 1991).

Most of the plasmids so far investigated show homology to nuclear sequences, but lack homology to the main mtDNA and their origin may be in the nucleus. In *Vicia faba*, sequences homologous to the nuclear DNA could be detected for all three mitochondrial plasmids (1.7, 1.7, and 1.5 kb) (Flamand et al. 1992).

In rice, two small, circular, mitochondrial plasmid-like DNAs, B1 and B2, were found in the male-sterile cytoplasm (CMS-Bo) (Sakamoto et al. 1991). The homologous nuclear sequences were genetically investigated by segregation analysis. The results indicate that the B1 and B2 homologous sequences are dispersed in the nuclear genome. The plasmid-like DNA B4 in the *japonica* rice variety Fujiminori also shows homology to several positions in the nuclear genome (Fukuchi et al. 1991). Structural analysis in rice is consistent with the hypothesis that the DNA sequences were transferred via RNA.

c) Introns Present in the mtDNA

In contrast to mammalian mitochondria several RNA processing events (for review see Gray et al. 1992) such as RNA-editing, *cis*- and *trans*-splicing can occur during maturation of a protein-coding transcript in plant mitochondria. Analysis of cDNA clones in *Petunia* revealed that RNA editing of pre-mRNA of *nad*1 can precede *cis*- and *trans*-splicing (Sutton et al. 1991).

For a number of plants, introns have been described for several mitochondrial protein-coding genes: *cox*II, *nad*1, *nad*4, *nad*5, *rps*3 (Fox and Leaver 1981; Wissinger et al. 1988; Hunt and Newton 1991; Sutton et al. 1993). Apart from *Marchantia polymorpha*, where group I introns are present (Oda et al. 1992), all investigated intervening sequences belong to group II introns (Michel and Dujon 1983; Michel et al. 1989). The size of the intron varies in different species due to insertions in domain IV of the predicted group II intron structure (Hanson and Folkerts 1992). The single intron present in the *cox*II gene in all monocotyledons examined (Lonsdale 1989) but not in all dicotyledons (Moon et al. 1985; Hanson et al. 1989; Hiesel et al. 1990; Mann et al. 1991) is *cis*-spliced. In contrast to the plant species investigated to date, the coding region of the *cox*II gene in *Daucus carota* is interrrupted by two group II introns (Lippok et al. 1992).

Genes like *nad*1 and *nad*5 contain several introns. The five exons of each of the two genes are dispersed over the whole mitochondrial genome. In the case of *nad*1 in *Petunia* one *cis*-splicing event and three *trans*-splicing events are necessary to create a mature *nad*1 transcript from four RNA precursors (Conklin et al. 1991). In wheat and maize, the exons for the *nad*5 gene are assembled into a 2.4 kb mRNA after two *cis*-splicing and two *trans*-splicing events (Souza et al. 1991). In broad bean, a maturase-like gene (*mat-r*) was found within the group II intron (Wahleithner et al. 1990).

The turnip rape *nad*4 gene contains four exons, which potentially encode a polypeptide of 495 amino acids and three large group II introns (Gass et al. 1992). Investigations of the *nad*4 organization in three distantly related dicotyledons reveals that introns 2 and 3 are not obligatory. In mung bean, the *nad*4 locus contains three introns as in turnip, whereas in spinach, intron 2 is not present, and in lettuce *nad*4, only intron 1 is detectable.

d) Plasmids

Autonomously replicated small nucleic acid molecules (linear and circular DNAs as well as single- and double-stranded RNAs) have been found in relatively high abundance in plant mitochondria (Hanson and Folkerts 1992; Brown 1993). Sequence data are available for a number of the extrachromosomal DNAs (Palmer et al. 1983; Wahleithner and Wolstenholme 1987; Crouzillat et al. 1989; De la Canal 1991). Apart from the only known functional copy of a $tRNA^{Trp}$ gene (Leon et al. 1989) on a 2.3 kb maize plasmid, no function could be assigned to the open reading frames on plasmids. In maize CMS-S, the two linear plasmids, S1 (6.4 kb) and S2 (5.4 kb), can recombine via a 187 nt repeated sequence at the terminus of these plasmids with the main mitochondrial chromosome. This recombination event leads to the integration and linearization of the main mitochondrial DNA (Schardl et al. 1984), assuming it normally exists as circles, which is still a point of some controversy (Bendich et al. 1993). A proof for the existence of the protein products encoded by ORF3 (S1 plasmid) and ORF1 (S2 plasmid) could be obtained by immunological investigations (Zabala et al. 1987).

In sunflower, three plasmids, P1 (1.4 kb), P2 (1.9 kb) and P3 (1.8 kb) have been described (Crouzillat et al. 1987; De la Canal et al. 1991). Transcripts could be detected for P1 and P2, but no evidence of translation products were provided so far (Crouzillat et al. 1989; De la Canal et al. 1991). As in maize CMS-S no correlation between the presence of any plasmids and cytoplasmic male sterility in sunflower could be found.

3. Nuclear-Mitochondrial Interactions

a) Cytoplasmic Male Sterility (CMS)

Plants exhibiting the cytoplasmic male sterility trait are unable to produce viable pollen. This phenomenon is commonly observed in higher plants, where it was described in over 140 plant species (Laser and Lersten 1972).

For several plant species, more than one type of CMS exists. Maize for example has three major male sterile cytoplasms: CMS-C, CMS-S, and CMS-T (see Braun et al. 1992). The investigations of CMS in *Brassica* have also focused on three male sterile cytoplasms: ogu, nap, and pol (Singh and Brown 1991). In sunflower, however, more than 30 male sterile cytoplasms have been described to date (Hahn and Friedt 1992; Zetsche and Horn 1993; Hahn 1993).

Most CMS types have occurred naturally or in inter- or intraspecific crosses. However, in the past few years, new methods have been introduced to produce male sterile plants, such as experimental mutagenesis (Christov 1992, cited in Spassova et al. 1992), asymmetric fusion of protoplasts (Melchers et al. 1992), or transformation of plants with anther-specific expressed genes (Mariani et al. 1990, 1992; van der Meer et al. 1992; Worrall et al. 1992). Although the CMS plants are phenotypically similar, as they are not able to produce functional pollen, the specific mechanisms differ among plant species (Warmke and Lee 1977, 1978) as well as within species. Holford et al. (1991) studied the structure of anther tissues during microsporogenesis in male sterile and fertile onions. They observed three types of abnormal tapetial behavior within a line, II/3ms, containing the CMS-S cytoplasm: type 1, premature breakdown of the tapetum at the tetrad stage, type 2, hypertrophy of the tapetum after the diad stage followed by its premature autolysis, and type 3, in which the tapetum remains in good condition but for an abnormally long period of time.

The first mitochondrial loci shown to encode CMS were described in maize CMS-T (T-*urf*13) and *Petunia* (*pcf*). Those CMS loci derived from multiple recombination events contain abnormal genes which are located upstream of apparently normal mitochondrial genes. These mutations, thought to cause CMS, have been the subject of many reviews (e.g., Braun et al. 1992; Hanson and Folkerts 1992; Williams and Levings 1992), so that we will focus on other species.

In rapeseed mtDNA, rearrangements located around the mitochondrial *atp*6 locus of Polima cytoplasm (pol CMS) are candidates for causing the male sterile phenotype. As a result of extensive rearrangements, a novel reading frame (*pol-urf*) was generated upstream of the *atp*6 gene only in pol CMS mitochondria which encoded 105 amino acids and might be cotranscribed with *atp*6 (Singh and Brown 1991; Handa and Nakajima 1992). These results are in conformity with the findings of Witt et al. (1991) where only hybridization with *atp*6 as probe led to different patterns of the mitochondrial transcripts of normal and pol CMS.

Sequence analysis of an Ogura-specific mtDNA fragment carrying Ogura CMS in rapeseed revealed a $tRNA^{fMet}$ sequence, a putative 138 amino acid open reading frame (*orf*138), and a 158 amino acid orf (*orf*158) previously observed in mitochondrial genomes of several other plant species. Fertile revertant plants lack transcripts of the *orf*138 sequence which is most likely to be responsible for CMS (Bonhomme et al. 1992).

In pearl millet (*Pennisetum glaucum*), mtDNA fragments were rearranged by spontaneous reversion from cytoplasmic male sterility to fertility. The correlation of reversion with the loss of fragments containing the *rrn*18, *rrn*5, and *cox*I gene suggests that those lost fragments, i.e., their gene contents, could be responsible for the expression of CMS (Smith and Chowdhury 1991).

In sunflower, the most thoroughly investigated CMS type is the PET1 cytoplasm, reported to be derived from *Helianthus petiolaris*. The mtDNA of this cytoplasm is rearranged in a region framed by the *atp*A gene and the *cob* gene including an 11 kb inversion and an about 5 kb insertion (Fig. 1). The rearrangements seem to be the result of recombination events within an inverted repeat of 261 bp resulting in the creation of a new open reading frame (*orfH522*) which is cotranscribed with *atp*A in sterile sunflower

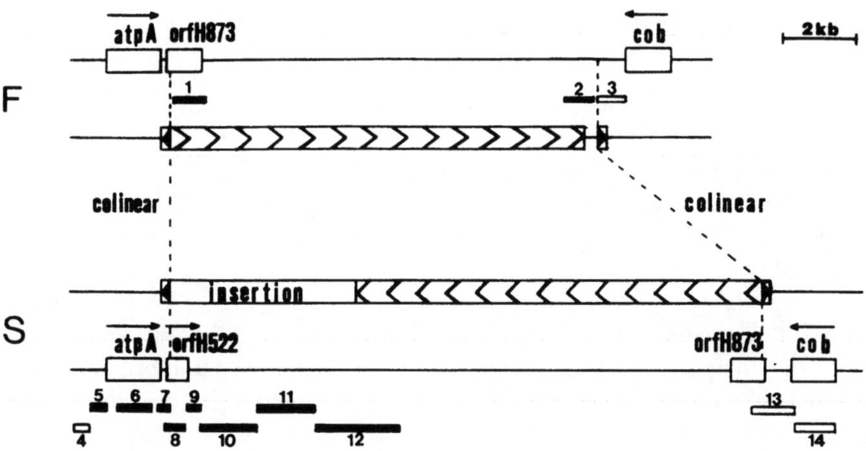

Fig. 1. Comparison of the rearranged mitochondrial DNA region of sunflower. The mtDNA of the fertile line Baso (*F*) and the male-sterile line CMSBaso (*S*) differs in an 11 kb inversion (*arrowheads*) and a 5 kb insertion. The rearranged mtDNA region is flanked by an inverted repeat (*solid arrows*) of 261 nucleotides. The *small arrows* above the *atp*A gene, the *cob* gene, and the *orfH522* (*open boxes*) indicate their direction of transcription. There are no transcripts of *orfH873* detectable

(Köhler et al. 1991; Laver et al. 1991). Southern analysis indicate that the 5 kb insertion is composed of sequences from mtDNA, the nuclear genome, and of sequences of yet unknown origin (Zetsche and Horn 1993). A homology to the 5 kb insertion could also be detected in another male sterile cytoplasm in sunflower, MAX1, but is not localized in the 3'-flanking region of the *atp*A gene (Hahn 1993). Further investigations may reveal if the homologous region plays a role in creating the CMS phenotype in the MAX1 cytoplasm. *In organello* labeling of mitochondrial translation products showed that a 16 kdalton protein is synthesized only by the mitochondria from male sterile sunflower (PET1 cytoplasm). The *orfH522* sequence possibly encodes the 16 kdalton protein but this has not yet been proved with certainty (Horn et al. 1991).

Cytoplasmic male sterility in common bean ist associated with the presence of a 3 kb unique mitochondrial sequence designated *pvs*. The *pvs* sequence, situated between a sequence repeated elsewhere in the CMS mitochondrial genome and a region with similiarity to the maize mitochondrial *atp*A gene (Chase and Ortega 1992), encodes at least two open reading frames (297 and 720 bp in length) with portions derived from the chloroplast genome (Johns et al. 1992). The organization of the *pvs* sequence between two different repeated sequences could lead to reversion by a mechanism involving recombination between like repeats, and selection for a recombinant form of the mitochondrial genome (Chase and Ortega 1992).

Despite the identification of the loci encoding CMS, the molecular mechanism of the disruption in pollen development is still an unsolved question. Current theories about the pollen disruption are reviewed by Hanson and Folkerts (1992).

b) Restoration of Male Fertility

It has been shown that size and number of plant mitochondrial transcripts are influenced by the nuclear background (for refs. see Iwabuchi et al. 1993). Specific nuclear genes termed nuclear fertility restorer genes (*Rf*) allow viable pollen formation by suppressing the CMS phenotype. *Rf* genes are known for many species. These genes might be involved in regulation of CMS gene expression, but it is not, however, understood how they restore male fertility.

In maize, two genes (*Rf*1, *Rf*2) are necessary to restore fertility. A 1.6 kb RNA species is seen in CMS-T genotypes containing the dominant allele of the *Rf*1 locus which is undetected in CMS-T plants homozygous rezessive for *Rf*1. The altered transcriptional profile is concurrent with an 80% reduction in the abundance of the URF13 protein. Preliminary evidence suggests that *Rf*1 modifies T-*urf*13 expression by an RNA processing event. Alternatively, the mode of action of the *Rf*1 gene may involve some form of transcriptional control such as premature termination of translation secondarily resulting in a 1.6 kb transcript. The dominant allele of *Rf*2 is also indispensible for fertility restoration, but it is uncertain in which way it contributes to pollen fertility (reviewed by Williams and Levings 1992).

In *Petunia* a single restorer gene affects the expression of *pcf*. In the presence of the restorer gene the 25 kdalton protein is significantly reduced (Nivison and Hanson 1989). However, it has been shown that the *Rf* gene does not affect the abundance of the *pcf* transcript (Young and Hanson 1987).

The restoration mechanism of the *Petunia* line 13-194, generated in the same protoplast fusion experiment that led to the definition of the S-*pcf* locus point to the existence of nuclear multiple fertility-restoration genes in this system, but does not seem to affect the arrangement or abundance of the S-*pcf* locus or its transcription. Now it has to be shown whether the protein level of the S-*pcf* genes is affected in these plants (Wolf-Litman et al. 1992).

In cytoplasmic male sterile rice carrying the CMS-bo cytoplasm, interaction of RNA processing and editing seems to be involved in controlling CMS expression and the restoration of fertility (Iwabuchi et al. 1993). In this mitochondrial genome, two *atp*6 genes were found. One (N-*atp*6) was identical to the normal cytoplasmic gene, while the second (B-*atp*6) was identified as a potential candidate for a CMS gene. Northern analysis showed that B-*atp*6 is transcribed into a 2.0 kb RNA in the absence of the *Rf*1 gene, whereas two discontinuous RNAs were detected in the presence of the *Rf*1 gene, presumedly generated from the 2.0 kb RNA by processing affected by the action of the *Rf*1 gene. Sequence analysis of cDNA clones showed that the processed B-*atp*6 RNAs are edited as efficiently as the N-*atp*6, whereas unedited and partially edited RNAs were detected among unprocessed RNAs. The nuclear *Rf*1 gene, which may code for an RNA processing enzyme, thus influences the sequential post-transcriptional editing of the B-*atp*6 RNAs. The unprocessed RNA is not efficiently edited and might be translated into an altered polypeptide, reducing ATPase activity by competing with normal ATP6 protein. In the presence of the *Rf*1 gene the processed RNA is efficiently edited and translated into the normal peptide (Iwabuchi et al. 1993).

The restoration of CMS in broad bean (*Vicia faba*) is unusual, because a restorer gene irreversibly converts the sterile cytoplasm (CMS-Sprite) to a fertile cytoplasm. Fertility restoration by the restorer gene *Fr* results in the loss of the *pvs* region. However, partial fertility in the F_1 generation implies that the process of restoration by *Fr* affords at least two generations. In situ hybridization experiments suggest that significant reduction or complete loss of the mitochondrial sterility-associated sequence occurred in fertile pollen of F_2 populations segregating for fertility (Johns et al. 1992).

c) Abnormal Plant Development

Nonchromosomal maize mutants (NCS) with abnormal development, including morphology, poor growth, and leaf striping, were shown to carry mtDNA alterations (Newton and Coe 1986; Lauer et al. 1990, Newton et al. 1990; Hunt and Newton 1991). At present it appears likely that essential mitochondrial genes are altered and mutant plants survive because they carry a mixture of wild-type and mutant mtDNAs (see Hanson and Folkerts 1992).

In the leaves of a tobacco plant resulting from fusion of protoplasts from tobacco with *Petunia*, a variegated maternally inherited mutation appeared. No defects were detected in the chloroplasts which were derived from *Petunia*. The mitochondria were very similar to the mitochondria of tobacco, although with some cybrid-specific fragments shown by the electrophoretic pattern of their mtDNA. In a second cycle of fusions, the *Petunia* chloroplasts of the variegated plants were replaced by tobacco chloroplasts. The results allowed exclusion of the chloroplasts, and confirm that the mitochondrial genome is associated with the mutant phenotype (Bonnett et al. 1993).

These nonchromosomal mutants in maize and tobacco now offer the opportunity to examine interactions between chloroplasts and mitochondria.

4. Rearrangements Induced by in Vitro Culture

In plant tissue, frequent modifications of the mitochondrial DNA as rearrangements and deletions can be observed in comparison to the mtDNA from seedlings or green leaves (Bears et al. 1989; Shirzadegan et al. 1989, 1991). However, stable cell cultures showing no detectable variation in the mitochondrial genome organization have also been described (Hanson 1984; Kool et al. 1985).

Rapid structural changes in the mitochondrial genome of *Brassica campestris* during in vitro culture seems to be the result of a preferential amplification and sorting of preexisting sublimons rather than de novo rearrangements (Shirzadegan et al. 1991).

Somaclonal variation was also observed in the mitochondrial genome of protoplast cultures in *Sorghum bicolor* using *atp*A, *atp*6, *cob*, and *cox*I as probes (Kane et al. 1992). Rearrangements in one of the two copies of the *atp*6 present in seedlings occurred during tissue culture. In addition, an amplification of a substoichiometric 3.3 kb EcoRI fragment containing the *cob* gene was observed.

Mutations in both the mitochondrial and nuclear genome are present in plants regenerated from *Nicotiana sylvestris* protoplast culture (Vitart et al. 1992). Due to the amplification of recombinant substoichiometric DNA sequences already present in the parental line, a new mitochondrial organization, called U, is observed. The recombination events involve two 404 bp repeats and lead to an alteration of the transcript pattern in this region although the sequences of the repeats are not changed by the recombination.

In *Nicotiana sylvestris* two cytoplasmic male-sterile plants were obtained by protoplast culture (Chetrit et al. 1992). The large deletions (up to 50 kb) in the mitochondrial genomes occurred via two homologous recombination events (rec.1 and rec.2) in the parental mitochondrial DNA. The sequences involved in rec.1 are located in a transcribed region. The rec.2 region is located next to the sequence homologous to *nad*5. In the CMS mutants either the lack of genes present in the deleted region or alterations of the transcription of genes next to the deletion may play a role in creating the CMS phenotype.

5. Somatic Hybrids

Electrofusion or PEG-mediated fusion of protoplasts in order to obtain somatic hybrids has become a useful tool in plant breeding if sexual crosses due to incompatibilities are not possible even by employing techniques like *"embryo rescue"* (Kräuter et al. 1991). Segregations and selections of fusion products allow stabilization of new genetic diversities in nucleus and cytoplasm which have arisen by recombinant events.

To introduce clubroot disease resistance of the Japanese radish (*Raphanus sativus*) into cauliflower (*Brassica oleracea*), protoplasts of the two species were electrofused (Hagimori et al. 1992). Three of the hybrids obtained contained mitochondria of the Japanese radish, with some modification, while seven hybrids had either parental or new patterns. All of the hybrid-type plants showed resistance to clubroot disease.

Intergener somatic hybrids of tall fescue (*Festuca arundinacea* Schreb.) and non-morphogenic Italian ryegrass (*Lolium multiflorum* Lam.) which were intermediate with respect to plant habitus, leaf, and inflorescence morphology were analyzed towards their organellar composition (Takamizo et al. 1991). The somatic hybrids showed additive or novel mtDNA patterns but in all cases the hybridization pattern of tall fescue was present when a chloroplast-specific probe was used.

Somatic hybrids were obtained by fusing hypocotyl protoplast from Broccoli (*Brassica oleracea* L. var. *italica*) with mesophyll protoplasts of two *B. napus* lines. One of these lines contained the Ogura (ogu) CMS cytoplasm while the other carried triazine-tolerance-conferring chloroplasts (ctr) in addition to the ogu mitochondria (Kao et al. 1992). No alterations in the ogu mtDNAs and ctr cpDNA could be detected in the two male-sterile somatic hybrids regenerated from the fusion of broccoli protoplasts with those of ogu/ctr cybrid *B. napus*. However, three of the four male-sterile plants recovered from the somatic hybridization of broccoli with ogu CMS *B. napus* appear to possess mitochondrial genomes that have resulted from recombination between the ogu and normal *B. oleracea* mtDNAs. Interestingly, all of these four plants had *B. oleracea* cpDNA, and none showed the seedling chlorosis usually associated with ogu chloroplasts. The chromosome number of most of the plants was 2n = 56 as could be expected for *B. oleracea* (2n = 18) + *B. napus* (2n = 38) hybrids.

In *Petunia*, analysis of recombinant mitochondrial genomes of somatic hybrids resulting from fusion of protoplasts from CMS and fertile lines were used to correlate the presence of the S-*Pcf* locus with the CMS trait in stable somatic hybrids. It could also be demonstrated that restoration of fertility by nuclear restorer genes does not affect the transcription of the S-*Pcf* locus (Wolf-Litman et al. 1992).

Fusion products of three different male-sterile tobacco cultivars were investigated to examine the influence of mitochondrial genes on stamen development in higher plants (Kofer et al. 1991). Cybrids showing recombined biparental floral morphology or novel male-sterile phenotypes carried mitochondrial genomes distinct from the parental mtDNAs.

The use of transgenic antibiotic-resistant *Nicotiana tabacum* as one fusion partner proved to be a simple, but effective selection system in order to produce fertile somatic hybrids between *N. tabacum* and *N. debneyi* (Sproule et al. 1991). A nonrandom pattern of organelle segregation occurred in the fusion products with preference towards *N. debeneyi* with regard to the chloroplasts and the mitochondria. Evidence for the presence of rearranged or recombinant cp and mtDNA was also obtained for some of the hybrids.

Segregation of a mixed plastid population in the fusion products can be directed by selective conditions which favour the multiplication of one of the parental plastid types (Malone et al. 1992). Protoplast fusion between *N. plumbaginifolia* and an albino and streptomycin-resistant *N. tabacum* plastid mutant allows a selection by different light intensities. High light intensity (1500 lux) gives a selective advantage to chloroplasts versus albino plastids, whereas on streptomycin-containing medium in the dark resistant plastids are preferentially multiplied.

The structure of the mitochondrial DNA of somatic hybrids derived from a fusion of *N. langsdorffii* and *N. glauca* was investigated by using the *atp*A as a probe in Southern hybridizations (Honda and Hirai 1992). In *N. langsdorffii* the four detected fragments which represent the *atp*A gene in four genomic environments can be explained by homologous recombination in the coding region of the *atp*A. In contrast, only one fragment can be detected in *N. glauca* whereas in the somatic hybrids obtained by fusion between protoplast from these two species new fragments, in addition to the parental fragments, were observed. The coding region of the *atp*A gene seems to be involved not only in intramolecular homologous recombination but also in intermolecular homologous recombination between the two parental mitochondrial genomes. Also in somatic hybrids of *N. tabacum* and *N. rustica* Southern analysis using *cox*I, *cox*II, 26S rDNA, and 5S-18S rDNA as probes provided evidence for recombination between the mitochondrial genomes of the two parents (Mukhopadhyay et al. 1991).

6. Transformation Using Mitochondrial Genes

A number of nuclear transformations of plants in order to obtain improved resistance against diseases or harmful insects have been described recently (Gasser and Farley 1989; Hayakawa et al. 1992; MacFarlane and Davies 1992), but to date no successful transformation system for higher plant mitochondria has been developed. An interesting alternative may be the direct cell-to-cell transfer of organelles by microinjection (Verhoeven and Blaas 1992).

In this chapter we deal only with transformations using mitochondrial genes or chimeric genes constructed to target the gene products into mitochondria.

The first success to induce male sterility by use of a transgenic mitochondrial gene was reported for the unedited *atp*9 mitochondrial gene from wheat (Bégu et al. 1990; Graves et al. 1990) introduced into tobacco (Hernould et al. 1993). Assuming that editing represents an essential process for producing active proteins, edited and unedited coding sequences of the *atp*9 gene from wheat fused to the coding sequence of a yeast *cox*IV transit peptide were used for transformation. Competition of the modified ATP9 with endogenous protein in constituting the ATP synthase complex should result in a mixed population of functional and nonfunctional complexes. The observed percentage of semifertile or male-sterile transgenic plants conforms to the proposed hypothesis.

Transformation of tobacco plants with the T-*urf*13 gene using Cauliflower Mosaic Virus S35 promoter confers methomyl and T-toxin sensitivity in various degrees to the plants. The targeting of URF13 into the mitochondria is not necessary to obtain the bleaching effect in the leaves exposed to the toxin (Von-Allem et al. 1991).

A new alternative to influence mitochondrial gene expression is reported by Small et al. (1992). Potato plants were transformed using a leucin tRNA gene from bean nuclear DNA carrying a 4 bp insertion in the anticodon loop. In the transgenic plants, the modified tRNA was also found inside mitochondria, indicating that the natural RNA import system allows the introduction of foreign RNA sequences into the mitochondrion.

7. Conclusions

The complexity of higher plant mitochondrial DNA is partly due to frequent recombination events across recombination repeats and rare intra- and intermolecular recombinations leading to subgenomic circles, isomeric forms, sublimons, and new master chromosomes. Integrations of sequences homologous to nuclear and chloroplast sequences were also shown to have an influence on the organization of the mtDNA. The presence of introns of various sizes in a number of plant mitochondrial genes requires *trans*- and *cis*-splicing processes for the maturation of the transcripts besides RNA-editing.

Maternally inherited traits such as CMS and nonchromosomal chlorophyll-stripe mutants allow to study interactions between nucleus, mitochondria, and chloroplast. Numerous models have been developed to explain rearrangements in the mitochondrial genome occurring, for example, during in vitro culture or due to fusion of protoplasts. However, the lack of in vitro systems for recombination, replication, and protein synthesis hampers the experimental verification of the theories.

References

Bears T, Curtis GJ, Lonsdale DM (1989) A specific rearrangement of mitochondrial DNA induced by tissue culture. Theor Appl Genet 77:620–624. **Bégu D,** Graves PV, Domec C, Arselin G, Litvak S, Araya A (1990) RNA editing of wheat mitochondrial ATP synthase subunit 9: direct protein and cDNA sequencing. Plant Cell 2:1283–1290. **Bendich AJ,** Loretz C, Monnat J (1993) The structure of the plant mitochondrial genome. In: Kück U, Brennicke A (eds) Plant mitochondria. Springer, Berlin Heidelberg New York. **Bonhomme S,** Budar F, Lancelin D, Small I, Defrance MD, Pelletier G (1992) Sequence and transcript analysis of the *Nco* 2.5 Ogura-specific fragment correlated with cytoplasmic male sterility in *Brassica* cybrids. Mol Gen Genet 235:340–348. **Bonnett HT,** Djurberg I, Fajardo M, Glimelius K (1993) A mutation causing variegation and abnormal development in tobacco is associated with an altered mitochondrial DNA. Plant J 3:519–525. **Braun CJ,** Brown GG, Levings CS III (1992) Cytoplasmic male sterility. In: Herrmann RG (ed) Cell organelles. Springer, Berlin Heidelberg New York, pp 219–245. **Brown GG** (1993) Regulation of mitochondrial gene expression. In: Verma DP (ed) Control of plant gene expression. CRC, Boca Raton, pp 175–198.

Chase CD, Ortega VM (1992) Organization of *atp*A coding and 3'-flanking sequences associated with cytoplasmic male sterility in *Phaseolus vulgaris* L. Curr Genet 22:147–153. **Chetrit P,** Rios R, Paepe R, Vitart V, Gutierres S, Vedel F (1992) Cytoplasmic male sterility is associated with large deletions in the mitochondrial DNA of two *Nicotiana sylvestris* protoclones. Curr Genet 21:131–137. **Conklin PL,** Hanson MR (1991) Ribosomal protein S19 is encoded by the mitochondrial genome in *Petunia hybrida*. Nucleic Acids Res 19:2701–2705. **Conklin PL,** Hanson MR (1993) A truncated recombination repeat in the mitochondrial genome of a *Petunia* CMS line. Curr Genet 23:477–482. **Conklin PL,** Wilson RK, Hanson MR (1991) Multiple trans-splicing events are required to produce a mature *nad*1 transcript in a plant mitochondrion. Genes & Dev 5:1407–1415. **Crouzillat D,** Leroy P, Perrault A, Ledoigt G (1987) Molecular analysis of the mitochondrial genome of *Helianthus annuus* in relation to cytoplasmic male sterility and phylogeny. Theor Appl Genet 74:773–780. **Crouzillat D,** Gentzbittel L, De La Canal L, Vaury C, Perrault A, Nicolas P, Ledoigt G (1989) Properties and nucleotide sequence of a mitochondrial plasmid from sunflower. Curr Genet 15:283–289.

De La Canal L, Crouzillat D, Flamand MC, Perrault A, Boutry M, Ledoigt G (1991) Nucleotide sequence and transcriptional analysis of a mitochondrial plasmid from a cytoplasmic male-sterile line of sunflower. Theor Appl Genet 81:812–818. **Dewey RE,** Levings CS III, Timothy DH (1986) Novel recombinations in the maize mitochondrial genome produce a unique transcriptional unit in the Texas male-sterile cytoplasm. Cell 44:439–449.

Fauron CM, Havlik M, Lonsdale D, Nichols L (1989) Mitochondrial genome organization of the maize cytoplasmic male-sterile type T. Mol Gen Genet 216:395–401. **Fauron CM,** Havlik M, Brettell RI (1990) The mitochondrial genome organization of a maize fertile CMS-T revertant line is generated through recombination between two sets of repeats. Genetics 124:423–428. **Fauron CM,** Havlik M, Casper M (1991) Organization and evolution of the maize mitochondrial genome. In: Herrmann RG, Larkins BA (eds) NATO ASI A, Life science, plant molecular biology 2. Plenum, New York, pp 345–363. **Fauron CM,** Casper M, Gesteland R, Albertsen M (1992) A multi-recombination model for the mtDNA rearrangements seen in maize CMS-T regenerated plants. Plant J 2(6):949–958. **Flamand MC,** Goblet JP, Duc G, Briquet M, Boutry M (1992) Sequence and transcription analysis of mitochondrial plasmids isolated from cytoplasmic male-sterile lines of *Vicia faba*. Plant Mol Biol 19(6):913–923. **Folkerts O,** Hanson MR (1989) Three copies of a recombination repeat occur on the 443 kb mastercircle of *Petunia hybrida* 3704 mitochondrial genome. Nucleic Acids Res 17:7345–7357. **Folkerts O,** Hanson MR (1991) The male sterility-associated *pcf* gene and the normal *atp*9-1 gene in petunia are located on different mitochondrial DNA molecules. Genetics 129:885–895. **Fox TD,** Leaver CJ (1981) The *Zea mays* mitochondrial gene coding cytochrome oxidase subunit II has an intervening sequence and does not contain TGA codons. Cell

26:315–323. **Fukuchi M,** Shikanai T, Kossykh VG, Yamada Y (1991) Analysis of nuclear sequences homologous to the B4 plasmid-like DNA of rice mitochondria, evidence for sequence transfer from mitochondria to nuclei. Curr Genet 20:487–494.

Gass DA, Makaroff CA, Palmer JD (1992) Variable intron content of the NADH dehydrogenase subunit 4 gene of plant mitochondria. Curr Genet 21:423–430. **Gasser CS,** Farley RT (1989) Genetically engineering plant for crop improvement. Science 244:1293–1299. **Graves PV,** Bégu D, Velours J, Neau E, Belloc F (1990) Direct protein sequencing of wheat mitochondrial ATP synthase subunit 9 confirms RNA editing in plants. J Mol Biol 214:1–6. **Gray MW,** Hanice-Joyce PJ, Covello PS (1992) Transcription, processing and editing in plant mitochondria. Annu Rev Plant Physiol Plant Mol Biol 43:145–175.

Hagimori M, Nagaoka M, Kato N, Yoshikawa H (1992) Production and characterization of somatic hybrids between the Japanese radish and cauliflower. Theor Appl Genet 84:819–824. **Hahn V** (1993) Molekulargenetische Untersuchungen über die Ursachen cytoplasmatischer Pollensterilität bei der Sonnenblume. Thesis Univ Gießen. **Hahn V,** Friedt W (1992) Nutzung von *Helianthus*-Wildformen zur Entwicklung neuer "Plasmen" für die Sonnenblumenzüchtung. Vortr Pflanzenzücht 22:145–151. **Handa H,** Nakajima K (1992) Different organization and altered transcription of the mitochondrial *atp*6 gene in the male-sterile cytoplasm of rapeseed (*Brassica napus* L.). Curr Genet 21:153–159. **Hanice-Joyce PJ,** Gray MW (1990) Processing of transfer RNA precursors in a wheat mitochondrial extract. J Biol Chem 265:13782–13792. **Hanson MR** (1984) Stability, variation and recombination in plant mitochondrial genomes via tissue culture and somatic hybridization. Oxford Surv Plant Mol Cell Biol 1:33–52. **Hanson MR,** Folkerts O (1992) Structure and function of the higher plant mitochondrial genomes. Int Rev Cytol 141:129–172. **Hanson MR,** Pruitt KD, Nivison HT (1989) Male sterility loci in plant mitochondrial genomes. Oxford Surv Plant Mol Cell Biol 6:61–85. **Hayakawa T,** Zhu Y, Itoh K, Kimura Y, Iwaza T, Shimamoto, Toriyama S (1992) Genetically engineered rice resistant to rice stripe virus, an insect-transmitted virus. Proc Natl Sci USA 89:9865–9869. **Hernould M,** Suharsono S, Litvak S, Araya A, Mouras A (1993) Male-sterility induction in transgenic tobacco plants with an unedited *atp*9 mitochondrial gene from wheat. Proc Natl Acad Sci USA 90:2370–2374. **Hiesel R,** Wissinger B, Brennicke A (1990) Cytochrome oxidase subunit II mRNA *Oenothera* mitochondria are edited at 24 sites. Curr Genet 18:371–375. **Holford P,** Croft J, Newbury HJ (1991) Structural studies of microsporogenesis in fertile and male-sterile onions (*Allium cepa* L.) containing the CMS-S cytoplasm. Theor Appl Genet 82:745–755. **Honda H,** Hirai A (1992) The gene for the alpha-subunit of ATPase: a site of homologous recombination in plant mitochondrial DNA also functions in somatic hybrid cells. Theor Appl Genet 84:33–38. **Horn R,** Köhler RH, Zetsche K (1991) A mitochondrial 16 kDa protein is associated with cytoplasmic male sterility in sunflower. Plant Mol Biol 17:29–36. **Hunt M,** Newton KJ (1991) the NCS3 mutation: genetic evidence for the expression of ribosomal protein genes in *Zea mays* mitochondria. EMBO J 10:1045–1052.

Iwabuchi M, Kyozuka J, Shimamoto K (1993) Processing followed by complete editing of an altered mitochondrial *atp*6 RNA restores fertility of cytoplasmic male sterile rice. EMBO J 12:1437–1446. **Izuchi S,** Terachi T, Sakamoto M, Mikami T, Sugita M (1990) Structure and expression of tomato mitochondrial genes coding for $tRNA^{Cys}$ (GCA), $tRNA^{Asn}$ (GUU) and $tRNA^{Tyr}$ (GUA): a native $tRNA^{Cys}$ gene is present in dicot plants but absent in monocot plants. Curr Genet 18:239–243.

Johns C, Lu M, Lyznik A, Mackenzie S (1992) A mitochondrial DNA sequence is associated with abnormal pollen development in cytoplasmic male sterile bean plants. Plant Cell 4:435–449. **Joyce PB,** Gray MW (1989) Chloroplast-like transfer RNA genes expressed in wheat mitochondria. Nucleic Acids Res 17:5461–5476.

Kane EJ, Wilson AJ, Chourey PS (1992) Mitochondrial genome variability in sorghum cell culture protoclones. Theor Appl Genet 83:799–806. **Kao HM,** Keller WA, Gleddie S, Brown GG (1992) Synthesis of *Brassica oleracea/Brassica napus* somatic hybrid plants with novel organelle DNA compositions. Theor Appl Genet 83:313–320. **Kofer W,** Glimelius K, Bonnett HT (1991) Modifications of the mitochondrial DNA cause changes in floral development in homeotic-like mutants of tobacco. Plant Cell 3(8):759–769. **Köhler RH,** Horn R, Lössl A, Zetsche K (1991) Cytoplasmic male sterility in sunflower is correlated with the co-transcription of a new open reading frame with the *atp*A gene. Mol Gen Genet 227:369–376. **Kool AJ,** de Haas JM, Mol JN, Van Marrewijk GA (1985) Isolation and physicochemical characterization of mitochondrial DNA from cultured cells of *Petunia hybrida.* Theor Appl Genet 69:223–233. **Kräuter R,** Steinmetz A, Friedt W (1991) Efficient interspecific hybridization in the genus *Helianthus* via "embryo rescue" and characterization of the hybrids. Theor Appl Genet 82:521–525.

Laser KD, Lersten NR (1972) Anatomy and cytology of microsporogenesis in cytoplasmic male sterile angiosperms. Bot Rev 38:425–454. **Lauer M,** Knudsen C, Newton KJ, Gabay-Laughnan S, Laughnan JR (1990) A partially deleted mitochondrial cytochrome oxidase gene in the NCS6 abnormal growth mutant of maize. New Biol 2:179–186. **Laver HK,** Reynolds SJ, Moneger F, Leaver CJ (1991) Mitochondrial genome organization and expression associated with cytoplasmic male sterility in sunflower (*Helianthus annuus*). Plant J 1:185–193. **Leon P,** Walbot V, Bedinger P (1989) Molecular analysis of the linear 2.3 kb plasmid of maize mitochondria: apparent capture of tRNA genes. Nucleic Acids Res 17:4089–4099. **Lippok B,** Brennicke A, Wissinger B (1992) The *cox*II gene in carrot mitochondria contains two introns. Mol Gen Genet 232(2):322–327. **Lonsdale DM** (1989) The plant mitochondrial genome. In: Stump PK, Conn EE (eds) The biochemistry of plant: a comprehensive treatise, vol 15. Academic Press, New York, pp 229–295. **Lonsdale DM,** Grienenberger JM (1992) The mitochondrial genome of plants. In: Herrmann RG (ed) Cell organelles. Springer, Vienna New York, pp 183–218.

Macfarlane SA, Davies J (1992) Plants transformed with region of the 201-kilodalton replicase gene from pea early browning virus RNA1 are resistant to virus infection. Proc Natl Acad Sci USA 89:5829–5833. **Malone R,** Horvath GV, Cseplo A, Buzas B, Dix PJ, Medgyesy P (1992) Impact of the stringency of cell selection on plastid segregation in protoplast fusion-derived *Nicotiana* regenerates. Theor Appl Genet 84:866–873. **Mann V,** Ekstein I, Nissen H, Hiser C, McIntosch L, Hirschberg J (1991) The cytochrome oxidase II gene in mitochondria of the sugar-beet *Beta vulgaris* L. Plant Mol Biol 17(3):559–566. **Mariani C,** de Beuckeleer M, Treuttner J, Leemans J, Goldberg RB (1990) Induction of male sterility in plants by a chimaeric ribonuclease gene. Nature 347:737–741. **Mariani C,** Goselle V, de Beuckeleer M, de Block M, Goldberg RB, de Greef W, Leemans J (1992) A chimaeric ribonuclease-inhibitor gene restores fertility to male sterile plants. Nature 357:384–387. **Melchers G,** Mohri Y, Watanabe K, Wakabayashi S, Harada K (1992) One-step generation of cytoplasmic male sterility by fusion of mitochondrial-inactivated tomato protoplasts with nuclear-inactivated *Solanum* protoplasts. Proc Natl Acad Sci USA 89:6832–6836. **Michel F,** Dujon B (1983) Conservation of RNA secondary structure features in two intron families including mitochondrial-, chloroplast-, and nuclear-encoded members. EMBO J 2:33–38. **Michel F,** Umesono K, Ozeki H (1989) Comparative and functional anatomy of group II catalytic introns – a review. Gene 82:5–30. **Moon E,** Kao TH, Wu R (1985) Pea cytochrome oxidase subunit II gene has no intron and generates two mRNA transcripts with different 5'-termini. Nucleic Acids Res 13:3195–3212. **Mukhopadhyay A,** Pradhan AK, Pentel D (1991) Mitochondrial DNA patterns are similar in gametosomatic and somatic hybrids of two *Nicotiana* species. Plant Cell Rep 10(10):522–524.

Newton KJ, Coe EH (1986) Mitochondrial DNA changes in abnormal growth (nonchromosomal stripe) mutants in maize. Proc Natl Acad Sci USA 83:7363–7366. **Newton KJ,** Knudsen C, Gabay-Laughnan S, Laughnan J (1990) An abnormal growth mutant in maize has a defective mitochondrial cytochrome oxidase gene. Plant Cell 2:107–113. **Nivison HT,** Hanson MR (1989) Identification of a mitochondrial protein associated with cytoplasmic male sterility in *Petunia.* Plant Cell 1:1121–1130. **Nugent JM,** Palmer JD (1988) Location, identity, amount and serial entry of chloroplast DNA sequences in crucifer mitochondrial DNAs. Curr Genet 14:501–509.

Oda K, Yamato K, Ohta E, Nakamura Y, Takemura M, Nozato N, Akashi K, Kanegae T, Ogur Y, Kohchi T, Ohyama K (1992) Gene organization from the complete sequence of liverwort *Marchantia polymorpha* mitochondrial DNA. J Mol Biol 223:1–7. **Orellano EG,** Carrillo N, Calcaterra

NB (1992) Evaluation of the extent of homologous chloroplast DNA sequences in the mitochondrial genome of cowpea (*Vigna unguiculata* L.) Plant Physiol 98(2):525–529.

Palmer JD (1992) Comparison of chloroplast and mitochondrial genome evolution in plants. In: Herrmann RG (ed) Cell organelles. Springer, Berlin Heidelberg New York, pp 99–133. **Palmer JD,** Herborn LA (1986) Tricircular mitochondrial genomes of *Brassica* and *Raphanus*: reversal of repeat configurations by inversion. Nucleic Acids Res 14:9755–9764. **Palmer JD,** Herborn LA (1987) Unicircular structure of the *Brassica hirta* mitochondrial genome. Curr Genet 11:565–570. **Palmer JD,** Shields CR, Cohen DG, Orton TJ (1983) An unusual mitochondrial DNA plasmid in the genus *Brassica*. Nature 301:725–728.

Sakamoto W, Kadowaki K, Kishimoto N, Yano M, Saito A, Tano S (1991) RFLP analysis of nuclear DNAs homologous with mitochondrial plasmid-like DNAs in cultivated rice. Theor Appl Genet 82:179–184. **Sangaré A,** Weil JH, Grienenberger JM, Fauron C, Lonsdale D (1990) Localization and organization of tRNA genes on the mitochondrial genomes of fertile and male-sterile lines of maize. Mol Gen Genet 223:224–232. **Schardl CL,** Lonsdale DM, Pring DR, Rose KR (1984) Linearization of maize mitochondrial chromosomes by recombination with linear episomes. Nature 310:292–296. **Schuster W,** Brennicke A (1986) Pseudocopies of the ATPase α-subunit gene in *Oenothera* mitochondria are present on different circular molecules. Mol Gen Genet 204:29–35. **Shirzadegan M,** Christey M, Earle ED, Palmer JD (1989) Rearrangement, amplification and assortment of mitochondrial DNA molecules in cultured cells of *Brassica campestris*. Theor Appl Genet 77:17–25. **Shirzadegan M,** Christey M, Earle ED, Palmer JD (1991) Patterns of mitochondrial DNA instability in *Brassica campestris* cultured cells. Plant Mol Biol 16:21–37. **Siculella L,** Palmer JD (1988) Physical and gene organization of mitochondrial DNA in fertile and male sterile sunflower CMS-associated alterations in structure and transcription of the *atp*A gene. Nucleic Acids Res 16:3787–3799. **Singh M,** Brown GG (1991) Suppression of cytoplasmic male sterility by nuclear genes alters expression of a novel mitochondrial gene region. Plant Cell 3:1349–1362. **Small I,** Isaac PG, Leaver CJ (1987) Stoichiometric differences in DNA molecules containing the *atp*A gene suggest mechanisms for the generation of mitochondrial genome diversity in maize. EMBO J 6:865–869. **Small I,** Marechal-Drouard L, Masson J, Pelletier G, Cosset A, Weil JH, Dietrich A (1992) In vivo import of a normal or mutagenized heterologous transfer RNA into the mitochondria of transgenic plants: towards novel ways of influencing mitochondrial gene expression. EMBO J 11(4):1291–1296. **Smith RL,** Chowdhury MKU (1991) Characterization of pearl millet DNA fragments rearranged by reversion from cytoplasmic male sterility to fertility. Theor Appl Genet 81:793–799. **Souza AP,** Jubier MF, Delcher E, Lancelin D, Lejeune B (1991) A trans-splicing model for the expression of the Tripartite *nad*5 gene in wheat and maize mitochondria. Plant Cell 3(12):1363–1378. **Spassova M,** Christov M, Bohorova N, Petrov P, Dudov K, Atanassov A, Nijkamp HJJ, Hille J (1992) Molecular analysis of a new cytoplasmic male sterile genotype in sunflower. FEBS Lett 297:159–163. **Sproule A,** Donaldson P, Dijak M, Bevis E, Pandeya R, Keller WA, Gleddie S (1991) Fertile somatic hybrids between transgenic *Nicotiana tabacum* and transgenic *N. debneyi* selected by dual-antibiotic resistance. Theor Appl Genet 82:450–456. **Stern DB,** Lonsdale DM (1982) Mitochondrial and chloroplast genomes of maize have a 12-kilobase DNA sequence in common. Nature 299:698–702. **Stern DB,** Palmer JD (1984) Extensive and widespread homologies between mitochondrial DNA and chloroplast DNA in higher plants. Proc Natl Acad Sci USA 81:1946–1950. **Stern DB,** Palmer JD (1986) Tripartite mitochondrial genome of spinach: physical structure, mitochondrial gene mapping, and locations of transposed chloroplast DNA sequences. Nucleic Acids Res 14:5651–5666. **Sutton CA,** Conklin PL, Pruitt KD, Hanson MR (1991) Editing of pre-mRNA can occur before *cis*- and *trans*-splicing in *Petunia* mitochondria. Mol Cell Biol 11:4274–4277. **Sutton CA,** Conklin PR, Pruitt KD, Calfee AJ, Cobb AG, Hanson MR (1993) Editing of *rps*3/*rpl*16 trancripts creates a premature truncation of the *rpl*16 open reading frame. Curr Genet 23:472–476.

Takamizo T, Spangenberg G, Suginobu KI, Potrykus I (1991) Intergeneric somatic hybridization in Gramineae: somatic hybrid plants between tall fescue (*Festuca arundinacea* Schreb.) and Italian ryegrass (*Lolium multiflorum* Lam.). Mol Gen Genet 231:1–6.

van der Meer IM, Stam Me, Zunen JA, Mol JN, Stuitje AR (1992) Antisense inhibition of flavonoid biosynthesis in petunia anthers results in male sterility. Plant Cell 4(3):253–262. **Verhoeven HA,** Blaas J (1992) Direct cell to cell transfer of organelles by microinjection. Plant Cell Rep 10(12):613–616. **Vitart V,** Paepe R, Mathieu C, Chetrit P, Vedel F (1992) Amplification of substoichiometric recombinant mitochondrial DNA sequences in a nuclear, male sterile mutant regenerated from protoplast culture in *Nicotiana sylvestris*. Mol Gen Genet 233:193–200. **Von-Allem JM,** Rottmann WH, Gengenbach BG, Harvey AJ, Lonsdale DM (1991) Transfer of methomyl and HmT-toxin sensitivity from T-cytoplasm maize to tobacco. Mol Gen Genet 229:405–412.

Wahleithner JA, Wolstenholme DR (1987) Mitochondrial plasmid DNAs of broad bean mitochondrial DNA. Curr Genet 12:55–67. **Wahleithner JA,** Macfarlane JL, Wolstenholme DR (1990) A sequence encoding a maturase-related protein in a group II intron of a plant mitochondrial *nad*1 gene. Proc Natl Acad Sci USA 87:548–552. **Warmke HE,** Lee SL (1977) Mitochondrial degeneration in T cytoplasmic male-sterile corn anthers. J Hered 68:213–222. **Warmke HE,** Lee SL (1978) Pollen abortion in T cytoplasmic male sterility corn (*Zea mays*): a suggested mechanism. Science 200:561–563. **Williams ME,** Levings CS (1992) Molecular biology of cytoplasmic male sterility. Plant Breed Rev 10:23–51. **Wintz H,** Grienenberger JR, Weie HJ, Lonsdale DR (1988) Location and nucleotide sequence of two tRNA genes and a tRNA pseudogene in the maize mitochondrial genome: evidence for the transcription of a chloroplast gene in mitochondria. Curr Genet 13:247–254. **Wissinger B,** Hiesel R, Schuster W, Brennicke A (1988) The NADH-dehydrogenase subunit 5 gene in *Oenothera* mitochondria contains two introns and is co-transcribed with the 5S rRNA gene. Mol Gen Genet 212:56–65. **Witt U,** Hanson S, Albaum M, Abel WO (1991) Molecular analysis of the CMS-inducing "Polima" cytoplasm of *Brassica napus* L. Curr Genet 19:323–327. **Wolf-Litman O,** Soferman O, Tabib Y, Izahr S (1992) Interaction of the mitochondrial S-*Pcf* locus for cytoplasmic male sterility in *Petunia* with multiple fertility-restoration genes in somatic hybrid plants. Theor Appl Genet 84:829–834. **Worrall D,** Hird DL, Hodge R, Paul W, Draper J, Scott R (1992) Premature dissolution of the microsporocyte callose causes male sterility in transgenic tobacco. Plant Cell 4(7):759–771.

Young EG, Hanson MR (1987) A fused mitochondrial gene associated with cytoplasmic male sterility is developmentally regulated. Cell 50:41–49.

Zabala G, O'Brien-Vedder C, Walbot V (1987) S2 episome of maize mitochondria encodes a 130-kilodalton protein found in male sterile and fertile plants. Proc Natl Acad Sci USA 84:7861–7865. **Zetsche K,** Horn R (1993) Molecular analysis of cytoplasmic male sterility in sunflower (*Helianthus annuus*). In: Kück U, Brennicke A (eds) Plant mitochondria. Springer, Berlin Heidelberg New York.

Professor Dr. Wolfgang Friedt
Dr. Renate Horn
Institut für Pflanzenbau
und Pflanzenzüchtung I
Justus-Liebig-Universität
Ludwigstraße 23
D-35390 Giessen, FRG

Dr. Volker Hahn
Present Address:
University Hohenheim
Landessaatzuchtanstalt
Versuchsstation Eckartsweier
D-77731 Willstätt, FRG

III. Mutation: Higher Plants

By Werner Gottschalk

1. Environmental Mutagenesis

The intensive studies on the mutagenicity of some luxuries have been continued during the past years. This is particularly valid with regard to tobacco. To discern the effects of cigarette smoke condensates or extracts, different human cells of smokers, especially lymphocytes, were evaluated, but also mice cells and rat urine. The frequency of micronuclei and of sister chromatid exchanges (SCEs) in the cells was used as parameter. Moreover, the mutagenicity of tobacco extracts on *Salmonella typhimurium* was studied. The following findings were obtained:

- A significant increase in the frequency of SCEs and in micronuclei in the cells of smokers was found by Jørgensen et al. (1991) and by Larramendy and Knuutila (1991).
- Cigarette smoke condensate was found to induce partial mitotic arrest in human lymphoid cell lines, increasing the frequency of metaphases and decreasing the proportion of anaphases. These effects are possibly due to the presence of antimitotic compounds in the condensate (Hsu et al. 1991).
- The mutagenicity of polyaromatic amines was enhanced through cigarette smoke condensates in *Salmonella* (Crebelli et al. 1991).
- The urinary mutagenicity of cigarette smoke condensates was increased in rats after injection. After inhalation of cigarette smoke, however, this effect was not observed (Doolittle et al. 1991).
- The effect of passive smoking was studied by Gorgels et al. (1992) evaluating the frequency of SCEs in blood cultures of nonsmokers who worked and lived together with smokers. A group of high passive smokers (70 h/week) was compared with low passive smokers (5 h/week) and no difference between the two groups was found.
- Aqueous extracts of smokeless tobaccos were found to be not mutagenic in *Salmonella*, but they obviously contain chemicals which become mutagenic under nitrosation conditions (Shirnamé-Moré 1991).
- In some states of India, a paste-like tobacco preparation called gudakhu is widely used. It was found to induce chromosome aberrations, micronuclei, and SCEs in mice (Dash and Das 1992).

Substances in betel quid seem to induce similar effects as tobacco. The frequency of micronuclei in oral mucosa cells of Indian chewers was significantly increased (Nair et al. 1991). This is a confirmation of former results obtained by other authors. The effect of cannabis was studied by Jørgensen et al. (1991) comparing the frequency of SCEs of 22 tobacco smokers with that of 22 persons smoking both tobacco and cannabis. Smoking itself caused a significant increase in SCEs; the added cannabis to tobacco, however, did not effect the frequency of SCEs further. It is therefore concluded that cannabis smoking

Progress in Botany, Vol. 55

has no genotoxic effects. Tannin extracts from green or black tea did not affect the rate of SCEs and chromosome aberrations in cultured Chinese hamster cells. After the use of mutagens (mitomycin C, UV), however, a posttreatment with high concentrations of tea tannin components produced an increase in the negative effects caused by the mutagens. Thus, the tannins seem to have a co-mutagenic effect. Interestingly, the mutagenic damage was suppressed by posttreatment with low concentrations of the tannin components (Imanishi et al. 1991). A modest increase in the number of SCEs, but not in chromosome aberrations was furthermore observed for theophylline (McFee 1991; Sinues et al. 1992). The genetic effects of ethanol were confirmed in male mice showing an increased rate of dominant lethal mutations after chronic ethanol treatment (Berryman et al. 1992).

Genotoxic effects of pharmaceuticals have been intensively studied, particularly in China, for almost 10 years. Extracts of 13 drugs out of 102 raw pharmaceuticals, used in traditional medicine, were found to be mutagenic. Some of the ingredients derive from the following species: *Astragalus mongholicus, Datura metel, Artemisia capillaris, Carthamus tinctorius, Paeonia suffruticosa, Cinnamomum mairei* (Xue-jun et al. 1991). A review on the genotoxic effects of cinnamaldehyde was given by Neudecker (1992). It induces point mutations, SCEs, fragementations, and chromosome aberrations. *Heliotropium curassavicum* is widely used in Argentinian medicine. It was found to cause an increase of mitotic anomalies and aberrations obviously due to pyrrolizidine alkaloids (Carballo et al. 1992). The species *Petiveria alliacea* is extensively used as a folk medicine in many developing countries, particularly in South America. It causes a dose-dependent increase of SCEs in human lymphocytes and mice bone marrow cells (Hoyos et al. 1992). Many hydroxyanthraquinone derivatives, produced by fungi and higher plants and used as purgatives and laxatives, proved to be mutagenic in the Ames *Salmonella* microsome system (Krivobok et al. 1992).

Mutagenicity was also demonstrated for other plant substances. Some examples are as follows:

- Pyrrolizidine alkaloids from several Italian *Senecio* species showed a weak mutagenic activity in the Ames test due to retrorsine, senecivernine, and seneciphylline (Rubiolo et al. 1992).
- A weak mutagenicity of smoke extracts of wood, used for food preservation and aromatization, was found in Nigeria, especially for white and red mangroves (Asita et al. 1991).
- Extracts from *Vicia faba* roots and leaves and from *Zea mays* leaves induced SCEs and chromosomee aberrations in human lymphocytes and Chinese hamster ovary cells (Kanaya et al. 1992).
- A significant increase in SCEs in human lymphocytes and in barley root tip cells was found after treatment with some amino acids. The two cell types, however, showed different reactions. Threonine, for instance, was effective in the barley cells but not in the lymphocytes (Zhang and Yang 1992).

2. Antimutagenic Plant Substances

Some findings on the antimutagenic action of distinct plant substances, already considered in volumes 49, 51 and 53 of *Progress in Botany*, were confirmed and supplemented:

- Pretreatment of rat and guinea pig cells with vitamins C and E caused an antimutagenic effect against bichromate (Chorvatovicová et al. 1991).
- The mutagenic activity of aflatoxin B_1 and some other mutagens was completely or almost completely inhibited by chlorophyllin. The substance was tested using the frequency of forward

mutations in *Salmonella*. More than 50% of the genetic damage, caused by tobacco snuff and black pepper extracts were prevented by chlorophyllin (Warner et al. 1991).

- The mutagenic efficiency of coffee has not yet been fully clarified. According to Abraham (1991), instant coffee was found to have even a protecting effect against some mutagens such as aflatoxin B_1, benzo[α]pyrene, urethane. The coffee was given to mice 2 and 20 h before the carcinogens and mutagens were injected; the frequency of micronuclei in bone marrow cells was evaluated.
- Low concentrations of tea tannin components, given in the form of a posttreatment to cultured mammalian cells, suppressed the frequency of SCEs and chromosome aberrations incudced by mitomycin C and UV (Imanishi et al. 1991). As already mentioned, high concentrations act as a comutagen.
- *β-carotene*, a natural food colorant and antioxidant, was found to act as an antimutagen and anticarcinogen in several test systems. A pretreatment 7 days before giving cyclophosphamide to mice inhibited the clastogenicity of this substance. Micronuclei and chromosomal aberrations in erythrocytes were used as parameters (Mukherjee et al. 1991).
- The amino acid lysine reduced the frequency of SCEs induced by bromodeoxyuridine in root tip cells of *Hordeum vulgare* (Zhang and Yang 1992). Some other amino acids, however, caused a significant increase in SCEs.

A concentration-dependent inhibition of the number of revertants in *Salmonella* was found for some herbs used in Chinese medicine. Aqueous extracts of *Oldenlandia diffusa, Scutellaria barbata, Astragalus membranaceus,* and *Ligustrum lucidum* were tested against aflatoxin B_1. The combination of *Oldenlandia* and *Scutellaria* gave an additive effect (Wong et al. 1992). A suppressing effect of vanillin on mitomycin C-induced chromosome aberrations was found in *Drosophila* (de Andrade et al. 1992). The resinous exudates of many plant species contain nordihydroguaiaretic acid, which is used as an antioxidant in fats and oils. This substance inhibited the mutagenicity of some mutagens such as benzo[α]pyrene, methyl methane sulfonate, and others in *Salmonella* (Z.y. Wang et al. 1991).

3. Gene Mutations

The development of mutant collections in allogamous species is essentially more difficult than in autogamous ones. A new method was used in *Lolium multiflorum*. Seeds from the spikes of mutagen-treated populations were sown in the form of isolated spike progenies. Only the plants within such a progeny were allowed to undergo natural crossing. The seeds of each of these populations were separately harvested and again sown as isolated progenies for selecting the mutants. This procedure is called CSP ("Crossing within spike progenies method") and was successfully used in Italian ryegrass, which had recurrently been treated with different mutagens over six generations (Ukai 1990). *Cucumis sativus* is a monoecious outcrossing species. The induction of mutations in seeds gives chimeras and is not efficient. Therefore, irradiated pollen were used for pollinations in the same cultivar. In this way, chimerism was avoided and good results were obtained (Iida and Amano 1990). In *Oryza sativa*, populations of experimentally produced autotetraploids were treated with gamma rays over several generations. Afterwards, diploid mutants were selected (Yamamoto et al. 1990). The authors regard this procedure as a new technique for obtaining mutants in this species. I do not understand the advantage of this circumstantial technique, as very effective methods in normal diploid rice have long existed.

A review on spontaneous mutagenesis with many references was given by Smith (1992). The derivation of maize from the Mexican wild grass teosinte by means of only a small number of gene mutations with large effects on adult morphology was already postulated by Beadle in 1939. This hypothesis is supported by modern molecular findings using RFLPs (Doebley 1992). The two forms are regarded to be subspecies now called *Zea mays* ssp. *mexicana* (teosinte) and *Zea mays* ssp. *parviglumis* (maize). A new genetic linkage map of *Secale cereale* was published by Melz et al. (1992). The latest linkage map of *Pisum sativum* contains 461 symbolized genes, 259 of them being assigned to the 7 linkage groups. Many more mutant genes have been identified; they are maintained in gene banks or collections. Unfortunately, not only 56 of the older genes, but also many recently obtained mutant genes of the species have been lost (Blixt et al. 1991).

a) Meiotic Mutants

From the great number of mutant genes available in many species, only a small number can be mentioned in this chapter. A review on meiotic mutants in *Zea mays* was given by Golubovskaya (1989). So far, more than 20 mutant genes, belonging to 10 different groups, are known to influence microsporogenesis negatively in this species. Examples for new genes controlling different meiotic processes are as follows:

- As-genes control the pairing behavior of the homologous chromosomes during the earliest stages of the first meiotic prophase resulting in asynapsis if they are present in the mutant condition. In the asynaptic mutant *sy-1* of *Secale cereale*, the synaptonemal complex is completely absent and only univalents are observed (Sosnikhina et al. 1992).
- Ds-genes control the chiasma formation in pachytene after normal pairing of the homologs. In the mutant condition, they cause desynapsis, i.e., a reduction of the number of chiasmata or their complete absence, respectively. A spontaneously arisen desynaptic mutant was found in *Zea mays* (Maguire et al. 1991; the paper contains fascinating photomicrographs of the synaptonemal complex). *Rhoeo spathacea* is a complex interchange hybrid with 2 n = 12 chromosomes present in a ring or a chain. A high frequency of univalents was found in a plant obviously due to partial desynapsis (Garcia-Velasquez 1991).
- Mutated ms-genes cause male sterility due to a complete breakdown of microsporogenesis whereas the macrosporogenesis is not influenced by these genes. Thus, the respective plants are female fertile. The developmental anomalies of male sterile soybean and maize mutants were compard by Palmer et al. (1992), those of pea and barley mutants by Kaul and Nirmala (1991). By means of trisomics, gene *ms1* of *Secale cereale* could be located on chromosome 4R, whereas the modifying genes *ms2* and *ms3* are located on chromosomes 3R and 6R (Melz and Adolf 1991). The fertility of EMS-induced male sterile rice plants was considerably increased when they were grown in a phytotron under 12/12 h photoperiod (Oard et al. 1991).

b) Mutants of Agronomic Interest

The number of commercial varieties developed by means of induced mutations is constantly increasing. In the lists published by the International Atomic Energy Agency (IAEA) in Vienna, a total of 1548 officially released mutant cultivars of crops and ornamentals was given until 1992. The true number is considerably higher because mutation breeding in ornamentals is widely performed in private firms and is not fully recorded by the IAEA. A review is given in Table 1 containing the most important crops of mutation breeding and the mutagens used. Problems of mutation breeding are discussed in the pro-

Table 1. Mutagens used for producing 1512 officially released mutant varieties of different crops and ornamentals. Twelve varieties more have been obtained by using electrons, beta rays, and combined treatment with rays and chemicals. (Data from Maluszynski et al. 1991, 1992)

Species	Gamma rays	X-rays	Neutrons	Laser	Chemicals	Colchicine	Crosses	Total
Sexually propagated species								
Rice	158	15	9	2	11	–	80	275
Barley	11	9	5	–	11	–	193	229
Bread wheat	72	3	7	2	5	–	16	105
Maize	6	–	1	–	–	–	26	33
Other species	149	44	16	2	48	3	96	358
Total	396	71	38	6	75	3	411	1000
Vegetatively propagated species								
Chrysanthemums	88	97	–	–	–	1	–	186
Other species	159	138	9	1	9	2	8	326
Total	247	235	9	1	9	3	8	512
Grand Total	643	306	47	7	84	6	419	1512

ceedings of symposia organized by the Institute of Radiation Breeding in Ohmiya-machi (1989) and IAEA (1991). Reviews on methods, results, future aims, and prospects of this branch in rice and pea were given by Rutger (1992) and Gottschalk (1990). The role of the phytotron for discerning the reactions of mutant genes to diverse environmental conditions was studied using a large number of mutants and recombinants of *Pisum sativum*. Especially photoperiod and temperature should be considered for judging the true value of genotypes for breeding purposes. Not only the flowering behavior, but also shoot structure and tolerance to unfavorable temperatures can be strongly influenced by these factors (Gottschalk 1990, 1991).

c) Somaclonal Variation

Somaclonal variation means the appearance of genetic changes in plants derived from cell cultures without having used mutagens. It is obviously partly due to genetic differences already existing in the original somatic cells or to distinct components of the culture media which act as mutagens. New examples are known for:

- *Lycopersicon esculentum* (van den Bulk et al. 1990),
- *Solanum tuberosum* (Rietveld et al. 1991),
- *Mentha arvensis* (Kukreja et al. 1991),
- *Triticum aestivum* (Guenzi et al. 1992; Qureshi et al. 1992).

A review on this problem was given by Evans (1989), demonstrating that somaclonal variation is a very complex phenomenon not only combined with alterations at the gene level but also with altered chromosome numbers, chromosome structures, and other anomalies. It is the opinion of many geneticists that somaclonal variation is a rich source of genetic variability and that it thus should be used in plant breeding besides the classical methods existing. Others, however, are very skeptical in this respect. The evaluation of wheat material for key agronomic traits gave no positive results (Qureshi et al. 1992). The authors interpret early published data in the literature as too optimistic with regard to the usefulness of somaclonal variation for cultivar improvement.

4. Genome Mutations

a) Haploids

Haploid plants of diploid species can be produced by means of two principally different methods:

- through parthenogenesis, i.e., gynogenesis;
- through androgenesis.

If these methods are applied to polyploid species, di- or polyhaploid plants arise.

Effective methods for producing dihaploid potatoes have been used for about 15 years by pollinating emasculated flowers of the autotetraploid species *Solanum tuberosum* ($2n = 4x = 48$) with diploid wild potatoes. In this way, parthenogenetic development of unfertilized eggs occurs, theoretically giving rise to euploid plants with $2n = 2x = 24$ chromosomes. Very often, however, aneuploids arise. After having used pollen of the diploid wild potato *Solanum phureja*, Clulow et al. (1991) obtained only 2 true dihaploids, whereas 10 plants were predominantly diploid with 25 and 26 chromosomes and 5 plants were around triploid. In the aneuploid cells, DNA fragments of *Solanum phureja* were found by RFLP analysis. The wild potato *Solanum acaule* is a segmental allotetraploid species. After having crossed 25 *acaule* plants with a "haploid inducer", two dihaploid plants with very low fertility were selected (Camadro et al. 1992). Dihaploids (SY) of *Elymus dolichatherus* (SSYY) and of *E. brevipes* (SSYY) were obtained by crossing these species with *E. tibeticus* and *E. panormitanus* (Lu 1992). Not only interspecific hybridization, but also chemical treatment of emasculated flower buds or high temperatures can induce parthenogenesis. This has been done in *Gossypium hirsutum* (Zhou et al. 1991).

Androgenetic development of haploids via microspore culture succeeded in *Brassica napus*. Their chromosome numbers were doubled by means of colchicine (Henderson and Pauls 1992). The same method was used in barley (Finnie et al. 1991). The androgenetic development of homozygous dihaploids can be reached directly by giving colchicine to the induction media. This very efficient method was used in *Triticum aestivum* (Barnabás et al. 1991).

b) Autopolyploids

So far, only a very small number of clear cases of autopolyploid origin of polyploid plant species is known. The tetraploid species *Hordeum depressum* belongs to this group (Petersen 1992). The European plum (*Prunus domestica*) was regarded as an allohexaploid species derived from *P. cerasifera* (2x) and *P. spinosa* (4x). According to new cytological findings, however, the plum has obviously arisen directly from *Prunus cerasifera*, which exists in diploid, tetra-, and hexaploid levels. Thus, it is obviously an autohexaploid species (Zohary 1992). The species *Hordeum brachyantherum, depressum*, and *jubatum* ($2n = 4x = 28$) are either autopolyploids or segmental allopolyploids with closely related homoeologous genomes (Salomon et al. 1991).

A new method for inducing polyploidy consists of using parafluorophenylalanine instead of colchicine. In this way, polyploid and aneuploid plants were obtained in *Zea mays, Lycopersicon esculentum*, and *Brassica napus* (Banks et al. 1991). The herbicides Pronamide and Amiprophos-methyl were found to induce chromosome doubling in high frequencies in anther-derived haploid maize cells. Callus growth is not inhibited by the two substances (Wan et al. 1991). In many leguminous species, the production of polyploids is very difficult. A new method was found by Andersen et al. (1991). Vegetative axillary meristems were excised and treated in vitro with colchicine, resulting in 44% fertile polyploid plants in *Trifolium pratense*. This method is obviously also suited for other species.

In the sweet orange (*Citrus sinensis*), crosses between diploid and tetraploid plants are often difficult. The triploids wanted are not only seedless, they show also a certain amount of heterosis. Gmitter et al. (1990) have regenerated triploid hybrid plants from endosperm-derived calli. As the endosperm cells are already triploid, the barriers to sexual hybridization can be overcome by this method.

Autotetraploid plants of *Hyoscyamus niger* (2 n = 4x = 68), produced by means of colchicine, are vigorous and have more than 90% seed set. Their unexpected good fertility is due to high bivalent formation in the first meiotic prophase, resulting in balanced gametes. The quadrivalent frequency decreased in three subsequent generations. The autotetraploids of this important medicinal plant yield considerably more tropane alkaloid than the original diploids (Lavania et al. 1991). Autohexaploids of *Ipomoea trifida*, arisen after having crossed diploid with tetraploid strains followed by colchicine treatment, are of interest for sweet potato breeding (Iwanaga et al. 1991).

c) Allopolyploids

α) Natural Allopolyploids, Genome Analyses

Sugarcanes are known to have high degrees of polyploidy. This has been confirmed and supplemented by cytogenetic studies of 31 wild *Saccharum* relatives including *S. robustum, S. spontaneum*, and representatives of the genera *Erianthus, Misanthus*, and *Narenga*. The highest chromosome number was found in *Saccharum robustum* with 2 n = 166. In all clones, predominantly bivalent formation was observed (Burner 1991). In eight North Indian populations of *Scirpus tuberosus*, cytotypes with 2 n = 94, 96, 104, 112, 114, 116 were found. The basic number is x = 7 or 8. Thus, clones with 2 n = 112 are regarded as 14- or 16-ploid, respectively; they have a vegetative mode of propagation (Bir et al. 1990).

The genomatic constitution of a number of allopolyploid species was analyzed. Especially favorable taxa for these studies are Gramineae because of their relatively easy interspecific or intergeneric crossability. Examples at the *tetraploid level* are as follows:

- In the genus *Elymus*, the four genomes S, H, Y, P have so far been identified. The S genome derives from *Pseudoroegneria libanotica* or *spicata*, the H genome from *Hordeum violaceum* (Salomon et al. 1991). The species *Elymus parviglumis, pseudonutans, caucasicus*, and *longearistatus* were found to have the constitution SSYY; the derivation of the Y genome is still unknown (Lu and von Bothmer 1990; Jensen and Wang 1991). One genome of the tetraploid species *Elymus glaucus, hystrix, multisetus, villosus*, and *virginicus* is homoeologous with one genome of the species *Hordeum brachyantherum, depressum, jubatum, secalinum* (Salomon et al. 1991).
- *Hordeum secalinum* has the constition XXHH (Salomon et al. 1991).
- The genomatic constitution of the tetraploid species *Triticum kotschyi* and *T. peregrinum* (= *Aegilops kotschyi* and *Ae. variabilis*) was analyzed by means of molecular genetic methods. One genome of both species was found to be identical to the U genome of *Triticum umbellatum*, the second to the S^1 genome of *Triticum longissimum* or *T. sharonense*. Thus, they have the composition S^1S^1UU (Zhang and Dvorák 1992a).
- Finger millet (*Eleusine coracana*, 2 n = 4x = 36) was crossed with five diploid *Eleusine* species. From the meiotic behavior of the triploid interspecific hybrids, it could be concluded that the A genome derives from *Eleusine indica*. The donor species of the B genome is still unknown (Hiremath and Salimath 1992).

- The species *Tinopyrum junceiforme* and *T. sartorii* (2 n = 4x = 28) have two closely related versions of the J genome, J^b and J^e; they show an allotetraploid meiosis. The J^b genome of *Tinopyrum sartorii* ($J^{b1}J^{b1}J^{e3}J^{e3}$) is very similar to the J^b genome of *T. bessarabicum*. The second genome is a modified J^e genome of *T. elongatum*. The genomic formula of *Tinpyrum junceiforme* is $J^{b2}J^{b2}J^{e2}J^{e2}$, both genomes being modified J^b and J^e genomes (Liu and Wang 1992). For the analyses, not only the meiotic behavior of interspecific hybrids, but also the structure of banded mitotic chromosomes was studied.
- The *Aranda Orchids* have arisen through intergeneric hybridizations between different species of the genera *Arachnis* and *Vanda* having the genomes A and V. Commercial *Aranda* cultivars from Singapore were found to have the following genomatic constitutions (Lee 1991):

  - Four cultivars are diploid hybrids AV.
  - Four cultivars are allotriploid hybrids AVV.
  - Two cultivars are unbalanced allotetraploid hybrids AVVV.

At the *hexaploid level*, the following allopolyploid species were genomatically analyzed:

- *Triticum syriacum* (= *Aegilops vavilovii*, DDMMSS) was crossed with autotetraploids of:
  Tr. longissimum (= *Ae. longissima:* $S^lS^lS^lS^l$
  Tr. bicorne (= *Ae. bicornis*): $S^bS^bS^bS^b$
  Tr. speltoides (= *Ae. speltoides*): S S S S.
  According to the meiotic behavior of the hybrids, none of the *syriacum* genomes is identical with S, S^l, and S^b (Yen and Kimber 1992). It is concluded that the species contains a substantially modified S genome. Zhang and Dvorák (1992b) studied this problem by means of molecular genetic methods investigating the variation of 27 repeated nucleotide sequences. They propose the genomatic formula $D^cD^cXX\ S^{se}S^{se}$. The D^c genome derives from ancient *Triticum tauschii*. The S^{se} genome is identical with the genome of *T. searsii*, whereas the origin of the X genome is not yet clear. The hexaploid species *Triticum crassum* (= *Aegilops crassa*) has the composition D^cD^cXXDD; its D genome is identical to the D genome of recent *Triticum tauschii*.
- It is assumed that the bread wheat *Triticum aestivum* (AABBDD) originated about 10 000 years ago near the Capsican Sea, presumably via hybridizations between *Triticum turgidum* (AABB) and *T. tauschii* (= *Aegilops squarrosa*, DD). The donor of the B genome is still under discussion. The diploid species *Triticum searsii, sharonense, speltoides, bicorne*, and *longissimum* are posssible donors (Gill et al. 1991; Zeller and Friebe 1991).
- The species *Elymus himalayanus* and *E. schrenkianus* (SSHHYY) are allohexaploids containing the three genomes in a slightly modified form as related to those in other *Elymus* species (Lu and von Bothmer 1992).
- *Phleum pratense* (2 n = 6x = 42) was formerly considered as one of the best examples for an autohexaploid mode of origin of a plant species. Its chromosome structure was compared to those of the diploid species *Ph. alpinum* (AA) and *Ph. bertolinii* (BB) using C-banding techniques (Cai 1991). From the karyotype analyses, it can be concluded that four sets derive from *Ph. alpinum* and two sets from *Ph. bertolinii*. Thus, it is obviously an auto-allohexaploid species of the constitution AAAABB.

β) Experimentally Produced Allopolyploids via Gametic Hybridization

Allopolyploids of this category can be produced by either interspecific or intergeneric hybridizations followed by colchicine treatment of the hybrids. Further subdivision of the plants ensues from the number of genomes present.

Allotetraploids through *interspecific hybridizations* were produced in the following genera:

- *Glycine latifolia/G. microphylla* (B_1B_1BB; 2 n = 4x = 80).
- *Glycine canescens/G. microphylla* (AABB; 2 n = 4x = 80, completely fertile).
- *Glycine latifolia/G. canescens* (B_1B_1AA; very sparse pod set; Singh et al. 1992).
- *Iris ensata/I. laevigata* (2 n = 4x = 56; Yabuya 1991).
- *Hordeum violaceum/H. bogdanii* (2 n = 4x = 28; Wang et al. 1991). This amphidiploid was produced via inflorescence culture. Callus formation was induced, later, the young plantlets were treated with cholchicine.

At the *hexaploid level*, *Glycine* plants of the constitution AAA^xA^xDD with 2 n = 6x = 120 chromosomes were produced by Singh and Hymowitz (1991a) in the following way: F_1 hybrids between *G. clandestina* and *G. canescens* (2 n = 2x = 40, A_1A) were crossed with the allotetraploid *G. tomentella* (2 n = 4x = 80, A^xA^xDD). The triploid hybrids AA^xD were colchicinized giving rise to the allohexaploids AAA^xA^xDD. They were backcrossed with *G. tomentella*, and allopentaploids of the constitution AA^xA^xDD with 2 n = 5x = 100 chromosomes arose. They resembled morphologically *G. tomentella* and were less vigorous than the hexaploids. In spite of manifold meiotic anomalies, they produced seeds.

Intergeneric hybridizations at the *tetraploid level* gave rise to the following amphiploids:

- *Hordeum chilense/Aegilops squarrosa* (2 n = 4x = 28; $H^{ch}H^{ch}DD$; Cabrera and Martin 1991). For their production, autotetraploid plants of the two species were used for the crosses. The allotetraploids proved to be vigorous and fertile. They are called *Tritordeum* because the *Aegilops* species are incorporated into the genus *Triticum* by many taxonomists.
- Thirteen forms of tetraploid *Triticale* were cytologically analyzed by Badaev et al. (1992) using the C-banding technique. All of them contained 14 rye chromosomes, whereas 69 different combinations with the A, B, and D genomes of wheat were found.

The following polyploid plants with *higher levels* via intergeneric hybridizations were produced:

- A resynthesis of the bread wheat *Triticum aestivum* (AABBDD) was carried out by Lange and Jochemsen (1992). They crossed *Triticum turgidum* ssp. *dicoccoides* (AABB) with *Aegilops squarrosa* (DD) and colchicinized the triploid hybrids ABD. Some of the allohexaploids arose spontaneously from the triploids. Another theoretical possibility for this resynthesis consists in crossing AABB with autotetraploids of *Aegilops squarrosa* (DDDD). This method, however, proved to be unsuccessful.
- Hexa- and octoploid *Tritordeum* (Alvarez et al. 1992). The diploid South American wild barley *Hordeum chilense* ($H^{ch}H^{ch}$) was crossed with *Triticum turgidum* ssp. *durum* (AABB) and *Triticum aestivum* (AABBDD) giving rise to hexaploid and octoploid *Tritordeums* ($H^{ch}H^{ch}AABB$ and $H^{ch}H^{ch}AABBDD$). This is a new synthetic crop with potential value for cereal breeding. The quality of the material available in Spain is similar to medium quality bread wheat varieties.
- *Triticum durum/Thinopyrum distichum* (2 n = 8x = 56).
- *Triticum aestivum/Thinopyrum distichum* (2 n = 10x = 70; Armstrong et al. 1991). The material was cytologically analyzed.

γ) Allopolyploids Through Somatic Hybridization

The fusion of somatic protoplasts leads directly to allopolyploids provided that viable plants develop from the calli. The examples given below demonstrate that this category represents – as in former years – a very inhomogeneous group of plants, not only with regard to their vitality and fertility, but also to their cytological stability and their potentialities for practical use. A review paper on the improvement of legumes by means of somatic hybridization was given by Kumar and Davey (1991). As in the allopolyploids arisen via gametic hybridization, we can subdivide the material existing into the following four groups:

- allotetraploid plants,
- plants with higher ploidy levels.

This is valid for both interspecific and intergeneric somatic hybrids. Moreover, we have to consider symmetric and asymmetric somatic hybrids.

The *symmetric somatic hybrids* contain in the ideal case the complete genomes of the two parental species. Thus, they are cytologically balanced and can be regarded as representatives of synthetic species. Examples for *interspecific hybrids* at the *allotetraploid level* are as follows:

- *Brassica oleracea/B. campestris* (AACC; Sundberg and Glimelius 1991). Some of these plants are fertile.
- Dihaploid *Solanum tuberosum/S. phureja*. To obtain these plants, the method of electrofusion was used. The technique applied was incorporated into the Dutch commercial potato breeding program (Mattheij and Puite 1992).
- *Solanum viarium/S. dulcamara*. Cotyledon protoplasts of *S. viarium* were electrofused with cell suspension protoplasts of *S. dulcamara*; the heterokaryons were isolated by means of a micromanipulator. Both parents have $2n = 24$ chromosomes. Root tip cells of the allotetraploids had mostly 46 chromosomes ranging between 40 and 48 (Chand et al. 1991).

The following allopolyploids of this group with *higher ploidy levels* were produced:

- *Brassica napus/B. nigra* (AABBCC; Sundberg and Glimelius 1991). Some of them were fertile, in others chromosomes of the B genome were eliminated.
- *Brassica napus/B. oleracea* (AACCCC; Sundberg and Glimelius 1991).
- *Brassica oleracea italica/B. napus* (AACCCC; Kao et al. 1992). Hypocotyl protoplasts of the broccoli were fused with mesophyll protoplasts of the second partner giving rise to six somatic hybrids. They had the expected chromosome number of $2n = 6x = 56$, but they were male sterile.
- *Medicago sativa* ($2n = 4x = 32$)/*M. coerulea* ($2n = 2x = 16$; Pupilli et al. 1992). Leaf protoplasts were electrofused with callus protoplasts resulting in four plants with $2n = 6x = 48$ chromosomes ranging between 45 and 50. They were fertile and produced seeds.
- *Brassica napus* (AACC)/*B. juncea* (AABB) → AAAABBCC. These auto-allo-octoploid forms were produced by Sundberg and Glimelius (1991).
- *Nicotiana tabacum/N. debneyi*. The two species already have an allotetraploid genomatic constitution. Thirteen somatic hybrids were obtained following mesophyll protoplast fusion. Their chromosome numbers ranged from $2n = 8x = 96$ till down to 60. They were partially male and female fertile (Sproule et al. 1991).

Examples for symmetric *allotetraploid intergeneric somatic hybrids* are as follows:

- *Raphanus sativus/Brassica oleracea* (RRCC; Hagimori et al. 1992). From the 40 regenerants obtained after electrofusion, 37 were found to be true allotetraploids with $2n = 4x = 36$ chromosomes. They were self-fertile and showed the resistance to clubroot disease transferred from *Raphanus sativus*.

- Dihaploid *Solanum tuberosum/Lycopersicon esculentum* (2 n = 4x = 48; Jacobsen et al. 1992). Balanced allotetraploid plants were obtained in high frequencies. They showed a normal diploid meiosis, but they were completely sterile.
- *Lycopersicon esculentum/Solanum muricatum* (2 n = 4x = 48; Sakomoto and Taguchi 1991). Mesophyll protoplasts of both species were electrofused. All the regenerants were euploid; no aneuploids were found. The plants flowered, but the pollen viability was less than 1%. Self-pollination gave no fruit set. Fruits were obtained after having backcrossed the allotetraploids with tomatoes, but their seeds did not germinate.
- *Citrus sinensis/Poncirus trifoliata.* Mesophyll protoplasts of *Poncirus* were chemically fused with nucellar protoplasts derived from excised ovules. The allotetraploids had the expected chromosome number of 2 n = 4x = 36. Five clones were grafted. Four years later, the trees produced fruits with seeds of normal germability. The material is of interest for *Citrus* breeding (Kobayashi and Ohgawara 1989; Ohgawara et al. 1991).

At the *hexaploid level* or with even *higher degrees of allopolyploidy*, the following examples can be given:

- Dihaploid *Solanum tuberosum/Lycopersicon esculentum.* Besides the allotetraploids mentioned above, one allohexaploid plant was obtained by Jacobsen et al. (1992). It produced berries, but their seeds were not fully developed.
- *Brassica napus* (AACC)/*Eruca sativa* (EE) → AACCEE.
- *Brassica napus* (AACC)/*Raphanus sativus* (RR) → AACCRR.
  These allohexaploids were female fertile but male sterile (Sundberg and Glimelius 1991). In the allotetra- and -hexaploid material studied by the authors, a correlation between the frequency of hybrids with eliminated chromosomes and the genetic distance of the species used was found. Somatic hybrids containing the A and C genomes proved to be more fertile than those containing the less related genomes A, C, and B, R, or E, respectively. In all the tetra- and hexaploid combinations mentioned, the DNA content of some somatic hybrids was equal to that of the sum of the parental genomes.

The somatic hybrids mentioned so far are genomatically symmetric, i.e., they contain the genomes of both parents. In other cases, some or all the chromosomes of one of the two species used are inactivated, for instance by gamma irradiation. In this way, *asymmetric somatic hybrids* arise. Some examples of this category are as follows:

- *Lycopersicon esculentum/L. pennellii.* Tomato leaf protoplasts were fused with irradiated suspension protoplasts of the second partner. About 50% of the regenerants were found to be asymmetric. At least in some of them, limited amounts of *pennellii* gene material were present (Melzer and O'Connell 1992).
- *Solanum tuberosum* (2 n = 4x = 48)/*S. brevidens* (2 n = 2x = 24). *Tuberosum* leaf protoplasts were fused with irradiated cell suspension protoplasts of *S. brevidens*. In all the 45 clones obtained, a high degree of aneuploidy was observed. In addition to the 48 potato chromosomes, at least 12 extra chromosomes were present. The somatic hybrids were genetically instable; some of them had very high chromosome numbers (Fehér et al. 1992).
- *Petunia hybrida/Nicotiana plumbaginifolia.* Twentyfour fertile regenerants were obtained from 86 hybrid clones. Some of them had only 20 or 40 *Nicotiana* chromosomes; in others, some additional fragments were found (Hinnisdaels et al. 1991).
- Dihaploid *Solanum tuberosum/Nicotiana plumbaginifolia.* Three different types of somatic hybrids arose as follows:

  - plants with 8–24 *Solanum* and more than 20 *Nicotiana* chromosomes,
  - plants with more than 24 *Solanum* and 1–20 *Nicotiana* chromosomes,
  - plants with diploid or subdiploid numbers of chromosomes of both species.

  Thus, an elimination of chromosomes of both parents had occurred, resulting in the three types of asymmetric hybrids (Gilissen et al. 1992).

- *Atropa belladonna* (2 n = 72)/*Nicotiana tabacum* (2 n = 48). Altogether, 447 colonies were obtained, most of them gave shoots. Three plants had 48 *Nicotiana* chromosomes and one *Atropa* fragment; they were highly fertile (Babiychuk et al. 1992).

d) Polyploids Through Unreduced Gametes

Very often, polyploid plants arise through fertilization of unreduced male or female gametes. This behavior was not only observed in diploid, but also in polyploid plants and in hybrids. In some cases, a genetic basis of this anomaly could be evidenced. The phenomenon is intensively under study, preferably in potatoes. Some new examples are as follows:

- In diploid wild potatoes (*Solanum chacoense, S. sparsipilum,* hybrids between *S. chacoense* and *S. infundibuliforme*), two different meiotic mechanisms for the production of unreduced diploid egg cells were clarified by Werner and Peloquin (1990). The second meiotic division of macrosprogenesis can be omitted, resulting in the production of two unreduced diploid megaspores. One of them disintegrates, the second one is functionable. The same result is obtained through failure of cytokinesis after the second meiotic division followed by the fusion of two haploid nuclei. Each of the two mechanisms was found to be controlled by a single mutant gene; the two genes are not allelic. These very interesting gene actions are of great importance for the evolution of polyploids. Later, five different meiotic mechanisms for the production of unreduced eggs in the diploid wild potatoes *Solanum berthaultii, brevicaule,* and *tarijense* were found (Werner and Peloquin 1991). These phenomena are obviously not rare in wild potatoes. They were observed in 17 diploid, 12 tetraploid, and 5 hexaploid species of this group with regard to unreduced pollen. Their frequency ranged between 2 and 10% in the material studied (Watanabe and Peloquin 1991).

- Similar meiotic anomalies have been found by Jongedijk et al. (1991) in dihaploid plants of *Solanum tuberosum*. In that material, the diploid eggs arise through omission of the second meiotic division followed by chromosome doubling after anaphase I. Moreover, they can originate through an equational division of univalents during anaphase I.
- In hybrids between dihaploid *Solanum tuberosum* and *S. chacoense*, micro- and macrosporogenesis were studied by Conicella et al. (1991). More than 50% of all the egg cells were found to be diploid whereas diploid pollen arose only rarely. The mechanisms could be clarified.
- *Ipomoea trifida*, a wild species closely related to the sweet potato, has a polyploid series with 2 n, 4n, and 6n cytotypes. The formation of unreduced pollen up to 20% in the diploid material is a common phenomenon in this taxon (Orjeda et al. 1990; Iwanaga et al. 1991).
- The high frequency of unreduced gametes in hybrids of *Triticum carthlicum* x *Aegilops squarrosa* ssp. *strangulata* is obviously genetically controlled (Fukuda and Sakamoto 1992).
- In diploid and tetraploid blueberry species (*Vaccinium* sect. *Cyanococcus*), the frequency of unreduced gametes was found to depend on the genotypic constitution of the comprehensive material studied, ranging between 1 and 29% in diploid clones. Derivatives of hybrids between *Vaccinium ashei* x *V. corymbosum* produce almost entirely diploid pollen (Ortiz et al. 1992a, b; Vorsa and Ortiz 1992).

e) Aneuploids

Aneuploids are a very heterogeneous group containing plants with lacking chromosomes (monosomics, nullisomics), but preferably with additional chromosomes (trisomics, tetrasomics) or with additional chromosome arms (telotrisomics, telotetrasomics). A review on wheat aneuploids was given by Nishikawa (1992). An efficient method for obtaining different types of aneuploids in maize consists of treating excised embryos or germinating seeds with para-fluoro-phenylalanine (Banks et al. 1989).

Monosomics (2 n – 1) are generally used for assigning distinct genes to specific chromosomes of the complement. If this has been reached, *monotelodisomics* (i.e., plants in which a telocentric chromosome instead of a normal one is present) can be used for analyzing whether the gene is located on the short or the long arm. This has been done with 34 mutant loci of *Gossypium hirsutum* (Endrizzi and Ray 1991).

A new method for obtaining plants with additional chromosomes was used by Wang and Iwata (1991) in *Oryza sativa*. They subjected 11 *primary trisomics* (2 n + 1 = 25) to anther culture and obtained 3732 plants. In the greenhouse, morphologically abnormal seedlings were selected and cytologically investigated. In this way, 43 *haplo-trisomics* with 13 chromosomes (n + 1) and 56 *tetrasomics* with 26 chromosomes (2 n + 2) were found. Further examples are as follows:

- Eight different types of primary trisomics were selected in *Oryza glaberrima* following pollination of autotriploids with diploids, the classical method (Ishiki 1991).
- Eight of the possible 20 primary trisomics of *Glycine max* are available. In 4 of them, the additional chromosome was identified by pachytene analysis (Singh and Hymowitz 1991b; Ahmad et al. 1992).
- Octoploid plants of *Elymus canadensis*, obtained from tissue cultures, were crossed with tetraploids. A hexaploid plant was backcrossed with tetraploids giving not only rise to a pentaploid but also to two trisomics and one tetrasomic (Park and Walton 1990).
- A tetrasomic plant of *Plantago lagopus* was obtained from a cross between two trisomics. The two additional chromosomes were identified (Bhan et al. 1990).
- Double trisomics with two different additional chromosomes were obtained in *Plantago lagopus* following crosses of two primary trisomics (Bhan and Koul 1989).

A special case of trisomy are the *monosomic addition lines*. In plants of this cytological constitution, the complements of a species are supplemented by a single chromosome of a foreign species. Nine different types of this category are available in *Beta vulgaris*. The additional chromosomes derive from the wild species *Beta webbiana* (Reamon-Ramos and Wricke 1992).

Telotrisomic plants contain not a normal chromosome but only one arm in addition to the diploid complement. They were cytologically and genetically analyzed in *Hordeum vulgare* (Shahla and Tsuchiya 1990; Wang and Tsuchiya 1990; Tsuchiya and Wang 1991). In *Triticum monococcum*, the extra telocentric chromosome was found to consist of the short arm of chromosome 5A (Kim and Kuspira 1992). From telotrisomics, *ditelotetrasomic plants* with two additional homologous telocentrics were obtained in the two species just mentioned (Wang and Tsuchiya 1990; Kim and Kuspira 1992).

Tertiary trisomics have a translocated chromosome in addition to the normal complements. They arose in the progeny of hybrids between *Lens culinaris* x *L. ervoides* (Ladizinsky et al. 1990). The transmission of such an extra chromosome was studied in

Secale cereale ranging from 0 to 36% in different inbred lines (Janse and Sybenga 1992). In *compensating trisomics*, finally, one chromosome of a bivalent is replaced by two translocated chromosomes which together contain at least the whole genetic material of the lacking chromosome. In *Secale cereale*, the meiosis of four telo-tertiary compensating trisomic plants was analyzed. Their complement was constructed from 4 different reciprocal translocations and three different telocentrics (de Vries and Sybenga 1989).

References

Abraham SK (1991) Mutat Res 262:109–114. **Ahmad F,** Singh RJ, Hymowitz T (1992) J Hered 83:221–224. **Alvarez JB,** Ballesteros J, Sillero JA, Martin LM (1992) Hereditas 116:193–197. **Anderson JA,** Mousset-Déclas C, Williams EG, Taylor NL (1991) Genome 34:1–5. **Armstrong KC,** Le H, Fedak G (1991) Theor Appl Genet 81:363–368. **Asita AO,** Matsui M, Nohmi T, Matsuoka A, Hayashi M, Ishidate M, Sofuni T, Koyano M, Matsushita H (1991) Mutat Res 264:7–14.

Babiychuk E, Kushnir S, Gleba YY (1992) Theor Appl Genet 84:87–91. **Badaev NS,** Badaeva ED, Dubovets NI, Bolsheva NL, Bormotov VE, Zelenin AV (1992) Genome 35:311–317. **Banks P,** Britten EJ, Gordon GH (1989) Genome 32:962–966. **Banks P,** Britten EJ, Gordon GH (1991) Genome 34:929–934. **Barnabás B,** Pfahler PL, Kovács G (1991) Theor Appl Genet 81:675–678. **Berryman SH,** Anderson RA, Weis J, Bartke A (1992) Mutat Res 278:47–60. **Bhan A,** Koul AK (1989) Nucleus 32:156–161. **Bhan A,** Koul AK, Langer A (1990) Nucleus 33:134–137. **Bir SS,** Cheema P, Sidhu M (1990) Nucleus 33:22–24. **Blixt S,** Folkeson D, Gottschalk W (1991) In: Tsuchiya T, Gupta PK (eds) Chromosome engineering in plants, vol B. Elsevier, Amsterdam, p 33. **Burner DM** (1991) Euphytica 54:125–133.

Cabrera A, Martin A (1991) Theor Appl Genet 81:758–760. **Cai Q** (1991) Genome 34:52–58. **Camadro EL,** Masuelli RW, Cortés MC (1992) Genome 35:431–435. **Carballo M,** Mudry MD, Larripa IB, Villamil E, D'Aquino M (1992) Mutat Res 279:245–253. **Chand PK,** Davey MR, Power JB, Cocking EC (1991) Nucleus 34:53–58. **Chorvatovicová D,** Ginter E, Kosinová A, Zloch Z (1991) Mutat Res 262:41–46. **Clulow SA,** Wilkinson MJ, Waugh R, Baird E, DeMaine MJ, Powell W (1991) Theor Appl Genet 82:545–551. **Conicella C,** Barone A, DelGuidice A, Frusciante L, Monti LM (1991) Theor Appl Genet 81:59–63. **Crebelli R,** Conti L, Fuselli S, Leopardi P, Zijno A, Carere A (1991) Mutat Res 259:29–36.

Dash BC, Das RK (1992) Mutat Res 280:45–53. **de Andrade HHR,** Santos JH, Gimmler-Luz MC, Correa MJF, Lehmann M, Reguly ML (1992) Mutat Res 279:281–287. **de Vries JN,** Sybenga J (1989) Theor Appl Genet 78:889–896. **Doebley J** (1992) Trends Genet 8:302–307. **Doolittle DJ,** Rahn CA, Lee CK (1991) Mutat Res 260:9–18.

Endrizzi JE, Ray DT (1991) J Hered 82:53–57. **Evans DA** (1989) Trends Genet 5:46–50.

Fehér A, Preiszner J, Litkey Z, Csanadi GY, Dudits D (1992) Theor Appl Genet 84:880–890. **Finnie SJ,** Forster BP, Chalmers KJ, Dyer AF, Waugh R, Powell W (1991) Genome 34:923–928. **Fukuda K,** Sakamoto S (1992) Hereditas 116:253–255.

Garcia-Velasquez A (1991) Genome 34:895–899. **Gilissen LJW,** van Staveren MJ, Verhoeven HA, Sree Ramulu K (1992) Theor Appl Genet 84:73–86. **Gill KS,** Lubbers EL, Gill BS, Raupp WJ, Cox TS (1991) Genome 34:362–374. **Gmitter FG,** Ling XB, Deng XX (1990) Theor Appl Genet 80:785–790. **Golubovskaya IN** (1989) Adv Genet 26:149–192. **Gorgels WJMJ,** van Poppel G, Jarvis MJ, Stenhuis W, Kok FJ (1992) Mutat Res 279:233–238. **Gottschalk W** (1990) In: Khan IA,

Farak SA (eds) Genetic improvement of pulse crops, vol 1. Premier Publ House, Hyderabad, p 329. **Gottschalk W** (1991) In: Plant mutation breeding for crop improvement, vol 2. Int Atomic Energy Agency, Vienna, p 55. **Guenzi AC,** Mornhinweg DW, Johnson BB (1992) Theor Appl Genet 84:952–957.

Hagimori M, Nagaoka M, Kato N, Yoshikawa H (1992) Theor Appl Genet 84:819–824. **Henderson CAP,** Pauls KP (1992) Theor Appl Genet 83:476–479. **Hinnisdaels S,** Bariller L, Mouras A, Sidorov V, Del-Favero J, Veuskens J, Negrutiu I (1991) Theor Appl Genet 82:609–614. **Hiremath SC,** Salimath SS (1992) Theor Appl Genet 84:747–754. **Hoyos LS,** Au WW, Heo MY, Morris DL, Legator MS (1992) Mutat Res 280:29–34. **Hsu TC,** Cherry LM, Bucana C, Shirley LR, Gairola CG (1991) Mutat Res 259:67–78.

Iida S, Amano E (1990) In: Proc Gamma Field Symp 29. Inst Radiation Breeding, Ohmiya-machi, Japan, p 95. **Imanishi H,** Sasaki YF, Ohta T, Watanabe M, Kato T, Shirasu Y (1991) Mutat Res 259:79–87. **Institute of Radiation Breeding** (ed) (1989) Gamma Field Symp 28: Production of Mutants in Tree Crops. Ohmiya-machi, Japan. **International Atomic Engergy Agency** (1991) Plant mutation breeding for crop improvement, vols 1, 2. Vienna. **Ishiki K** (1991) Euphytica 55:7–13. **Iwanaga M,** Freyre R, Orjeda G (1991) Genome 34:201–208.

Jacobsen E, Reinhout P, Bergervoet JEM, de Looff J, Abidin PE, Huigen DJ, Ramanna MS (1992) Theor Appl Genet 85:159–164. **Janse J,** Sybenga J (1992) Theor Appl Genet 84:487–493. **Jensen KB,** Wang RR-C (1991) Genome 34:860–867. **Jongedijk E,** Ramanna MS, Sawor Z, Hermsen JGTh (1991) Theor Appl Genet 82:645–656. **Jørgensen K,** Wulf HC, Husum B, Niebuhr E (1991) Mutat Res 261:193–195.

Kanaya N, Takehisa S, Nicoloff H, Nicolova T, Damianova V (1992) Mutat Res 281:47–54. **Kao HM,** Keller WA, Gleddie S, Brown GG (1992) Theor Appl Genet 83:313–320. **Kaul MLH,** Nirmala C (1991) Nucleus 34:32–39. **Kim N-S,** Kuspira J (1992) Genome 35:849–854. **Kobayashi S,** Ohgawara T (1989) In: Gamma Field Symp 28, Inst Radiation Breeding, Ohmiya-machi, Japan, p 1. **Krivobok S,** Seigle-Murandi F, Steiman R, Marzin DR, Betina V (1992) Mutat Res 279:1–8. **Kukreja AK,** Dhawan OP, Mathur AK, Ahuja PS, Mandal S (1991) Euphytica 53:183–191. **Kumar V,** Davey MR (1991) Euphytica 55:157–169.

Ladizinsky G, Weeden NF, Muehlbauer FJ (1990) Euphytica 51:179–184. **Lange W,** Jochemsen G (1992) Euphytica 59:197–220. **Larramendy ML,** Knuutila S (1991) Mutat Res 259:189–195. **Lavania UC,** Srivastava S, Sybenga J (1991) Genome 34:190–194. **Lee YH** (1991) Euphytica 54:251–254. **Liu Z-W,** Wang RR-C (1992) Genome 35:758–764. **Lu B-L** (1992) Theor Appl Genet 83:997–1002. **Lu B-R,** von Bothmer R (1990) Hereditas 113:109–119. **Lu B-R,** von Bothmer R (1992) Genome 35:230–237.

Maguire MP, Paredes AM, Riess RW (1991) Genome 34:879–887. **Maluszynski M,** Sigurbjörnsson B, Amano E, Sitch L, Kamra O (1991) Mutat Breed Newslett 38:16–49. **Maluszynski M,** Sigurbjörnsson B, Amano E, Sitch L, Kamra O (1992) Mutat Breed Newslett 39:14–33. **Mattheij WM,** Puite KJ (1992) Theor Appl Genet 83:807–812. **McFee AF** (1991) Mutat Res 264:219–224. **Melz G,** Adolf K (1991) Theor Appl Genet 82:761–764. **Melz G,** Schlegel R, Thiele V (1992) Theor Appl Genet 85:33–45. **Melzer JM,** O'Connell MA (1992) Theor Appl Genet 83:337–344. **Mukherjee A,** Agarwal K, Aguilar MA, Sharma A (1991) Mutat Res 263:41–46.

Nair U, Obe G, Nair J, Maru GB, Bhide SV, Pieper R, Bartsch H (1991) Mutat Res 261:163–168. **Neudecker T** (1992) Mutat Res 277:173–185. **Nishikawa K** (1992) Wheat Inf Serv 74:1–3.

Oard JH, Hu J, Rutger JN (1991) Euphytica 55:179–186. **Ohgawara T,** Kobayashi S, Ishii S, Yoshinaga K, Oiyama I (1991) Theor Appl Genet 81:141–143. **Orjeda G,** Freyre R, Iwanaga M (1990) J Hered 81:462–467. **Ortiz R,** Bruederle LP, Laverty T, Vorsa N (1992a) Euphytica 61:241–246. **Ortiz R,** Vorsa N, Bruederle LP, Laverty T (1992b) Theor Appl Genet 85:55–60.

Palmer RG, Albertsen MC, Horner HT, Skorupska H (1992) Nucleus 35:1–18. **Park CH,** Walton PD (1990) Genome 33:389–394. **Petersen G** (1992) Hereditas 116:101–105. **Pupilli F,** Scarpa GM, Damiani F, Arcioni S (1992) Theor Appl Genet 84:792–797.

Qureshi JA, Hucl P, Kartha KK (1992) Euphytica 60:221–228.

Reamon-Ramos SM, Wricke G (1992) Theor Appl Genet 84:411–418. **Rietveld RC,** Hasegawa PM, Bressan RA (1991) Theor Appl Genet 82:430–440. **Rubiolo P,** Pieters L, Calomme M, Bicchi C, Vlietinck A, Vanden Berghe D (1992) Mutat Res 281:143–147. **Rutger JN** (1992) Mutat Breed Newslett 8:1–23.

Sakomoto K, Taguchi T (1991) Theor Appl Genet 81:509–513. **Salomon B,** von Bothmer R, Jacobsen N (1991) Hereditas 114:35–39. **Shahla A,** Tsuchiya T (1990) J Hered 81:127–130. **Shirnamé-Moré L** (1991) Mutat Res 259:37–42. **Singh RJ,** Hymowitz T (1991a) Genome 34:751–756. **Singh RJ,** Hymowitz T (1991b) J Hered 82:75–77. **Singh RJ,** Kollipara KP, Ahmad F, Hymowitz T (1992) Genome 35:140–146. **Sinues B,** Broto A, Suarez MA, Duce F, Martinez-Berganza A, Bernal ML (1992) Mutat Res 280:271–277. **Smith KC** (1992) Mutat Res 277:139–162. **Sosnikhina SP,** Fedotova YuS, Smirnov VG, Mikhailova EI, Kolomiets OL (1992) Theor Appl Genet 84:979–985. **Sproule A,** Donaldson P, Dijak M, Bevis E, Pandeya R, Keller WA, Gleddie S (1991) Theor Appl Genet 82:450–456. **Sundberg E,** Glimelius K (1991) Theor Appl Genet 83:81–88.

Tsuchiya T, Wang S (1991) Theor Appl Genet 83:41–48.

Ukai Y (1990) In: Gamma Field Symp 29. Inst Radiation Breeding, Ohmiya-machi, Japan, pp 55–69.

van den Bulk RW, Löffler HJM, Lindhout WH, Koornneef M (1990) Theor Appl Genet 80:817–825. **Vorsa V,** Ortiz R (1992) J Hered 83:346–349.

Wan Y, Duncan DR, Rayburn AL, Petolino JF, Widholm JM (1991) Theor Appl Genet 81:205–211. **Wang RR-C,** Marburger JE, Hu C-J (1991) Theor Appl Genet 81:151–156. **Wang S,** Tsuchiya T (1990) Theor Appl Genet 80:145–152. **Wang ZX,** Iwata N (1991) Theor Appl Genet 83:12–16. **Wang ZY,** Agarwal R, Zhou ZC, Bickers DR, Mukhtar H (1991) Mutat Res 261:153–162. **Warner RJ,** Nath J, Ong T-M (1991) Mutat Res 262:25–30. **Watanabe K,** Peloquin SJ (1991) Theor Appl Genet 82:621–626. **Werner JE,** Peloquin SJ (1990) J Hered 81:371–374. **Werner JE,** Peloquin SJ (1991) Genome 34:975–982. **Wong BYY,** Lau BHS, Tadi PP, Teel RW (1992) Mutat Res 279:209–216.

Xue-jun Y, De-xiang L, Hechuan W, Yu Z (1991) Mutat Res 260:73–82.

Yabuya T (1991) Euphytica 55:85–90. **Yamamoto K,** Fukuoka H, Kageyama Y, Nashimoto S, Takeda G (1990) In: Gamma Field Symp 29. Inst Radiation Breeding, Ohmiya-machi, Japan, pp 27–44. **Yen Y,** Kimber G (1992) Genome 35:709–713.

Zeller FJ, Friebe B (1991) BIUZ 21:248–254. **Zhang H-B,** Dvorák J (1992a) Genome 35:182–191. **Zhang H-B,** Dvorák J (1992b) Genome 35:806–814. **Zhang Z,** Yang J (1992) Mutat Res 280:279–283. **Zhou S-Q,** Qian D-Q, Cao X-Y (1991) Genome 34:255–260. **Zohary D** (1992) Euphytica 60:75–77.

Professor Dr. Werner Gottschalk
Institut für Genetik
der Universität Bonn
Kirschallee 1
D-53115 Bonn, FRG

IV. Function of Genetic Material, Regulation of Genes Encoding Seed Storage Proteins

By Rolf Blaich

1. Introduction: A Model System for Gene Regulation

The development of storage tissues in seeds (endosperm or cotyledons) involves the multiplication, enlargement, and differentiation of cells with consequent accumulation of different storage products. The biosynthesis of carbohydrates, fats, or other nonprotein components of cells involves a complex regulatory network with a large number of genes. In contrast, storage proteins – such as the maize zeins, the vicilins and legumins of pea, the hordeins of barley, the wheat gliadins – are directly encoded by single genes. At well-defined times, not only the proteins but also the corresponding primary transcripts and mRNAs are present in high concentrations – up to 50% of the cellular RNA – and can be isolated, analyzed, and used for the gene analysis.

Although being integrated in complex regulation systems which are not always well understood, even by specialists, the control of storage protein genes can be considered independently, making this subsystem well suited to exemplify the actual state of knowledge on gene function, particularly to nonexperts.

There are, however, a number of other reasons why storage protein genes are among the best studied: due to their high concentrations (in cereals about 10%, in legumes up to 30%) storage proteins determine the nutritional value of seeds which might be enhanced by the introduction of essential amino acids via gene manipulation. Unlike enzyme proteins, the function of storage proteins would seldom be disturbed by such exchanges.

This latter fact makes storage protein genes particularly well suited for the study of evolutionary events: a large variety of mutations may be found in populations, gene duplication is facilitated, and large multigene families have evolved comprising not only many functional genes but also a large number of pseudogenes which have lost their activity due to mutations in the regulatory regions (for further discussion see Thompson et al. 1991).

2. Glossary: Definition of Frequent Expressions

A definition of some frequently used expressions will avoid repetition in the following text and may facilitate its lecture, although it is, of course, not intended to repeat basic textbook knowledge; for further information, see also previous reviews in this series (Blaich 1989, 1992).

Progress in Botany, Vol. 55

A *box* is a short DNA sequence that is evolutionarily conserved, i.e., found in homologous genes of many organisms.

The *TATA box* (TATAA, TATAT) is one of the control signals that determine the site of transcription initiation in eukaryotes, probably as the recognition site for RNA polymerase II. The start of transcription is, however, around 20–50 nucleotides downstream from the TATA box. For interesting details on "proteins saddling up the TATA box" see Philips (1993).

The *(C)CAAT box* is also involved in stimulating transcription, but is not indispensable. In zeins it is usually replaced by an AGGA box.

Upstream and *downstream* are related to the direction of transcription of a gene, i.e., towards (or before) the 5' end and towards (or after) the 3' end, respectively, of the DNA.

A *reading frame* is a DNA coding sequence delimited by a translation start codon (AUG) and a stop codon (TAG, TGA, TAA).

An *intron* is a region of the transcribed part of a gene which is removed from the primary transcript (hnRNA) by "splicing" and hence not contained in the mature mRNA, which is then translated during protein synthesis.

It is not yet clear whether introns are useless evolutionary relicts, in any case they may be removed from genes by gene manipulation. The remaining *mini genes* are functional. The studies of Chee et al. (1991) may serve as an example: removal of the introns of a phaseolin (legume storage protein) gene did not affect its normal developmental expression.

3. Overview: Regulation of Storage Protein Synthesis

Most storage protein genes are synthesized only during seed development as pre-proteins with an N-terminal signal peptide that allows them to enter the endoplasmatic reticulum. After removal of the signal (transit peptide) and other modifications (e.g., glycosylation) they are deposited in protein bodies derived from the endoplasmic reticulum or the vacuole. While the storage proteins of legumes are accumulated mainly in the embryo (cotyledons), in cereals they are synthesized in the endosperm which does not belong to the embryo. This is important for functional studies of transferred genes.

Storage protein synthesis involves mainly transcriptional regulation of their genes, although there are many examples of posttranscriptional regulation.

Bagga et al. (1992), for example, found that constitutive expression (abnormal continuous transcription in all tissues due to a modified promoter) of a phaseolin gene in transgenic alfalfa plants did not lead to accumulation of phaseolin in nonseed tissue. The protein in nonseed organs underwent correct post-translational processing but was not stable in the nonseed environment.

Several evolutionarily conserved sequence motives have been found in the upstream (regulatory) regions of genes encoding storage proteins. (Kreis et al. 1985; Casey and Domoney 1987). The terms "vicilin box" or "legumin box" (Gatehouse et al. 1986; Chamberland et al. 1992) indicate such sequences, which have sometimes been shown to be associated with DNA regions that might form stem-loop structures (Maier et al. 1987). Stemloops are formed by base pairing within a single strand of the DNA due to short palindromic base sequences.

One or more of these boxes is involved in regulating the expression of storage protein genes. For example, about 300 base pairs upstream from the ATG-codon of wheat α-gliadin and barley B-hordein genes is a sequence ATGTAAAGTGAATAAG. A similar sequence is repeated further upstream, and another conserved 15 bp sequence CACATGTGTAAAGGT, with related features, is found upstream of the TATA box of most maize zein genes (prolamin box). Such "enhancer" core sequences, which stimulate gene expression, are known also from viruses and animals.

The expression of glycinin genes in legumes is modulated by a CATGCAT element (Lelievre et al. 1992). Napins are seed storage gene from *Brassica napus* (rape) genes. They also contain conserved sequence motives which are frequently repeated and seem to have regulatory functions (TACACAT and CATGCA). One copy of the CATGCA motif occurs near the TATA box and overlaps with the octamer sequence ATGCAAAT, known from many other eukaryotic promoters (Ericson et al. 1991).

4. Control Structures of Zein Genes

The best-studied storage proteins are the zeins of maize, which belong to the alcohol-soluble prolamins and are either rich in glutamin (z1 zeins) or proline, cysteine, or methionine (z2 zeins). They are restricted to the endosperm of seeds (not present in the embryo) and encoded by large multigene families comprising about 60 genes for z1 zeins (further grouped in z1A, z1B, z1C, z1D) but far less for z2 genes. The proteins fall into different molecular weight classes (e.g., 10, 15, 22, 27 kDa; for more details see the review of Heidecker and Messing 1986). The majority of these genes are highly clustered and have between 98 and 85% homology. A model for the evolution of the families involving unequal crossing-over and gene conversion may be found in Heidecker et al. (1991).

Zein genes show a number of peculiarities: They have no introns and have an "AGGA box" instead of the "CAAT box" (Heidecker and Messing 1986). The studies of Thompson et al. (1990) give a good example for the analysis of gene control with a 19-kDa, a zein gene. The promoter was fused to the GUS reporter gene and assayed by transient expression in carrot protoplasts. The regulatory effect of different upstream sequences was then tested in various combinations. Understandably enough, CaMV (cauliflower mosaic virus) activating sequences did not activate transcription of the GUS reporter with the zein promoter. It is, however, more interesting to note that the upstream sequences of the maize gene did not promote the transcription of the reporter gene in somatic maize protoplasts, whereas they are functional in the carrot protoplast system: the control mechanism which limits the gene activity to endosperm cells may be incompatible in both plant genera. The activities of some constructs obtained by deletions of parts of the native sequence are given in Table 1.

Another unique structure of most zein genes is the presence of two or more promoters. A second promoter is found 900 base pairs upstream from the first one. Consequently sometimes two types of mRNA have been observed, one of them being 900 pb longer than the other. Brown and Feix (1990) found 18 short reading frames with stop codons between the first promoter (P1), and the AUG codon (translation initiation). A splicing site seems to exist 40 nucleotides upstream of the coding region. A splicing event removes all of the short reading frames from the mRNA transcribed from the P1 promoter, thus making this mRNA translatable (Brown and Feix 1990). In a more recent paper, Liu

Table 1. (1) Part of the regulatory sequence of a maize zein gene according to Thompson et al. (1990). Nucleotide positions are related to the translation start. Scheme of the native control region is indicated on the left (lengths of subregions not true to corresponding scale). (2) Schematic diagram of (1) with four parts arbitrarily numbered 1–4. (3)–(6) Activity of some artificial regulatory regions with different deletions

1	2	3	4	5	6	
5' (Upstream end)						
!						
– 483	4	4	←			High activity
!	4	4				independent of
– 347	4	4				the 222 region
!	4	4				(444 may be orie-
TGTAAAG	4	4				ted in either
/	4	4				direction)
/	4	4				
/	4	4				
– 309	3	3	3	←		Also high activity:
!	3	3	3			TGTAAAG consensus
!	3	3	3			sequence is not
– 226	3	3	3			indispensable
!	3	3	3			
– 189	2	1	2	2	←	Double activity
!	2	1	2	2		as compared to
!	2	1	2	2		111 alone
– 114	2	1	2	2		
!	1	1	1	1	1 ←	Low
!	1	1	1	1	1	activity
TATA	1	!	1	1	1	
/	1	!	1	1	1	
!	1	ZP	1	1	1	
!	1	!	1	1	1	
– 8	1	GUS	1	1	1	
!	!	3'	!	!	!	
ZP	ZP		ZP	ZP	ZP	(Zein promoter)
!	!		!	!	!	
GUS	GUS		GUS	GUS	GUS	(Reporter gene)
3'	3'		3'	3'	3'	(Downstream end)

and Rubenstein (1992) describe genes in the zein subfamily 4 (SF4, about 15 members) which lack such early in-frame stop codons.

The transcription of zein genes is tissue-specific and developmentally regulated. Contradictory results on the function of the promoters P1 and P2 could be explained by a different specifity. Giovinazzo et al. used a GUS construct to test such effects. They found that in tobacco protoplasts, GUS expression was activated only by P2, while in maize endosperm cells only P1 gave significant activity.

The spatial distribution of transcripts representing each gene family during endosperm development was analyzed by Dolfini et al. (1992) by in situ hybridizations in different wild-type and mutant lines. The zeins are expressed in all endosperm cells (except for the aleurone layer) but each mRNA type accumulates at a different level in the various endosperm regions: there are specific territories of expression for each mRNA within the tissue. The typical expression patterns appear early in development and are maintained during endosperm development. The wild-type specific distribution of the transcripts may be changed by mutations.

The specificity of zein genes is maintained in transgenic systems: Ueda and Messing (1991) detected no transcripts in leaf tissue-derived suspension cultures, whereas zeins were expressed in endosperm cell cultures (although somewhat diminished as compared to developing endosperm tissue) and Quayle et al. (1991) reported that an endosperm derived tissue culture synthesized zein protein and responded to a zein gene regulatory element whereas zein proteins lacked in embryo-derived tissue culture.

5. Trans-Activation of Genes by Regulatory Proteins

While the regulating sequences discussed above are *cis*-acting (acting within the same molecule), *trans*-acting factors are encoded by genes situated elsewhere in the genome. For storage protein genes, a number of protein factors have been identified that bind specifically to conserved regions (boxes) of the *trans*-regulatory regions and may activate transcription:

Kim and Wu (1990) describe a protein-binding sequence in the promoter of a rice glutelin gene (TGAGTCA) which binds two different transcription factors. In maize endosperm nuclei "high mobility group proteins" have been identified that bind specifically to boxes of the 19-kdalton zein gene (Grasser et al. 1990). Heidecker et al. (1991) report that the A30 and A20 subfamilies contain at least three repeats of a consensus sequence ACGAACAAta/gG, but on the other hand, Matzke et al. (1990) describe a promoter region of a Z4 zein gene which lacks the complete 15 bp consensus sequence. The most interesting system, however, is the O2 proteins and their target sites, which have been found only in controlling regions of zein genes of the 22-kdalton class (Schmidt et al. 1992).

Opaque-2 (O2) is a regulatory locus in maize that controls expression of 22-kdalton α-zein genes. This locus produces a regulatory protein synthesized in the cytoplasm and then targeted into the nucleus. The O2 gene protein (O2p) contains a domain that can interact with zein gene promoters (bZIP = basic/leucine zipper domain). Zipper domains are found in all mammalian, fungal, and plant DNA binding proteins of this class. They are thought to act on the DNA like the separator of a zip-fastener.

Although protein targeting is well known from animal and yeast systems, the opaque-2/zein system was the first to be studied in plants (Varagona et al. 1991). O2p is present in maize endosperm nuclei known to produce α-zeins. Neither embryo tissue from wild-type kernels nor endosperm harboring an inactive O2 allele contained the protein. Upon transformation of the O2 gene (equipped with a constitutive promoter) into tobacco, O2p could be detected which was transported into the tobacco nuclei indicating compatibility of nuclear transport systems in monocots and dicots.

More recently, these finding could be corroborated in another plant system. Oleosins are small, hydrophobic, abundant proteins localized in the oil bodies of plant seeds. A maize oleosine gene was transferred into *Brassica napus*. The maize oleosin mRNA and protein were present in the oil bodies of maturing *Brassica* seeds together with *Brassica* oleosin (Lee et al. 1991).

The binding site for O2p in the promoter of the 22-kDa zein genes is similar to other bZIP proteins: it contains an ACGT core that is necessary for DNA binding about –300 bp upstream of the translation start near the prolamin box (Schmidt et al. 1992).

This control mechanism is not only important for cereals but may be generalized: O2p is only one out of a group of endosperm proteins that can bind to the target site. In yeast cells, O2p can activate expression of a lacZ gene the promoter of which contains a multimer of the O2 target sequence (Schmidt et al. 1992). The protein shows autoregulation: it is capable of binding to, and activating, its own promoter (Lohmer et al. 1991). These authors have also shown that the promoter of an abundant maize seed albumin (b-32) is activated in vivo by O2p. Five binding sites (sequence: GATGAPyrimidinePurineTGPurine) reside in a 440-bp promoter fragment, two of them enclosed in copies of the "endosperm box" known from 22-kDa zein promoters.

The system could be further characterized by modifications of both the O2 protein and its target site: Ueda et al. (1992) report that a single nucleotide substitution in the target sequence abolished O2p binding in vitro and transactivation in vivo. On the other hand, slightly different sequence motifs in promoters of zein genes of other molecular weight classes (insensitive to O2 activation) could be "reactivated" by site-directed mutagenesis. Aukerman et al. (1991) described a mutant O2 protein which lost its recognition specificity for zein promoter fragments because an arginine residue was replaced by a lysine. This particular arginine is conserved within the bZIP domains of all mammalian, fungal and plant DNA binding proteins of this class.

6. Conclusions

It is evident that research on gene function has now entered a stage where many results are obtained by specific experimental intervention in regulating mechanisms rather than by mere conclusions from comparative descriptions. Within the complex gene systems controlling growth and differentiation, the genes for storage proteins constitute only a small part. Yet the examples given above show that this part is sufficient as model to obtain an idea of the whole system, demonstrating the existence of conserved mechanisms common for most organisms, as well as of species-specific phenomena.

References

Aukerman MJ, Schmidt RJ, Burr B, Burr FA (1991) Genes Dev 5:310–320.

Bagga S, Sutton D, Kemp JD, Sengupta-Gopalan C (1992) Plant Mol Biol 19:951–958. **Blaich R** (1989) Prog Bot 51:225–236. **Blaich R** (1992) Prog Bot 53:207–223. **Brown J WS,** Felix G (1990) Nucleic Acids Res 18:111–118.

Casey R, Domoney C (1987) Plant Mol Biol Rep 5:261–285. **Chamberland S,** Daigle N, Bernier F (1992) Plant Mol Biol 19:937–945. **Chee PP,** Jones JM, Slightom JL (1991) J Plant Physiol 137:402–408.

Dolfini SF, Landoni M, Tonelli C, Bernard l, Viotti A (1992) Dev Genet 13:264–276.

Ericson ML, Muren E, Gustavsson HO, Josefsson LG, Rask L (1991) Eur J Biochem 197:741–746.

Gatehouse JA, Evans IM, Croy RRD, Boulter D (1986) Philas Trans R Soc Lond Ser B 314:343–353. **Giovinazzo G,** Manzocchi LA, Binachi MW, Corragio I, Viotti A (1992) Plant Mol Biol 19:257–263. **Grasser KD,** Maier UG, Haass MM, Feix G (1990) J Biol Chem 265:4185–4188.

Heidecker G, Messing J (1986) Annu Rev Plant Physiol 37:439–466. **Heidecker G,** Chaudhuri S, Messing J (1991) Genomics 10:719–732.

Kim SY, WU R (1990) Nucleic Acids Res 18:6845–6852. **Kreis M,** Shewry PR, Forde BG, Forde J, Miflin BJ (1985) Oxford Surv Plant Mol Cell Biol 2:253–317.

Lee WS, Tzen JTC, Kridl JC, Radke SE, Huang AHC (1991) Proc Natl Acad Sci USA 88:6181–6185. **Lelievre JM,** Oliveira LO, Nielsen NC (1992) Plant Physiol 98:387–391. **Liu CN,** Rubenstein I (1992) Mol Gen Genet 234:244–253. **Lohmer S,** Maddaloni M, Motto M, Di Fonzo N, Hartings H, Salamini F, Thompson RD (1991) Eur Mol Biol Organ J 10:617–624.

Maier UG, Brown JWS, Toloczyki C, Feix G (1987) EMBO J 6:17–22. **Matzke AJM,** Stoger EM, Schernthaner JP, Matzke MA (1990) Plant Mol Biol 14:323–332.

Philips SEV (1993) Curr Biol 3:112–114.

Quayle TJA, Hetz W, Feix G (1991) Plant Cell Rep 9:544–548.

Schmidt RJ, Ketudat M, Aukerman MJ, Hoschek G (1992) Plant Cell 4:689–700.

Thompson GA, Boston RS, Lyzik LA, Hodges TK, Larkins BA (1990) Plant Mol Biol 15:755–764. **Thompson AJ,** Bown D, Yaish S, Gatehouse JA (1991) Biochem Physiol Pflanz 187:1–12.

Ueda T, Messing J (1991) Theor Appl Genet 82:93–100. **Ueda T,** Waverczak W, Ward K, Sher N, Ketudat M, Schmidt RJ, Messing JRR (1992) Plant Cell 4:701–709.

Varagona MJ, Schmidt RJ, Raikhel NV (1991) Plant Cell 3:105–114.

Professor Dr. Rolf Blaich
Bundesanstalt für Züchtungsforschung
an Kulturpflanzen
Institut für Rebenzüchtung Geilweilerhof
D-76833 Siebeldingen, FRG

V. Extranuclear Inheritance: Plastid Genetics

By Rudolf Hagemann and Monika M. Hagemann

1. Introduction

This chapter connects with our articles in *Progress in Botany* 47, 49, and 51 (1985, 1987, and 1989).

We shall first report on the present status of complete sequencing of plastid genomes and unravelling of the coding capacity of the plastid DNA. Then we characterize the differentiation between plastidal "genetic system genes" and "photosynthesis genes". This leads to the report on plastid DNA deletions in nonphotosynthetic plants. An interesting new finding (during the past 2 years) is the occurrence of RNA editing in plastids, which is dealt with later. Finally, gene transfer into plastids by different methods is reported.

During the past 2 years, several books and review articles have been published which contain relevant results on our topic. In the book *Cell Organelles* (Herrmann 1992) many articles (written by Boynton, Gillham, Hagemann, Harris, Herrmann, Link, Newman, Ohyama, Palmer, Rochaix, Sugiura, Whittier) deal with numerous aspects of genetics and molecular biology of plastids. Also the Symposium Volume *Regulation of Chloroplast Biogenesis* (Argyroudi-Akoyunoglou 1992) contains many papers relevant for plastid genetics. Sugiura (1992) has given a very interesting review on the chloroplast genome.

2. Present Status of Complete Sequencing of Plastid Genomes

The complete sequencing of the plastid DNAs of *Marchantia polymorpha* (Ohyama et al. 1986) and of *Nicotiana tabacum* (Shinozaki et al. 1986) marks an exciting breakthrough in the field of molecular biology and genetics of plastids. With these results, a reference system has been established to which all research groups dealing with plastid DNA sequences can refer (Fig. 1).

In the meantime, the complete nucleotide sequence of the plastid DNA of *Oryza sativa* (Hiratsuka et al. 1989), of *Zea mays* (Kössel et al. 1992), and of *Epifagus virginiana* (Wolfe et al. 1992a, b) was determined. The plastid DNA of the gymnosperme *Pinus thunbergii* is presently being sequenced (Tsudzuki et al. 1992). The sequencing of the plastid genomes of other species is under way. We may predict that the sequencing of the plastid genomes of taxonomically divergent taxa of other lower and higher plants may considerably increase our knowledge about the structure and coding capacity of the pla-

Progress in Botany, Vol. 55

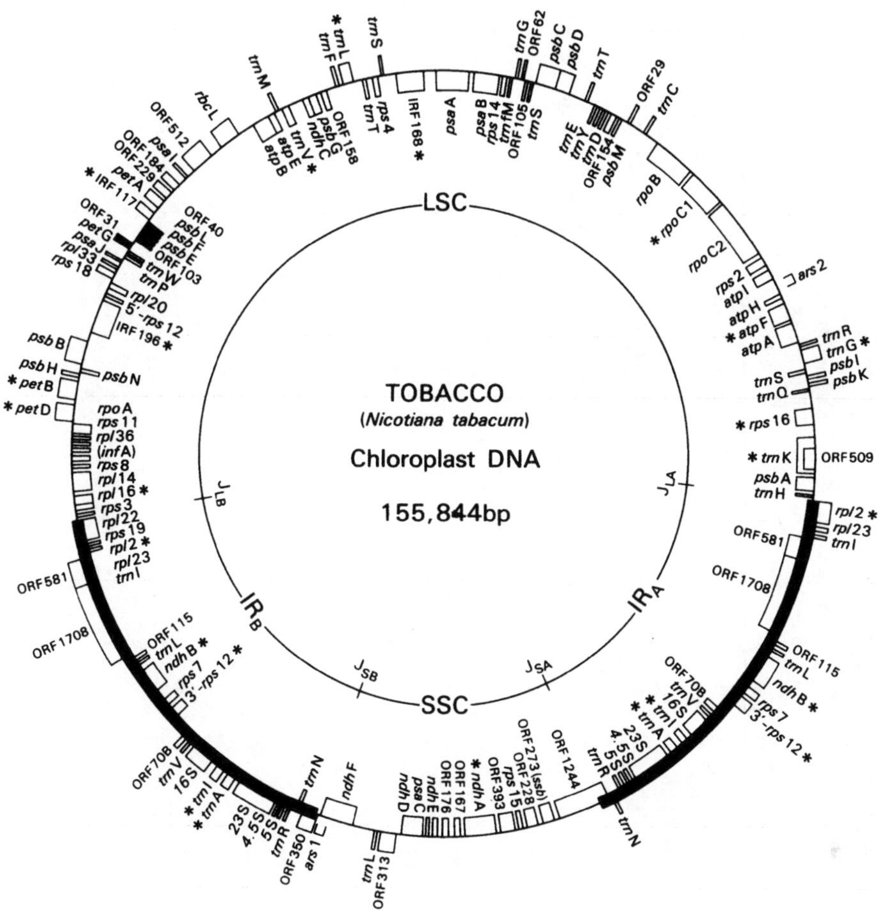

Fig. 1. Physical map of the plastid genome of *Nicotiana tabacum*, drawn to scale. Genes shown on the *outside of the circle* are encoded on the A strand and transcribed counterclockwise. Genes *on the inside* are encoded on the B strand and transcribed clockwise. *Asterisks* denote genes with introns. *LSC* Large single copy region; *SSC* small single copy region; *IR* inverted repeat. (Whittier and Sugiura 1992 and Sugiura, pers. comm. 1992)

stid DNAs and special features of the transcription and translation in plastids and the processing following both steps of gene expression. Moreover, interesting insights into the processes of plastid evolution are to be expected. Table 1 gives an overview on the general characteristics of the five completely sequenced plastid genomes.

A survey of the plastid genes and the polypeptides encoded by these genes is given in Tables 2 and 3 (based on lists published by Shinozaki et al. 1988; Sugiura 1992; Whittier and Sugiura 1992; Hagemann 1993).

Table 1. Characteristics of the completely sequenced plastid genomes

	Marchantia polymorpha	*Nicotiana tabacum*	*Oryza sativa*	*Zea mays*	*Epifagus virginiana*
	Nature 322 (1986) 572–574	EMBO J. 5 (1986) 2043–2049	MGG 217 (1989) 185–194	Igloi, Maier, Kössel (pers. commun. 1992)	PNAS 89 (1992) 10648–10652
Total size	121 024 bp	155 844 bp	134 525 bp	140 377 bp	70 028 bp
LSC	81 095 bp	86 684 bp	80 592 bp	82 346 bp	19 799 bp
SSC	19 813 bp	18 482 bp	12 335 bp	12 535 bp	4 759 bp
IR (2x)	10 058 bp	25 339 bp	20 799 bp	22 748 bp	22 735 bp
tRNA genes	32	30	30 +3 pseudogenes		17 +5 pseudogenes
rRNA	4	4	4		4
Protein genes					
Identified	44	44			21
Predicted	11	11			
Total	55	55	56		21
Unknown ORFs	ca. 30	ca. 30	ca. 35		

Table 2. RNA coding plastid genes

Genes	Gene products
Ribosomal RNAs	
23S rDNA	23S rRNA
16S rDNA	16S rRNA
5S rDNA	5S rRNA
4.5S rDNA	4.5S rRNA
Transfer RNAs	
trnA-UGC[a]	tRNA ^{Ala} (UGC)
trnR-ACG	tRNA ^{Arg} (ACG)
trnR-UCU	tRNA ^{Arg} (UCU)
trnN-GUU	tRNA ^{Asn} (GUU)
trnD-GUC	tRNA ^{Asp} (GUC)
trnC-GCA	tRNA ^{Cys} (GCA)
trnO-UUG	tRNA ^{Gln} (UUG)
trnE-UUC	tRNA ^{Glu} (UUC)
trnG-GCC	tRNA ^{Gly} (GCC)
trnG-UCC[a]	tRNA ^{Gly} (UCC)
trnH-GUG	tRNA ^{His} (GUG)
trnI-GAU[a]	tRNA ^{Ile} (GAU)
trnI-CAU	tRNA ^{Ile} (CAU)
trnL-UAA[a]	tRNA ^{Leu} (UAA)
trnL-CAA	tRNA ^{Leu} (CAA)
trnL-UAG	tRNA ^{Leu} (UAG)
trnK-UUU[a]	tRNA ^{Lys} (UUU)
trnfM-CAU	tRNA ^{fMet} (CAU)
trnM-CAU	tRNA ^{Met} (CAU)
trnF-GAA	tRNA ^{Phe} (GAA)
trnP-UGG	tRNA ^{Pro} (UGG)
trnS-GGA	tRNA ^{Ser} (GGA)
trnS-UGA	tRNA ^{Ser} (UGA)
trnS-GCU	tRNA ^{Ser} (GCU)
trnT-GGU	tRNA ^{Thr} (GGU)
trnT-UGU	tRNA ^{Thr} (UGU)
trnW-CCA	tRNA ^{Trp} (CCA)
trnY-GUA	tRNA ^{Tyr} (GUA)
trnV-GAC	tRNA ^{Val} (GAC)
trnV-UAC[a]	tRNA ^{Val} (UAC)

[a] Genes with introns.

Table 3. Plastid genes for proteins

Gene	Gene product	Transcription	Amino acids (in *Nicotiana*)
Ribosomal proteins 30S subunit			
rps 2	CS 2		236
rps 3	CS 3	Polycistronic	218
rps 4	CS 4		201
rps 7	CS 7	Polycistronic	155
rps 8	CS 8	Polycistronic	134
rps 11	CS 11	Polycistronic	138
rps 12[a]	CS 12	Polycistronic	123
rps 14	CS 14		100
rps 15	CS 15		87
rps 16[a]	CS 16		86
rps 18	CS 18	Polycistronic	101
rps 19	CS 19	Polycistronic	92
50S subunit			
rpl 2[a]	CL 2	Polycistronic	274
rpl 14	CL 14	Polycistronic	123
rpl 16[a]	CL 16	Polycistronic	134
rpl 20	CL 20	Polycistronic	128
rpl 21	CL 21		116 (Marchantia)
rpl 22	CL 22	Polycistronic	155
rpl 23	CL 23	Polycistronic	93
rpl 32	CL 32		55
rpl 33	CL 33	Polycistronic	66
rpl 36	CL 36	Polycistronic	37
RNA polymerase and maturase genes			
rpo A	α-Subunit		337
rpo B	β-Subunit		1070
rpo C1[a]	β'-Subunit		687
rpo C2	β''-Subunit		1392
mat K	Intron maturase (Epifagus)		
Plastidal NADH dehydrogenase			
ndh A[a]	NADH dehydrogenase subunit 1		364
ndh B[a]	Subunit 2		387
ndh C	Subunit 3		120
ndh D	Subunit 4		509
ndh E	Subunit 4L		101
ndh F	Subunit 5		710
ndh G	Subunit 6		176

Table 3 (continued)

Gene	Gene product	Transcription	Amino acids (in *Nicotiana*)
Plastidal NADH dehydrogenase			
ndh I	"psb G" 27-kDa polypeptide		284
ndh H	ORF 393; 49-kDa subunit		393
frx B	ORF 167; 18-kDa Fe-S protein		167
	ORF 158		158
Stromal polypeptides			
rbc L	Large subunit of rubisco	Monocistronic	477
inf A	Initiation factor 1		96
tuf A	Elongation factor EF-Tu (*Chlamydomonas*)		
clp P	ATP-dependent protease		
Photosystem I			
psa A	P700 apoprotein Ia	Polycistronic	750
psa B	P700 apoprotein Ib	Polycistronic	734
psa C	9-kDa Fe-S protein	Polycistronic	81
psa I	I-polypeptide		
psa J	J-polypeptide		
Photosystem II			
psb A	D_1-reaction center polypeptide	Monocistronic	353
psb B	47-kDa chlorophyll apoprotein	Polycistronic	508
psb C	43-kDa chlorophyll apoprotein	Polycistronic	473
psb D	D_2-reaction center polypeptide	Polycistronic	453
psb E	Cytochrome b_{559} α-subunit	Polycistronic	83
psb F	Cytochrome b_{559} β-subunit	Polycistronic	39
psb G	G-protein	Polycistronic	284
psb H	10-kDa phosphoprotein	Polycistronic	73
psb I	4.8-kDa I-polypeptide	Polycistronic	52

Table 3 (continued)

Gene	Gene product	Transcription	Amino acids (in *Nicotiana*)
Photosystem II			
psb K	3.9-kDa K-polypeptide		
psb L	L-polypeptide of PS II		
psb M	M-polypeptide of PS II		
psb N	N-polypeptide of PS II		
Cytochrome b/f complex			
pet A	Cytochrome f	Polycistronic	320
pet B[a]	Cytochrome b_6	Polycistronic	215
pet D[a]	Subunit IV	Polycistronic	160
pet G	Subunit V; ORF 37		
H^+ ATPase			
atp A	α-Subunit	Polycistronic	507
atp B	β-Subunit	Polycistronic	498
atp E	e-Subunit	Polycistronic	133
atp F[a]	Subunit I	Polycistronic	184
atp H	Subunit III	Polycistronic	81
atp I	Subunit IV	Polycistronic	247

[a] Genes with introns.

3. Classification of the Plastid Genes: "Genetic System Genes" and "Photosynthesis Genes"

Shimada and Sugiura (1991) have recently made a detailed comparison of the fine structural features of the three plastid genomes of tobacco, rice and liverwort, and came to interesting general results. The analysis of homology, GC content, and codon usage of the plastid genes allowed their classification into at least two groups, the "photosynthesis genes" and the "genetic system genes".

The group of the photosynthesis genes consists of the plastid genes for the photosystem I and photosystem II thylakoid proteins, for the subunits of the cytochrome b/f and the H^+-ATPase complex and the gene for the large subunit of rubisco.

To the group of the genetic system genes belong the plastid genes for the plastidal rRNAs (23S, 16S, 5S, 4,5S) and tRNAs, and moreover the plastid protein genes for the

Table 4. Different groups of plastid genes

	Genetic system genes		NADH dehydrogenase genes		Photosynthesis genes	
			G+C content of genes			
Rice	G+C	38.3%	G+C	36.5%	G+C	40.0%
Tobacco	G+C	38.0%	G+C	36.3%	G+C	40.1%
Liverwort	G+C	29.5%	G+C	27.0%	G+C	33.3%
			Codon usage in the third position			
Rice	A+U	4599	A+U	2207	A+U	4517
	G+C	2099 = 31.3%	G+C	876 = 28.4%	G+C	2149 = 32.2%
Tobacco	A+U	4430	A+U	2265	A+U	4642
	G+C	1714 = 27.9%	G+C	839 = 27.0%	G+C	2078 = 30.9%
Liverwort	A+U	5617	A+U	2706	A+U	5668
	G+C	666 = 10.6%	G+C	347 = 11.4%	G+C	949 = 14.2%

ribosomal proteins (for the 30S and the 50S subunits), the RNA polymerase subunits, and the initiation factor IF. Obviously, the plastid genes for the plastidal NADH dehydrogenase subunits belong to this group (although their characteristics are given separately in Table 4).

The comparison of the protein coding genes of these two groups reveals interesting differences (Table 4):

The genetic system genes are characterized by a lower GC content of the genes (compared with the photosynthesis genes). The value of the NADH dehydrogenase genes is even lower than that of the other genetic system genes.

Nuclear genes of higher plants very often have G or C in their third codon position, monocots more than 80%. In contrast, plastid genes in general have a high usage of A or U in the third codon position. Moreover, again the genetic system genes have a still higher frequency of A and U in the third codon position (and vice versa a lower GC content) than the photosynthesis genes; this is especially striking for liverwort (Table 4).

These criteria allow one to assign still unidentified open reading frames (ORFs) to one of these two groups; this may indicate the function of the gene products of the ORFs.

The fact that within the plastid genome two distinct groups of genes exist is interesting per se; moreover, it leads to questions about the evolutionary basis of these differences.

4. Plastid DNA Deletions in Nonphotosynthetic Plants

Very interesting results have been obtained in studies on colorless flagellates and nonphotosynthetic parasitic angiosperms. In three taxa the plastid DNA underwent large deletions, which caused the loss of many genes.

The colorless euglenoid flagellate *Astasia longa* has a plastid DNA of only 73 kilobases, compared with 146.4 kb of green *Euglena gracilis* cells. A detailed analysis demonstrated the presence of plastid genes coding for the 16S and 23S ribosomal RNAs (in three tandem repeats and one extra copy of 16S rRNA), for the large subunit of rubisco (rbcL) and elongation factor Tu (tufA). In contrast DNA sequences for psaA, psbA, psbD, psbE, and atpA could not be found (Siemeister and Hachtel 1989, 1990). Interestingly, the genes still present in this truncated plastid DNA are genetic system genes.

Comparable findings resulted with the nonphotosynthetic parasitic plants *Epifagus virginiana* and *Conopholis americana*, two species of the Orobanchaceae.

The nonphotosynthetic root parasite *Epifagus virginiana* has a plastid DNA of only 71 kb, far smaller than any previously characterized land plant plastid genome (dePamphilis and Palmer 1990). The comparison with the tobacco plastid chromosome (156 kb) reveals extensive deletions. The *Epifagus* plastid DNA lost most, if not all, of the photosynthetic genes and several ndh genes, which are coding for constituents of a plastidal respiratory chain. But the plastid genome retained transcribed ribosomal RNA genes and ribosomal protein genes, which are genetic system genes. (Only one of the 22 photosynthetic genes examined, psbA, was detected in *Epifagus* DNA; but it seems to be incorporated into mitochondrial DNA, not present in plastid DNA).

Further studies revealed the presence of seven plastidal tRNA genes (trnD-GUC, trnE-UUC, trnfM-CAU, trnO-UUG, trnS-GCU, trnS-UGA, and trnY-GUA). Surprisingly, the sequences that are completely absent include genes for three subunits of the plastidal RNA polymerase, four tRNAs, and one ribosomal protein (Morden et al. 1991; cf. Wolfe et al. 1992a, b, and Table 1).

Similar results were obtained in studies of *Conopholis americana*; the plastid genome of this species also has large deletions. The *Conopholis* plastid genome has lost numerous photosynthesis genes. In contrast, the 16S and 23S ribosomal genes are transcribed and processed. Interestingly, *Conopholis* lacks the functional plastidal rpo genes and a functional trnL gene, like *Epifagus* (Taylor et al. 1991; Wimpee et al. 1991, 1992).

The preferred loss of photosynthesis genes, but the preservation of genetic system genes in the nonphotosynthetic plants hints at the idea that the genetic system genes of the plastids may fulfill housekeeping functions for the life of the plant cell as a whole, and are therefore retained in the cell beyond the process of photosynthesis.

Both *Epifagus* and *Conopholis* have lost the plastidal genes for three subunits of the plastidal RNA polymerase. These results suggest the hypothesis that – at least in these cases – a nuclear encoded RNA polymerase (and nuclear-encoded tRNAs) are imported into the plastid and enable transcription (and translation) within the plastid.

The findings are not the only indication for such transport processes. Many years ago, we described the nuclear mutant albostrians of barley, *Hordeum vulgare*, which induces plastid mutations and thus leads to the occurrence of pure white and green-white striped seedlings (Hagemann and Scholz 1962). The white plastids lack plastid ribosomes (Börner, Schumann, Hagemann 1976). Recently, Hess et al. (1993) reported that, nevertheless, in these plastid ribosome-deficient mutant cells there is an intense transcription of the plastidal rpo genes (rpoB/C1/C2) and the rps 15 gene. These findings, also, are arguments in favor of the hypothesis that a nonplastid-encoded RNA polymerase is imported into the organelle and functions within the (genetically deficient) plastids.

5. RNA Editing in Plastids

RNA editing is defined as a process that causes nucleotide sequence differences between a mature RNA and its corresponding DNA template. It was first described for kinetoplast (mitochondrial) mRNAs of protozoans (*Trypanosoma, Leishmania, Crithidia*), where uridine insertions and deletions occur.

Many editing events have been described in mitochondria of higher plants; the most frequent change in the process of RNA editing is the transition of C into U (much more seldom is the change of U to C).

Various other types of editing have been found in other organisms: insertions of C into mitochondrial RNAs of the slime mold *Physarum polycephalum*; C to U conversion in the mammalian nuclear mRNA for apolipoprotein B; A to G conversions in the nuclear mRNA for glutamate receptor of the mouse, and pyrimidine conversions in the nuclear selenocysteine tRNA (bovine) and aspartate tRNA (rat) (reviews: Cattaneo 1991; Wissinger et al. 1992; Chan 1993; Gray and Covello 1993; Hajduk et al. 1993).

In most cases, the primary nucleotide sequence of the pre-RNA would lead to a non-functional polypeptide; only the edited secondary RNA sequence codes for the functional protein. The essential experimental procedure is the comparison of the sequence of the genomic DNA with that of the cDNA obtained by reverse transcription of the corresponding mature RNA; the changes caused by RNA editing thus become evident.

Until August 1991, processes of RNA editing had been unknown for plastids; but since that date, several examples of RNA editing by C to U conversions in plastids of higher plants have been elucidated (at first in the laboratory of H. Kössel, Freiburg i. Br.).

The editing events so far reported in plastids, and their positions in the plastid DNA and RNA, respectively, are summarized in Table 5. The processes of RNA editing in plastids are characterized by the following features:

- Editing may create initiation codons for translation as well as occur in several internal codon positions of a mRNA.
- Editing results in restoration of codons for conserved amino acid residues (Maier et al. 1992a).
- So far only one editing event was found in the first codon position (ndhB: CAU to UAU); all other events converted the second codon position (Table 5).
- The possible codon transitions occur in different frequencies (Maier et al. 1992b): high frequency of Ser to Phe, Ser to Leu and Pro to Leu, so far no case of Ala to Val and Thr to Ile (both also not found in plant mitochondria).
- There are genes, the RNA of which is edited (Table 5); but there are other genes whose transcripts – according to our present knowledge – are not edited (e.g., maize rpoA, rpoCl, petD, and rps4; Kössel et al. 1992).
- Transcripts which are edited in one species may contain the "correct" nucleotide already on DNA level in another species; e.g., in contrast to spinach the psbF/L transcript of *Antirrhinum* has no editing site (Kudla et al. 1992; Bock et al. 1993).
- There is no direct correlation between angiosperm plants phylogeny and editing pattern of plastid transcripts: there are differences in editing pattern of the psbF/L transcript between *Oenothera* and *Pelargonium* (both Rosidae) and between *Nicotiana*

Table 5. All editing events and positions so far reported for plastids

Species	Gene	Editing event		Nucleotide position in mRNA	Reference
Zea mays	rpl 2	ACG	→ AUG (Start codon, F-Met)	2	Hoch et al. (1991)
	ndh A	UCG (Ser)	→ UUG (Leu)	49	Maier et al. (1992a)
		UCA (Ser)	→ UUA (Leu)	473	
		UCA (Ser)	→ UUA (Leu)	1586	
		UCC (Ser)	→ UUC (Phe)	2093	
	ndh B	CCA (Pro)	→ CUA (Leu)	467	Maier et al. (1992b)
		CAU (His)	→ UAU (Tyr)	586	
		UCA (Ser)	→ UUA (Leu)	611	
		CCA (Pro)	→ CUA (Leu)	637	
		UCA (Ser)	→ UUA (Leu)	1534	
		CCA (Pro)	→ CUA (Leu)	2184	
	rpo B	UCG (Ser)	→ UUG (Leu)		Kössel et al. (1992)
		UCA (Ser)	→ UUA (Leu)		
		UCG (Ser)	→ UUG (Leu)		
		CCG (Pro)	→ CUG (Leu)		
	pet B	CCA (Pro)	→ CUA (Leu)	1366	Kössel et al. (1992) Freyer et al. (1993)
	IRF 170	UCC (Ser)	→ UUC (Phe)		Kössel et al. (1992)
		ACG (Thr)	→ AUG (Met, internal)		

Nicotiana tabacum	psb L	ACG	→ AUG (Start codon, F-Met)	2	Kudla et al. (1992)
Spinacia oleracea	psb L	ACG	→ AUG (Start codon, F-Met)	2	Bock et al. (1993)
	psb F	UCU (Ser)	→ UUU (Phe)	77	Bock et al. (1993)
Capsicum annuum	psb L	ACG	→ AUG (Start codon, F-Met)	2	Kuntz et al. (1992)
Solanum tuberosum	psb L	ACG	→ AUG (Start codon, F-Met)	2	Bock and Kudla (pers. comm. 1993)
Oenothera hookeri	psb F	UCU (Ser)	→ UUU (Phe)	77	Bock and Kudla (pers. comm. 1993)
Pelargonium zonale (plastome I)	psb L	ACG	→ AUG (Start codon, F-Met)	2	Bock and Kudla (pers. comm. 1993)
	psb F	UCU (Ser)	→ UUU (Phe)	77	Bock and Kudla (pers. comm. 1993)

and *Antirrhinum* (both Scrophulariales, Asteridae), respectively, but consistency between *Spinacia* (Caryophyllidae) and *Pelargonium* (Rosidae) (see Table 5).
- A first example of organ- and tissue-specific differences of RNA editing in plastids has been found (spinach: psbF/L transcript): in roots and seeds both sites are only partially edited; in plastids of all other tissues and developmental stages, editing is obviously complete (Bock et al. 1993).
- RNA editing in plastids seems to be an early step of pre-mRNA processing which takes place independently of splicing and cleavage of the polycistronic into monocistronic mRNAs (Freyer et al. 1993).
- Editing takes place in angiosperms, both dicots and monocots. In contrast, there are no indications for RNA editing in plastids (and mitochondria) of mosses (*Marchantia polymorpha*).
  It occurs in mRNAs of both photosynthesis genes and genetic system genes (there is so far no indication for editing in rRNAs or tRNAs either in mitochondria or in plastids).
- Editing sites of plastids and mitochondria may share common sequence motifs (Maier et al. 1992b).
- It seems likely that editing of plastid RNAs is not as frequent as in mitochondria. Nevertheless, the data obtained in 1992/93, show that editing is a regular process of RNA modification in plastids.

The well-established procedures of plastid transformation (gene transfer) and chloroplast in vitro systems may yield suitable approaches for investigating the mechanism and function of the editing processes in plastids, which are at present still unclear.

6. Direct Gene Transfer into Plastids of Higher Plants

In our review in *Progress in Botany* 51 (1989), we reported on the successful experiments with the particle gun to directly introduce genes into the chloroplast of *Chlamydomonas reinhardtii* (Boynton et al. 1988). In the meantime, these studies have been extended; they yielded interesting new results and have led to models of the recombination events following the introduction of genes into the plastid of *Chlamydomonas* (reviews: Gillham et al. 1991; Boynton et al. 1992).

Soon afterwards, experiments were started to test whether a direct transfer of plastid genes into plastids of higher plants is also possible. In fact, two experimental strategies proved to be successful.

The research group of P. Maliga (Maliga, Hajdukiewicz, Staub, Svab; Rutgers, Piscataway) used the bombardment of leaves or leaf discs with DNA-coated tungsten particles for the transfer of plastid genes into plastids of tobacco (= "biolistic" gene transfer).

In the first series of experiments (Svab et al. 1990), a plasmid vector (pZS148) of the following genetic structure was used: the DNA molecule of the vector contained a 3.7-kb plastid DNA fragment with the 16S rDNA of tobacco. In the 16S rDNA, a spectinomycin resistance mutation is flanked on the 5' side by a streptomycin resistance mutation and on the 3' side by a PstI site generated by ligating an oligonucleotide in the intergenic region. Calli were regenerated from the bombarded leaves. Transgenic plants were selected by spectinomycin and distinguished from spontaneous mutants by the flanking, cotransformed streptomycin resistance and PstI markers. The transgenic lines carried the spectinomycin and the PstI marker; but there was a segregation for streptomycin resistance/sensitivity. The transgenic plastid traits were inherited in an extranuclear mode (uniparental maternal inheritance). Obviously, the transgenic plastid genomes were the product of a multistep process involving DNA recombination, copy correction and sorting-out of plastid DNA copies within the organelle, and sorting-out transformed and nontransformed plastids.

In a following series of experiments, Svab and Maliga (1993) used another transforming plasmid (pZS197) which contains a chimeric aadA gene encoding aminoglycoside 3"-adenyltransferase, which was cloned between rbcL and an open reading frame ORF512. The gene aad4 confers resistance to spectinomycin and streptomycin. After the biolistic transfer of pZS197 DNA into tobacco leaf cells with the particle gun, selection was made for spectinomycin resistance. This transformation and transfer system gave a 100-fold increased frequency of plastid transformation (compared with the experiments of the first series) and approaches the frequency of nuclear gene transformation (Staub and Maliga 1992; Maliga 1993).

Thus, these experiments with the well suited model object tobacco represent a breakthrough for higher plants similar to that in the experiments with *Chlamydomonas*, mentioned earlier.

Another experimental approach has been used by the research group of Koop (Koop, Golds; Munich; in collaboration with Maliga). Koop's group used the same plasmid vector as in Maliga's first experiments (pZS148); but the transformation protocol was based on polyethylene glycol (PEG) treatment of tobacco leaf protoplasts in the presence of transforming DNA. Clones with transformed plastid DNA were selected by spectinomycin resistance encoded by the genetically modified mutant 16S ribosomal RNA gene. Successful plastid gene transfer was confirmed by the two unselected markers, streptomycin resistance and a novel PstI site in the plasmid pZS148 (mentioned above). Also this experimental approach for plastid gene transfer gave positive results.

Thus, so far two experimental strategies (use of the particle gun and PEG treatment) have led to a stable transfer of plastid genes into the plastids of higher plants. Some other experimental procedures have been tested, including *Agrobacterium*-mediated transfer, and opening temporary holes in the cell or protoplast by a UV laser microbeam; but so far there has been no report on stable transformation (cf. Golds et al. 1993; Maliga 1993).

References

Argyroudi-Akoyunoglou JH (ed) (1992) Regulation of chloroplast biogenesis (NATO ASI Series). Plenum, New York.

Bock R, Hagemann R, Kössel H, Kudla J (1993) Mol Gen Genet 240:238–244. **Börner T,** Schumann B, Hagemann R (1976) Biochemical studies on a plastid ribosome-deficient mutant of *Hordeum vulgare.* In: Bücher T, Neupert W, Sebald W, Werner W (eds) Genetics and biogenesis of chloroplasts and mitochondria. Elsevier/North-Holland, Amsterdam, pp 41–48. **Boynton JE,** Gillham NW, Harris EH, Hosler JP, Johnson AM, Jones AR, Randolph-Anderson BL, Robertson D, Klein TM, Shark KB, Sanford JC (1988) Science 240:1534–1538. **Boynton JE,** Gillham NW, Newman SM, Harris EH (1992) Organelle genetics and transformation of *Chlamydomonas*. In: Herrmann RG (ed) Cell organelles. Springer, Vienna New York, pp 3–64.

Cattaneo R (1991) Annu Rev Genet 25:71–88. **Chan L** (1993) BioEssays 15:33–41.

de Pamphilis CW, Palmer JD (1990) Nature 348:337–339.

Freyer R, Hoch B, Neckermann K, Maier RM, Kössel H (1993) Plant J 4:(4) (in press).

Gillham NW, Boynton JE, Harris EH (1991) Transmission of plastid genes. In: Bogorad L, Vasil JK (eds) The molecular biology of plastids. Academic Press, New York, pp 55–92. **Golds T,** Maliga P, Koop H-U (1993) Bio/Technology 11:95–97. **Gray MW,** Covello PS (1993) FASEB J 7:64–71.

Hagemann R (1993) Biol Zentralbl 112:244–287. **Hagemann R,** Scholz F (1962) Züchter (TAG) 32:50–59. **Hajduk SL,** Harris ME, Pollard VW (1993) FASEB J 7:54–63. **Herrmann RG** (ed) (1992) Cell Organelles (Plant Gene Research). Springer, Vienna New York. **Hess WR,** Prombona A, Fieder B, Subramanian AR, Börner Th (1993) EMBO J 12:563–571. **Hiratsuka J,** Shimada H, Whittier R, Ishibashi T, Sakamoto M, Mori M, Kondo C, Honji Y, Sun CR, Meng BY, Li YQ, Kanno A, Nishizawa Y, Hirai A, Shinozaki K, Sugiura M (1989) Mol Gen Genet 217:185–194. **Hoch B,** Maier RM, Appel K, Igloi GL, Kössel H (1991) Nature 353:178–180.

Kössel H, Hoch RM, Maier RM, Igloi GL, Kudla J, Freyer R, Neckermann K, Zeltz P, Ruf St (1992) In: Brennicke A (ed) Plant mitochondria. Berlin-Spandau. HFSP Workshop, Abstract, p 6. **Kudla J,** Igloi GL, Metzlaff M, Hagemann R, Kössel H (1992) EMBO J 11:1099–1103. **Kuntz M,** Camara B, Weil J-H, Schautz R (1992) Plant Mol Biol 6:1185–1188.

Maier R, Hoch B, Zeltz P, Kössel H (1992a) Plant Cell 4:609–616. **Maier RM,** Neckermann K, Hoch B, Akhmedov NB, Kössel H (1992b) Nucleic Acids Res 20:6189–6194. **Maliga P** (1993) Trends Biotechnol 11:101–107. **Morden CW,** Wolfe KH, de Pamphilis CW, Palmer JD (1991) EMBO J 10:3281–3288.

Ohyama K, Fukuzawa H, Kohchi T, Shirai H, Sano T, Sano S, Umesono K, Shiki Y, Takeuchi M, Chang Z, Aota S-I, Inokuchi H, Ozeki H (1986) Nature 322:572–574.

Shimada H, Sugiura M (1991) Nucleic Acids Res 19:983–995. **Shinozaki K,** Ohme M, Tanaka M, Wakasugi T, Hayashida N, Matsubayashi T, Zaita N, Chunwongse J, Obokata J, Yamaguchi-Shinozaki K, Ohto C, Torazawa K, Meng BY, Sugita M, Deno H, Kamogashira T, Yamada K, Kusada J, Takaiwa F, Kato A, Tohdoh N, Shimada H, Sugiura M (1986) EMBO J 5:2043–2049. **Shinozaki K,** Sugiura M (1986) Adv Biophys 21:57–78. **Shinozaki K,** Hayashida N, Sugiura M (1988) Photosynth Res 18:7–31. **Siemeister G,** Hachtel W (1989) Curr Genet 15:435–441. **Siemeister G,** Hachtel W (1990) Plant Mol Biol 14:825–833. **Staub JM,** Maliga P (1992) Plant Cell 4:39–45. **Stuart K** (1991) Annu Rev Microbiol 45:327–344. **Sugiuara M** (1992) Plant Mol Biol 19:149–168. **Svab Z,** Maliga P (1993) Proc Natl Acad Sci USA 90:913–917. **Svab Z,** Hajdukiewicz P, Maliga P (1990) Proc Natl Acad Sci USA 87:8526–8530.

Taylor GW, Wolfe KH, Morden CW, dePamphilis CW, Palmer JF (1991) Curr Genet 20:515–518. **Tsudzuki J,** Nakashima K, Tsudzuki T, Hiratsuka J, Shibata M, Wakasugi T, Sugiura M (1992) Mol Gen Genet 232:206–214.

Whittier RF, Sugiura M (1992) Plastid chromosomes from vascular plants – genes. In: Herrmann RG (ed) Cell organelles. Springer, Vienna New York, pp 164–182. **Wimpee CF,** Wrobel RL, Garvin DK (1991) Plant Mol Biol 17:161–166. **Wimpee CF,** Morgan R, Wrobel RL (1992) Curr Genet 21:417–422. **Wissinger B,** Brenicke A, Schuster W (1992) Trends Genet 8:322–328. **Wolfe KH,** Morden CW, Palmer JD (1992a) Proc Natl Acad Sci USA 89:10648–10652. **Wolfe KH,** Morden CW, Ems SC, Palmer JD (1992b) J Mol Evol 35:304–317.

Professor Dr. Rudolf Hagemann
Institut für Genetik
Fachbereich Biologie
Martin-Luther-Universität
Domplatz 1
O6108 Halle (Saale), FRG

Dr. Monika M. Hagemann
Institut für Humangenetik
und Medizinische Biologie
Medizinische Fakultät
Universitätsplatz 7
06097 Halle (Saale), FRG

VI. Population Genetics

By Jürgen Tomiuk and Klaus Wöhrmann

1. Introduction

During the last decades, the conservation of species, animals as well as plants, has become an issue which is being discussed with increasing intensity. The legislated protection of species has become indispensible as the natural resources have been increasingly used and consumed by the fast-growing human population, resulting in severe changes in the condition of our biosphere. The changes have become most obvious through the increasing worldwide reduction of forests; Thorpe (1978), for example, demonstrated the increase of forest fragmentation in Warwickshire in England from 400 to 1960. In the last years, however, the destruction of forests has gone on, as Curtis (1956) impressively showed for the forests in Wisconsin, USA. Such alarming situations are also observed in Madagascar (Green and Sussman 1990) and in many other parts of our world.

At the time when humans were hunting and collecting their food, the size of the human population was in a state of equilibrium with the environment, i.e., the size of the human population was regulated by the carrying capacity of the existing environment. The carrying capacity of an environment is defined as the possible maximum number of individuals of a species which are able to survive in the respective environment. The carrying capacity can be altered in two different ways: (1) a change in the genotypic structure (in genotypes and their frequencies) can result in a more efficient exploitation of the available resources and therefore in a higher number of individuals, and/or (2) in the case of humans, the carrying capacity has been steadily enlarged by inventions or changes in the way of life.

Organisms other than humans cannot actively increase their carrying capacity. They can only passively adapt to the man-made changes in the environment through altering the genetic structure of their populations.

In Central Europe, the endemic forests and their associated communities have been drastically reduced by human activities to extend the agricultural areas, and the remaining ecosystems are still experiencing this strong influence. In contrast to ancient times, however, the present changes are very radical and fast, which implies that the period of time is too short for the reconstruction of the species' genotypic composition. The lack of time for adaptation and possibly the deficiency of appropriate genetic information, both connected with the reduction of area sizes, is the cause of the increasing rate of extinction of species (Frankel and Soulé 1981; Soulé 1983).

Progress in Botany, Vol. 55

At any time during the evolution of organisms, new species evolve and other species become extinct. On the basis of records on fossil species, it is assumed that 99% of all species which ever evolved on earth have disappeared, and becoming extinct is therefore quite natural. In the very recent decades, however, an increasing number of extinctions can be observed which can never be qualified as "normal". If the destructive activities of mankind proceed in this way, it is assumed that 10–20% of all species will be threatened with extinction in the next few decades. Although less is known about the extinction risk of plants, it is estimated that about 10% of the world flowering plants are endangered.

Living creatures and their genotypes, with few exceptions, are unique. One goal of conservation biology is the identification of the factors which cause loss of diversity on the level of species, genes, and habitats, and to apply this knowledge to stem the loss of diversity or, at least, to mitigate the effects of declining diversity. For achieving this aim, the knowledge of all biological disciplines is necessary, especially those from ecology (the interaction of species with the biotic and abiotic environment), from population dynamics (growth and size of populations), and from population genetics (e.g., genetic variation, mating systems, adaptive and nonadaptive processes). All are traditional fields of research, but only very recently have they been seen in the light of conservation biology.

In particular, conservation biology focuses on the following aims:

1. Areas or parks have to be established for protecting the endangered species.
2. Endangered species should be fostered in zoological or botanical gardens. The purpose of their preservation is the later release of individuals into suitable habitats.
3. The variability of genetic information should be preserved by adequate sampling procedures and storage of organisms or parts of them. The last point holds particularly for plants in seed banks.

This chapter shows to what extent population genetics and population biology contribute decisive arguments to conservation biology.

The importance of genetics for the conservation of species is pointed out by Templeton (1991). The author emphasizes four fields of research in which genetic methods are inevitable to solve the respective problems. (1) The use of methods of molecular genetics (Flavell 1991) and especially DNA "fingerprintings", for example, to identify the geographical source of tusk material as part of an effort to curb poaching of African elephants; to identify the origin of eagle owls (*Bubo bubo*); or to study gene flow between populations (Seitz 1991). (2) The application of molecular genetic methods (determination of DNA structures of enzyme patterns) in systematics for the identification of closely related taxa in order to know what it is we are trying to conserve. (3) The monitoring of hybridization between related species (hybridization has been greatly increased by human activities through introduction of, e.g., exotics). Hybridization might cause genotypic extinction through extensive gene flow of nonadaptive genes into a species. (4) Genetic studies are necessary for both the short-term health and the long-term flexibility of captive populations (see Sect. 4).

2. The Significance of Genetic Variability for the Plasticity of a Population/Species

The polymorphisms of various characters have been studied and the data are available for a large number of different species. Before the use of protein electrophoresis in genetics, morphological, physiological, and chromosomal polymorpisms were considered. The use of protein electrophoresis for identifying genetic variants has paved the way to determining the individual genotypes in large samples (Lewontin and Hubby 1966; Lewontin 1974). The results of such investigations are summarized, e.g., by Brown and Weir (1983), Hamrick (1983), and Nevo (1984). Nowadays, DNA polymorphisms are increasingly used for describing genetic differences among and within populations (Flavell 1991). In general, the hypothesis is confirmed that self-fertilizing organisms show a lower degree of genetic variability than cross-fertilizers, and sexual species are more polymorphic than asexual ones (Nevo 1984; Ledig 1986).

Genetic variability is generated by mutation, enlarged by recombination, and changed by adaptive and nonadaptive processes such as selection, genetic drift, and mating systems. These mechanisms are inherent to the organisms (e.g., breeding system), caused by the environment (e.g., selection), or random (genetic drift). Ignoring random effects and assuming a highly constant environment (and infinite population sizes), the distribution of genotypes/phenotypes of populations is in a state of equilibrium, i.e., there is no change of the genetic constitution of the population from one generation to the next. Under these assumptions, each genotype reproduces a specific number of offspring, but if the environmental conditions change, the reproductive ability of the genotypes can also change, which causes a shift of genotypic frequencies. However, it is of importance that all individuals of the population reproduce a minimum number of offspring to prevent the extinction of the population. The adaptation by the shifting of genotype frequencies is denoted population homeostasis (Lerner 1954), whereas individual homeostasis is applied to describe the physiological adaptation of an individual to the experienced environment. Homeostatic individuals have in general a low plasticity. High plasticity or a "wide" reaction norm indicates a high genotype-environment interaction. In the case of high plasticity when fitness-relevant characters are affected, populations are extremely endangered. This was shown for the freshwater pearl mussel (*Margaritifera margaritifera*; Bauer 1991). Both systems, the adaptation by the shifting of genotype frequencies and homeostasis, are able to guarantee the essential number of progeny. However, they can act only within certain limits, which are given in the first case by the variability of the available genetic information in the population, and in the second by the genetic constitution of the individual itself. All populations exhibit a certain degree of genotypic variability which, in cross-breeding species, is represented by the existence of heterozygotes and homozygotes, whereas in selfing organisms the variability is mostly represented by a multitude of different homozygous lineages. Very simplified, it is thought that genetically polymorphic species (more than one allele at a locus) are potentially better adapted to varying environment than monomorphic ones.

3. Population Size and Genetic Variability

In two different ways, man can alter "nature": (1) the species or populations are deprived of essential resources, e.g., the draining of wet habitats has destroyed the living conditions of frogs, whereby storks are deprived of part of their prey; or by the destruction of specific weeds using herbicides, the resources of the larvae of monophagous insects are lost; (2) the habitat size is reduced, e.g., because of industrial purposes or road construction which decrease the population size. The reduction in the population size can cause two different effects: strong random genetic drift and inbreeding.

a) Random Genetic Drift

A "small population" can be characterized by its number of individuals and the intensity of selection: the smaller the number of individuals, the stronger selection forces must be to compensate drift effects. Consequently, genetic information which would otherwise be maintained by selection can be lost by chance. In such a small population, genes having a negative effect on reproductive ability which would be eliminated by selection in a large population can become randomly fixed.

Many populations consist of a set of subpopulations. The genetic structure of each subpopulation is determined by the strength of selection and genetic drift. In the extreme case, different subpopulations can be fixed for different alleles, but the total population is polymorphic (see also metapopulations). Those population structures can be of importance for the conservation concept of a species.

Assuming that a bottleneck (extreme low population size) lasted in the past for some generations, formulas are given which calculate the relative proportion of the genetic variance that can be maintained in a population with constant effective population size, N_e (Franklin 1980; Falconer 1981; Loeschcke 1988). For example, in a population of $N_e = 2$ individuals, the genetic variance after one generation of bottleneck is still 75% of the total variance, but after 50 generations the total variance is much less than 0.01%. When N_e is larger than 64 individuals, the genetic variance is about 99% after one generation and still 68% after 50 generations (Loeschcke 1988). A surprising amount of genetic variation is maintained in the population and there seems to be no serious danger of a short-term reduction of population size. Additionally, procedures are given which estimate the change of allele numbers at a locus. The number of alleles which remain in the population depends on the number of individuals during the bottleneck, the number of generations of the bottleneck, and on the initial allele frequencies. Again, the changes for the number of alleles are similar to those of the genotypic variability. This kind of consideration is obviously important in the case of species conservation in botanical and zoological gardens (see Sect. 4). Franklin (1980) stated that the population size should not decrease below 50 individuals when conservation is performed for a short period, but it should be larger than 500 individuals for a long-term conservation. For several reasons, these magic numbers regarding viable population sizes may be simplistic, but are still considered to be about the right order of magnitude for application (Gilpin and Soulé 1986). Gilpin and Soulé (1986) offered an analysis of population vulnerability. Their investigations based on earlier papers of the minimum viable population which were provided by MacArthur and Wilson (1967), Richter-Dyn and Goel (1972), and Leigh (1975). Gilpin and Soulé proposed a model which, in the opinion of the authors, is "comprehensive but far from being a universal protocol for the estimation of minimum viable populations sizes (MVPS) in species cases". Reasons for the lack of precision are that the species differ in their vulnerability because of their different life histories including longevity, reproduction mode, and body size. Thomas (1990) discussed what the real population dynamics tells us about MVPS, and Menges (1991) applied the theory to plant populations.

Besides the absolute numbers of individuals which determine density effects, the effective number of individuals, N_e, is of importance for the reproductive ability of a population. The effective population size is given by the number of those individuals which genetically contribute to the next generation. Nonreproductive individuals are not considered. This implies that the number of individuals which are of importance for the survival of the population can be much less than the number of existing individuals. The concept of the effective population size was first introduced by Wright (1931) in order to calibrate the amount of random genetic drift in populations with a nonbinomial (or non-Poisson) distribution of progeny numbers. In conservation biology, the effective population size is obviously an important idea which has been studied in several publications (Franklin 1980; Lande and Barrowclough 1987; Burgman et al. 1993). Lande and Barrowclough (1987) showed that the genetic effective population size is the primary variable of importance for monitoring or managing the genetic variation in a population. Therefore, they suggested methods to measure N_e as well as to monitor the level of genetic variation. For many taxa, it may be easier to maintain genetic variation for quantitative characters than single-locus variation. Therefore, it is argued that every effort should be made to preserve evolutionary important amounts of quantitative genetic variation by maintaining effective population sizes of at least several hundreds of individuals. However, it is stated that populations on the verge of extinction should not be considered hopeless because, if the population regains larger numbers, genetic variation can be regenerated by mutations, thus restoring the necessary genetic variation for their future evolution.

Natural populations of the endangered species *Salvia pratensis* and *Scabiosa columbaria* were studied in the Netherlands. During the last 30 years, the number of populations has considerably decreased in these species (Bijlsma et al. 1991). Using enzyme markers, Bijlsma et al. showed evidence for a significant correlation between population size and the proportion of polymorphic loci and the average number of alleles per locus. Clearly, small populations are less variable than larger ones. The same results were found for some considered morphological traits. Drift effects caused higher differentiation among smaller populations than among large ones. Similar results are reported by Moran and Hopper (1983) for *Eucalyptus caesia* and by Karron (1987) for some other species. Although it is assumed that most allozyme loci are selectively neutral and are therefore mainly influenced by genetic drift whereas morphological characters are of selective significance (Turner et al. 1979; Wolff 1988), both groups of characters show similar correlations with the population size (Allendorf et al. 1982; Lagercrantz and Ryman 1990).

b) Inbreeding

In a population with only few individuals, the number of mating partners is reduced. Consequently, the frequency of mating between relatives, i.e., inbreeding, increases. This results in an increasing amount of homozygotes, which increases the probability that recessive alleles with negative effects are combined. In the beginning, inbreeding decreases the population fitness and is disadvantageous as long as deleterious recessive alleles remain in the population.

In populations which do not usually inbreed, the heterozygotes often exhibit a higher reproductive rate than the homozygotes. The lower reproductive ability of homozygotes is balanced by the higher one of the heterozygotes and, therefore, the population size remains stable and genetic polymorphism can be maintained (Foley 1992; Burgman et al. 1993). Consequently, inbreeding reduces the reproductive ability of the total population.

Soulé (1980) discussed the relation between the fitness and the genetic heterozygosity from the point of view of maintaining the fitness and the evolutionary potential of populations. From data obtained from natural populations, he concluded that heterozygote individuals have greater viability and, in some cases, higher fecundity than homozygotes. Moreover, within the same population/species, even with a small amount of inbreeding, heterozygotes can have higher growth rates and higher survival rates. A 10% increase in genetic homozygosity in the population can reduce the reproduction performance by as much as 25%. Most inbred lines become extinct after 3 to 20 generations. It is a rule of thumb that the allowable rate of inbreeding is 1% and that this corresponds to an effective population size of about 50 individuals. Soulé, however, points out that in the "world of real populations several times this numbers may be needed" and that "this century will see the end of significant evolution of larger plants and terrestrial vertebrates in the tropics. Nature reserves are too small for the necessary isolation required by allopatric species" (Soulé 1980).

There are many examples of the effect of inbreeding and heterosis/superdominance, e.g., in plant breeding. The successful breeding of, e.g., corn (*Zea mays*) is based on heterosis effects. A striking example is also the genetic variation within populations of the cheetah (*Acinonyx jubatus*) in Africa. There are two remaining populations of the cheetah of which the southern one consists of about 1500 individuals and which is almost devoid of genetic variation. Male fertility, juvenile mortality, and susceptibility to diseases are adversely changed in this population. It is assumed that this is due to the lack of variation caused by a bottleneck in the recent history of the population (O'Brien et al. 1985, 1987; O'Brien 1989). Another example is the small perennial shrub (*Isotoma petraea)* where 90% of the ovules fail to develop into seeds because of the expression of recessive lethals. The shrub is restricted to rocky habitats in Western Australia, which causes breeding between relatives, resulting in homozygosity causing reduced fertility, survivorship, and disease resistance (James 1982).

In a contribution on heterozygosity, heterosis, and fitness in outbreeding plants, Ledig (1986) states that the maintenance of heterozygosity by several forms of selection and selection for heterozygosity is a common explanation. However, overdominance is seldom documented, and heterosis is rare in interspecific hybrids or in individuals which are offspring of two individuals both coming from different locally adapted populations. Ledig concluded that "the bulk of evidence suggests that the relation between fitness, or its surrogates, and heterozygosity is the result of a correlation between homozygosity and inbreeding depression. Overdominance probably occurs at some loci, but is relatively less important". In his opinion, extensive pollen flow and a spatially and temporally heterogeneous habitat might provide an explanation for genetic diversity. Furthermore, mutation might offer a better explanation than migration and selection, especially in trees, which have a higher genetic load than short-living plants. His lesson for conservation biology is to maintain larger populations of plants to keep the level of inbreeding low and therefore to keep a high level of heterozygosity for deleterious alleles.

4. Conservation of Genes

It is not always possible that a sufficiently large area for the protection and conservation of species is available. The very last resort are zoological and botanical gardens, where, although in small numbers, endangered species can be managed. This type of conservation has two different goals: (1) to keep the species for demonstration and/or (2) to breed them for a later release into nature. In both cases genetic problems are involved.

a) Conservation and Breeding

In animals, the genetic management of breeding groups is necessary in order to minimize undesired effects. Detailed breeding progams aim at maintaining a maximum of genetic variability within vital and fertile groups. Of course, it is not possible to preserve all endangered species in zoological or botanical gardens. Therefore, it is first necessary to find detached arguments for the choice of which species, species complexes, or even parts of an ecosystem should be preserved for a long time. Consequently, we have to desist from the collection of nice specimens and must focus on a population biological point of view. To date, there are only few studies on these problems. This "agony of choice" was first overcome by the use of the evolutionary distinctiveness of taxa (Vane-Wright et al. 1991). The authors used the topology of the phylogeny of species groups to decide objectively and to preserve a maximum of genetic diversity. The proposed method has been improved by Crozier (1992), who additionally considered the accumulation of genetic divergence along branches of the evolutionary tree.

As already pointed out in Section 3, genetic drift and inbreeding act in finite populations and result in a loss of genetic information which possibly causes reduced vitality and fertility. Such effects can be minimized by pedigree breeding (Wright 1969; Crow and Kimura 1970). The construction of the pedigree requires the knowledge of kinship (relatedness) within the breeding group, which is very often based only on observations of the mating behavior. Nevertheless, Smith (1982) pointed out that observations of the behavior sometimes do not yield reliable information on the relationships in breeding groups. This was also supported by genetic studies on platyrrhine species (Vanderberg et al. 1990).

The construction of pedigrees, however, was improved by the application of molecular genetic techniques. For example, protein electrophoresis, the analysis of mt-DNA and highly variable nuclear genes have been important in documenting the extent of genetic variation (Amos and Hoelzel 1992; Hedrick and Miller 1992). Especially in small populations, where no variation at the enzyme level is observed, DNA analysis can be applied to identify genetic differences. The study of genetic variation in cheetahs is a well-known and convincing example (O'Brien et al. 1983, 1985, 1987).

Furthermore, populations of the same species can differ in their genotypic structure as a result of local adaptation. Such differences can also be detected by protein electrophoresis or, in the case of low variation, by DNA analysis, e.g., the subspecies of the North American *Crytobranchus alleganiensis* (Templeton et al. 1990). The hybridization of individuals from locally adapted populations can produce a vigorous breeding group, but can also cause maladaptation to any area after the release of individuals (Greig 1979).

The analysis of pedigrees is also useful for maintaining a maximum number of founder alleles. This approach is based on the idea that by saving the original genotypic variability, the adaptability of a population is preserved and, therefore, the probability of a later successful reintroduction of the species to its original habitat is high.

Several methods of pedigree analysis can be used, but peeling and gene dropping give the most information for this purpose (Hedrick and Miller 1992). The technique of peeling a pedigree (Cannings et al. 1978) gives exact probabilities for the possible genotypes in the set of current individuals in the pedigree given a particular distribution of founder genotypes. Gene dropping (MacCluer et al. 1986) is a Monte Carlo simulation that applies Mendelian segregation, and calculates which founder allele is randomly passed to each offspring.

b) Conservation by Gene Banks

A special problem of practical importance is the conservation of species related to our cultivated plants. As a consequence of breeding and cultivation of high-performance varieties, which yield a high harvest in a wide range of environments, the number of varieties and of their more primitive local forms is drastically reduced. This situation becomes much more severe as most high-performance varieties are very closely related. Both facts, the wide spread of only few closely related high-performance varieties and the replacement of more primitive forms, can result in a reduction of genetic variation which is then lost for future breeding programs. In order to counter this development, gene banks have been established where wild relatives of our cultigens are stored, usually as seeds.

Conservation collections are only as good as the diversity they contain. Thus the forethought and methods that go into sampling procedures play a critical role. After the decision which species should be collected, the following questions have to be answered: (1) how many populations should be sampled per species? (2) how many individuals should be sampled per population? (3) how many propagules should be collected from each individual? In the answers to these questions, statistical considerations are involved (Brown and Briggs 1991) and, of course, knowledge of the biology, e.g., breeding systems (Karron 1991) and especially the genetics of the desired traits or markers, has to be available. The genetics of quantitative characters is very complex and in general not clearly understood. It is therefore of importance to know to what extent a correlation between the variation in quantitative traits and in enzyme markers exists (Hamrick et al. 1991).

The viability of, e.g., seeds is limited even under optimal conditions during their storage. Consequently, the stored seeds must be regularly cultivated after a certain period, and all of the above-mentioned questions about sample size and the loss of genetic variability are again involved (Roberts 1975; Pernes 1984; I.B.P.G.R. 1985; Ford-Lloyd and Jackson 1986; Brown et al. 1989).

The storage of seeds in gene banks is termed a "static" conservation. The collected sample experiences no evolutionary change besides the loss of information whereas in nature, the evolution of competitors, parasites, and predators continues. In the case of a future release, the stored populations will meet different biotic environments than at the time of their sampling (van Valen 1973). On this basis, Henry et al. (1991) propose a method of "dynamic conservation" which is no strict alternative but rather a complement to gene banks. This method requires the cultivation of the populations for many years at sites of different conditions to use the effect of natural selection. The idea is mostly based on the concept of metapopulations (Gilpin 1987). A metapopulation is a structured population with a patchy distribution. The subpopulations are "small". In contrast to the total population they are homogeneous by drift. The subpopulations can become extinct and the sites can again be colonized. Migration between the subpopulations is antagonistic to drift. Such a system is able to maintain polymorphism without any heterozygote advantage (see above). The problem of environmental stochasticity is discussed by Goodman (1987), who concludes that pratical models must take environmental variation into account and serious attention must be given to field programs of estimation of the relevant parameters of environmental variation and, last but not least, the strategies for population management and reserve design must be re-evaluated in the light of the results of models which incorporate small environmentally driven variation.

A brief history of the concept of metapopulation is given by Hanski and Gilpin (1991). Models of structured populations are developed by Gyllenberg and Hanski (1992), Hastings (1991), Nee (1992), Stacey and Taper (1992), and the understanding of dynamics of metapopulations is discussed by Hanski (1989).

As enzyme loci are in general neutral, and therefore the frequencies are influenced by drift, fitness-correlated characters are mainly considered in metapopulation dynamics. Wheat is the model organism studied by Henry et al. (1991). These authors monitored variation for the protein content of seeds, resistance to powdery mildew, dwarfing, and several other quantitative traits like spike emergency, lodging, grain weight, height, spike features, etc. First results show a significant divergence among the populations within the experimental period. Some alleles have been lost in some populations, but genetic diversity is still present in the total population. There are some similar long-term experiments (about 40 years) which use barley (Kahler et al. 1975; DeSmet et al. 1985).

Some books on conservation biology not cited in this chapter are available (McCabe 1988; Simberloff 1988; Orians 1990; Decker 1991; Saunders and Hobbs 1991; Fielder and Jain 1992; Kessel 1992).

Acknowledgment. We are grateful to Dr. Dave Parker for comments on the manuscript.

References

Allendorf FW, Knudsen KL, Blake GM (1982) Genetics 100:497–504. **Amos B,** Hoelzel A (1992) Biol Conserv 61:133–144.

Bauer G (1991) In: Seitz A, Loeschcke V (eds) Species conservation: a population-biological approach. Birkhäuser, Basel, pp 103–111. **Bijlsma R,** Ouborg NJ, van Teuren R (1991) In: Seitz A, Loeschcke V (eds) Species conservation: a population-biological approach. Birkhäuser, Basel, pp 89–101. **Brown AHD,** Briggs JD (1991) In: Falk DA, Holsinger KE (eds) Genetics and conservation of rare plants. Oxford University Press, Oxford, pp 99–119. **Brown AHD,** Weir BS (1983) In: Tansley SD, Orton TJ (eds) Isozymes in plant genetics and breeding. Elsevier, Amsterdam, pp 219–239. **Brown AHD,** Frankel OH, Marshall DR, Williams JT (1989) The use of plant genetic resources. Cambridge University Press, Cambridge. **Burgman MA,** Ferson S, Akcakaya HR (1993) Risk assessment in conservation biology. Chapman and Hall, London.

Cannings C, Thomson EA, Skolnick (1978) Adv Appl Probab 10:26–61. **Curtis J** (1956) In: Thomas WL (ed) Man's role in changing the face of the earth. University of Chicago Press, Chicago. **Crow JF,** Kimura M (1970) An introduction to population genetics. Harper and Row, New York. **Crozier RH** (1992) Biol Conserv 61:11–15.

Decker DJ (1991) Westview special studies in natural resources and energy management challenges in the conservation of biological resources: a practitioner's guide. Westview, Boulder, Col. **DeSmet GMW,** Scharen A, Hockett EA (1985) Euphytica 34:265–272.

Falconer DS (1981) Introduction to quantitative genetics, 2nd edn. Longman, Harlow. **Fielder PL,** Jain SK (1992) Conservation biology: the theory and practise of nature conservation, preservation, and management. Chapmann and Hall, New York. **Flavell RB** (1991) Biol J Linn Soc 43:73–80. **Foley P** (1992) Evolution 46:763–774. **Ford-Lloyd B,** Jackson M (1986) Plant genetic resources: an introduction of their conservation and use. Edward Arnold, London. **Frankel OH,** Soulé ME (1981) Conservation and evolution. Cambridge University Press, Cambridge. **Franklin IR** (1980) In: Soulé ME, Wilcox BA (eds) Conservation biology. An evolutionary ecological perspective. Sinauer, Sunderland, pp 135–150.

Gilpin ME (1987) In: Soulé ME (ed) Viable populations for conservation. Cambridge University Press, Cambridge, pp 125–139. **Gilpin ME,** Soulé ME (1986) In: Soulé ME (ed) Conservation biology. The science of scarcity and diversity. Sinauer, Sunderland, pp 19–34. **Goodman D** (1987) In: Soulé ME (ed) Viable populations for conservation. Cambridge University Press, Cambridge, pp 11–34. **Green GM,** Sussman RW (1990) Science 248:212–215. **Greig JC** (1979) S Afr J Wildl Res 9:57–78. **Gyllenberg M,** Hanski I (1992) Theor Popul Biol 42:35–61.

Hamrick JL (1983) In: Schonewald CM, Chambers SM, MacBryde B, Thomas L (eds) Genetics and conservation: a reference for managing wild animal and plant populations. Benjamin-Cummings, Menlo Park, pp 335–348. **Hamrick JL,** Godt MJW, Murawski DA, Loveless MD (1991) In: Falk DA, Holsinger KE (eds) Genetics and conservation of rare plants. Oxford University Press, Oxford, pp 75–86. **Hanski I** (1989) Trends Ecol Evol 4:113–114. **Hanski I,** Gilpin M (1991) Biol J Linn Soc 42:3–16. **Hastings A** (1991) Biol J Linn Soc 42:57–72. **Hedrick PW,** Miller PS (1992) Ecol Appl 2:30–46. **Henry JP,** Pontis C, David J, Gouyon PH (1991) In: Seitz A, Loeschcke V (eds) Species conservation: a population-biological approach. Birkhäuser, Basel, pp 185–198.

I.B.P.G.R. (1985) Cost-effective long-term seed stores. International board for plant genetic resources. Rome.

James SH (1982) In: Groves RH, Ride WDL (eds) Species at risk: research in Australia. Australian Academy of Science, Canberra, pp 63–71.

Kahler AL, Clegg MT, Allard RW (1975) Proc Natl Acad Sci (USA) 72:943–946. **Karron JD** (1987) Evol Ecol 1:47–58. **Karron JD** (1991) In: Falk DA, Holsinger KE (eds) Genetics and conservation of rare plants. Oxford University Press, Oxford, pp 87–98. **Kessel RV** (1992) Monogr Biol 0:69–90.

Lagercrantz U, Ryman N (1990) Evolution 44:38–53. **Lande R,** Barrowclough GF (1987) In: Soulé M (ed) Viable populations for conservation. Cambridge University Press, Cambridge, pp 86–123. **Ledig FT** (1986) In: Soulé ME (ed) Conservation biology. The science of scarcity and diversity. Sinauer, Sunderland, pp 77–104. **Leigh EG** (1975) In: Cody ML, Diamond JM (eds)

Ecology and evolution of communities. Harvard University Press, Cambridge, pp 51–73. **Lerner IM** (1954) Genetic homeostasis. Oliver and Boyd, Edinburgh. **Lewontin RC** (1974) The genetic basis of evolutionary change. Columbia University Press, New York. **Lewontin RC,** Hubby JL (1966) Genetics 54:595–609. **Loeschcke V** (1988) Naturwiss Rundsch 41:310–314.

MacArthur RH, Wilson EO (1967) The theory of island biogeography. Princeton University Press, Princeton. **MacCluer JW,** Vanderberg JL, Read B, Ryder OA (1986) Zool Biol 5:148–160. **McCabe RE** (1988) Transactions of the fifty-third North American wildlife and natural resources conference, No. 53, Conference on new approaches in managing natural resources. Louisville, Kentucky. **Menges ES** (1991) In: Falk DA, Holsinger KE (eds) Genetics and conservation of rare plants. Oxford University Press, Oxford, pp 45–61. **Moran GF,** Hopper SD (1983) Aust J Bot 31:161–172.

Nee S (1992) J Anim Ecol 61:37–40. **Nevo E** (1984) Lect Notes Biomath 53:13–65.

O'Brien SJ (1989) Reply. Trends Ecol Evol 4:178. **O'Brien SJ,** Wildt DE, Goldman D, Merril CR, Bush M (1983) Science 221:459–462. **O'Brien SJ,** Roelke ME, Marker L, Newman A, Winkler CA, Meltzer D, Colly L, Everman JF, Bush M, Wildt DE (1985) Science 227:1428–1434. **O'Brien SJ,** Wildt DE, Bush M, Cavo TM, Fitzgibbon C, Aggundey I, Leakey RE (1987) Proc Natl Acad Sci (USA) 84:508–545. **Orians GM** (1990) The preservation of biological resources. Workshop Seattle Washington. University of Washington Press, Seattle.

Pernes J (1984) Gestion des ressources génétiques des plantes. Tome 2: Manuel. Agence de Coopération culturelle et technique, Paris.

Richter-Dyn N, Goel NS (1972) Theor Popul Biol 3:406–433. **Roberts EH** (1975) In: Frankel OH, Hawkes JG (eds) Crop genetic resources for today and tomorrow. Cambridge University Press, Cambridge, pp 269–295.

Saunders DA, Hobbs RJ (1991) Nature conservation 2: the role of corridors. Surrey Beatty and Sons, Chipping Norton, New South Wales. **Seitz A** (1991) In: Seitz A, Loeschcke V (eds) Species conservation: a population-biological approach. Birkhäuser, Basel, pp 1–13. **Simberloff D** (1988) Annu Rev Ecol Syst 19:473–512. **Smith DG** (1982) Lab Anim Sci 32:540–545. **Soulé ME** (1980) In: Soulé ME, Wilcox BA (eds) Conservation biology. An evolutionary ecological perspective. Sinauer, Sunderland, pp 111–151. **Soulé ME** (1983) In: Schonewald-Cox CM, Chambers SM, McBryde B, Thomas WL (eds) Genetics and conservation: A reference for managing wild animal and plant populations. Benjamin Cummings, Menlo Park, pp 115–132. **Stacey PB,** Taper M (1992) Ecol Appl 2:18–29.

Templeton AR (1991) In: Seitz A, Loeschcke V (eds) Species conservation: a population-biological approach. Birkhäuser, Basel, pp 15–29. **Templeton AR,** Shaw K, Routman E, Davis SK (1990) Ann MO Bot Gard 77:13–27. **Thomas CD** (1990) Conserv Biol 4:324–327. **Thorpe H** (1978) In: Hawkes JG (ed) Agriculture and conservation. Duckworth, London, pp 52–86. **Turner JRG,** Johnson MS, Eanes WF (1979) Proc Natl Acad Sci (USA) 76:1924–1928.

Vanderberg JL, Aivaliotis MJ, Williams LE, Abee CR (1990) Biochem Genet 28:41–56. **Vane-Wright RI,** Humphries CJ, Williams PH (1991) Biol Conserv 55:235–254. **van Valen L** (1973) Evol Theor 1:1–30.

Wolff K (1988) Natural selection in *Plantago* species: a genetical analysis of ecologically relevant morphological variability. PhD Thesis, University of Groningen. **Wright S** (1931) Genetics 16:97–159. **Wright S** (1969) Evolution and the genetics of populations, vol 2. The theory of gene frequencies. University of Chicago Press, Chicago.

PD Dr. Jürgen Tomiuk
Abteilung für Klinische Genetik
Universität Tübingen
Wilhelmstrasse 27
D-72074 Tübingen, FRG

Professor Dr. Klaus Wöhrmann
Biologisches Institut
Universität Tübingen
Auf der Morgenstelle 28
D-72076 Tübingen, FRG

D. Taxonomy

I. Lichenized and Lichenicolous Fungi 1991–1992

By H. J. M. Sipman

1. Introduction

In 1991–1992 considerable progress was made in the classification of the lichen diversity on the earth. At least 28 genera and over 330 species, or ca. 2% of the presently known number, were newly described. To illustrate the significance of this figure, the number of new species described per year for the African phanerogam flora (ca. 30000–40000 species) was 133 in 1976 and gradually declining (Léonard 1979).

The progress is strongly dependent on three developments of the last decades. (1) The consequent application of TLC methods for chemical analysis of secondary products. TLC data often enable a first segregation of superficially similar specimens, and then it is easier to look for less apparent characters to support the divisions. (2) The use of iodine staining for investigation of the ascus apex structure. (3) The intensified investigation of structures for vegetative reproduction, where, e.g., schizidia, pustules, and polysidiangia have been added to the two traditional categories, isidia and soralia (Fig. 1).

The new taxa were not discovered mainly in the tropics, as one might expect. Slightly more than one-third of the new species was detected in the northern temperate to subtropical zones, the best investigated part of the earth. About one third is from the tropics and slightly less than one third from the cooler southern areas. A surprisingly high number was detected during an investigation of sorediate and isidiate, corticolous crustose lichens in Norway (Tønsberg 1992). Of the 128 species present, 19 appeared to be undescribed, and another 13 unrecorded for Europe, Scandinavia, or Norway. This type of lichen is usually lacking sexual organs and is thus difficult to classify. Consequently, it has been neglected for a long time.

2. Character Investigation

a) Morphology, Anatomy, and Ontogeny of the Mycobiont

Ascus structures are now widely accepted as a base for generic delimitations, and are treated in many taxonomic papers. Rambold and Triebel (1992) present a survey of types in Lecanorales (Fig. 2), in which they draw consequences for the classification. More

Progress in Botany, Vol. 55

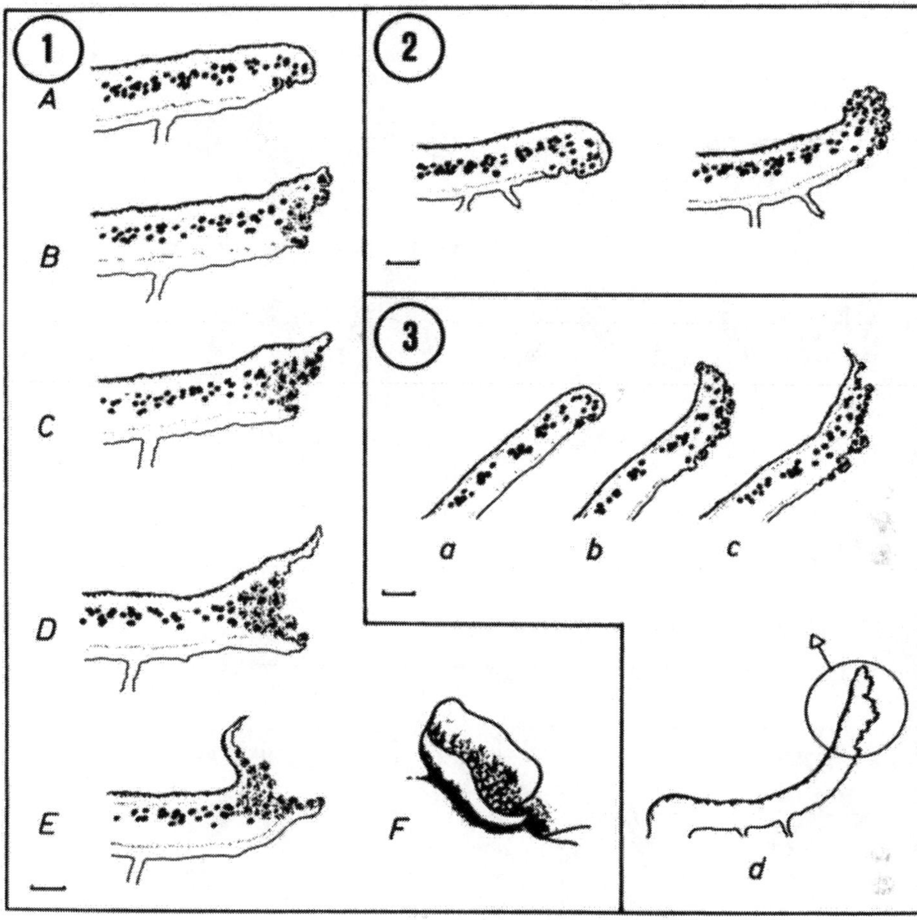

Fig. 1. Different ways of vegetative diaspore production in *Xanthoria fallax* agg. (all except 1F thallus transverse sections) Bar = 0.1 mm. (Poelt and Petutschnig 1992)

subtle differences promise to bring a better subdivision of cetrarioid Parmeliaceae (Kärnefelt et al. 1992).

A different type of paraphyses with inflated, oil(?)-containing cells, called oil paraphyses, is reported by Giralt et al. (1992) in *Caloplaca aegatica* and *C. alnetorum* (*C. flavorubescens* agg.).

Apothecium ontogeny was studied in *Bacidia bagliettoana* by Lumbsch (1991). The development appeared to be gymnocarpic, the paraphysoids becoming replaced by true paraphyses. Henssen (1992a) describes apothecial development in *Omphalodium* and *Omphalora*, confirming their affinities with Parmeliaceae. *Lecanora opiniconensis* is shown to be similar to Parmeliaceae as to apothecium development (Lumbsch et al. 1991).

SEM investigation of the thallus surface of *Coccocarpia* and Pannariaceae was done by Lumbsch and Kothe (1992). They found a similar pored epicortex in *Coccocarpia* and

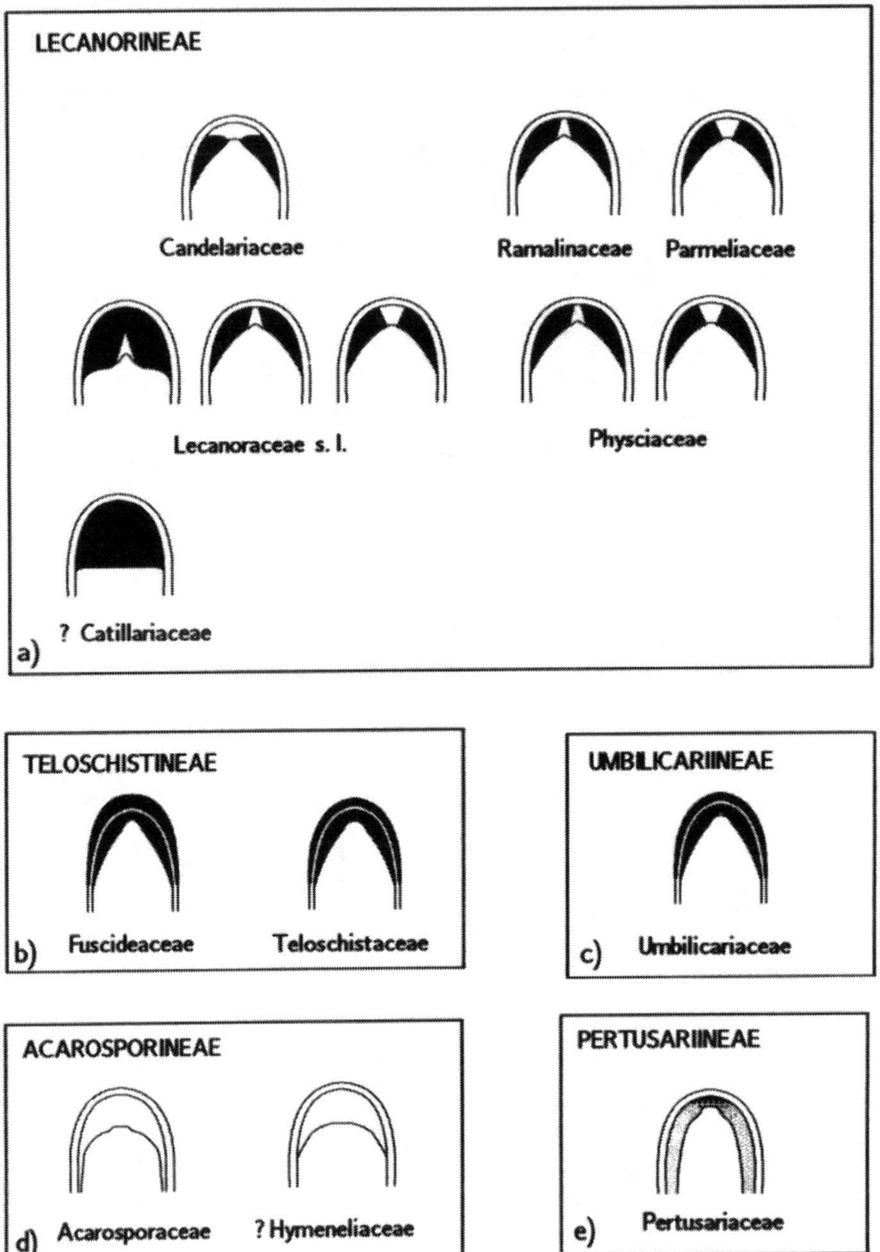

Fig. 2 a–e. A proposed classification of Lecanorales s.l., with pertinent ascus types. (Rambold and Triebel 1992)

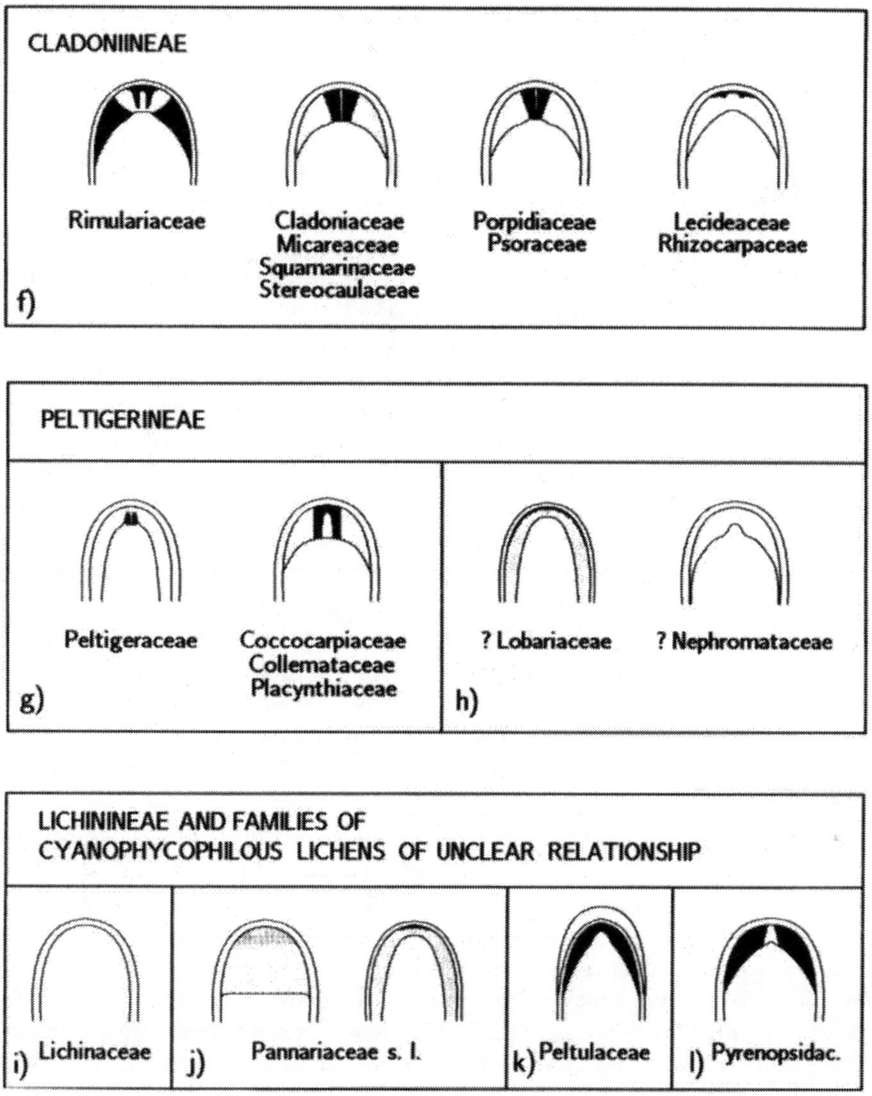

Degelia. The epicortex of *Psiloparmelia* appeared to be only rudimentarily pored (Lumbsch et al. 1992a). TEM studies were done of spore ontogeny in *Sphaerophorus* (Wedin 1991), of the three-dimensional ultrastructure of the bionts in *Lasallia hispanica* and *Umbilicaria spodochroa* (Valladares and Ascaso 1992), and of the thalloconidia in eight species of *Umbilicaria* (Ascaso et al. 1992). Honegger and Bartnicki-Garcia (1991) analyzed the cell wall structure and composition of three lichenized ascomycetes and concluded that they are similar to nonlichenized euascomycetes.

The notorious plasticity of lichen thalli was studied by Sanders (1992), who analyzed the morphological variation in the perforated thallus laciniae of *Ramalina menziesii* and found that the thallus form is to some extent correlated with environmental factors such as wind exposure. Rosentreter and McCune (1992) report that vagant thalli of *Dermatocarpon* in North America belong to two, not one, species. Apparently, adaptation to a vagant existence is of limited taxonomic value.

b) Chemotaxonomy

"Lichen substances" turn out not to be a speciality of the lichen symbiosis, but a product of the fungal component. Culberson and Armaleo (1992) observed that the isolated mycobiont of *Cladonia grayi*, cultivated on a solid medium, produced grayanic acid and the same related substances as found in the natural lichen. Alga-free cultures grown from ascospores of the *Ramalina siliquosa* complex appeared to produce the same characteristic secondary products as the intact thalli (Culberson et al. 1992). However, their chemistry was more complex: in addition to the usual set of substances, biogenetically explainable precursors were found.

Numerous new secondary substances with taxonomic significance were isolated and structurally elucidated. Notable progress was made with xanthones, investigated by, e.g., Elix et al. (1991a, b), Elix and Crook (1992). Leuckert and Knoph (1992) present a standardized method to identify the xanthone chemosyndromes of European *Lecidella*. An improved computer program for the identification of lichen substances, wintabolites, was developed by Mietzsch et al. (1992). Common (1991) developed a simple test for lichenan; it appeared to be widespread in Parmeliaceae. Czeczuga continued his analyses of carotenoids in lichens, without finding clear correlations with taxonomy or distribution (Czeczuga et al. 1991).

The localization of lichen substances in the thallus was investigated by Modeseni and Lajolo (1992: gyrophoric acid in *Umbilicaria rufidula*) and Modeseni and Debarbieri (1991: lobaric acid in *Parmelia saxatilis*; determined by fluorescence and SEM). The biogenesis of lichen substances, in this case lecanoric acid and related substances, in lichen thalli was investigated by Yoshimura and Kurokawa (1991). Clayden (1992) presents chemical evidence that usnic acid-deficient populations of *Arctoparmelia centrifuga* in North America developed independently from European ones. Manrique and López (1991) found different concentrations of lichen substances in different parts of *Pseudevernia furfuracea* thalli.

c) Molecular Biology

Attempts to derive phylogenetic information from the structure of macromolecules are also made with lichens. Efforts are mainly directed at the selection of phylogenetically meaningful substances. Hageman and Fahselt (1992a) used enzyme electromorphic data to trace taxonomic relationships in Umbilicariaceae. None of the investigated species appeared significantly different from the others, and *Lasallia* not particulary different from *Umbilicaria*. An isoenzyme analysis of *Umbilicaria mammulata* (Hageman and Fahselt 1992b) showed that populations at close distance may or may not be more similar than more distant ones. Mattson (1991) analyzed protein content and phosphatase activity in thallus extracts of five *Cetraria* species and found rather large differences within populations. *C. pinastri* showed remarkable differences between its European and North America populations. Apparently, European *C. pinastri* is genetically more distinct from the North American populations than morphology suggests. Environmentally induced enzymatic differences were found by Fahselt (1992). She detected differences in esterase and alkaline phosphatase production between samples of *Cladonia mitis* exposed to and sheltered from geothermal fumes in Japan.

Armaleo and Clerc (1991) analyzed DNA extracts fo find out whether the phycotypes involved in lichen chimeras have really the same mycobiont. They found a great similarity between the mycobionts of both thallus types of *Pseudocyphellaria murrayi/rufovirescens* and *Sticta dufourii/canariensis*, the significance of the differences remaining unclear.

d) Biology

Hestmark (1991a, b) discusses the role of thalloconidia in the life cycle of *Umbilicaria* species. Armstrong (1992) investigated the dispersal of soredia of *Hypogymnia physodes*. Sanders and Rico (1992) discovered that the rhizomorphs of a squamulose lichen (*Aspicilia crespiana*, ined.) which form a network linking the squamules, capture algae and form new squamules, and are thus comparable to a prothallus.

e) Phylogeny

The application of cladistic analyses in lichens faces considerable difficulties. The number of characters available for evaluation is mostly very small. Therefore many alternative, ± equally parsimonious cladograms can be expected. The suprageneric taxonomy of pyrenocarps was investigated by Aptroot (1991a, c). Kärnefelt and Thell (1992) and Kärnefelt et al. (1992) present cladograms for groups of Parmeliaceae; the latter conclude that some of the present generic subdivisions, based mainly on morphology, cannot be maintained.

3. Classification

a) Lichenized Ascomycetes

An updated and improved list of accepted orders, families, and genera of ascomycetes, including lichenized taxa, is presented by Eriksson and Hawksworth (1991).

α) Pyrenocarpous Orders

A revision of the taxa included in *Leptorhaphis* (Aguirre-Hudson 1991) showed them to belong to six genera in four different orders: *Leptorhaphis*, *Arthopyrenia* (Dothideales); *Celothelium, Rhaphidicyrtis* (Pyrenulales?); *Sarcopyrenia* (Verrucariales?); and the new genus *Cresporhaphis* (Trichosphaeriales). Aptroot (1991a) proposed a new arrangement for the Melanommatales, with 11 new genera: *Clypeopyrenis, Distopyrenis, Mazaediothecium, Pyrenowilmsia* (Pyrenulaceae); *Granulopyrenis, Lacrymospora, Parapyrenis, Pyrenographa* (Requienellaceae); *Architrypethelium, Megalotremis, Ornatopyrenis* (Trypetheliaceae); the genus *Lithothelium* is much expanded. The question whether the perithecia found on *Normandina pulchella* squamules belong to the squamules or to a different, lichenicolous fungus, is answered by Aptroot (1991b) in favor of the first hypothesis. An unusual species of *Monoblastia*, deviating by immersed perithecia with noncarbonized wall, is described by Aptroot (1991e). Orange (1991) presents a key for *Thelidium* in NW Europe, accepting only 12 species.

β) Discocarpous Orders

Lecanorales, Parmeliaceae (incl. *Alectoriaceae, Hypogymniaceae*) Elix and Nash (1992) present a synopsis of *Psiloparmelia* (12 species in the mountains of South America and southern Africa). Kurokawa (1991b) established the new genus *Rimeliella* for the *Parmelia subcaperata* group (eight species). Henssen described two subcrustose Parmeliaceae genera from Patagonia: *Omphalodiella* (Henssen 1991) and *Placoparmelia* (Henssen 1992b). A new genus, *Allocetraria*, is described by Kurokawa and Lai (1991) to accommodate a few East Asiatic *Cetraria* species; Randlane and Saag (1992b) expanded it with among others the widespread *C. cucullata* and *C. nivalis*. A synopsis of *Nephromopsis*, with chemical data for all species, is presented by Randlane and Saag (1992a). The new genus *Coelopogon* is proposed, among others, for the southern temperate *C. epiphorellum* (Brusse and Kärnefelt 1991).

Physciaceae. Two crustose lichens with *Physcia*-type spores from coastal California and Baja California are placed in an new genus, *Mobergia* (Mayrhofer et al. 1992). Trass (1992) presents a list of all *Heterodermia* species, with principal characters, distribution, and a key. Sheard (1992) treats the genus *Hafellia* in North America, recognizing five species; with notes on *H. callispora* and related species.

Teloschistaceae. A detailed analysis of the soredium production provided criteria for a revised taxonomy of the *Xanthoria candelaria* group (Poelt and Petutschnig 1992; cf. Fig. 1). Anomalous spores and oil paraphyses in *Caloplaca* are reported by Navarro-Rosines and Hladun (1992) and Giralt et al. (1992), respectively.

Other Families of Lecanorales s.l. The first accounts of Australian Cladoniaceae, Collemataceae, and Pannariaceae were published in George (1992). Hafellner and Casares-Porcel (1992) investigated the ascus structure in *Acarospora* and *Biatorella* and decided that the latter belongs to a different family, for which the name Biatorellaceae is proposed. Timdal (1991a) describes the new genus *Anamylopsora* (Trapeliaceae) for a widespread species of arid areas on the northern hemisphere. Henssen et al. (1992) describe *Austropeltum* (Stereocaulaceae?) from Tasmania and New Zealand. Poelt and

Obermayer (1991) present a treatment of *Bryonora* (Lecanoraceae) in the Himalaya, with a revision of all 11 species known worldwide. Hafellner (1991c) proposes *Lobothallia* as separate genus for the *"Lecanora radiosa"* group. Moreno and Egea investigated the family Lichinaceae, especially in the western Mediterranean: they give a detailed treatment of morphology, anatomy, and ecology (Moreno and Egea 1991), a description of *Digitothyrea* gen. nov. (Moreno and Egea 1992a), a survey of the complex *Anema-Thyrea-Peccania* (16 species) with a list of name changes for *Thyrea* s.l. (Moreno and Egea 1992b), and a redefinition of *Lichinella* (expanded with 5 *Gonohymenia* species, Moreno and Egea 1992c). The North American corticolous representatives of *Ochrolechia* were revised by Brodo (1991; 19 species, 6 new). A key to Australian Pertusariae, including numerous new species, was presented by Archer (1991). Brako (1991) monographed the neotropical species of *Phyllopsora*. Galloway (1992a) revised 53 species of *Pseudocyphellaria* in southern S. America. Feuerer (1991) revised European *Rhizocarpon* with nonyellow thallus and pluricellular spores (18 species). The genus *Thamnochrolechia* is described from Papua New Guinea by Aptroot and Sipman (1991). Hafellner and Vezda (1992) propose the genus *Tibellia* (Bacidiaceae) for a byssoid lichen from Australia and present a key to all known lichens with byssoid thallus. Timdal (1991b) presents a worldwide revision of *Toninia*, accepting 48 species, including a number of lichenicolous and crustose taxa previously placed in *Kiliasia*, *Lobiona* and *Lecidea*. Posner et al. (1992) present the aromatic compounds for 33 *Umbilicaria* species.

γ) Other Discocarpous Orders

Thor (1991) shows that the well-known lichen which has been known as *Herpethallon sanguineum* or *Chiodecton sanguineum* most likely belongs in *Cryptothecia*. Sérusiaux (1992) reintroduces the genus *Eremothecella* for a group of *Arthoniae* with exceptionally long conidia (up to 450 x 2–3 μm). Tehler (1992) investigated the delimitation of *Lecanactis* and transferred eight taxa from *Schismatomma*. Torrente and Egea (1991) describe the new genus *Llimonaea* from coastal rocks in Morocco and the Iberian Peninsula, intermediate between Opegraphaceae and Roccellaceae. Vezda and Poelt (1991) present a survey with key of *Gyalidea*. Kantvilas et al. (1992) describe the new genus *Siphulella* (Baeomycetaceae) from Tasmania.

b) Incertae Sedis

Eriksson (1992) describes the new genus *Flakea*, widespread in the tropics, somewhat similar to *Psoroglaena* in thallus structure; ascomata unknown. Chemical investigation leads to an improved classification of *Lepraria* species (Kümmerling et al. 1991; Leuckert and Kümmerling 1991; Laundon 1992).

c) Lichenicolous Fungi

Rambold and Triebel (1992) present an account of the lichenicolous lichenized and non-lichenized fungi, especially those belonging to the order Lecanorales in a redefined, wide sense; introductory chapters present information on various aspects, such as taxonomy of hosts, ecology, and chemistry. Hawksworth (1991) includes *Charcotia* in *Arthonia* and treats two lichenicolous species of this genus, on *Umbilicaria*. Harada (1991) considers *Heterocarpon ochroleucum* (Verrucariales) as a lichenicolous fungus.

4. Floristics

a) General

The floristic literature, so important to document the diversity of the lichen flora and the distribution of the taxa, largely exceeds that on other aspects of lichens. Here, only a very small selection of titles is presented, with apologies to the investigators whose enormous efforts are thus much undervalued.

Several important floras and comparable publications came out. Harris (1990) produced keys for selected groups (ca. 450 species) of Florida lichens. Awasthi (1991) published keys to most microlichens (1150 taxa in 163 genera) of India, Nepal, and Sri Lanka (not keyed: *Graphina, Helminthocarpon, Lecanora, Lecidea*). The first lichen volume of the *Flora of Australia* (George 1992) appeared, with keys to some 230 species, mainly Cladoniaceae, Collemataceae, Hypogymniaceae, and Pannariaceae. Purvis et al. (1992) produced an up to date flora for all ca. 1700 species known from the British Isles.

b) The Northern Extratropics

α) Europe

Larger floristic reports include: Hafellner (1991a), rare or otherwise interesting lichens from Carinthia, Austria; Schreiner and Hafellner (1992), sorediate crustose epiphytes from the eastern Alps; Diederich et al. (1991), lichens and lichenicolous fungi from Belgium and surroundings; Ekman et al. (1991), 317 taxa from Estonia (74 new for the country); Houmeau and Roux (1991), 322 taxa from the Pyrenean Mts.; Hauck (1992), species list for Niedersachsen, Germany; Litterski (1992a), lichens from Rügen, Germany; Kümmerling (1991), 249 species from the Hohe Meissner and surroundings, Germany; John (1991), lichen atlas of Rheinland-Pfalz, Germany; Jacobsen (1992), survey of lichens in Schleswig-Holstein and Hamburg, Germany; Gilbert and Coppins (1992), 322 lichens from eastern Scotland; Christensen et al. (1991), 43 taxa from Attica, Greece; Vitikainen (1991), 428 species from SW. Finland; Aptroot and Alstrup (1991), 119 species from Edgeøya, Spitzbergen; Thor (1992), 558 species from Åsele Lappmark, Sweden; Egea and Llimona (1991), selected mediterranean silicicolous lichens with their mediterranean distribution. Regional treatments: *Lecania* in the Benelux (van den Boom 1992); key to *Caloplaca* in Italy (Nimis 1992); Cladoniaceae of Spain (Burgaz and Ahti 1992).

β) North America

Changes to the *Fifth Checklist of Lichen-Forming etc. Fungi of the Continental United States and Canada* were published by Egan (1991). Lager floristic reports include: St. Clair et al. (1991), 318 species from Utah; Thomson and Weber (1992), 372 species from the Canadian Arctic; Goward and Ahti (1992), 293 macrolichens from British Columbia; Wong and Brodo (1992), 482 taxa from Southern Ontario, Canada, with some keys; Triebel et al. (1991), 58 lichenicolous fungi; Talbot et al. (1992), 217 lichens from Tuxedni Wilderness Area, Alaska. The following generic revisions appeared: *Hafellia* (Sheard 1992); corticolous *Ochrolechia* (Brodo 1991); *Staurothele* (Thomson 1991).

γ) Other Areas

Wei (1991) presents a checklist of 1766 species in 232 genera reported in the literature from China, with indication of distribution, habitat, and Chinese names; Ahti (1991) treats Chinese *Cladoniae*. Poelt and Obermayer (1991) deal with *Bryonora* from the Himalaya, Poelt and Petutschnig (1992) with *Xanthoria* and *Teloschistes*, and Poelt and Grube (1992) with *Protoparmelia*. Kurokawa (1991a) lists Japanese Parmeliaceae. Kalb and Hafellner (1992) report 159 lichen species from Madeira, and Hafellner (1992) presents a checklist. Huneck et al. (1992) report 120 species from Mongolia, Zhurbenko (1992) 141 taxa from W. Taymir, Siberia. John (1992a, b) presents a bibliography and reports 20 species from Turkey.

c) The Tropics

General. A Bundle of presentations from the Tropical Lichens symposium held in London 1990 was published (Galloway 1991a), with floristic contributions on East Africa, Ecuador, the Guianas, and SE Brazil. *America*. Ahti (1992) reports 50 Cladoniaceae from the paramos of northern S America; Sipman and Aptroot (1992) 273 species from Upper Mazaruni district, Guyana; Sipman (1992a) a list of lichens from the Colombian paramos with discussion of the origin of this flora; Sipman (1992b) 214 lichens from Cerro Guaiquinima, Venezuela; Fleig and Riquelme (1991) 72 species from southern Brazil; Marcelli (1992) 289 taxa in SE Brazilian mangroves, with notes on distribution and ecology; Barillas and Lücking (1992) 104 foliicolous species from Guatemala; Lücking (1992a) keys and illustrations for over 200 foliicolous lichens from Costa Rica. Regional revisions: Sipman and Topham (1992), *Umbilicaria* in Colombia; Tibell and Kalb (1992), *Calicium* from tropical and subtropical America. *Africa*. Ahti and Aptroot (1992) and Stenroos (1991) report on Cladoniaceae, Aptroot (1991d) on Parmeliaceae, from Madagascar. SE Asia: Makhija et al. treat the genus *Graphina* (1992a) and the further Graphidaceae (1992b) for the Andaman islands; Aptroot and Sipman (1991) present 112 taxa from New Guinea, with notes on floristic affinities and altitudinal zonation of the lichen flora.

d) The Southern Extratropics

(Sub)Antartica. Inoue (1991), reports on 200 species from a continental and a maritime site; Søchting and Øvstedal (1992) treat 14 species of *Caloplaca.* Kantvilas and Seppelt (1992) present a checklist (141 taxa) of Macquarie Island. *South America.* Larger floristic contributions deal with: Argentina (Adler and Elix 1992), southern Chile (Galloway 1992c), 28 Cladoniaceae from the Falkland Islands (Stenroos and Ahti 1992). Galloway and Marticorena (1991) list 662 titles of books and papers on Chilean lichens 1787–1991. Regional revisions: *Pseudocyphellaria* (Galloway 1992a); Cladoniaceae of Tierra del Fuego (Stenroos et al. 1992); *Teloschistes* (Almborn 1992). *Australia and New Zealand.* Hayward and Lumbsch (1992) mention 199 lichens from Stewart Island, New Zealand, Kantvilas and Jarman (1991) 141 lichens from Mount Sprent, Tasmania, Hayward et al. (1991) 251 lichens from Little Barrier Island (Hauturu), northern New Zealand, with notes on habitats and biogeographical relations, Kantvilas and Elix (1992) 24 lichens from Tasmania. McCarthy updated the checklist of Australian lichens (1991b) and the bibliography (1992), Galloway (1992b) produced a checklist of New Zealand lichens. Regional treatments: *Endocarpon* in Australia (McCarthy 1991a); key to *Pertusaria* in Australia (Archer 1991).

5. Chorology

a) Mapping in Europe

After a mapping scheme for lichens on the British Isles has become so successful (cf. Seaward 1992), a similar project is being undertaken on an European scale. For this purpose a meeting was held in Stuttgart (Germany), where 31 presentations on the theme were held, mostly on the state of lichen mapping in various European countries, and future joint activities planned (Wirth and Oberhollenzer 1990). Litterski (1992b) gives the distribution of 14 lichen species in Europe. Wirth (1992) summarizes the state of lichen mapping in Germany. John (1991) presents an atlas of lichens in Rheinland-Pfalz.

b) Intercontinental Floristic Affinities

Examples of amphi-pacific distributions in Umbilicariaceae are given by Wei and Biazrov (1991). Galloway (1991b) analyzes the biogeographical relationships of the tropical lichen flora in the Pacific area. Ahti (1992) analyzed the floristic composition of the Cladoniaceae from the paramos of northern South America, and found a wide-temperate element (36%), an endemic element (18%), and other neotropical elements (30%). Galloway (1991c) reviewed the phytogeography of the southern temperate and cold zones. Kantvilas (1991) reports three species with southern Australia-East Africa disjunct ranges.

6. Ecology

a) Analysis of Vegetation Structures

The description of lichen vegetation units following the Braun-Blanquet system is currently making most progress in the West Mediterranean area. Examples are: Bricaud and Roux (1991), Ménard and Roux (1991), Roux and Bricaud (1991), Roux and Egea (1992). In Central Europe, mainly saxicolous habitats and impoverishment are studied. Daniëls and Harkema (1992) describe epilithic lichen communities on man-made substrate in The Netherlands. Litterski (1992a) treats the lichen associations of Rügen, Germany. Mohr (1992) describes impoverished epiphytic lichen communities from northern Germany.

Lawrey (1991) found that competitive exclusion is rare in lichen community development, especially on stable substrates. Lichen succession seems more driven by addition of colonists. A comparison of species-area curves of disturbed and undisturbed lichen communities with randomized ones (Lawrey 1992) showed that the latter are more similar to the disturbed communities. Consequently, natural communities are not regulated by random processes.

b) External Relations

A remarkable case of habitat selection was found by Green et al. (1992): two lichen species in the dry Taylor valley, Antarctica, appeared to be restricted to a band only a few cm wide. *Cladonia coniocraea* on decaying stumps was found to be particularly polymorphic, depending on microhabitats (Schwerdtner and Cordes 1992). Schöller and Jahns (1992), Schöller (1992), Halonen et al. (1991), Canters et al. (1991) and Hyvärinen et al. (1992) investigated habitat selection by lichens in fores and analyzed various parameters such as tree species and age, light, humidity. Relevant microclimatic factors seem to be of usually little importance to higher plants (Canters et al. 1991).

Interactions with other organisms were studied by, e.g., Hätscher et al. (1991), who found feeding deterrency and growth retardation of extracted lichen substances on *Spodoptera littoralis* (Lepidoptera). Pfleiderer and Winkler (1991) investigated the feeding of gastropods on limestone lichens. They observed which animals were found associated with which lichens and noted several lichen species which were eaten. Wessels and Wessels (1991) investigated the role of lichenophagous moth larvae in sandstone weathering in South Africa. The larvae use loosened quartz grains for the construction of the bags in which they live, and contribute thus to pedogenesis. Kinder et al. (1992) found germination retardment and growth reduction after application of lichen extract and thallus fragments of *Cladonia furcata* on three phanerogams in a greenhouse experiment.

Crittenden (1991) investigated necromasses of mat-forming *Cladonia* and *Stereocaulon* and suggested that they contribute considerably to the growth, and thus competitiveness, of the mats. Knops et al. (1991) treated the role of epiphytic lichens in the mineral cycles of ecosystems. Tenhunen et al. (1992) studied the role of lichens in a tundra.

Various other themes were addressed. Hafellner (1991a) studied the lichen flora of high-altitude serpentinite in the eastern Alps, Europe, and found that in contrast to low-altitude sites, no substrate-specific lichens were present. Ott and Scheidegger (1992) investigated the relations between *Peltula euploca* and *Glyphopeltis ligustica*, which develops on its thallus. Prussia and Killingback (1991) measured the concentration of ten elements in *Flavoparmelia baltimorensis* and *F. caperata* on different substrates, and found significant differences between thalli growing on rock and those on bark, sometimes also between those on bark of different trees. Gehrmann et al. (1992) investigated and classified the pitting resultant from endolithic lichens on ancient stone monuments. Green and Lange (1991) and Green et al. (1991) measured ecophysiological adaptations of temperate rainforest lichens.

The ecology of tropical lichens has received very little attention so far. Lücking (1992b, c) investigated the ecology and distribution of foliicolous lichens in Costa Rica. He found that whole families tend to share ecological preferences and that floristic composition varies with forest type and climatic conditions. Marcelli (1992) studied distribution and ecology of lichens in mangroves in SE Brazil.

7. Applied Aspects

a) Pollution Monitoring

The tendency of lichens to accumulate substances is much used now to monitor air-transported pollution. Deruelle (1992) measured lead accumulation near a motorway and found that lead does not only accumulate, but its content may decrease under certain conditions. Seaward (1991) measured radionuclide contents in lichens to trace effects of the Chernobyl nuclear reactor accident. Namakura and Inoue (1991) demonstrated active iron uptake by Antartic lichens, with a selective effect on isotopes.

The use of the lichen vegetation as a cheap means to map air contamination continues to be the dominant application, which has been used, e.g., in Siberia (Oksanen et al.

1991), Greece (Diamantopoulos et al. 1992), Germany (Pfeiffer and Barclay-Estrup 1992; Woesner 1992), The Netherlands (Sloof and Wolterbeek 1991), Finland (Kauppi and Halonen 1992). Improvement of air quality is also observed: e.g., in Paris (Seaward and Letrouit-Galinou 1991; Letrouit-Galinou et al. 1992) and surroundings (Letrouit 1992).

Transplant experiments were done by Palomäki et al. (1992) to monitor fluoride and sulfur deposition in Finland. Lackovicová and Kubinská (1991) used *Cladonia rangiformis* transplants to measure fluorine content. Chen et al. (1992) used a computer to measure growth rates of transplanted *Parmelia sulcata.*

Fumes from industrial sources and other fossil fuel burning are not the only airborne cause of lichen decline. Alstrup (1992a) reported effects of four pesticides on ca. 50 lichen species in different habitats. Another probable effect of air contamination, by fertilizing substances, is reported by Søchting et al. (1992), who analyzed the epiphyllic growth on spruce needles, including three lichens: the increase in this growth threatens the commercial use of young spruce as Christmas trees.

b) Conservation

The awareness of a dramatic mundial decline in biodiversity is causing an increasing interest in conservation. In the pertinent research, lichens should play an important role because they are among the first groups where extinction can be expected. Many species appear to be restricted to habitats with long ecological continuity, habitats which cannot easily be regenerated. Diederich (1991) presents lists of species which characterize forests with a long historical continuity in Luxemburg. *Texosporium sancti-jacobi*, a terricolous lichen of dry grassland, which cannot stand any change in its habitat, is known only from four general locations worldwide (McCune and Rosentreter 1992).

Decline of the lichen flora is often signalized. Ernst (1992) reports a loss of 42 species on erratic blocs in northern Germany. Ketner-Oostra (1992) describes a strong decline of lichens on coastal dunes in The Netherlands. In areas where the lichen flora is sufficiently known, "red lists" are prepared, indicating which species are endangered and to what extent: Hauck (1992) for Niedersachsen, northern Germany; Siebel et al. (1992) for The Netherlands. The latter indicate that lichens are the most threatened plant group in The Netherlands (Table 1).

Table 1. Dutch Red List lichens. (After Siebel et al. 1992)

	Total number of species	633	100.0%
0	Extinct	91	14.4%
1	Endangered	115	18.2%
2	Very vulnerable	54	8.5%
3	Vulnerable	35	5.5%
4	Rare	72	11.4%
	Total number of listed species	367	58.0%

Several papers deal with practical aspects of conservation. Alstrup (1992b) investigated the effect of grazing on stone and soil lichens; ceased grazing is claimed to be responsible for the decline of about 22% of the endangered lichens in Denmark. Garty (1991) investigated recovery after bush fire. Gilbert (1991) transplanted *Lobaria amplissima* in Great Britain, in an attempt to improve its survival possibilities. Gustafsson et al. (1992) studied factors of importance for 19 "old woodland" lichens which prefer broadleaved, deciduous woodlands of long continuity, and give recommendations for their conservation. Lipnicki (1991) presents plans for the conservation of lichen stations in Poland. Wirth (1992) stresses the need to keep collecting of uncommon lichens in Germany at a minimum. Wolseley and Aguirre-Hudson (1991) discuss the role of lichens as indicators of environmental change in the tropical forests of Thailand.

The preceding treats only a small selection of all publications. For additional titles three sources are especially recommended: for literature on all subjects, the series *Recent Literature on Lichens* in the journal *The Bryologist* (alphabetically arranged and *Eléments de bibliographie lichénologique récente* in the *Bulletin of the Association Française de Lichénologie* (arranged according to subjects, in French); for pollution monitoring the series *Lichens and air pollution*, published in *The Lichenologist*.

References

Adler MT, Elix JA (1992) Mycotaxon 43:283–288. **Aguirre-Hudson B** (1991) Bull Br Mus Nat Hist (Bot) 21(2):85–192. **Ahti T** (1991) Mycosystema 4:59–64. **Ahti T** (1992) In: Balslev H, Luteyn JL (eds) Páramo, an Andean ecosystem under human influence. Academic Press, London, pp 111–117. **Ahti T**, Aptroot A (1992) Cryptogam Bryol Lichénol 13:117–124. **Almborn O** (1992) Nordic J Bot 12:361–364. **Alstrup V** (1992a) Bryonora (Praha) 9:2–4. **Alstrup V** (1992b) Int J Mycol Lichenol 5:1–2. **Aptroot A** (1991a) Bibl Lichenol 44:178. **Aptroot A** (1991b) Willdenowia 21:263–267. **Aptroot A** (1991c) In: Galloway DJ (ed) Tropical lichens: their systematics, conservation, and ecology. System Assoc Spec Vol 43, Oxford, pp 253–273. **Aptroot A** (1991d) Cryptogam Bryol Lichénol 12:149–154. **Aptroot A** (1991e) Bryologist 94:404–406. **Aptroot A**, Alstrup V (1991) Graphis Scripta 3:73–75. **Aptroot A**, Sipman H (1991) Willdenowia 20:221–256. **Archer AW** (1991) Mycotaxon 41:223–269. **Armaleo D**, Clerc P (1991) Exp Mycol 15:1–10. **Armstrong RA** (1992) Environ Exp Bot 32:55–63. **Ascaso C**, Sancho LG, Valladares F (1992) Cryptogam Bryol Lichénol 13:335–340. **Awasthi DD** (1991) Bibl Lichenol 40, 340 pp.

Barillas R, Lücking R (1992) Cryptogam Bryol Lichénol 13:297–314. **Brako L** (1991) Flora Neotropica. Monograph 55. The New York Botanical Garden, New York, 67 pp. **Bricaud O**, Roux C (1991) Bull Soc Linn Provence 42:79–90. **Brodo IM** (1991) Can J Bot 69:733–772. **Brusse FA**, Kärnefelt I (1991) Mycotaxon 42:35–41. **Burgaz AR**, Ahti T (1992) Nova Hedwigia 55:37–53.

Canters KJ, Schöller H, Ott S, Jahns HM (1991) Lichenologist 23:237–252. **Chen T**, Kreeb K-H, Röper B (1992) Int J Mycol Lichenol 5:3–12. **Christensen SN**, Alstrup V, Voreadou C (1991) Graphis Scripta 3:97–103. **Clayden SR** (1992) Bryologist 95:1–4. **Common RS** (1991) Mycotaxon 41:67–112. **Crittenden PD** (1991) Lichenologist 23:323–331. **Culberson CF**, Armaleo D (1992) Exp Mycol 16:52–63. **Culberson CF**, Culberson WL, Johnson A (1992) Mycologia 84:705–714. **Czeczuga B**, Stenroos S, Christensen SN, Ahti T (1991) Ann Bot Fenn 28:123–130.

Daniëls FJA, Harkema M (1992) Phytocoenologia 21:209–235. **Deruelle S** (1992) Bull Soc Bot Fr 139, Act Bot:99–109. **Diamantopoulos J**, Pirintsos S, Laundon JR, Vokou D (1992) Lichenologist

24:63–71. **Diederich P** (1991) Bull Soc Nat Luxemb 92:31–39. **Diederich P**, Sérusiaux E, van den Boom P (1991) Lejeunea 136, 47 pp.

Egan RS (1991) Bryologist 94:396–400. **Egea JM**, Llimona X (1991) Bot Chron 10:179–198. **Ekman S**, Fröberg L, Kärnefelt I, Sundin R, Thor G (1991) Folia Cryptogam Estonica 28:5–25. **Elix JA**, Crook CE (1992) Bryologist 95:52–64. **Elix JA**, Nash TH (1992) Bryologist 95:377–391. **Elix JA**, Bennet SA, Jiang H (1991a) Aust J Chem 44:1157–1162. **Elix JA**, Chappell H-M, Jiang H (1991b) Bryologist 94:304–307. **Eriksson O** (1992) Systema Ascomycetum 11:11–27. **Eriksson OE**, Hawksworth DL (1991) Systema Ascomycetum 9:39–271. **Ernst G** (1992) Int J Mycol Lichenol 5:13–36.

Fahselt D (1992) Lichenologist 24:181–192. **Feuerer T** (1991) Bibl Lichenol 39, 218 pp. **Fleig M**, Riquelme I (1991) Acta Bot Bras 5:3–12.

Galloway DJ (1991a) (ed) Tropical lichens: Their systematics, conservation, and ecology. System Assoc Spec Vol 43. Clarendon, Oxford, 302 pp. **Galloway DJ** (1991b) In: Galloway DJ (ed) Tropical lichens: Their systematics, conservation, and ecology. System Assoc Spec Vol 43: Clarendon, Oxford, pp 1–16. **Galloway DJ** (1991c) In: Nimis PL, Crovello TJ (eds) Quantitative approaches to phytogeography. Kluwer, Dordrecht, pp 233–262. **Galloway DJ** (1992a) Bibl Lichenol 46, 275 pp, 44 pl. **Galloway DJ** (1992b) DSIR Land Res Sci Rep 26:1–58. **Galloway DJ** (1992c) Glob Ecol Biogeogr Lett 2:37–45. **Galloway DJ**, Marticorena C (1991) Gayana Bot 48 (1–4):17–66. **Garty J** (1991) Can J Bot 70:301–312. **Gehrmann CK**, Krumbein WE, Petersen K (1992) Int J Mycol Lichenol 5:37–48. **George AS** (ed) (1992) Flora of Australia 54. Aust Govt Publ Ser, Canberra, 349 pp. **Gilbert OL** (1991) Lichenologist 23:73–76. **Gilbert OL**, Coppins BJ (1992) Lichenologist 24:143–163. **Giralt M**, Nimis PL, Poelt J (1992) Cryptogam Bryol Lichénol 13:261–273. **Goward T**, Ahti T (1992) Acta Bot Fenn 147, 60 pp. **Green TGA**, Lange OL (1991) Lichenologist 23:267–282. **Green TGA**, Kilian E, Lange OL (1991) Oecologia 85:498–503. **Green TGA**, Seppelt RD, Schwarz A-MJ (1992) Lichenologist 24:57–61. **Gustafsson L**, Fiskesjö A, Ingelög T, Pettersson B, Thor G (1992) Lichenologist 24:255–266.

Hafellner J (1991a) Carinthia II 181/101:507–527. **Hafellner J** (1991b) Mitt Naturwiss Ver Steiermark 121:95–106. **Hafellner J** (1991c) Acta Bot Malacitana 16:133–140. **Hafellner J** (1992) A new checklist of lichenized and lichenicolous fungi of the Madeira Archipelago, Graz (priv. publ.) 29 pp. **Hafellner J**, Casares Porcel M (1992) Nova Hedwigia 55:309–323. **Hafellner J**, Vezda A (1992) Nova Hedwigia 55:183–193. **Hagemann C**, Fahselt D (1992a) Lichenologist 24:91–100. **Hagemann C**, Fahselt D (1992b) Bryologist 95:316–323. **Halonen P**, Hyvärinen M, Kauppi M (1991) Lichenologist 23:61–72. **Harada H** (1991) Systema Ascomycetum 10:1–6. **Harris RC** (1990) Some Florida lichens. Published by the author, New York. 109 pp. **Hätscher I**, Veit M, Proksch P, Lange OL, Zellner H (1991) Planta Med 57 Suppl 2:A26. **Hauck M** (1992) Informationsdienst Natursch Niedersachsen 1/92, 43 pp. **Hawksworth DL** (1991) Systema Ascomycetum 10:127–134. **Hayward BW**, Wright AE, Hayward GC (1991) Rec Auckland Inst Mus 28:185–199. **Hayward BW**, Lumbsch HT (1992) N Z Nat Sci 19:69–78. **Henssen A** (1991) Lichenologist 23:333–342. **Henssen A** (1992a) Lichenologist 24:27–41. **Henssen A** (1992b) Lichenologist 24:133–142. **Henssen A**, Döring H, Kantvilas G (1992) Bot Acta 105:457–467. **Hestmark G** (1991a) Lichenologist 23:343–359, 361–380. **Hestmark G** (1991b) Sommerfeltia (Oslo) 3, 47 pp. **Honegger R**, Bartnicki-Garcia S (1991) Mycol Res 95:905–914. **Houmeau J-M**, Roux C (1991) Bull Soc Bot Centre-Ouest N 22:545–556. **Huneck S**, Ahti T, Cogt U, Poelt J, Sipman H (1992) Nova Hedwigia 54:277–308. **Hyvärinen M**, Halonen P, Kauppi M (1992) Lichenologist 24:165–180.

Inoue M (1991) Proc NIPR Symp Polar Biol 4:91–106

Jacobsen P (1992) Mitt Arbeitsgem Geobot Schleswig-Holstein Hamb 42:234. **John V** (1991) Beitr Landespfl Rheinland-Pfalz 13A, B. Landesamt für Umweltschutz und Gewerbeaufsicht Rheinland-Pfalz, Oppenheim, pp 1–276; 1–272. **John V** (1992a) Flechten der Türkei I. Pollichia,

Bad Dürkheim, 14 pp. **John V** (1992b) Flechten der Türkei III. Pollichia, Bad Dürkheim, 20 unnumbered pp.

Kalb K, Hafellner J (1992) Herzogia 9:45–102. **Kantvilas G** (1991) Nord J Bot 11:369–373. **Kantvilas G**, Elix JA (1992) Muelleria 7:507–517. **Kantvilas G**, Jarman SJ (1991) In: Banks MR et al. (eds) Aspects of Tasmanian Botany – a tribute to Winifred Curtis. Roy Soc Tasm, Hobart, pp 149–162. **Kantvilas G**, Seppelt RD (1992) Anare Res Notes (Australia) 87, 20 pp. **Kantvilas G**, Elix JA, James PW (1992) Bryologist 95:186–191. **Kärnefelt I**, Thell A (1992) Plant Syst Evol 180:181–204. **Kärnefelt I**, Mattson J-E, Thell A (1992) Plant Syst Evol 180:113–160. **Kauppi M**, Halonen P (1992) Ann Bot Fenn 20:1–9. **Ketner-Oostra R** (1992) Int J Mycol Lichenol 5:63–66. **Kinder M**, Vagts I, Müller J (1992) Int J Mycol Lichenol 5:67–76. **Knops JMH**, Nash TH, Boucher VL, Schlesinger WH (1991) Lichenologist 23:309–321. **Kümmerling H** (1991) Bibl. Lichenol 41, 315 pp. **Kümmerling H**, Leuckert C, Wirth V (1991) Nova Hedwigia 53:507–517. **Kurokawa S** (1991a) J Jpn Bot 66:152–159. **Kurokawa S** (1991b) Ann Tsukuba Bot Gard 10:1–14. **Kurokawa S**, Lai M-J (1991) Bull Natl Sci Mus Tokyo B 17:59–65.

Lackovicova A, Kubinská A (1991) Biológia 46:781–789. **Laundon JR** (1992) Lichenologist 24:315–350. **Lawrey JD** (1991) Lichenologist 23:205–214. **Lawrey JD** (1992) Bryologist 95:137–141. **Léonard J** (1979) In: Kunkel G (ed) Taxonomic aspects of African economic botany. Proc IX Plenary Meet AETFAT, Las Palmas de Gran Canaria, pp 163–167. **Letrouit-Galinou M-A** (1992) Bull Assoc Fr Lichénol, Mém 1:35–48. **Letrouit-Galinou M-A**, Seaward MRD, Deruelle S (1992) Bull Soc Bot Fr 139, Lett Bot (2) 115–126. **Leuckert C**, Knoph J-G (1992) Lichenologist 24:383–397. **Leuckert C**, Kümmerling H (1991) Nova Hedwigia 52:17–32. **Lipnicki L** (1991) Parki Narodowe i Rezerwaty Przyrody 10:137–150. **Litterski B** (1992a) Gleditschia 20:123–152. **Litterski B** (1992b) Herzogia 9:149–166. **Lücking R** (1992a) Beih Nova Hedwigia 104, 179 pp. **Lücking R** (1992b) Nova Hedwigia 54:309–353. **Lücking R** (1992c) Herzogia 9:181–212. **Lumbsch HT** (1991) Nova Hedwigia 52:419–426. **Lumbsch HT**, Kothe HW (1992) Mycotaxon 42:277–282. **Lumbsch HT**, Schmitz KE, Feige CB (1991) Mycotaxon 42:77–87. **Lumbsch HT**, Schuster G, Elix JA, Nash TH (1992) Mycotaxon 45:489–494.

Makhija U, Adawadkar B, Patwardhan PG (1992a) Biovignyanam 18:21–32. **Makhija U**, Kekre S, Patwardhan PG (1992b) Biovignyanam 18:12–20. **Manrique E**, López F (1991) Bryologist 94:207–212. **Marcelli MP** (1992) Bibl Lichenol 47, 288 pp. **Mattson J-E** (1991) Bryologist 94:261–269. **Mayrhofer H**, Sheard JW, Matzer M (1992) Bryologist 95:436–442. **McCarthy PM** (1991a) Lichenologist 23:27–52. **McCarthy PM** (1991b) Checklist of Australian lichens, 4th edn, National Herbarium of Victoria, Melbourne, 177 pp. **McCarthy PM** (1992) Bibliography of Australian lichens: 1807–1991. National Herbarium of Victoria, Melbourne, 62 pp. **McCune B**, Rosentreter R (1992) Bryologist 95:329–333. **Ménard T**, Roux C (1991) Bull Soc Linn Provence 42:91–116. **Mietzsch E**, Lumbsch HT, Elix JA (1992) Wintabolites – users manual. Essen, priv publ. **Modeseni P**, Debarbieri A (1991) Nova Hedwigia 52:231–238. **Modeseni P**, Lajolo L (1992) Nova Hedwigia 54:271–276. **Mohr K** (1992) Int J Mycol Lichenol 5:81–97. **Moreno PP**, Egea JM (1991) Biología y taxonomía de la familia Lichinaceae, con especial referencia a las especies del S.E. Español y Norte de Africa. Secretariado de Publicaciones, Universidad de Murcia, 87 pp. **Moreno PP**, Egea JM (1992a) Lichenologist 24:215–228. **Moreno PP**, Egea JM (1992b) Acta Bot Barc 41, 66 pp. **Moreno PP**, Egea JM (1992c) Cryptogam Bryol Lichenol 13:237–253.

Namakura A, Inoue M (1991) Proc NIPR Symp Polar Biol 4:82–90. **Navarro-Rosines P**, Hladun NL (1992) Cryptogam Bryol Lichénol 13:227–235. **Nimis PL** (1992) Not Soc Lichenol Ital 5:9–28.

Oksanen J, Läärä E, Zobel K (1991) Lichenologist 23:167–180. **Orange A** (1991) Lichenologist 23:99–106. **Ott S**, Scheidegger C (1992) Symbiosis 12:159–172.

Palomäki V, Tynnyrinen S, Holopainen T (1992) Ann Bot Fenn 29:25–34. **Pfeiffer HN**, Barclay-Estrup P (1992) Bryologist 95:38–41. **Pfleiderer J**, Winkler S (1991) Jahresh Ges Naturkd Württemb 146:89–113. **Poelt J**, Grube M (1992) Nova Hedwigia 55:381–395. **Poelt J**, Petutschnig W (1992) Nova Hedwigia 54:1–36. **Poelt J**, Obermayer W (1991) Nova Hedwigia 53:1–26. **Posner B**,

Feige GB, Huneck S (1992) Z Naturforsch 47c:1–9. **Prussia CM**, Killingback KT (1991) Bryologist 94:135–142. **Purvis OW**, Coppins BJ, Hawksworth DL, James PW, Moore DM (1992) The Lichen Flora of Great Britain and Ireland. Natural History Museum Publications in association with the British Lichen Society, London, 710 pp.

Rambold G, Triebel D (1992) Bibl Lichenol 48, 201 pp. **Randlane T**, Saag A (1992a) Mycotaxon 44:485–489. **Randlane T**, Saag A (1992b) Mycotaxon 44:491–493. **Rosentreter R**, McCune R (1992) Bryologist 95:15–19. **Roux C**, Bricaud O (1991) Cryptogam Bryol Lichénol 12:95–110. **Roux C**, Egea JM (1992) Cryptogam Bryol Lichénol 13:105–115.

Sanders WB (1992) Bryologist 95:192–204. **Sanders WB**, Rico VJ (1992) Bot Acta 105:449–456. **Schöller H** (1992) Int J Mycol Lichenol 5:113–119. **Schöller H**, Jahns HM (1992) Herzogia 9:19–44. **Schreiner E**, Hafellner J (1992) Bibl Lichenol 45, 291 pp. **Schwerdtner H**, Cordes H (1992) Int J Mycol Lichenol 5:121–136. **Seaward MRD** (1991) Proc 1st Symp Environmental Pollution and Control 1:80–89. **Seaward MRD** (1992) Lichenologist 24:104. **Seaward MRD**, Letrouit-Galinou MA (1991) Lichenologist 23:181–186. **Sérusiaux E** (1992) Systema Ascomycetum 11:39–47. **Sheard JW** (1992) Bryologist 95:79–87. **Siebel HN**, Aptroot A, Dirkse GM, van Dobben JH, van Melick HMH, Touw A (1992) Gorteria 18:1–20. **Sipman HJM** (1992a) In: Balslev H, Luteyn JL (eds) Páramo, on Andean ecosystem under human influence. Academic Press, London, pp 95–109. **Sipman HJM** (1992b) Trop Bryol 6:1–31. **Sipman H**, Aptroot A (1992) Trop Bryol 5:79–107. **Sipman H**, Topham P (1992) Nova Hedwigia 54:63–75. **Sloof JE**, Wolterbeek HT (1991) Lichenologist 23:139–165. **Søchting U**, Øvstedal DO (1992) Nord J Bot 12:121–134. **Søchting U**, Jensen B, Unger L (1992) Epifylfloraen på Rødgran. Miljøministeriet, Skor-og Naturstyrelsen, Hørsholm, 44 pp. **St. Clair LL**, Newberry CC, Nebeker GT (1991) Mycotaxon 40:199–264. **Stenroos S** (1991) Ann Bot Fenn 28:107–110. **Stenroos S**, Ahti T (1992) Ann Bot Fenn 29:67–73. **Stenroos S**, Ferraro LI, Ahti T (1992) In: Guarrera SA, Gamundi de Amos I, Materi CM (eds) Flora Criptogámica de Tierra del Fuego 13(7). Consejo Nacional de Investigaciones Cientificas y Técnicas de la República Argentina, Buenos Aires, 111 pp.

Talbot SS, Looman Talbot S, Thomon JW (1992) Bryologist 95:20–30. **Tehler A** (1992) Willdenowia 22:201–214. **Tenhunen JD**, Lange OL, Hahn S, Siegwolf R, Oberbauer SF (1992) In: Chapin FS et al. (eds) Arctic ecosystems in a changing climate. Academic Press, San Diego, pp 213–237. **Thomson JW** (1991) Bryologist 94:351–367. **Thomson JW**, Weber WA (1992) Bryologist 95:392–405. **Thor G** (1991) Bryologist 94:278–283. **Thor G** (1992) Graphis Scripta 4:19–29. **Tibell L**, Kalb K (1992) Nova Hedwigia 55:11–36. **Timdal E** (1991a) Mycotaxon 42:249–254. **Timdal E** (1991b) Opera Bot 110, 137 pp. **Tønsberg T** (1992) Sommerfeltia (Oslo) 14:331 pp. **Torrente P**, Egea JM (1991) Nova Hedwigia 52:239–245. **Trass H** (1992) Folia Cryptog Estonica 29:2–24. **Triebel D**, Rambold G, Nash TH (1991) Mycotaxon 42:263–296.

Valladares F, Ascaso C (1992) Lichenologist 24:281–297. **Van den Boom PPG** (1992) Nova Hedwigia 54:229–254. **Vezda A**, Poelt J (1991) Nova Hedwigia 53:99–113. **Vitikainen O** (1991) Graphis Scripta 3:35–43.

Wedin M (1991) In: Galloway DJ (ed) Tropical lichens: their systematics, conservation, and ecology. System Assoc Spec Vol 43, Clarendon, Oxford, pp 245–251. **Wei J-C** (1991) An enumeration of lichens in China. International Academic Publishers, Beijing, 278 pp. **Wei J-C**, Biazrov LG (1991) Mycosystema 4:65–72. **Wessels DCJ**, Wessels L-A (1991) Lichenologist 23:283–291.

Wirth V (1992) Flor Rundbriefe Göttingen Beih 2:119–126. **Wirth V**, Oberhollenzer H (1990) Stuttg Beitr Naturkd A456, 199 pp. **Woesner E** (1992) Int J Mycol Lichenol 5:137–146. **Wolseley PA**, Aguirre-Hudson A (1991) Glob Ecol Biogeogr Lett 1:170–175. **Wong PY**, Brodo IM (1992) Can Mus Nature Syllogeus (Ottawa) 69, 79 pp.

Yoshimura I, Kurokawa T (1991) Bull Kochi Gakuen College 22:663–672.

Zhurbenko MP (1992) Graphis Scripta 4:1–4.

Dr. H. J. M. Sipman
Botanischer Garten und Botanisches Museum
Königin-Luise-Straße 6–8
D-14191 Berlin, FRG

II. Molecular Systematics: 1991–1993

By Kenneth J. Sytsma and William J. Hahn

1. Introduction

The impact of the DNA revolution on plant systematics has increased dramatically during the last few years. These effects are noticeable in obvious ways such as the amount of grant funding, published papers, and presented talks and posters dealing with macromolecules. A quick survey of six important and broadly encompassing journals with systematic coverage, at least in part (*Taxon, American Journal of Botany, Systematic Botany, Canadian Journal of Botany, Nordic Journal of Botany,* and *Australian Journal of Botany*), reveals more than a 100% increase in molecular studies published during the last five years relative to the previous five. More subtly, these effects are seen in (1) the increasing number of studies that attempt (correctly) to fully integrate molecular results with those from more traditional studies (i.e., "morphological" broadly defined), (2) the collaboration of different "traditional" and "molecular" research programs working on the same set of systematic problems, and (3) the training in molecular systematic methods at all professional levels. Not surprising, then, is the increase in the number of taxonomic revisions based in part on molecular data (see below).

The 2 years covered in this review (mid-1991 to mid-1993) represent a very important stage in this integration of molecular systematics, as one area of systematic research, into the broader and more multifaceted field of plant biosystematics. This time period has seen the movement to rapid DNA sequencing and thus of large-scale attempts to uncover relationships at all taxonomic levels, from those involving issues of "intraspecific phylogeography" to questions with "phylogeny of life". No longer is molecular systematic research to be viewed as a separate endeavor or as an ad hoc tool to be used when "morphology" presents problems. Rather, molecular systematics as a field stands poised, with its incredibly large (but often not fully analyzed) DNA information bank, to provide significant contributions to these diverse issues either singly or in concert with traditional data. Which way molecular systematics goes will be decided in the not too distant future.

2. Progress from 1991–1993

Did we make progress in the last 2 years? Consider our list of ten of the most noteworthy papers in the field published during this time (not a trivial exercise, try it). (1) Chase et

Progress in Botany, Vol. 55

al. (1993) lays claim to this title by its sheer audacity in presenting a phylogenetic analysis of roughly 1200 bp of *rbc*L sequence for 499 species, representing most of the seed plant families, that 42 authors could agree upon! This largest phylogenetic analysis ever undertaken, although it has its limitations, places plant molecular systematics on a well-deserved pedestal. (2) The work of Rodman et al. (1993) is a culmination of a research program (largely without outside funding) that should be an exemplary model of how persistance and broad training in biosystematics (on the part of the first author) pays off. The payoff in terms of settling long-standing issues regarding evolution of glucosinolate-producing plants (and the characters carried with them) was elegantly obtained using sequence data but was preceded by years of earlier, careful phenetic and cladistic analyses of morphological, anatomical, and chemical data. (3) The work of Gaut et al. (1992) deserves recognition for its introduction of the maximum likelihood relative rate test on the presence of an *rbc*L molecular clock. The fivefold rate difference seen just among grasses and palms should be a matter of concern and a warning for those continually trying to date the rise of angiosperms (see below). (4) and (5) Raubeson and Jansen (1992b) and Doyle et al. (1992a) present chloroplast genome rearrangement data that bear directly on two long-standing issues: the relationships of major land plant lineages and the origin of the grass family, respectively. Both, however, present tantalizing hypotheses that can be further tested by casting out a larger net of taxonomic sampling. (6) D. E. Soltis et al. (1991a) introduces to the botanical systematic community the concept of "intraspecific phylogeography", an important interface of biogeography and phylogenetic analysis of populations long recognized and studied by animal molecular systematics. (7) Rieseberg and Soltis (1991) present a capstone paper on one of the most perplexing, but yet most exciting, areas of systematic biology. Cytoplasmatic gene flow among closely related species, a phenonenon made public largely by the efforts of molecular systematists (such as these two pioneering authors) is placed in its varied evolutionary and, hence, systematic contexts. (8) Morden et al. (1992) present a synthesis hypothesis on another, but more far-reaching, case(s) of genome transfer – the endosymbiotic origin(s) of plastids. (9) Knox et al. (1993) elegantly solve the long-standing issue surrounding the origin and relationships of the pantropical, giant lobeliads using an ordered series of chloroplast genome rearrangements. (10) The work of Spooner et al. (1993) is a superior example of a growing list of molecular phylogenetic studies which, in conjunction with morphological data, propose nomenclatural changes. Their results advocate the controversial, but previously suggested, treatment of *Lycopersicon* as congeneric with *Solanum* in the subgenus *Potatoe*.

Did we have disagreements in the last 2 years? Four examples of papers generating controversary will suffice to answer this in the affirmative. (1) Donoghue and Sanderson (1992) raise the stakes in the ongoing debate of what exactly are morphology and molecular data telling us, with their strong urging for (indiscriminate?) use of "total evidence". (2) Doyle (1992) argues for a radical approach to scoring different data sets in the related and ongoing debate of "gene" versus "species" trees. (3) Martin et al. (1993) continue the saga on the attempt to deduce the timing of the origin of angiosperms and other noteworthy past events using a molecular clock assumption, a method presently not advocated by many in the business and perhaps suspect if one considers relative rate tests. (4) Steele et al. (1991) close the book, hopefully, on the extended argument concerning the phylogenetic reliability of 5S rRNA sequences; sequences clearly too small and too rapidly evolving to satisfactorily answer questions of land plant relationships. The extended arguments have covered everything from alignment problems to choice of discrete or distance analyses (and to whether funding agencies should be supporting molecular systematics at their present level!).

Books, texts, and other notable publications? Possibly reflecting the large number of plant molecular papers published during these last 2 years (30 pages of single-spaced, pica font references collected for this review), over 15 books touching on the field were printed. *Molecular Systematics of Plants* (P.S. Soltis et al. 1992b) fills a nice niche for many Molecular Systematic/Evolution courses as the most up-to-date synthesis of the topic, but is rapidly (and not unexpectly) going out of date. *Molecular Techniques in Taxonomy* (Hewitt et al. 1991) is far-reaching in scope, although heavily biased towards animal systematics. The phylogenetic and ecological perspectives impacting on molecular systematics are evident in Brooks and McLennan (1991), Harrison (1993), Harvey and Pagel (1991), and Wagner and Funk (1993). For a more molecular point of view, try Selander et al. (1991), Miyamoto and Cracraft (1991), and Li and Graur (1991). Several "how to" books are worth considering or investing in: *Molecular Techniques in Taxonomy* (Hewitt et al. 1991); *Molecular Evolution: Producing the Biochemical Data* (Zimmer et al. 1993); *Molecular Genetic Analysis of Populations: A Practical Approach* (Hoelzel 1992). Additional reviews can be found in the more applied *DNA-Based Markers in Plants* (Phillips and Vasil 1993), *The Molecular Biology of Plastids* (Bogorad and Vasil 1991), and *Organelles* (Herrmann 1992). Not surprisingly, biosystematic compendia (including a good amount of molecular systematics) of several angiosperm groups are now available: Caryophyllales (Behnke and Mabry 1993), Solanaceae (Hawkes et al. 1991), and Labiatae (Harley 1992). Can we look forward to monocots, Onagraceae, legumes, and Asteraceae?

Another vaccuum has been partially filled with the emergence of several new journals in the last 2 years devoted, exclusively or in part, to the dissemination of the expected voluminous molecular evolutionary, systematic, genetic, and ecological information: *Molecular Phylogenetics and Evolution* (1992), *Plant Molecular Evolution Newsletter* (1991), *Molecular Ecology* (1992), *Current Opinions in Genetic and Development* (1991). Hopefully, the trend toward more specialized trade journals will not keep important molecular systematic information away from colleagues interested or versed only in "traditional" systematics. Also of interest is the renaming of *Systematic Zoology* to *Systematic Biology*, a change designed to permit exposure of botanical studies with topical or broad coverage. Does this suggest that our zoological counterparts do not view the explicitly botanical systematic journals with much regard?

3. Advances in Methodology

a) Tissue Preservation/DNA Extraction

The phenomenal increase in tissue collection and subsequent DNA extraction of species worldwide fortunately has been accompanied by several innovations in tissue preservation and DNA extraction from recalcitrant plants (reviewed in Doyle 1991a; Smith et al. 1991; Milligan 1992; Stein 1993; Sytsma 1993; Sytsma et al. 1993). Foremost is the now widely tested and fairly successful use of silica gel for drying leaf tissue in the field (Chase and Hills 1991). High salt/CTAB liquid preservation of leaf tissue needs further testing, but looks promising (Rogstad 1992). Specific remedies for the extraction of DNA (or its purification) from variously preserved tissues of some recalcitrant plants are now available (e.g., Do and Adams 1991; Doyle and Doyle 1991; Fang et al. 1992; Gawal and Jarret 1991; Jhingan 1992; Sytsma 1993). These advances have certainly played a pivotal role in expanding the taxon sampling now seen in most molecular studies. Additionally, with the increased number of molecular studies using herbarium material (e.g., Liston et al. 1992; Mishler et al. 1992; Chase et al. 1993; Rodman et al. 1993), molecular studies may soon come close to morphological studies in taxon coverage, a feature of morphological studies often cited as an inherant blessing of the latter and a curse of the former (Donoghue and Sanderson 1992). The use of fossil DNA, however, has not kept pace with the use of herbarium DNA. Only a handful of additional sequences are available since the pioneering work of Golenberg et al. (1990) on a Miocene *Magnolia* (P.S. Soltis et al. 1992c, 1993; E. Golenberg, pers. comm.); there is no Jurassic sequence as yet.

b) PCR/Sequencing Innovations

The single most important advance that has been embraced by plant molecular systematics over the last 2 years is certainly, no question about it, the polymerase chain reaction (PCR) to amplify specific DNA stretches. We can largely say farewell to the genome library construction, subcloning and then sequencing from days of old, except of course to get those first (oh so important!) sequences that allow the construction of universal primers. Gone also are the unnecessary steps to clone the PCR product for sequencing (e.g., Les et al. 1991; Donoghue et al. 1992), with the now quite routine methods of asymmetric PCR and DNA sequencing (e.g., Olmstead et al. 1992a) and direct double-stranded DNA sequencing (e.g., Conti et al. 1993), and newer innovative procedures (Ruanto and Kidd 1991; Douglas et al. 1993). A number of streamlined DNA extraction protocols have been designed specifically for PCR studies involving hundreds of plants (Berhomieu and Meyer 1991; Edwards et al. 1991; Langridge et al. 1991; Deragon and Landry 1992; Hoelzel and Green 1992; Luo et al. 1992; Stewart and Via 1993). These methods allow for rapid tissue processing of many plants and greatly reduce the possibility of cross-contamination by elimination of most tube-to-tube transfers and reliance on few buffers. Whether DNA sequencing will largely displace restriction site mapping studies (of the chloroplast genome, for example), as has been proclaimed for animal mtDNA and nuclear rDNA (Wilson et al. 1989), is a contentious point. Restriction site analysis of cpDNA is still very informative at the population to genus level (see chapters in P.S. Soltis et al. 1992b and Rieseberg et al. 1992 for an innovative PCR-based method for restriction site analysis), and we predict that it will continue to be a well-used and cost-effective systematic tool for some time to come.

c) Genome Analysis

Important insights into the evolution of all three plant genomes have accumulated during these last 2 years. The relatively untapped resource of the nuclear genome for phylogenetic purposes, aside from rDNA analysis (Hamby and Zimmer 1992), has attracted considerable attention on a number of fronts. The review of Whitkus et al. (1993) lays out the pros and cons of various methods for tapping this resource. "Restriction fragment length polymorphisms" (RFLPs), an unfortunate designation of a term to a subset of restriction fragment studies, are now routinely being used in studies of cultivated plants and their wild relatives. (Kesseli et al. 1991; Melchinger et al. 1991; Demeke et al 1992; Dias et al. 1992; Hosaka and Spooner 1992; Song and Osborn 1992; Wang et al. 1992). Despite warnings to the contrary (Whitkus et al. 1993), most of the phylogenetic results using nuclear RFLPs in *Solanum*, *Lactuca*, and *Brassica* are consistent with those obtained from previous molecular studies. For example, the demonstration of multiple origins of *Brassica napus* (Brassicaceae) using phylogenetic analysis of RFLPs is verified by analysis of both the mitochondrial and chloroplast genomes (Song and Osborn 1992). A few studies have attempted to address the important issue of the molecular basis for these RFLPs, either restriction site mutations or indels (insertion-deletions), in order to appease the critics who insist that problems with homology rule out any confidence in the phylogenetic application of nuclear RFLPs. Comparative genome mapping, for later phylogenetic use, has been done on *Brassica* (Song et al. 1991), *Sorghum* (Whitkus et al. 1992), and *Solanum/Lycopersicon* (Tanksley et al. 1992).

In a similar vein, the use of RAPDs (random amplified polymorphic DNAs) in phylogenetic studies has been viewed with suspicion (Whitkus et al. 1993). These criticisms rest on three perceived or real problems: (1) the dominant nature of RAPDs, thus making heterozygotes indistinguishable from dominant homozygotes; (2) nonrepeatability of obtaining the same amplified bands, often due to influences of DNA quality and quantity (but see Fritsch et al. 1993 for another view); and (3) nonhomology of comigrating bands. Tests of the applicability of RAPD technology to phylogenetic studies at a variety of levels must be done using well-characterized plant groups whose relationships are based on a wealth of other molecular and morphological data. RAPD analyses within *Microseris elegans* (Asteraceae) are largely congruent with isozyme analysis, although the latter is limited by the number of informative polymorphisms (van Heusden and Bachmann 1992). More of these kinds of studies are sorely needed. Nevertheless, RAPD analyses have been used successfully in the documentation of hybridization or the assessment of genetic variation (Brauner et al. 1993; Crawford et al. 1993).

Phylogenetic studies of specific nuclear genes or noncoding regions have been limited to the rRNA coding genes (5S, 5.8S, 18S, 26S) (Buchheim and Chapman 1991; Chapman and Buchheim 1991; Troitsky et al. 1991; Hamby and Zimmer 1992; Bult and Zimmer 1993), their internal, transcribed spacers (ITS1, ITS2) (Baldwin 1992, 1993; Baum et al. 1993; Kim and Jansen 1993; Suh et al. 1993; Wojciechowski et al. 1993), and a handful of other genes (e.g., Doyle 1991b; Martin et al. 1993). The potential dangers of using nuclear-encoded genes, especially with reference to orthologous versus paralogous comparisons, are lucidly explained by Doyle (1991b). As usual, we plant systematists are trailing our zoological counterparts who have identified at least 14 promising nuclear genes for phylogenetic studies (Friedlander et al. 1992). Has the occurrence of the versatile chloroplast in plants made us too unwilling to expand our horizons? Work in progress on *rbc*S (Meagher et al. 1989; Martin and Dowd 1991), *pgi* (Thomas et al. 1992), *gap*C (Martin et al. 1993), and *adh* (Gaut and Clegg 1993) might do the trick.

The spread and evolution of retrotransposable elements in the nuclear genome has been tackled from a phylogenetic perspective in plants. Phylogenetic analyses of *copia*-like retrotransposons in *Gossypium* (Malvaceae) indicate that nine sharply differentiated lineages exist, that these lineages are incongruent with an organismal phylogeny of *Gossypium*, and that horizontal (recent) transfer has not occurred (VanderWiel et al. 1993). A consideration of processes that obscure phylogenetic reconstruction of multigene families demonstrated that such incongruity is expected if vertical processes are the sole means of transmission.

The mitochondrial (mt) genome remains elusive in terms of its potential applicability to systematic studies (Lonsdale 1992; Oda et al. 1992; Palmer 1992 a, c; Palmer et al. 1992). Although well characterized in animals and fungi, the mt genome of land plants is beset with problems from a practical point of view: mtDNA is more difficult to extract, not abundant in leaves relative to cpDNA, few genes have been sequenced, the genome structure is complex and changes rapidly, it contains substantial portions of the cp genome, and nucleotide silent substitutions are three to four times slower than in cpDNA genes (Palmer 1992a). Hirai and Nakazono (1993) report that 6% of the mitochondrial genome in rice comes originally from the chloroplast genome. Twenty-two kb of cpDNA (about 19% of the rice genome excluding one copy of the inverted repeat) occurs in the rice mt genome in 16 fragments ranging from 32 to 6.8 kb. Restriction site studies at the interspecific level will certainly not be successful, and no higher level sequence study of mt genes has been published. The situation in algae might be different. Coleman and Goff (1991) report that the mt genomes of algae so far examined (admittedly few) are much smaller and possibly less complex than land plant mt genomes.

Our knowledge of the structural and sequence evolution of the chloroplast (cp) genome (Palmer 1991, 1992c; Clegg and Zurawski 1992; Downie and Palmer 1992a; Clegg 1993), well characterized over the last 10 years with complete sequences known from a bryophyte, a monocot, and a dicot, was further enhanced in the last 2 years by the complete sequencing and structural analysis of the 70 028 bp plastid genome from the obligately parasitic and nonphotosynthetic *Epifagus virginiana* (Orobanchaceae) (Morden et al. 1991; Wolfe et al. 1992a, b). This model system is providing valuable information on nonphotosynthetic functions operating in plastids and on rates of molecular divergence in retained genes. Also of critical importance during these last 2 years has been the examination of algal cpDNA and gene sequences (Gray 1991; Palmer 1991; see below), information that should prove invaluable in addressing questions of polarity in studies of early land plant radiations. Complete chloroplast genome sequences of some algae will soon be available (Palmer 1991).

Other firsts during this time period were the explicit, phylogenetic use of noncoding, spacer sequences (Golenberg et al. 1993) and of gene and/or intron loss information (Downie et al. 1991; Gantt et al. 1991; Downie and Palmer 1992a). The transfer of the plastid *rpl*22 gene to the nucleus in the immediate ancestor of all legume families is the most recent evolutionary gene transfer yet described (Gantt et al. 1991). Importantly, the transfer event was shown to greatly precede its loss from the chloroplast and also involved the gain of an intron. This example and phylogenetic analysis of intron distributions suggest that all nuclear introns and most group I and II introns were inserted late in eukaryotic evolution and that the role of exon-shuffling in the formation of primordial genes can be disregarded (Palmer and Logsdon 1991).

d) Data Analysis

Has data accumulation, especially in the form of nucleotide sequences, outpaced our efforts to analyze (let alone publish!) them? Several important advances have occurred in the last 2 years that are slowing the widening of this gap. These advances fall into three groups, irrespective of whether distance or discrete analytical procedures are used: (1) computing distance measures or weighting characters/character states, (2) estimating "best" trees with numerical algorithms, and (3) comparing trees from one or more data sets.

Simulation studies (reviewed in Nei 1991) using different rates of sequence divergence have demonstrated that the simplistic Jukes-Cantor method (one-parameter model) is effective in estimating actual nucleotide divergence between two sequences when the substitution rate (K) is less than about 0.3. At higher levels of divergence but less than $K = 1.0$, or when transition bias is high, the Kimura method (two-parameter model) should be used. Various other models have been proposed and are effective under certain other assumptions of sequence evolution. A maximum-likelihood based, relative rate test has been proposed by Muse and Weir (1992) and used by Gaut et al. (1992). Various weighting schemes (generalized parsimony) have been suggested for discrete character analysis of either restriction site or nucleotide sequence data (Albert et al. 1992a, 1993; Albert and Mishler 1992; Holsinger and Jansen 1993; Swofford 1993). The empirically determined stepmatrices designed for *rbc*L sequence analysis (Albert and Mishler 1992; Albert et al. 1993), however, greatly increase the computational load of the tree-building algorithms to such an extent that the ability to find most parsimonious solutions is (often?) severely handicapped (e.g., Chase et al 1993). Various approaches to remedy this problem are being explored (Olmstead et al. 1993; Page 1993b).

New insights into the pros and cons of tree-building algorithms strongly suggest a need for molecular systematics to expand its repertoire to include other methods rather than relying solely on maximum parsimony. Despite intense pressure from some groups, distance and maximum likelihood methods have been used successfully over the last couple of years (e.g., Clegg and Zurawski 1992; Gaut et al. 1992; Clegg 1993; Duvall et al. 1993a, b; Golenberg et al. 1993). Explicit strengths and weaknesses of various distance methods such as distance-Wagner, neighbor-joining, Fitch-Margoliash, and UPGMA and discrete character methods such as maximum parsimony, evolutionary parsimony, and maximum likelihood have been examined based on theoretical grounds (Stewart 1993) and, more importantly, using simulation studies (reviewed in Nei 1991). The latter review is invaluable in presenting information (although much more needs to be done!) on the power and reliability of various methods under different parameters of sequence evolution, time of divergence, and number of nucleotides examined. The following quote concerning maximum parsimony is sufficient to illustrate the problem of a sole endorsement of this method in molecular systematic studies: "One of the general conclusions is that the maximum parsimony method ... is as efficient as the ME [minimum evolution], NJ [neighbor-joining], and ML [maximum likelihood] methods only when the number of nucleotide substitutions per site is very small and the number of nucleotides examined is very large. Otherwise, the latter methods are better than the MP [maximum parimony] method. This seems to be true whether or not the transition/transversion bias is taken into account in the parsimony method ..." (Nei 1991). Three new versions of the most versatile and user-friendly phylogenetic computer programs, as a group, incorporate most of these methods: PHYLIP version 3.5 (Felsenstein 1993), PAUP/Mac version 3.1.1 (Swofford 1993; debugged replacement for the just released version 3.1), and MacClade version 3.0 (Maddison and Maddison 1992). PAUP has also been available and run on PC, Unix, and Cray computers. Ongoing attempts to better reconstruct phylogenetic trees are summarized in Penny et al. (1992).

Although still in debate, methods of tree comparison from within or especially between data sets are being examined and used (Barrett et al. 1991; Swofford 1991; Baum 1992; Caputo et al. 1992; Donoghue and Sanderson 1992; Doyle 1992; Källersjö et al. 1992; Ragan 1992; Sytsma and Smith 1992; DeQueiroz 1993; Page 1993a). Two basic approaches have been followed: combining all data sets together without weighting to arrive at a "total evidence" view of organismal relationships (Barrett et al. 1991; Donoghue and Sanderson 1992), or initial and separate analyses of different data sets prior to combining all data (Swofford 1991; Sytsma and Smith 1992; Kim and Jansen 1993). The "gene" versus "species" tree debate prompted Doyle (1992) to devise a weighting scheme for organelle data, a procedure perhaps difficult to justify (why not do the same for nuclear data?). Metric approaches to tree comparison (comparing and quantifying differences in shapes of trees) have also been suggested by Baum (1992), Caputo et al. (1992), and Ragan (1992) as a means of phylogenetic inference without combining the actual data behind the trees. Hidden behind these debates on handling different data sets is the debate on the phylogenetic utility of various data types. Donoghue and Sanderson (1992) and Swofford (1991) present good arguments for considering all data. The explicit claim of Donoghue and Sanderson (1992) that morphological and molecular data have the same internal consistency, when the number of taxa are factored in, is not supported by a larger survey of plant restriction site and sequence studies (Givnish and Sytsma 1992). This survey shows that restriction site data have higher internal consistencies than do sequence data, and the latter in turn is more consistent than morphological studies. Unfortunately, only a few systems have been phylogenetically examined with a diversity of data types including cladistic analyses of morphology (e.g. *Krigia*, Asteraceae, Asteridae, Onagraceae, seed plants).

4. Systematic Progress in Algae

More molecular and molecular phylogenetic studies have been done on the chlorophytes than any other algal or protistan group. The extensive sequencing of nuclear rRNA (and later rDNA) in the late 1980s is summarized in Buchheim and Chapman (1991) and Chapman and Buchheim (1991). The taxonomic implications of these data are just beginning to be evaluated. The monophyly (or even paraphyly) of the large genus *Chlamy-*

domonas is certainly not supported, and a major revision of the genus is in order (Chapman and Buchheim 1991). Likewise, the monophyly of the Chlorophyceae is not supported using both DNA and amino acid sequence analysis of *rbc*L (Manhart and VonderHaar 1991). The nuclear rRNA data, interestingly, support the Charales (*Chara* and *Nitella*) as the sister group to plants and not *Coleochaete*; the latter being implicated as closest to land plants based on cladistic analysis of morphological and ultrastructure characters (Graham et al. 1991).

Although chlorophytes have been the focus of most molecular phylogenetic work, recent data on the plastid genomes of nonchlorophyte algae have accumulated (e.g., Ariztia et al. 1991; Gray 1991; Palmer 1991; Scherer et al. 1991; Shivji 1991; Bourne et al. 1992; Howe et al. 1992; Shivji et al. 1992; Reith and Munholland 1993). These and other studies bear directly on the precise nature of the endosymbiotic events that gave rise to plastids and their genome (Gray 1991; Wolfe et al. 1991; Howe et al. 1992). Two endosymbiotic routes for the acquisition of plastids by eukaryotes are possible: "primary" involving a eukaryote and a bacterium, and "secondary" involving the uptake of a eukaryote by another eukaryote. Phylogenetic analysis of two 18S rRNA genes from *Cryptomonas* clearly indicates secondary endosymbiosis (Douglas et al. 1991; Douglas 1992). One gene, inferred to come from nucleomorph (vestigial nucleus) DNA, is placed in the rhodophyte lineage as expected from the pigment composition of plastids in *Cryptomonas*. The nuclear 18S rRNA gene, however, is more closely related to chromophytes and chlorophytes. Evidence supporting secondary endosymbiosis as the origin of the plastid in *Euglena* (and perhaps all chromophytes) is also accumulating (Douglas et al. 1991; Morden and Golden 1991; Wolfe et al. 1991).

The interrelationships of eukaryotic algae (chlorophytes, rhodophytes, chromophytes, etc.), cyanobacteria, and prochlorophytes (chlorophyll b-containing bacteria) are still poorly known. Evidence for a monophyletic origin of chloroplasts is suggested by the presence in all chloroplasts but not in cyanobacteria of two groups of genes that are clustered, *rpo*/*rps*2/atp and *psb*B/N/H (Reith and Munholland 1993). Howe et al. (1992) present alternative scenarios for the evolution of the entire group, but stress that polyphyletic origins of plastids and re-occurring endosymbiosis are likely events that are confounding the issue. More than one lineage of bacteria is implicated as the source of plastid genes in other studies (Gray 1991; Wolfe et al. 1991). Phylogenetic analysis of both the large and small subunit plastid genes of rubisco (*rbc*L and *rbc*S, respectively) indicates that nonphotosynthetic b-purple bacteria are closer relatives to rhodophytes and chromophytes than they are to chlorophytes and cyanobacteria (Morden and Golden 1991). In contrast, rhodophytes, chromophytes, and chlorophytes are a monophyletic group and fall within the cyanobacteria lineage based on sequences from plastid 16S rRNA and other photosynthetic genes. Morden et al. (1992) expand on these results by examining relationships among purple bacteria, cyanobacteria, and the plastids of various algal groups and land plants using six different gene sequences. All six phylogenetic trees are consistent in (1) having the chlorophytes derived from cyanobacteria and not from prochlorophytes (supporting Palenik and Haselkorn 1992; Urbach et al. 1992) and (2) the clustering of the nongreen algae. Incongruity of the *rbc*L and *rbc*S sequence results with the other four gene results suggests that a single bacterial endosymbiotic event occurred leading to all plastids, followed by the lateral transfer of the rubisco gene operon from a purple bacterium to a rhodophyte. The relationships of the algal lines and the precise nature of the endosymbiotic event(s) are clearly not yet settled.

Surprising evidence for a green (possibly dinoflagellate) ancestry of malarial parasites has accumulated (reviewed in Palmer 1992b). *Plasmodium* and related protists contain two extrachromosomal genomes, one of which is suggestively similar to the plastid genome of dinoflagellates but which has been reduced, a likely outcome due to the host's present nonphotosynthetic life style (Depamphilis and Palmer 1990; Morden et al. 1992; Wolfe et al. 1992a, b).

16S-like sequences from a diversity of protists indicate that the Oomycete fungi are placed with the chromophyte algae (Ariztia et al. 1991). This result supports the hypothesis that protists with tripartite hairs form a natural assemblage.

An emerging area in algal biosystematics is at the level of intraspecific relationships. Chloroplast, mitochondrial, and nuclear DNA sequences and restriction site analyses have been used to test species concepts in a wide variety of algae (Coleman and Goff 1991; Steane et al. 1991; Maggs et al. 1992; Manhart and McCourt 1992).

5. Systematic Progress in Bryophytes

Molecular studies in bryophytes have largely been confined to answering the question of the interrelationships of the three major groups within this obviously paraphyletic grouping. A cladistic analysis of cytoplasmic rRNA sequences from eight bryophytes and various tracheophyte taxa indicates that the mosses and hornworts are a monophyletic clade and represent the sister group to tracheophytes (Waters et al. 1992), whereas the liverworts are the sister group to all other bryophytes + tracheophytes. These molecular results are contrary to most morphologically based cladistic studies placing the mosses as the sister group to tracheophytes; a relationship seen, however, in one-step longer trees using these molecular sequence data.

Mishler et al. (1992) analyzed relationships of the bryophytes with chloroplast rDNA sequences of both the small and large subunits. Individually, the two sequences provided no resolution for relationships or generated clearly unacceptable relationships based on previous morphological studies. The combined data set placed the mosses sister to the tracheophytes, as expected based on morphological evidence, with the liverworts more basal. Clearly, larger numbers of taxa and characters are needed to solve the phylogenetic relationships of the three lineages of bryophytes and their relationship to tracheophytes. One wonders whether this question is not unlike a zoological counterpart: the holy trilogy of gorilla, chimpanzee, and human.

6. Systematic Progress in Nonseed Vascular Land Plants

Like the bryophytes, the early vascular land plants have been generally neglected in molecular systematic studies. A concerted and multilaboratory effort is unquestionably needed to examine the genomes and phylogenetic relationships of this group. Chloroplast genome mapping and gene localization studies indicate that a good number of gene rearrangements occur within ferns. The distribution of a 30 kb inversion in the cp genome found in a liverwort and a moss but not in seed plants was surveyed in other vascular plants (Raubeson and Jansen 1992b). Three lycopods (*Lycopodium*, *Selaginella*, and *Isoetes*) share the inversion orientation with the bryophytes, identifying the lyocopods as the basal lineage in the vascular land plants. *Psilotum*, sometimes considered primitive, is shown to be a derived member of the tracheophytes. Perhaps the old notion must be reconsidered that bryophytes are secondarily reduced tracheophytes?

The four families Adiantaceae, Cyatheaceae, Dennstaedtiaceae, and Dryopteridaceae possess cpDNA having an inverted repeat that is rearranged with respect to both *Osmunda* (Osmundaceae) and angiosperms (Hasebe and Iwatsuki 1992; Stein et al. 1992). The *psb*A gene, usually on the large single copy flanking region of the inverted repeat, has been "captured", therefore duplicated, by the inverted repeat in these four families. This unique rearrangement indicates a monophyletic origin of the four families relative to Osmundaceae, and is consistent with some, but not all, fern classification schemes. A broader survey of fern families is clearly in order.

Maximum-likelihood analysis of amino acid sequences inferred from *rbc*L nucleotide sequences indicates that the eusporangiate family Ophioglossaceae is derived from within the leptosporangiate families (Hasebe et al. 1993). These results do not support the hypothesis that eusporangiate ferns are linked with the progymnosperm – seed plant lineage (e.g., Kato 1988).

Restriction site analysis of cpDNA in tropical tree ferns (Cyatheaceae) (Conant et al. 1993) supports one of three previous classification schemes for the family. Subsequent morphological character analysis strongly suggests that simple, emarginate scales are derived from complex, marginate ones; the evolution of indusial structures, however, are ambiguous using molecular data as a guide.

7. Systematic Progress in Gymnosperms

Interestingly, two *rbc*L sequence-based studies generate different conclusions concerning the monophyly of extant anthophytes (angiosperms + Gnetales). The large-scale analysis of Chase et al. (1993) supports a paraphyletic gymnosperm clade with the Gnetales as the sister group to angiosperms. In the more limited study of Hasebe et al. (1992b), the gymnosperms (with Gnetales) are monophyletic. The latter study, however, differs from the former in a number of respects: first, only six gymnosperms and four angiosperms are used (versus 31 and 450+ in the former); second, implied amino acid sequence data were used rather than nucleotide sequences because of variable GC content at the third position in all codons; and third, neighbor-joining and maximum-likelihood analyses were employed (rather than maximum parsimony in the former).

In a related paper, Hasebe et al. (1992a), indicate that *Welwitschia* and *Gnetum* are sister genera within the Gnetales using *rbc*L nucleotide sequences. Analysis of the implied protein sequence, however, indicates that *Welwitschia* and *Ephedra* are sister genera. They suggest that the former should be accepted (as in most studies, including

Chase et al. (1993)) because in this case the third position in the codon is not variable with respect to GC ratios, and because few differences are seen at the amino acid level.

The chloroplast gene (*chl*L = *frx*C) is present in all chlorophyll-containing organisms except for angiosperms, Gnetales, and *Psilotum* (Burke et al. 1993). This distribution provides additional evidence for the monophyly of angiosperms + Gnetales and implies a secondary loss in the lineage to *Psilotum*.

The structural loss of the inverted repeat in cpDNA, a landmark feature of almost all land plant lineages (previously known to be absent from Fabaceae and Pinaceae), is shown to be absent from all conifer families (Raubeson and Jansen 1992a). The fact that the Taxaceae share the loss with all remaining conifers discounts Florin's contention four decades ago that the groups separated prior to the evolution of the seed.

Sequence analysis of *rbc*L within the Cupressaceae sensu stricto provides support for recognition of a basically northern subfamily (Cupressoideae) and a southern subfamily (Callitroideae) (Gadek and Quinn 1993). These molecular results agree with previous anatomical and seed protein data, but are strongly contrary to the widely accepted intrafamilial classification scheme. The consensus of these different approaches indicates that this family deserves a new classification system.

8. Systematic Progress in Angiosperms

Relationships among flowering plants have been examined using a diverse set of methods including sequencing of *rbc*L, rDNA coding genes (or rRNA), and the internal transcribed spacers in rDNA (ITS), rearrangements of the cp genome, and restriction site analyses of both cpDNA and rDNA. For the first time, sequence analysis of DNA has been performed in an attempt to rigorously resolve higher level relationships. A good start (despite being based on a single gene) is seen in the large *rbc*L survey of flowering plants (Chase et al. 1993) and the accompanying papers (at the family, order, and class levels) in the special volume of the *Annals of the Missouri Botanical Garden* dedicated to phylogenetic analyses of *rbc*L variation.

a) Major Lineages of Flowering Plants

Seven primary lineages of flowering plants are identified in Chase et al. (1993): Ceratophyllaceae, Magnoliales, Laurales, Nymphaeales, Piperales (including "paleoherbs"), monocots, and eudicots. The enigmatic Ceratophyllaceae is shown to be the most basal lineage of flowering plants (Les et al. 1991; Chase et al. 1993). Unfortunately, none of the global morphological analyses have included Ceratophyllaceae (e.g., Doyle and Donoghue 1992), so its placement at the base of angiosperms in neither rejected nor supported with other lines of evidence. Four woody families lacking vessels are strongly supported members of the now expanded Nymphaeales: Austrobaileyaceae, Amborellaceae, Illiciaceae, and Schisandraceae (Chase et al. 1993; Qui et al. 1993).

The monocots from a well-supported monophyletic clade with a number of major lineages recognized (Chase et al. 1993; Clark et al. 1993; Duvall et al. 1993a, b; Smith et al. 1993); these relationships are similar in many respects to those advocated by Dahlgren. Of considerable interest is the identification of *Acorus*, now in its own family Acoraceae, as the most basal monocot (Duvall et al. 1993c). The implications of this result, if verified with other lines of evidence, will be immense for understanding the characterization of early monocot evolution. Exclusive of *Acorus*, the Aranae/Alismatanae are most basal. Higher up, the Lilianae form a paraphyletic grade of three lineages roughly corresponding in Dioscoreales, Liliales, and Aparagales, with certain other taxa included (e.g., Cyclanthaceae, Pandanaceae). Next comes the Arecanae, followed by the Zingiberales plus a clade of Commelinaceae, Pontederiaceae, Philydraceae, and some Haemodoraceae. The Zingiberales is one of the best-supported orders of flowering plants based on morphological and molecular data (Smith et al. 1993). The Bromeliaceae and Rapateaceae form a well-supported clade that is the sister group to the Poales, Cyperales, and Typhales.

The eudicots are a monophyletic group characterized as the higher dicots with triaperturate or triaperturate-derived pollen and, with *rbc*L sequence data, as one of the best-supported clades among the angiosperms (Qui et al. 1993). The two orders of Mangoliidae with triaperturate pollen, the Ranunculales and Papaverales, form the sister group to all other eudicots. Within the other eudicots, two large clades are identified that roughly correspond to the Rosidae and Asteridae. The groupings indicated by *rbc*L sequence variation reflect the schemes of Dahlgren best, then Thorne, and those of Cronquist/Takhtajan the worst. The ever elusive Dilleniidae is fragmented into the Rosidae with some elements (Ericalean lineage, Kron and Chase 1993) going into the Asteridae. The Hamamelidae is shown, not surprisingly, to be a polyphyletic grouping (Chase et al. 1993). Trochodendraceae and Tetracentraceae form a basal clade within eudicots, the Hamamelidaceae allies with portions of the Saxifragaceae, the Urticales is closest to Rhamnaceae, and the remainder (Fagales, Juglandales, etc.) forms a fourth clade within the Rosidae.

A number of studies using *rbc*L (Chase et al. 1993; Michaels et al. 1993; Olmstead et al. 1992a, b, 1993) and one study using restriction site mapping of the inverted repeat in cpDNA (Downie and Palmer 1992b) have extensively examined the subclass Asteridae. Although not all in complete agreement, six major groups are identified that correspond well to traditionally recognized orders: Asterales, Boraginales, Dipsacales, Gentianales, Scrophulariales (with Lamiales), and Solanales. The Asteridae is only monophyletic if additional families are included, most notably Apiaceae, Araliaceae, Cornaceae, Hydrangeaceae, and Loasaceae. It is noteworthy that these families have been previously suggested as asterid relatives, the strongest evidence coming from the much maligned field of serology. Additional families, especially Ericaceae and relatives, can be easily placed within the enlarged Asteridae.

b) Ordinal Studies

Molecular data has had an impact on the circumscriptum of a number of orders, but the three most interesting include the Capparales, Caryophyllales, and Poales. Glucosinolate (mustard oil) compounds are known from 15 families, with 14 of these families exclusively glucosinolate-producing. Cladistic analyses of all but one of the families (Gadek et al. 1992; Rodman et al. 1993) indicate that nearly all glucosinolate producers form a monophyletic clade (an expanded Capparales?). The one exception is *Drypetes*, the only mustard oil producer in the Euphorbiaceae, which is placed with *Euphorbia* elsewhere in the Rosidae. These results, which tie together such disparate families as Limnanthaceae, Caricaceae, Moringaceae, Brassicaceae, and Tovariaceae, are contrary to Cronquist and others but congruent in the large part with Dahlgren and recent morphological analyses of Rodman (1991a, b).

Four different *rbc*L analyses (Giannasi et al. 1992; Rettig et al. 1992; Chase et al. 1993; Manhart and Rettig 1993) and one cpDNA restriction site analysis (Downie and Palmer 1993) have been done on the Caryophyllales. Surprisingly, these analyses conflict with each other in many aspects of the resulting trees, a problem attributed to the poor sampling of some studies (Manhart and Rettig 1993). The Phytolaccaceae are not basal, contrary to Cronquist, nor are the nonbetalain-producing taxa. Four lineages apparently diverged early in the history of the order: (1) Chenopodiaceae and Amaranthaceae; (2) Basellaceae, Portulacaceae, Cactaceae, and Didiereaceae; (3) Caryophyllaceae; and (4) Phytolaccaceae, Petiveriaceae, and Nyctaginaceae (Manhart and Rettig 1993). A broader sampling within the families, especially Phytolaccaceae, and incorporation of other genes or genomes are clearly in order.

Two quite different studies have shed light on the relationships and circumscription of the grass alliance. Analyses of *rbc*L sequences (Chase et al. 1993; Duvall et al. 1993a) and the distribution of three cpDNA inversions (Doyle et al. 1992a) strongly implicate the separation of the Poaceae and Cyperaceae as two distinct but related alliances. The nongrass families Restionaceae, Ecdeiocoleaceae, and Joinvilleaceae group with Poaceae. The Flagellariaceae, Anarthriaceae, Typhaceae, and Juncaceae are further removed from Poaceae, or are allied with Cyperaceae. Better resolution is needed in this portion of the commelinoid monocot lineage.

c) Familial Studies

The saga continues on understanding generic/tribal relationships in a number of model families including Asteraceae, Leguminosae, Onagraceae, Saxifragaceae, and Solanaceae. A large number of studies have examined generic relationships within the Asteraceae, making it probably the best studied of the larger families from a molecular perspective (Jansen et al. 1991a, b, 1992; Kim et al. 1992a; see Sect. 9, below).

Molecular phylogenetic analysis of the Leguminosae, recently reviewed in Doyle et al. (1992b), continues with higher level analysis of tribes and subfamilies using cpDNA characters such as the loss of the inverted repeat, inversions, and restriction site variation (Doyle and Doyle 1991, 1993; Lavin and Doyle 1991; Lavin et al. 1991; Delgado-Salinas et al. 1993). The tribe Phaseoleae, probably the most extensively studied, is not monophyletic (Doyle and Doyle 1993), a result congruent with recent morphological studies.

The Onagraceae is one of the most thoroughly studied families of moderate size, but molecular evidence is continuing to give surprises. Three quite different attempts to address the question of tribal relationships were published: morphology (Hoch et al. 1993), nuclear rRNA sequences (Bult and Zimmer 1993), and *rbc*L sequences (Conti et al. 1993). Although only 17 informative characters were available from nuclear rRNA, relationships based on that data were very similar to that obtained from the more robust *rbc*L data, with only the relative positions of *Hauya* and *Lopezia* in conflict. It is no surprise that *Ludwigia* is consistently shown to be the sister group to the remainder of the family, but the very close relationships of temperate *Circaea* to largely tropical *Fuchsia*, seen in both molecular studies, is of some biogeographical and evolutionary interest. A summary of the congruence or lack of it between DNA, isozymes, and morphology in *Clarkia* and *Fuchsia* is provided in Sytsma and Smith (1992) (also see Sect. 9, below).

The Saxifragaceae have been the focus of several large surveys involving *rbc*L sequence comparisons, loss of the intron in the chloroplast gene *rpl*2, and cpDNA restriction site analysis (D. E. Soltis et al. 1991a, b, 1992c, 1993; Morgan and Soltis 1993). The Saxifragaceae sensu lato is a diverse polyphyletic entity (Chase et al. 1993; Morgan and Soltis 1993), probably the most unnatural family yet examined. Within the core Saxifragaceae, a number of poorly resolved basal lineages are seen (D. E. Soltis et al. 1993), suggesting a rapid and early diversification of the group. The genus *Saxifraga* is also shown not to be monophyletic (Soltis et al. 1993).

Phylogenetic relationships of 12 of the currently recognized 14 tribes in the Solanaceae, based on cpDNA restriction site analysis, indicate that the subfamily Cestroideae is ancestral but paraphyletic in the family and that the monophyletic subfamily Solanoideae is derived from the Cestroideae (Olmstead and Palmer 1991, 1992). The advanced nature of the Solanoideae is contrary to the tradional view. Analysis of character evolution using the cpDNA tree suggests that floral zygomorphy and reduction in stamen number occurred more than once; this makes the tribe Salpiglossideae, defined by these characters, polyphyletic.

9. Morphological/Molecular Integration

The last few years have seen the systematic community accepting molecular data, either alone or in conjunction with cladistic analysis of morphological data, as a basis for classificatory and nomenclatural changes. A much needed reevaluation of the classification of the Asteraceae has been presented (Bremer et al. 1992) with the erection of a third subfamily, the Barnadesioideae, along with the Asteroideae and Cichorioideae (Bremer and Jansen 1992). These taxonomic changes were based on detailed and rigorous cladistic analysis of both morphological and molecular data sets (Jansen et al. 1991a) and are a pleasant foretaste of where plant systematics is headed. Two genera within the Onagraceae, shown to be embedded within larger genera based on molecular data from both the nuclear and chloroplast genomes, have been formally submerged: *Boisduvalia* in *Epilobium* (Hoch and Raven 1992; Baum et al. 1993) and *Heterogaura* in *Clarkia* (Sytsma and

Gottlieb 1985; Lewis and Raven 1992; Sytsma and Smith 1992). The situation with *Heterogaura*, exhibiting reduced indehiscent nutlets relative to many-seeded capsules seen in its sister species in *Clarkia*, is also reflected in *Roemeria* (Papaveraceae) and *Astragalus* (Fabaceae). Chloroplast DNA restriction site data unambiguously place the elongated, loculicidal-dehiscent *Roemeria* squarely within a well-defined portion of the poricidal *Papaver* (Kadereit and Sytsma 1992). *Astragalus gambelianus* differs from all species of subsection *Californici*, in which it is placed using cpDNA restriction site variation, by its orbicular pods that separate into one-seeded indehiscent halves (Liston 1992).

Sequence analysis of *rbc*L supports previous claims based on cladistic analysis of morphological characters that *Sambucus*, *Viburnum*, and *Adoxa* form a natural group separate from Caprifoliaceae sensu stricto, with nomenclatural changes to follow (Donoghue et al. 1992). Molecular, chemical, and morphological data support, although to different degrees, the placement of *Ulmus rubra* in a new section *Fulva* (Wiegrefe et al. 1993). As has been painfully obvious for many years (even to Linnaeus), tomatoes are really potatoes. The recent submerging of *Lycopersicon* into *Solanum* (Spooner et al. 1993), based on chloroplast DNA and other lines of evidence, hopefully is the end to this controversy. The Hawaiian endemic *Rollandia* has been shown to be a part of the larger Hawaiian endemic *Cyanea* using cpDNA evidence (Givnish et al. 1993). With support from morphological studies *Rollandia* has been submerged, making *Cyanea* the largest genus in the Hawaiian islands (Lammers et al. 1993).

CpDNA restriction site analyses of the *Machaeranthera* complex (Asteraceae) support an alliance of *Psilactis* with some species of *Aster* (Morgan and Simpson 1992; Morgan 1993) and nomenclatural changes were proposed to remove *Psilactis* from *Machaeranthera* (Morgan 1993). The reluctance in not proposing nomenclatural changes for the related species of *Aster* with which *Psilactis* is allied will not settle well with some. Is there a "Russian doll" problem here? The chloroplast genome of *Iostephane* (Heliantheae: Astereaceae) is virtually identical to *Viguiera* sect. *Maculatae* (Schilling and Panero 1991), but the authors were reluctant (perhaps wisely so) to make formal taxonomic changes until all related genera were examined. The subtribe Brassicinae (*Brassica* and ten other genera) has similar generic problems. Chloroplast DNA analyses (Warwick and Black 1991; Warwick et al. 1992) indicate that drastic expansion of the genus *Brassica* is necessary to include related genera in the subtribe, as proposed by earlier workers. A different approach to integrate molecular and morphological data is illustrated in B. B. Bremer's (1992) study in Rubiaceae. Restriction site analysis of cpDNA defined a monophyletic assemblage that included a well-defined tribe Chiococceae characterized a posteriori by several flower and fruit characters. Retaining Chiococceae in the narrow sense, however, would cause all remaining taxa in the broader monophyletic assemblage to become a paraphyletic cluster without a single morphological character uniting this cluster. B. B. Bremer (1992) proposed, as the best taxonomic solution to this problem, a redefinition and widening of the tribe Chiococceae to include the entire monophyletic assemblage. We predict a flood of similar taxonomic changes, and a backlash of opposition. Food for thought: what impact might these changes have on conservation biology, status of endangered species, and apparent distinctiveness of certain floras (Hawaii, Guayana Shield)?

An increasing number of studies are attempting to directly quantitate the congruency (or lack of it) between different data sets. The largest published suite of different data sets that have been analyzed in this fashion is found in *Krigia* (Asteraceae), and include morphology, 45S rDNA restriction sites, ITS sequence, and cpDNA restriction sites (Kim and Mabry 1991; Kim et al. 1992b, c; Kim and Jansen 1993). Multiple data sets exist for assessing tribal relationships within Onagraceae (e.g., Bult and Zimmer 1993; Conti et al. 1993; Hoch et al. 1993) and for familial relationships within Zingiberales (see Smith et al. 1993), but the overall congruency tests have not been published.

Bremer and Struwe (1992) point out that most incongruency between molecules and morphology in the Rubiaceae is due to procedural problems in how traditional morphological data have been analyzed (i.e., apparent but not real incongruency, sensu Sytsma 1990). When both data sets are handled in a cladistic fashion, very little incongruity is seen; the authors, however, do not provide any explicit measure of incongruency.

10. Hybridization/Introgression

The molecular perspective of reticulate and/or polyploid evolution is reviewed in detail by Ritland and Eckenwalder (1992), Rieseberg and Ellstrand (1993), Rieseberg and Morefield (1993), and P. S. Soltis et al. 1992a). Several reviews of the phenomena of and problems associated with hybridization and introgression from a molecular phylogenetic perspective are noteworthy. Rieseberg and Soltis (1991), Nason et al. (1992), Rieseberg and Brunsfeld (1992), and Rieseberg and Wendel (1993) present the evidence for introgression from cpDNA, nuclear rDNA, and isozymes, categorize the different modes and possible mechanisms of introgression, and discuss the evolutionary and systematic consequences of gene or genome exchange. The in-depth survey of examples of introgression in plants by Rieseberg and Wendel (1993) documents the relative frequency of cytoplasmic versus nuclear introgression in plants. All but two of the cases involve cytoplasmic introgression and most of these occurring in the absence of simultaneous nuclear introgression. More than half of the known cases of cytoplasmic introgression were not suspected based on morphological, isozymic, or biogeographic evidence. Only two cases of known differential nuclear introgression have been reported including the recent (and soon to be classic) studies of *Iris* (Iridaceae) (Arnold et al. 1991; Arnold 1993; Cruzan et al. 1993).

Introgression has been described in a variety of plant groups, the most detailed in the last 2 years being sunflowers (Rieseberg 1991; Rieseberg et al. 1991a, b; Dorado et al. 1992), Saxifragaceae (Soltis and Soltis 1991b; D. E. Soltis et al. 1991a, b, 1992a), oaks (Whittemore and Schaal 1991; Nason et al. 1992), willows (Brunsfeld et al. 1992), and cotton (Wendel et al. 1991, 1992; Wendel and Albert 1992).

The absence of hybridization and/or introgression was demonstrated in three wild relatives of potato (Spooner et al. 1991b), a case previously used as exemplary of diploid species hybridization. Multiple origins of the allotetraploid *Tragopogon mirus* (Asteraceae) were shown by Soltis and Soltis (1991b), the latest in a series of similar cases illustrating polyphyly of a species. Likewise, nuclear rDNA and cpDNA indicate that several asexual *Taraxacum* (Asteraceae) taxa are of polyphyletic origin (King 1993). How are these polyploid and asexual populations/taxa to be recognized in a cladistic sense? In contrast, Kim et al. (1992b) provide cpDNA evidence that suggests that all populations surveyed of the autotetraploid *Krigia virginica* originated from a common ancestor.

Using a wealth of information from isozymes, DNA, morphology, cytology, and biogeography, Darbyshire et al. (1992) conclusively demonstrate that *Poa labradorica* (Poaceae) is of intergeneric hybrid origin between *Dupontia fisheri* subsp. *psilosantha* and *Poa eminens*. Based on RAPD markers, Crawford et al. (1993) document the intergeneric hybrid origin of *Margyracaena skottsbergii* (Rosaceae) on the Juan Fernandez Islands and identify the parents as species of *Margyricarpus* and *Acaena*.

The role of the cytoplasm in the evolution of *Brassica* (Brassicaceae) amphidiploid species was investigated by Song et al. (1993). In an elegantly devised system, a species of synthetic amphidiploids (analogs of the three natural amphidiploids) were developed. Tracking the cytoplasm (cpDNA and mtDNA; strictly maternally inherited) and the nuclear genome (RFLPs) for both synthetic and natural amphidiploids, indicates that substantial divergence in relative proportions of two parental nuclear genomes has occurred in the natural amphidiploids.

11. Biogeography

The merging of molecular systematics and biogeography, previously based almost solely on isozyme data, has been occurring even more rapidly in the last 2 years (Crawford et al. 1992). Most of these studies have dealt with "island" systems (tropical mountains, oceanic islands, etc.), but an increasing number are examining intra- or intercontinental systems.

The combined analysis of cpDNA restriction sites and rearrangements permitted Knox et al. (1993) to demonstrate that the pantropical group of giant lobeliads most likely originated paedomorphically from herbaceous ancestors as suggested earlier by Carlquist (1962) using wood anatomy. The molecular data, however, contradict the explicit biogeographic hypothesis of Mabberley (1975) suggesting independent dispersals to Hawaii in order to account for the supposed occurrence of two sections of *Lobelia*. The South American giant lobeliads are shown to be a mixed assemblage, indicating a number of transatlantic dispersal events between Africa and South America.

Dispersal patterns in *Cyanea* (Lobelioideae), the largest endemic genus in the Hawaiian islands, were estimated using phylogenetic analyses of cpDNA restriction site variation in *Cyanea* and nested outgroup genera (Givnish et al. 1993). As has been demonstrated with other plant and animal groups (Wagner and Funk 1993), the origin of *Cyanea* most likely occurred on the oldest island Kauai (or an older but now submerged seamount) and most dispersal events ensued from older to younger islands.

The North American (presumably Californian) origin of the spectacular Hawaiian silversword alliance (Asteraceae) was demonstrated using both cpDNA restriction site and ITS sequence analyses (Baldwin et al. 1991; Baldwin 1992). The Californian tarweed ancestor implicated in the derivation of the Hawaiian silverswords overcame a dispersal barrier of at least 3900 km and the breeding barrier of self-incompatibility.

A rapid radiation of the subgenera of *Dendroseris* (Asteraceae: Lactuaceae) on the Juan Fernandez Islands is suggested by cpDNA restriction site data (Crawford et al.

1992b). The tremendous diversity in habit, leaf form, and capitulum seen in the genus arose in less than 3 million years, using estimated divergence times calculated from cpDNA divergence rates. This is well after the main island of Masatierra was available for colonization (ca. 3.7–4.4 Ma ago).

In one of the most detailed biogeographic studies using molecular data, Wendel and Albert (1992) use an area cladogram of *Gossypium* and various dating methods (geological, palynological, molecular clock) to suggest an origin for the genus during the mid to upper Oligocene in either Australia of Africa. Two colonizations of the New World are indicated, an ancient and long dispersal from Africa and a later dispersal of the maternal ancestor of the allopolyploids.

The combination of cpDNA and nuclear rDNA restriction site data and known fossil distributions suggests that the unusual and Old World restricted *Fuchsia* sect. *Skinnera* dispersed first to New Zealand ca. 30 Ma ago (using the then interconnected continents of Antarctica and Australia as stepping stones), and then secondarily dispersed to Tahiti (Sytsma et al. 1991). The amount of genetic divergence seen between the Tahitian and related New Zealand species and the relatively young age of Tahiti suggest that the secondary dispersal occurred either (1) well after the species had already diverged on New Zealand or (2) to an older, now submerged seamount northwest of Tahiti.

An explicit biogeographic scenario for the evolution of the diploid, tetraploid, and hexaploid species in *Solanum* sect. *Petota* is presented (Spooner et al. 1991a; Spooner and Sytsma 1992). An origin of diploid species in Mexico is inferred, followed by dispersal to South America and subsequent polyploidization. A secondary dispersal of a tetraploid to Central America, hybridization with a diploid, and the formation of hexaploid clade are also supported by the cpDNA data. The biogeographic scenario mirrors that postulated earlier based on the inferred relationships of different genome types.

An important addition to molecular systematics and biogeography is the concept of "intraspecific phylogeography" (Avise et al. 1987). Once restricted to zoological examples, a number of intriguing plant studies have been published (Liston et al. 1992; D. E. Soltis et al. 1991, a, 1992a, b). North-south geographic partitioning of chloroplast DNA variation in *Datisca glomerata* in western North America is suggested to be in response to Pleistocene glaciation and cytoplasmic bottlenecks (Liston et al. 1992). Similar examples of north-south distributions in plastome type have been reported in two species of Saxifragaceae: *Tellima grandiflora* and *Tolmiea menziesii* (D. E. Soltis et al. 1991a, 1992b). Whether these examples from western North America are unusual (due to an interesting interaction of Pleistocene glaciation and plant migration) or exemplary of what can be found elsewhere, remains to be determined.

12. Interfacing Ecology and Systematics

An important trend that is occurring at the interface of ecology and systematics is the use of molecular phylogenies to assess ecological character evolution. This approach allows both ecologists and systematists to circumvent the problem of circularity that accompanies the construction of a phylogeny and then using the phylogeny to understand the evolution of the very characters upon which it is based (Philbrick and Jansen 1991; Sytsma et al. 1991; Givnish et al. 1993).

Phylogenetic analyses of nucleotide sequence data from *rbc*L indicate that carnivory has arisen independently a number of times within angiosperms, as expected (Albert et al. 1992b). Of more interest is the finding that structurally similar flypaper traps are polyphyletically derived and that they share close common ancestry with all other trap mechanisms.

A cpDNA restriction study of the pantropical, woody *Erythrina* (subfamily Papilionoideae) examined the evolution of ornithophilous pollination systems (Bruneau and Doyle 1993). Perching birds (order Passeriformes) pollinate all Old World species and 16 of the New World species, while hummingbirds pollinate the remaining 55 New World species. *Erythrina* is most closely related to Old World genera, indicating that passerine pollination is the ancestral state. The switch to hummingbird pollination evolved convergently in at least three distinct lineages, and possible reversals to passerine pollination syndromes are implicated by the molecular phylogeny.

A cpDNA-based phylogenetic analysis of the Hawaiian *Cyanea* demonstrates that fruit color (purple vs. orange) is a reliable guage of evolutionary history, but that floral features are largely homoplastic (Givnish et al. 1993). Thorns and juvenile leaf dimorphisms, highly correlated in distribution, appear to have arisen a number of times independently in the genus. The two characters, the first for defense and the second for visual deception, presumably evolved in response to selective pressures exerted by now extinct large flightless birds (Givnish et al. 1993).

Character evolution in Callitrichaceae, a largely aquatic family with highly reduced and/or specialized flowers, was examined using a cpDNA phylogeny (Philbrick and Jansen 1991). Problems with selecting an appropriate outgroup or with lack of resolution prevented unambiguous assessment of polarity of growth habit (terrestrial vs. aquatic), fruit development (geocarpic vs. aerial), or chromosome number. The molecular tree conclusively demonstrated, however, that internal geitonogamy is a synapomorphic state for one clade within *Callitriche*.

The probable course of evolution to various breeding systems in *Fuchsia* sect. *Skinnera* was ascertained by cpDNA restriction site analysis (Sytsma et al. 1991). Gynodioecy (controlled by a dominant gene) evolved in the ancestor of the Old World section and is retained by two of the three species in New Zealand. Subdioecy is a derived condition in the third New Zealand species, while the Tahitian species presumably regained the hermaphroditic condition following a long-distance dispersal and founding event of a homozygous recessive individual of gynodioecious ancestry.

A cpDNA-based phylogenetic analysis of the ditypic *Datisca* (Datiscaceae) indicates that the Old World *D. cannabina* is ancestral (i.e., paraphyletic) to *D. glomerata* of western North America (Liston et al. 1992). Surprisingly, the androdioecious condition seen in *D. glomerata* is thus apparently derived from the dioecious condition seen in *D. cannabina* (Rieseberg et al. 1992).

Molecular studies of heterostyly will be successful if genes governing the floral polymorphisms are identified and their DNA sequences obtained. Sequences of S-glycoproteins in *Nicotiana* and *Brassica* show little similarity, supporting the notion of independent origins for the two systems of homomorphic incompatibility represented by these two genera (Barrett 1992, p. 9). Phylogenetic analysis of heterostylous families in which specific evolutionary hypotheses have been proposed concerning breeding-system variation would be particulary worthwhile (Plumbaginaceae, Rubiaceae, Boraginaceae, Lythraceae, Oxalidaceae) (Barrett 1992, p. 23). Such information is important for understanding how and why heterostyly has evolved in particular families and provides the necessary historical background for microevolutionary studies of the selection process.

13. Molecular Clocks

There have been an increasing number of attempts to estimate divergence times of major plant groups using cpDNA and nuclear DNA and applying the principle of the molecular clock (and its implicit assumption of approximate constancy of evolutionary rate over time). The most explicit, molecular clock-driven study is that of Martin et al. (1993) a re-resurrection of an earlier hypothesis and study again based on the molecular clock. Using sequences from *rbc*L (cpDNA) and *gap*C (nuclear DNA) and calibrating the clock using various bryophytes and gymnosperms, they indicate that both sequences suggest that the separation of monocot and dicot lineages took place in late Carboniferous (ca. 300 Ma ago). They suggest that this finding might again "evoke vehement criticism"; they are probably right, but this time criticism of their methods and lack of sampling as well.

Four other papers explicitly demonstrate, however, that a molecular clock assumption cannot be invoked for *rbc*L sequences (Bosquet et al. 1992; Gaut et al. 1992; Clegg 1993; Hasebe et al. 1993). Extensive variation in the rates of substitution was seen among 50 *rbc*L sequences of seed plants with annual plants in angiosperms evolving more rapidly (Bosquet et al. 1992). In a survey of 35 *rbc*L sequences within monocots, a five fold difference in the substitution rate was seen (Gaut et al. 1992). The pattern of rate variation is consistent with the generation-time-effect hypothesis, with grasses and palms exhibiting the fastest and slowest rates, respectively. The synonymous rate of change in the lineage leading to *Adiatum* is more than four times that seen relative to other lineages of ferns, with the null hypothesis of constancy in rate being rejected at the 1% level (Hasebe et al. 1993).

14. Future Prospects and Problems

Four areas can be highlighted for future work due to either inherent problems now faced by molecular systematics or to rapidly advancing lines of research. Foremost among the problems to tackle is the analysis of DNA data, the "soft underbelly" of molecular systematics (reviewed in Doyle 1993). It is perhaps inevitable that consideration must be given to alignment (e.g., Thorne and Kishino 1992), character choice and character state transformations (e.g., Albert and Mishler 1992), choice of analytical procedure (e.g., Nei 1991), searching strategies (Penny et al. 1992; Page 1993b), confidence in trees obtained (Källersjö et al. 1992), and gene vs. species trees (e.g., Doyle 1992; Sanderson and Doyle 1992; Page 1993a). A second problem, with no good solutions or prospects yet, involves the search for alternative genes or DNA regions in the nuclear genome for phylogenetic use, a strategy already employed by zoologists (Friedlander et al. 1992).

Third, the prospect of detailed coevolutionary studies using molecules to track the evolution of the different organisms involved is certainly bright. Activity in this field is continuing and we can look forward soon to some interesting results (Armbruster 1992; Mitter et al. 1991). Fourth, we can look forward to seeing the explicit interfacing of developmental genetics and phylogenetic systematics to understand morphological evolution. Is the genetic analysis of the differences between maize and teosinte (Doebley and Stec 1991) a prelude of more to come? Can we expect similar studies now that genetic linkage maps of *Solanum* (Tanksley et al. 1992), *Brasscia* (Song et al. 1991), and *Helianthus* (Rieseberg et al. 1993) are available? These and other studies, perhaps in *Arabidopsis* and *Antirrhinum*, might explore the use of genes controlling morphological features (e.g., floral evolution), phenotypic plasticity versus genome variability, and developmental and modifying genes.

References

Albert VA, Mishler BD (1992) Cladistics 8:73–83. **Albert VA**, Chase MW, Mishler BD (1992a) Character state weighting for restriction site date in phylogenetic reconstruction, with an example from chloroplast DNA. In: Solits PS, Soltis DE, Doyle JJ (eds) Molecular systematics of plants. Chapman & Hall, New York, pp 369–403. **Albert VA**, Williams SE, Chase MW (1992b) Science 257:1491–1495. **Albert VA**, Chase MW, Mishler BD (1993) Ann Mo Bot Gard 80:752–766. **Ariztia EV**, Andersen RA, Sogin ML (1991), J Phycol 27:428–436. **Armbruster SW** (1992) BioScience 42:12–20. **Arnold ML** (1993), Am J Bot 80:577–583. **Arnold ML**, Buckner CM, Robinson JJ (1991) Proc Natl Acad Sci USA 88:1398–4102. **Avise JC**, Arnold J, Ball RM et al. (1987) Annu Rev Ecol Syst 18:489–522.

Baldwin BG (1992) Mol Phylogen Evol 1:3–16. **Baldwin BG** (1993) Am J Bot 80:222–238. **Baldwin BG**, Kyhos DW, Dvorak J, Carr GD (1991) Proc Natl Acad Sci USA 88:1840–1843. **Barrett M**, Donoghue MJ, Sober E (1991) Syst Zool 40:486–493. **Barrett M**, Donoghue MJ, Sober E (1993) Syst Zool 42:216–217. **Barrett SCH** (1992) Heterostylous genetic polymorphisms: model systems for evolutionary analysis. In: Barrett SCH (ed) Evolution and function of heterostyly. Springer, Berlin Heidelberg New York, pp 1–29. **Baum BR** (1992) Taxon 41:3–10. **Baum DA**, Sytsma KJ, Hoch PC (1993) Syst Bot (submitted). **Behnke H-D**, Mabry TJ (1993) Evolution and systematics of the Caryophyllales. Springer, Berlin Heidelberg, New York (in press). **Berthomieu P**, Meyer C (1991) Plant Mol Biol 17:555–557. **Bogorad L**, Vasil K (1991) Cell culture and somatic cell genetics of plants, 7A, the molecular biology of plastids. Academic Press, San Diego. **Bosquet J**, Strauss SH, Doerksen AH, Price RA (1992) Proc Natl Acad Sci USA 89:7844–7848. **Bourne CM**, Palmer JD, Stoermer EF (1992) J Phycol 28:347–355. **Brauner S**, Crawford DJ, Stuessy TF (1993) Am J Bot 79:1436–1439. **Bremer BB** (1992) Ann Mo Bot Gard 79:380–387. **Bremer BB**, Struwe L (1992) Am J Bot 79:1171–1184. **Bremer K** (1992) Syst Zool 41:436–445. **Bremer K**, Jansen RK (1992) Ann Mo Bot Gard 79:414–415. **Bremer K**, Jansen RK, Karis PO, Kallerjso M, Keeley SC, Kim K-J, Michaels HJ, Palmer JD, Wallace RS (1992) Nord J Bot 12:141–148. **Brooks DR**, McLennan DA (1991) Phylogeny, ecology, and behavior. University of Chicago Press, Chicago. **Bruneau A**, Doyle JJ (1993) Syst Bot 18:229–247. **Brunsfeld SJ**, Soltis DE, Soltis PS (1992) Syst Bot 17:239–256. **Buchheim MA**, Chapman RL (1991) BioSystems 25:85–100. **Bult CJ**, Zimmer EA (1993) Syst Bot 18:48–63. **Burke DH**, Raubeson LA, Alberti M et al. (1993) Plant Syst Evol (in press).

Caputo P, Stevenson DW, Moretti A (1992) Brittonia 44:376–382. **Carlquist S** (1962) Phytomorphology 12:30–45. **Chapman RL**, Buchheim MA (1991) Crit Rev Plant Sci 10:343–368. **Chase MW**, Hills HG (1991) Taxon 40:215–220. **Chase MW**, Soltis DE, Olmstead RG et al. (1993) Ann Mo Bot Gard 80:528–580. **Clark WD**, Gaut BS, Duvall MR, Clegg MT (1993) Ann Mo Bot Gard (in press). **Clegg MT** (1993) Proc Natl Acad Sci USA 90:363–367. **Clegg MT**, Zurawski G (1992) Chlorplast DNA and the study of plant phylogeny: present status and future prospects. In: Soltis PS, Soltis DE, Doyle JJ (eds) Molecular systematics of plants. Chapman & Hall, New York, pp 1–13. **Coleman AW**, Goff LJ (1991) J Phycol 27:463–473. **Conant DS**, Stein DB, Valinski AEC et al. (1993) (submitted). **Conti E**, Fischbach A, Sytsma KJ (1993) Ann Mo Bot Gard 80:672–685. **Crawford DJ**, Brauner S, Cosner MB, Stuessy TF (1993) Am J Bot 80:89–92. **Crawford DJ**, Lee NS, Stuessy TF (1992a) Aliso 13:395–409. **Crawford DJ**, Stuessy TF, Cosner MB, Haines DW (1992b) Syst Bot 17:676–682. **Cruzan MB**, Arnold ML, Carney SE, Wollenberg KR (1993) Am J Bot 80:344–350.

Darbyshire SJ, Cayouette J, Warwick SI (1992) Plant Syst Evol 181:57–76. **Delgado-Salinas A**, Bruneau A, Doyle JJ (1993) Syst Bot 18:6–17. **Demeke T**, Adams RP, Chibbar R (1992) Theor Appl Genet 84:990–994. **Depamphilis CW**, Palmer JD (1990) Nature 348:337–339. **DeQueiroz A** (1993) Syst Biol (in press). **Deragon J-M**, Landry BS (1992) PCR Meth Appl 1:175–180. **Dias JS**, Lima MB, Song KM et al. (1992) Euphytica 58:221–229. **Do N**, Adams RP (1991) BioTechnology 10:162–166. **Doebley J**, Stec A (1991) Genetics 129:285–295. **Donoghue MJ**, Sanderson MJ (1992) The suitability of molecular and morphological evidence in reconstructing plant phylogeny. In: Soltis PS, Soltis DE, Doyle JJ (eds) Molecular systematics of plants. Chapman & Hall, New York, pp 340–368. **Donoghue MJ**, Olmstead RG, Smith JF, Palmer JD (1992) Ann Mo Bot Gard 79:333–345. **Dorado O**, Rieseberg LH, Arias D (1992) Evolution 45:566–572. **Douglas AM**, Georgalis AM, Atchison BA (1993) BioTechnology 14:824–828. **Douglas SE** (1992) Biosystems 28:57–68. **Douglas SE**, Turner S (1991) J Mol Evol 33:267–273. **Douglas SE**, Murphy CA, Spencer DF, Gray MW (1991) Nature 350:148–151. **Downie SR**, Palmer JD (1992a) Use of chloroplast DNA rearrangements in reconstructing plant phylogeny. In: Soltis PS, Soltis DE, Doyle JJ (eds) Molecular systematics of plants. Chapman & Hall, New York, pp 14–35. **Downie SR**, Palmer JD (1992b) Ann Mo Bot Gard 79:266–283. **Downie SR**, Palmer JD (1993) Phylogenetic relationships of the Caryophyllales using restriction site variation of the chloroplast DNA inverted repeat. In: Behnke H-D, Mabry TJ (eds) Evolution and systematics of the Caryophyllales. Springer Berlin Heidelberg New York (in press). **Downie SR**, Olmstead RG, Zurawski G, Soltis DE, Soltis PS, Watson JC, Palmer JD (1991) Evolution 45:1245–1259. **Doyle JA**, Donoghue MJ (1992) Brittonia (in press). **Doyle JJ** (1991a) DNA protocols for plants. In: Hewitt G, Johnson AWB, Young JPW (eds) Molecular techniques in taxonomy. NATO ASI Series H, Cell Biology, vol 57, Springer Berlin Heidelberg New York, pp 283–293. **Doyle JJ** (1991b) Mol Biol Ecol 8:366–377. **Doyle JJ** (1992) Syst Bot 17:144–163. **Doyle JJ** (1993), BioScience 43:380–389. **Doyle JJ**, Doyle JL (1991) DNA and higher plant systematics: some examples from the legumes. In: Hewitt G, Johnson AWB, Young JPW (eds) Molecular techniques in taxonomy. NATO ASI Series H, Cell Biology, vol 57, pp 101–115. **Doyle JJ**, Doyle JL (1993) Syst Bot 18:309–327. **Doyle JJ**, Davis JI, Soreng RJ et al. (1992a) Proc Natl Acad Sci USA 89:7722–7726. **Doyle JJ**, Lavin M, Bruneau A (1992b) Contributions of molecular data to papilionoid legume systematics. In: Soltis DE, Doyle JJ (eds) Molecular systematics of plants. Chapman & Hall, New York, pp 223–251. **Duvall MR**, Chase MW, Soltis DS, Clegg MT (1993a) A phylogeny of seed plants resulting from analysis of DNA sequence variation among the *rbc*L loci of 475 species with particular emphasis on alliances among monocotyledons. In: Hoch PC (ed) Experimental and molecular approaches to plant biosystematics. Monographs in systematic botany. Missouri Botanical Garden, St. Louis (in press). **Duvall MR**, Clegg MT, Chase MW, Clark WD et al. (1993b) Ann Mo Bot Gard 80:607–619. **Duvall MR**, Learn GH, Clegg MT (1993c) Proc Natl Acad Sci USA (in press).

Edwards K, Johnstone C, Thomson C (1991) Nucleic Acids Res 19:1349.

Fang G, Hammar S, Grumet R (1992) BioTechnology 13:52–56. **Felsenstein J** (1993) PHYLIP (phylogeny inference package) version 3.5. Department of Genetics, University of Washington, Seattle. **Friedlander TP**, Regier JC, Mitter C (1992) Syst Zool 41:483–490. **Fritsch P**, Hanson MA, Spore CD, Pack PE, Rieseberg LH (1993) Plant Mol Biol Rep (in press).

Gadek PA, Quinn CJ (1993) Ann Mo Bot Gard 80:581–586. **Gadek PA**, Quinn CJ, Rodman JE, Karol KG, Conti E, Price RA, Fernando ES (1992) Aust Syst Bot 5:717–724. **Gantt JS**, Baldauf SL, Calie PJ, Weeden NF, Palmer JD (1991) EMBO J 10:3073–3078. **Gaut BS**, Clegg MT (1993) Proc Natl Acad Sci USA (in press). **Gaut BS**, Muse SV, Clark WD, Clegg MT (1992) J Mol Evol 35:292–303. **Gawal NJ**, Jarret RL (1991) Plant Mol Biol Rep 9:262–266. **Giannasi DE**, Zurawski G, Learn GH, Clegg MT (1992) Syst Bot 17:1–15. **Givnish TJ**, Sytsma KJ (1992) Am J Bot 79(6):145. **Givnish TJ**, Sytsma KJ, Smith JF, Hahn WJ (1993) Molecular evolution, adaptive radiation, and geographic speciation in *Cyanea* (Campanulaceae), a species-rich genus of Hawaiian plants. In: Wagner WL, Funk VL (eds) Origin and adaptive radiation of the Hawaiian biota. Smithsonian Institution Press, Washington DC (in press). **Graham LE**, Delwiche CF, Mishler BD (1991) Adv Bryol 4:213–244. **Golenberg EM**, Giannasi DE, Clegg MT et al. (1990) Nature 344:656–658. **Golenberg EM**, Clegg MT, Durbin ML, Doebley J, and Ma DP (1993) Mol Phylogen Evol (in press). **Gray MW** (1991) Origin and evolution of plastid genomes and genes. In: Bogorad L, Vasil IK (eds) Cell culture and somatic cell genetics of plants, 7A. The molecular biology of plastids. Academic Press, San Diego, pp 303–330.

Hamby RK, Zimmer EA (1992) Ribosomal DNA as a phylogenetic tool in plant systematics. In: Soltis PS, Soltis DE, Doyle JJ (eds) Molecular systematics of plants. Chapman & Hall, New York, pp 50–91. **Harley R** (1992) Advances in labiate science. Royal Botanic Gardens, Richmond, UK. **Harrison R** (1993) Hybrid zones and the evolutionary process. Oxford University Press, New York. **Hasebe M**, Iwatsuki K (1992) Bot Mag (Tokyo) 105:413–419. **Hasebe M**, Ito M, Kofuji R et al. (1992a) Bot Mag (Tokyo)105:385–391. **Hasebe M**, Ito M, Kofuji R et al. (1992b) Bot Mag (Tokyo) 105:673–679. **Hasebe M**, Ito M, Kofuji R et al. (1993) J Mol Evol (in press). **Hawkes JG**, Lester RN, Nee M, Estrada N (1991) Solanaceae 3: taxonomy – chemistry – evolution. Royal Botanic Gardens, Richmond, UK. **Herrmann RG** (1992) Plant gene research, vol 6. Organelles. Springer, Berlin Heidelberg New York. **Hewitt GM**, Johnston AWB, Young JPW (1991) Molecular techniques in taxonomy. NATO ASI Series H, Cell Biology, vol 57. Springer, Berlin Heidelberg New York. **Hirai A**, Nakazono MM (1993) Plant Mol Biol Rep 11:98–100. **Hoch PC**, Raven PH (1992) Phytologia 73:456–459. **Hoch PC**, Crisci J, Tobe H, Berry PE (1993) Syst Bot 18:31–47. **Hoelzel AR** (1992) Molecular genetic analysis of populations: a practical approach. Oxford University Press, New York. **Hoelzel AR**, Green A (1992) Analysis of population-level variation by sequencing PCR-amplified DNA. In: Hoelzel AR (ed) Molecular genetic analysis of populations: a practical approach. Oxford University Press, New York, pp 159–188. **Holsinger KE**, Jansen RK (1993) Methods Enzymol (in press). **Hosaka K**, Spooner DM (1992) Theor Appl Genet 84:851–858. **Howe CJ**, Beanland TJ, Larkum AWD, Lockhart PJ (1992) TREE 7:378–383.

Jansen RK, Michaels HJ, Palmer JD (1991a) Syst Bot 16:98–115. **Jansen RK**, Wallace RS, Kim K-J, Chambers KL (1991b) Am J Bot 78:1015–1027. **Jansen RK**, Michaels HJ, Wallace RS, Kim K-J et al. (1992) Chloroplast DNA variation in the Asteraceae: phylogenetic and evolutionary implications. In: Soltis PS, Soltis DE, Doyle JJ (eds) Molecular systematics of plants. Chapman & Hall, New York, pp 252–279.

Kadereit J, Sytsma KJ (1992) Nord J Bot 12:205–217. **Källersjö M**, Farris JS, Kluge AG, Bult C (1992) Cladistics 8:275–287. **Kato M** (1988) Taxon 37:381–386. **Kesseli R**, Ochoa O, Michelmore R (1991) Genome 34:430–436. **Kim K-J**, Jansen RK (1993) Plant Syst Evol (in press). **Kim K-J**, Mabry TG (1991) Plant Syst Evol 177:53–69. **Kim K-J**, Jansen RK, Wallace RS, Michaels HJ, Palmer JD (1992a) Ann Mo Bot Gard 79:428–445. **Kim K-J**, Jansen RK, Turner BL (1992b) Am J Bot 79:708–715. **Kim K-J**, Turner BL, Jansen RK (1992c) Syst Bot 17:449–469. **King LM** (1993) Evolution 47:136–151. **Knox EB**, Downie SR, Palmer JD (1993) Mol Biol Ecol 10:414–430. **Kron KA**, Chase MW (1993) Ann Mo Bot Gard 80:735–741.

Lammers TJ, Givnish TJ, Sytsma KJ (1993) Novon (in press). **Langridge U**, Schwall M, Langridge P (1991) Nucleic Acids Res 19:6954. **Lavin M**, Doyle JJ (1991) Syst Bot 16:162–172. **Lavin M**, Mathews S, Hughes C (1991) Am J Bot 78:1576–1585. **Les DH**, Garvin DK, Wimpee CF (1991) Proc Natl Acad Sci USA 88:10119–10123. **Lewis H**, Raven PH (1992) Madroño 39:163–169. **Li W-H**, Graur D (1991) Fundamentals of molecular evolution. Sinauer, Sunderland, Mass. **Liston A** (1992) Am J Bot 79:953–961. **Liston A**, Rieseberg LH, Hanson MA (1992) Plant

Syst Evol 181:121–132. **Lonsdale DM** (1992) Structure and expression of mitochondrial genomes. In: Herrmann RG (ed) Plant gene research, vol 6. Organelles. Springer, Berlin Heidelberg New York. **Luo G**, Hepburn AG, Widholm JM (1992) Plant Mol Biol Rep 10:319–323.

Mabberley DJ (1975) New Phytol 74:365–374. **Maddison WP**, Maddison DR (1992) MacClade: analysis of phylogeny and character evolution. Sinauer, Sunderland, Mass. **Maggs CA**, Douglas SE, Fenety J, Bird CJ (1992) J Phycol 28:214–232. **Manhart JR**, McCourt RM (1992) J Phycol 28:730–737. **Manhart JR**, Rettig JH (1993) Gene sequence data. In: Behnke H-D, Mabry TJ (eds) Evolution and systematics of the Caryophyllales. Springer, Berlin Heidelberg New York (in press). **Manhart JR**, VonderHaar RA (1991) J Phycol 27:613–617. **Martin PG**, Dowd JM (1991) Ann Mo Bot Gard 78:296–337. **Martin W**, Lydiate D, Brinkmann H, Forkmann G et al. (1993) Mol Biol Evol 10:140–162. **Meagher RB**, Berry-Lowe S, Rice K (1989) Genetics 123:845–863. **Melchinger A**, Messmer K, Lee M, Woodman W, Lamkey K (1991) Crop Sci 31:669–678. **Michaels HJ**, Scott KM, Olmstead RG, Szaro T et al. (1993) Ann Mo Bot Gard 80:742–751. **Milligan BG** (1992) Plant DNA isolation. In: Hoelzel AR (ed) Molecular genetic analysis of populations: a practical approach. Oxford University Press, New York, pp 59–88. **Mishler BD**, Thrall PH, Hopple JS, De Luna E, Vilgalys R (1992) Bryologist 95:172–180. **Mitter C**, Farrel B, Futuyama DJ (1991) TREE 6:290–293. **Miyamoto MM**, Cracraft J (eds) (1991) Phylogenetic analysis of DNA sequences. Oxford University Press, New York. **Morden CW**, Golden SS (1991) J Mol Evol 32:3ö79–395. **Morden CW**, Wolfe KH, dePamphilis CW, Palmer JD (1991) EMBO J 10:3281–3288. **Morden CW**, Delwiche CF, Kuhsel M, Palmer JD (1992) BioSystems 28:75–90. **Morgan DR** (1993) Syst Bot 18:290–308. **Morgan DR**, Simpson BB (1992) Syst Bot 17:511–531. **Morgan DR**, Soltis DE (1993) Ann Mo Bot Gard 80:631–660. **Muse SV**, Weir BS (1992) Genetics 132:269–276.

Nason JD, Ellstrand NC, Arnold ML (1992) Am J Bot 79:101–111. **Nei M** (1991) Relative efficiencies of different tree-making methods for molecular data. In: Miyamoto MM, Cracraft J (eds) Phylogenetic analysis of DNA sequences. Oxford University Press, New York, pp 90–128. **Nelson G** (1993) Syst Biol 42:215–216. **Nickrent DL** (1992) Plant Mol Evol Newslett 1:27–31.

Oka K, Yamato K, Ohta E, et al. (1992) J Molec Biol 223:1–7. **Olmstead RG**, Palmer JD (1991) Chloroplast DNA and systematics of the Solanaceae. In: Hawkes JG, Lester RN, Nee M, Estrada N (eds) Solanaceae 3: taxonomy – chemistry – evolution. Royal Botanic Gardens, Richmond, UK, pp 161–168. **Olmstead RG**, Palmer JD (1992) Ann Mo Bot Gard 79:346–360. **Olmstead RG**, Michales HJ, Scott KM, Palmer JD (1992a) Ann Mo Bot Gard 79:249–265. **Olmstead RG**, Scotland R, Wagstaf S, Sweere J, Reeves P (1992b) Plant Mol Evol Newslett 2:27–30. **Olmstead RG**, Scott KM, Palmer JD (1992c) A chloroplast DNA phylogeny for the Asteriade: implications for the Lamiales. In: Harley R (ed) Advances in labiate science. Royal Botanic Gardens, Richmond, UK, pp 19–26. **Olmstead RG**, Bremer B, Scott KM, Palmer JD (1993) Ann Mo Bot Gard 80:700–722.

Page RDM (1993a) Syst Biol 42:77–84. **Page RDM** (1993b) Syst Biol 42:200–210. **Palenik B**, Haselkorn R (1992) Nature 355:265–267. **Palmer JD** (1991) Plastid chromosomes: structure and evolution. In: Bogard L, Vail IK (eds) Cell culture and somatic cell genetic plants, 7A. The molecular biology of plastids. Academic Press, San Diego, pp 5–53. **Palmer JD** (1992a) Mitochondrial DNA in plant systematics: applications and limitations. In: Soltis PS, Soltis DE, Doyle JJ (eds) Molecular systematics plants. Chapman & Hall, New York, pp 36–49. **Palmer JD** (1992b) Curr Biol 2:318–320. **Palmer JD** (1992c) In: Herrmann RG (ed) Plant gene research, vol 6. Organelles. Springer Berlin Heidelberg New York, pp 99–133. **Palmer JD** (1993) Chloroplast and mitochondrial genome evolution in land plants. In: Osborn TC, Beckmann JS (eds) Plant genomes: methods for genetic and physical mapping. Kluwer, Dordrecht (in press). **Palmer JD**, Logsdon JM (1991) Curr Opinion Genet Dev 1:470–477. **Palmer JD**, Soltis DE, Soltis PS (1992) Curr Genet 21:125–129. **Penny D**, Hendy MD, Steel MA (1992) Trends Evol Ecol Biol 7:73–79. **Philbrick CT**, Jansen RK (1991) Syst Bot 16:478–491. **Philipps RL**, Vasil IK (1993) DNA-based markers in plants. Kluwer, Dordrecht (in press). **Price RA**, Palmer JD (1993) Ann Mo Bot Gard 80:661–671.

Qui YL, Chase MW, Les DH, Parks CR (1993) Ann Mo Bot Gard 80:587–606.

Ragan MA (1992) Mol Phylogen Evol 1:53–58. **Raubeson LA**, Jansen RK (1992a) Biochem Syst Ecol 20:17–24. **Raubeson LA**, Jansen RK (1992b) Science 255:1697–1699. **Reith M**, Munholland J (1993) Plant Cell 5:465–475. **Rettig JH**, Wilson HD, Manhart JR (1992) Taxon 41:201–209. **Rieseberg LH** (1991) Am J Bot 78:1218–1237. **Rieseberg LH**, Brunsfeld SJ (1992) Molecular evidence and plant introgression. In: Soltis PS, Soltis DE, Doyle JJ (eds) Molecular systematics of plants. Chapman & Hall, New York, pp 151–201. **Rieseberg LH**, Ellstrand NC (1993) Crit Rev Plant Biol (in press). **Rieseberg LH**, Morefield JD (1993) IOPB (in press). **Rieseberg LH**, Soltis DE (1991) Evol Trends Plants 5:65–84. **Rieseberg LH**, Wendel JF (1993) Introgression and its consequences. In: Harrison RG (ed) Hybrid zones and the evolutionary process. Oxford University Press, New York, pp 70–110. **Rieseberg LH**, Beckstrom-Sternberg S, Liston A, Arias DM (1991a) Syst Bot 16:50–76. **Rieseberg LH**, Choi H, Ham D (1991b) J Hered 82:489–493. **Rieseberg LH**, Hanson M, Philbrick CT (1992) Syst Bot 17:324–336. **Ritland K**, Eckenwalder JE (1992) Polymorphism, hybridization, and variable evolutionary rate in molecular phylogenies. In: Soltis PS, Soltis DE, Doyle JJ (eds) Molecular systematics of plants. Chapman & Hall, New York, pp 404–428. **Rodman J** (1991a) Syst Bot 16:589–618. **Rodman J** (1991b) Syst Bot 16:619–629. **Rodman J**, Price R, Karol KG, Conti E et al. (1993) Ann Mo Bot Gard 80:686–699. **Rogstad SH** (1992) Taxon 41:701–708. **Ruanto G**, Kidd KK (1991) Proc Natl Acad Sci USA 88:2815–2819.

Sanderson MJ, Doyle JJ (1992) Syst Biol 41:4–17. **Sanderson MJ**, Doyle JJ (1993) Syst Bot 18:395–408. **Scherer S**, Herrmann G, Hirschber J, Böger P (1991) Curr Genet 19:503–507. **Schilling EE**, Panero JL (1991) Am J Bot 78:1054–1062. **Selander RK**, Clark AG, Whittam TS (1991) Evolution at the molecular level. Sinauer, Sunderland, Mass. **Shivji MS** (1991) Curr Genet 19:49–54. **Shivji MS**, Li N, Cattalico RA (1992) Mol Gen Genet 232:65–73. **Smith JF**, Kress WJ, Zimmer EA (1993) Ann Mo Bot Gard 80:620–630. **Smith JF**, Sytsma KJ, Shoemaker JS, Smith RL (1991) Phytochem Bull 23:2–9. **Soltis DE**, Mayer MS, Soltis PS, Edgerton M (1991a) Am J Bot 78:1379–1390. **Soltis DE**, Soltis PS, Collier TG, Edgerton ML (1991b) Am J Bot 78:1091–1112. **Soltis DE**, Soltis PS, Kuzoff RK, Tucker TL (1992a) Plant Syst Evol (in press). **Soltis DE**, Soltis PS, Milligan BG (1992b) Intraspecific chloroplast DNA variation: systematic and phylogenetic implications. In: Soltis PS, Soltis DE, Doyle JJ (eds) Molecular systematics of plants. Chapman & Hall, New York, pp 117–150. **Soltis DE**, Soltis PS, Thompson JN, Pellmyr O (1992c) Syst Bot 17:607–619. **Soltis DE**, Grable A, Morgan DR, Soltis PS, Kuzoff RK (1993) Am J Bot (in press). **Soltis PS**, Soltis DE (1991a) Syst Bot 16:407–413. **Soltis PS**, Soltis DE (1991b) Aliso 13:215–223. **Soltis PS**, Soltis DE (1992) Plant Syst Evol 181:11–20. **Soltis PS**, Soltis DE, Wolf PG (1991) Syst Bot 16:245–256. **Soltis PS**, Doyle JJ, Soltis DE (1992a) Molecular data and polyploidy in plants. In: Soltis PS, Soltis DE, Doyle JJ (eds) Molecular systematics of plants. Chapman & Hall, New York, pp 177–201. **Soltis PS**, Soltis DE, Doyle JJ (1992b) Molecular systematics of plants. Chapman & Hall, New York. **Soltis PS**, Soltis DE, Smiley CJ (1992c) Proc Natl Acad Sci USA 89:449–451. **Soltis PS**, Soltis DE, Novak SJ, Schultz JL, Kuzoff RK (1993) Ann Mo Bot Gard (in press). **Song KM**, Osborn TC (1992) Genome 35:992–1001. **Song KM**, Suzuki JY, Slocum MK, Williams PH, Osborn TC (1991) Theor Appl Genet 82:296–304. **Song KM**, Tank K, Osborn TC (1993) Theor Appl Genet (in press). **Spooner DM**, Sytsma KJ (1992) Syst Bot 17:432–448. **Spooner DM**, Sytsma KJ, Conti E (1991a) Am J Bot 78:1354–1366. **Spooner DM**, Sytsma KJ, Smith JF (1991b) Evolution 45:757–764. **Spooner DM**, Anderson GJ, Jansen RK (1993) Am J Bot 80:676–688. **Stabile JE**, Wurtzel ET, Gallagher JC (1992) J Phycol 28:90–94. **Steane DA**, McClure BA, Clarke AE, Kraft GT (1991) J Phycol 27:758–762. **Steele K**, Holsinger KE, Jansen RK, Taylor D (1991) Mol Biol Evol 8:240–248. **Stein DB** (1993) Isolating and comparing nucleic acids from land plants: nuclear and other organellar genes. In: Zimmer EA, White TJ, Cann RL, Wilson AC (eds) Molecular evolution: producing the biochemical data. Methods Enzymol 224 (in press). **Stein DB**, Conant DS, Ahearn ME, Jordan ET et al. (1992) Proc Natl Acad Sci USA 89:1856–1860. **Stewart C-B** (1993) Nature 361:603–607. **Stewart CN**, Via LE (1993) BioTechnology 14:748–750. **Suh Y**, Thien LB, Zimmer EA (1993) Am J Bot (in press). **Swofford DL** (1991) When are phylogeny estimates from molecular and morphological data incongruent? In: Miyamoto MM, Cracraft J (eds) Phylogenetic analysis of DNA sequences. Oxford University Press, New York, pp 295–333. **Swofford DL** (1993) PAUP: phylogenetic analysis using parsimony, version 3.1.1. Illinois Natural History Survey, Champaign, Il. **Sytsma KJ** (1990) TREE

5:104–110. **Sytsma KJ** (1993) DNA extraction from recalcitrant plants: long, pure, and simple? In: Adams RP, Miller JS, Golenberg EM, Adams JE (eds) Conservation of plant genes II: intellectual property rights and DNA utilization. Missouri Botanical Garden, St. Louis (in press). **Sytsma KJ**, Gottlieb LD (1986) Evolution 40:1248–1261. **Sytsma KJ**, Smith JF (1992) Molecular systematics of Onagraceae: examples from *Clarkia* and *Fuchsia*. In: Soltis PS, Soltis DE, Doyle JJ (eds) Molecular systematics of plants. Chapman & Hall, New York, pp 295–323. **Sytsma KJ**, Smith JF, Berry PE (1991) Syst Bot 16:257–269. **Sytsma KJ**, Givnish TJ, Smith JF, Hahn WJ (1993) Obtaining and storing land plant samples for macromolecular comparisons. In: Zimmer EA, White TJ, Cann RL, Wilson AC (eds) Molecular evolution: producing the biochemical data. Methods Enzynol 224:23–37.

Tanksley SD, Ganal MW, Prince JP et al. (1992) Genetics 132:1141–1160. **Thomas BR**, Laudencia-Chingcuanco D, Gottlieb LD (1992) Plant Mol Biol 19:745–757. **Thorne JL**, Kishino H (1992) Mol Biol Evol 9:1148–1162. **Troitsky AV**, Melekhovets YF, Rakhimova GM, Bobrova VK et al. (1991) J Mol Evol 32:253–261.

Urbach E, Robertson DL, Chisholm SW (1992) Nature 355:267–270.

van Heusden AW, Bachmann K (1992) Plant Syst Evol 179:221–233. **VanderWiel PL**, Voytas DF, Wendel JF (1993) J Mol Evol 36 (in press).

Wagner WL, Funk VL (1993) Origin and adaptive radiation of the Hawaiian biota. Smithsonian Institution Press, Washington DC (in press). **Wang Z**, Second G, Tanksley S (1992) Theor Appl Genet 83:565–581. **Warwick SI**, Black LD (1991) Theor Appl Genet 82:81–92. **Warwick SI**, Black LD, Aguinagalde I (1992) Theor Appl Genet 83:839–850. **Waters DA**, Buchheim MA, Dewey RA, Chapman RL (1992) Am J Bot 79:459–466. **Webb DM**, Knapp SJ (1990) Plant Mol Biol Rep 8:180–185. **Wendel JF**, Albert VA (1992) Syst Bot 17:115–143. **Wendel JF**, Stewart JM, Rettig JW (1991) Evolution 45:694–711. **Wendel JF**, Brubaker CL, Percival AE (1992) Am J Bot 79:1291–1310. **Whitkus R**, Doebley J, Lee M (1992) Genetics 132:1119–1130. **Whitkus R**, Doebley J, Wendel JF (1993) Nuclear DNA markers in systematics and evolution. In: Phillips RL, Vasil IK (eds) DNA-based markers in plants. Kluwer, Dordrecht (in press). **Whittemore AT**, Schaal BA (1991) Proc Natl Acad Sci USA 88:2540–2544. **Wiegrefe SJ**, Sytsma KJ, Guries RP (1993) Syst Bot (submitted). **Wilson AC**, Zimmer EA, Prager EM, Kocher TD (1989) Restriction mapping in the molecular systematics of mammals: a retrospective salute. In: Fernhom B, Bremer K, Jornvall H (eds) The hierarchy of life, Elsevier, Amsterdam, pp 407–419. **Wolfe KH**, Morden CW, Palmer JD (1991) Curr Opinion Genet Dev 1:523–529. **Wolfe KH**, Morden CW, Palmer JD (1992a) J Mol Biol 223:95–104. **Wolfe KH**, Morden CW, Palmer JD (1992b) Proc Natl Acad Sci USA 89:10648–10652. **Wojciechowski MF**, Sanderson MJ, Baldwin BG, Donoghue MJ (1993) Am J Bot 80:711–722.

Zimmer EA, White TJ, Cann RL, Wilson AC (1993) Molecular evolution: producing the biochemical data. Methods enzymology, vol 224. Springer, Berlin Heidelberg New York

Dr. Kenneth J. Sytsma
Dr. William J. Hahn
Department of Botany
University of Wisconsin
430 Lincoln Drive
Madison, WI 53706, USA

III. Karyology and Cytogenetics

By Martin Röser and Wilfried Morawetz

1. Introduction

Many recent karyological and karyosystematic studies have made successful attempts to incorporate not only chromosome counts, but also more detailed data on karyomorphology, chromosome banding, interphase nuclear organization, and DNA hybridization tests. They employ methods which have been or are being developed in a limited number of karyologically well-known genera such as *Vicia*, *Allium*, or *Secale*. Especially the banding methods which use dyes specific to DNA base pairs, partly combined with in situ hybridization techniques, now become applicable to a wider range of organisms. They undoubtedly prove to be most useful in comparative systematic studies. In the near future, this approach most probably will succeed to fill the gap between classical karyological studies which use various methods of microscopy and modern DNA sequencing.

Moreover, karyologists are trying to interpret karyological characters in terms of phylogeny, especially when unusual chromosomal characters are detected for the first time or are compared with those of supposedly related taxa.

Numerous new studies from all continents deal with the evolutionary and geographic differentiation of plant groups. Special emphasis is placed on investigations of wild monocots which have been neglected for some time in favor of the monocotyledonous crop plants.

2. A Novel Approach: Genomic in Situ Hybridization

Very promising innovations come from the improvement and the combination of DNA in situ hybridization and fluorochrome staining techniques. Such "chromosome painting" first was used to identify whole parental genomes in the cells of hybrids between *Secale africanum* and *Hordeum chilense* (Schwarzacher et al. 1989; Anamthawat-Jónsson et al. 1990; Leitch et al. 1990), and in *Triticale* and in wheat-rye translocation lines (Le et al. 1989; Heslop-Harrison et al. 1990). The apparently first successful application of this method in wild species was performed by Bennett et al. (1992) in the grass genus *Milium*. Previous karyological investigation of *Milium* suggested that *M. montianum* (2n = 22), with its distinctive bimodal karyotype consisting of 8 large and 14 small chromosomes, was allopolyploid (Bennett and Thomas 1991). Although karyotype analyses permitted

Progress in Botany, Vol. 55

only speculative arguments as to the origin of the small chromosomes, it was proposed that the large ones were donated by the diploid *M. vernale* (2n = 8), or a closely related genotype. Attempts to produce hybrids remained unsuccessful, and thus it was impossible to test the origin of the larger chromosomes using meiotic chromosome pairing behavior in the interspecific hybrid as an indicator of genome homology. In situ hybridization of biotinylated total genomic DNA of *M. vernale* to root tip chromosomes of *M. montianum* showed that the *M. vernale* probe hybridized preferentially to all large chromosomes of *M. montianum*, but not to the small chromosomes. The results confirmed the allopolyploid nature of *M. montianum*, supported the theory that the large chromosomes of *M. montianum* were donated by *M. vernale* or a closely related genotype, and showed that the large chromosomes have largely retained their genomic integrity in the new allopolyploid background.

It can be reasonably assumed that genomic in situ hybridization methods will become a powerful tool for biosystematic and evolutionary studies. Significantly, it may allow the analysis of species interrelationships where conventional methods have failed. It is applicable when reproductive barriers prevent the production of artificial test hybrids. Moreover, it is applicable in hybrids that fail to flower, or when no meiotic chromosome pairing is seen, or in situations where genes controlling chromosome pairing prevent firm conclusions regarding homology being drawn from chromosome behavior at first meiotic metaphase.

3. Cold-Sensitive or Cold-Induced Chromosome Regions (CSRs, CIRs)

Comparatively few plant gerena have chromosomal regions which react to low temperature of about 0 °C by a reversible reduction of chromatid diameter and reduced staining capacity. In monocotyledons, such regions were, for example, reported in *Trillium*, *Paris* incl. *Kinagusa*, *Tulbaghia*, *Hyacinthus*, *Scilla*, *Hordeum*, and *Secale*. In dicotyledons, CSRs even seem to occur more rarely and are known in the genera *Vicia*, *Cestrum*, *Adoxa*, but recently have also been discovered in *Sambucus* and *Viburnum*. Closer investigations of CSRs including fluorochrome staining, C- and N-banding, and $AgNO_3$-staining had previously been lacking, but have been just recently presented by Berg and Greilhuber (1992, 1993a, b) in the genus *Cestrum* (Sonacaceae), and by Benko-Iseppon and Morawetz (1993) in the genera *Sambucus* and *Viburnum* of the Sambucaceae.

In *Cestrum* it was found that CSRs correspond to highly repetitive AT-rich sequences, and were interpreted to have phylogenetic homology, which also seems be true for a part of the other, very complicated heterochromatin band patterns observed in this genus. The occurrence of CSRs in two sections of *Cestrum* and all species investigated shows that this type of heterochromatin may be widespread in the genus, perhaps ubiquitous. CSRs are unknown so far in other genera of the Cestreae, and the rest of Solanaceae. In view of the current arbitrary internal taxonomic structure of *Cestrum*, extended studies could be rewarding.

In contrast to *Cestrum*, CSRs in *Sambucus* and *Viburnum* were GC-rich, showing that the cold-induced mitotic undercondensation of chromosomes is not coupled with a uniform DNA base-pair composition. In addition to the CSRs, Benko-Iseppon and Morawetz (1993) discussed further karyological characters of *Sambucus* and *Viburnum*, and the families and orders in which they had been included previously. They argued that the karyomorphological data coroborate the opinion that (1) *Sambucus* and *Viburnum* are closely related, (2) should be excluded from Caprifoliaceae s.str., and (3) stand near the Adoxaceae. Furthermore, *Sambucus*, *Viburnum*, and *Adoxa* should be considered as isolated groups which do not necessarily belong to be Dipsacales.

4. Karyotypes, Polyploidy, DNA Amounts, and Cytogeographic Distribution

a) Dicotyledoneae

In the Piperaceae a comparison of chromosomal races in *Piper nigrum* from India was presented by Nair et al. (1993). The study was based mainly on chromosome numbers and discussed previous conflicting statements on basic chromosome numbers in the Piperaceae.

In the related, entirely herbaceous genus *Peperomia*, Samuel and Morawetz (1992) measured DNA amounts in several dipoloid (2n = 22) and polyploid species (2n = 33, 2n = 44, 2n = 66). Diploids significantly varied in chromosome size and DNA content. In most species a positive correlation between chromosome length of the 2n complement and DNA amount was found. Additionally, the polyploids showed a proportionate increase in DNA amount per nucleus. An exception was the triploid *P. metallica* (2n = 33) whose DNA content was much higher than expected regarding its chromosome length. In this case the authors discussed the possibility of polynemy, a lateral replication of chromatids. As a whole, DNA values in *Peperomia* are low, but fit well into the results already obtained in *Piper* and other Magnoliidae.

Chromosomes of *Coptis teeta* (Ranunculaceae), a perennial medicinal herb from the eastern Himalayas, were investigated for the first time (Pandit and Babu 1993). *Coptis teeta* has 2n = 18 with a basic number of x = 9. As this number is uncommon in Ranunculaceae, the authors suggest that it was derived from either x = 8 or 7 and may further justifiy a separate tribe Coptidae in Ranunculaceae.

Diosdado and Pastor (1993) investigated seven orophytic species of *Ranunculus* from the Iberia peninsula (all diploid with 2n = 16). They confirmed the view that the karyotype of *R. aconitifolius* originated from that of *R. platanifolius* by translocations. *Ranunculus alpestris* and *R. seguieri*, though morphologically distinct, have very similar karyotypes. The high number of orophyte taxa in *Ranunculus*, their constant diploidy, and their symmetric karyotypes suggest the origin of the genus in mountain locations.

Taxa of the aquatic *Ranunculus* subgen. *Batrachium* were studied from the Iberian peninsula (Diosdado et al. 1993) and the Aegean islands (Dahlgren 1991). The latter study gave the first and, at the same time, most comprehensive chromosome counts for aquatic ranunculi in the Greek archipelago. It is of special interest for the investigation of speciation and evolution on isolated islands and it represents the first complete investigation of aquatic plants in the Aegean where terrestrial plants are much better known. Most taxa of subgen. *Batrachium* were polyploid (4x, 5x, 6x), only two species (*R. rionii*, *R. shaerospermus*) and one subspecies (*R. penicillatus* subsp. *pseudofluitans*) of altogether eight taxa found in the area were constantly diploid. Chromosomes of *R. sphaerospermus* and *R. penicillatus* subsp. *pseudofluitans* additionally were considerably larger than in the other taxa. This was considered a primitive trait. Combined with morphological characters, e.g., the production of nectaries, both taxa appear to belong to the older taxa of the subgenus. Dahlgren (1991) stressed that all diploids are species with divided leaves only, while the taxa with also laminate leaves are polyploids. She reconstructed a basic haploid karyotype formula of 2m 3sm 3st for subgen. *Batrachium* which proved to be applicable also to the Iberian aquatic ranunculi.

In the Iberian peninsula, diploid taxa with 2n = 16 chromosomes prevail. *Ranunculus tripartitus* is tetraploid here (2n = 32), while both numbers occur in *R. ololeucus*, *R. peltatus*, and *R. penicillatus*. Chromosomes are rather asymmetric and short in most cases and, therefore, resemble those of sect. *Flammula* and sect. *Hecatonia* from subgen. *Ranunculus*. Subgen. *Batrachium* can be considered as a very advanced group of the genus *Ranunculus* due to its extraordinary phenotypic plasticity and adaptation to unstable or extremely ephemeral habitats linked with a high degree of chromosomal variation.

The comparison of both studies shows that the Greek archipelago, although influenced much less by glacial climatic changes, harbors a higher percentage of polyploid aquatic ranunculi than the Iberian peninsula. Effects of geographical isolation combined with a subsequent lack of competition may have played some role here.

Ullucus tuberosus (Basellaceae) is one of the minor tuber crops in the Andes, and has been cultivated for at least 7000 years. Larkka et al. (1992) studied wild and cultivated clones with respect to karyotype and meiosis. In contrast to the previously known wild ullucos from northern Argentina with 2n = 24, the clones from Bolivia and Peru showed 2n = 36 chromosomes. Cultivated clones had 2n = 24, except a line with 2n = 48. Karyotype structures, especially the presence of large and short marker chromosomes, and meiotic chromosome pairing lead to the suggestion that x = 6 and not x = 12 is the basic chromosome number of *Ullucus* and the Basellaceae as a whole.

A karyological contribution to the classification of the controversial genera *Fallopia*, comprising annual and perennial, herbaceous to woody climbers, and *Reynoutria* comprising erect and perennial herbs with a rhizome, and *Polygonum* (all Polygonaceae) was presented by Bailey and Stace (1992). *Fallopia* and *Reynoutria*, with a center of diversity in Asia, contained diploid to octoploid cytotypes, with base numbers of x = 10 and x = 11. The readiness with which hybridization occurred between taxa of differing base chromosome number and ploidy level, and similiarities revealed in chromosome morphology, meiotic pairing, and DNA amount suggested to the authors that these two genera are best united under *Fallopia*.

Various species of the Brassicaceae are widely used as vegetable, oilseed, or spice crops. The great morphological diversity present already in wild taxa is continued or even exceeded in cultivated species. This is most obvious in the genus *Brassica*, where *B. oleracea* contains different cabbage and kale varieties together with broccoli and kohlrabi. However, chromosomes appear to be conservative since the pronounced morphological differentiation is not accompanied by major karyotypic changes or an establishment of effective crossing barriers both of which have obviously led to a low degree of speciation. Olin-Fatih and Heneen (1992) found comparable C-banding patterns in *B. campestris* (2n = 20 or, according to their Fig. 1c, possibly also 2n = 40), *B. oleracea* (2n = 18), and *B. napus* (2n = 38). As stated by the authors, chromosome morphology of the probably allopolyploid *B. napus* could not give clear evidence of its ancestry. This result is not too surprising since the two compared diploid species are not the parents of *B. napus*.

Neuenschwander (1991) dealt with the cytogeography and systematics of *Arabis pumila* s.l. in the Swiss Alps. He proposed to discriminate tetraploid *A. pumila* s.str. from diploid *A. stellulata* at a specific level, and provided many morphological, biometric, phytocoenological, ecological, and chorological arguments.

A similarly thorough cytogeographic study was performed by Clot (1992) in yellow-flowered *Erysimum* species of the northern Iberian peninsula. Apart from diploids (2n = 14), numerous hypotetraploids (2n = 26), hypohexaploids (2n = 38), and higher polyploids (between 2n = 55 and 2n = 88) were found. Karyological variation was shown to be based mainly on two mechanisms observed in extant *Erysimum* taxa of the investigation area: (1) irregular segregation in the first meiotic division leading to aneuploidy, and (2) the production of viable nonreduced gametes. The wild hypotetraploids have special karyotype features and thus appear to be long established. Hypohexaploids probably arose from crosses between diploids and hypotetraploids. The karyotypes of hypohexaploids could not be investigated in detail, but because of their localization in overlapping areas of different species groups, it was concluded that they were allopolyploids. The author convincingly discussed the multiple occurrence of such chromosomal events in taxonomically disparate groups of *Erysimum*. Apparently, at the moment, the chromosomal data do not contribute very much to the infrageneric classification. However, although most karyologically investigated microspecies of *Erysimum* were parapatric, many populations localized between two taxa were intermediate in karyology and morphology. Further studies on the multiple origin of polyploids in *Erysimum* would be rewarding.

Arctostaphylos is a small holarctic genus of Ericaceae. It is poorly represented in Europe, but is well diversified in Asia and western North America. Ecologically, the woody species of *Arctostaphylos* are variously adapted. They not only inhabit subshrub communities above the timberline in the Alps and the Himalayas, but are also present in lowland areas of the California floristic province with a number of species. Interrelations between the American species of *Arctostaphylos* were analyzed by Schierenbeck et al. (1992). The central Sierran tetraploid species *A. mewukka* was purported to originate from diploid species via a polyphyletic manner from *A. viscida* and *A. patula*.

In Rosaceae subfam. Maloideae, Krügel (1992a, b) summarized the chromosomal knowledge of this subfamily and also gave a comprehensive list of all previous publications. Chromosomal variation in the genus *Cotoneaster* is low. Although definite chromosomal evidence was lacking, the author regarded some infraspecific cytotypes as neopolyploids.

The small eastern Asian genus *Stephanandra* of Rosaceae subfam. Spiraeoideae was studied by Iwatsubo and Naruhashi (1993). The authors gave first chromosome counts in *Stephanandra* and discussed comparisons of chromosomal affinities of subfam. Spiraeoideae to that of other Rosaceae.

A study on DNA amounts in all subfamilies of Rosaceae was performed by Dickson et al. (1992). It was shown that the Maloideae have the largest diploid genomes within the whole family, consistent with their higher chromosome number and their presumed polyploid origin. C-values of genera within the subfamilies Spiraeoideae and Rosoideae are among the smallest of flowering plants so far reported.

Rare and new species of *Vicia* (Fabaceae) from Syria, Turkey, and Armenia were investigated with respect to chromosome numbers, chromosome morphology, and DNA amounts (Maxted et al. 1991). The lower chromosome numbers (2n = 10, 12) most probably originiated from 2n = 14. Reduction of chromosome number showed a certain correlation with increasing morphological advancement of the taxa concerned, but the distribution of DNA amount among the species studied seemed to be independent of taxonomic position.

Akiyama et al. (1992) discussed chromosome morphology and evolution in 16 species of *Impatiens* (Balsaminaceae) from central and eastern Nepal and found chromosomal numbers of 2n = 14, 18, 20, 28. In contrast to the taxa of *Impatiens* from the western Himalayas, 2n = 18 was most frequent. These species very often had bimodal karyotypes.

In the genus *Angelica* (Apiaceae), the diploid chromosome number of 2n = 22 prevailed, but sometimes also 2n = 44 was found within the same species (Vasil'eva and Pimenov 1991). With the exception of *A. sachaliensis*, no completely polyploid species is known so far. Among the diploid species, *A. miqueliana* had a distinct karyotype with many acrocentric chromosomes, while the remaining species showed rather symmetric karyotypes. *Angelica gmelinii* had the peculiar chromosome number of 2n = 28.

Pringle and Murray (1992a, b, 1993) discussed the karyology of 11 species of the genus *Cyphomandra* (Solanaceae), and gave special attention to the tamarillo (*C. betacea*), a softwood tree with edible fruits, and to C-band patterns within the genus. All species were diploid with 2n = 24. The two Brazilian species studied differed from those of Andean origin, suggesting a major evolutionary divergence between these two groups. Centromeric C-bands are lacking in *Cyphomandra*, and *C. luteoalba* was characterized by a special deviant banding style. Noticeably, this species stands near *Solanum*.

Centromeric C-bands occur constantly in perennial and annual species of *Capsicum* (Moscone et al. 1993). The perennial and woody *C. chacoense* and *C. parvifolium* showed low amounts of heterochromatin, while most annuals had higher amounts of heterochromatin. The taxonomic grouping of white and purple-flowered species was discussed critically, and chromosomal evidence for the origin of taxa with x = 13 chromosomes from progenitors with x = 12 was given.

Luque (1992) surveyed the chromosome numbers of ca. 15 Spanish taxa of Boraginaceae tribe Eritrichieae. She confirmed that x = 12 is most probably the primary base chromosome number in this tribe.

Most chromosome counts in eight genera of Argentinean Acanthaceae presented by Piovano and Bernardello (1991) were new. Polyploidy and aneuploidy were shown to be the most important factors of chromosomal differentiation in Acanthaceae. Diploids with 2n = 14 are uncommon in the family.

Lammers and Hensold (1992) studied chromosome numbers in Campanulaceae subfam. Lobelioideae, followed by a comprehensive compilation of all available chromosome counts in this group (Lammers 1993). The Chilean species of the morphologically somewhat isolated *Lobelia tupa* complex were all hexaploids. Taking the karyological diversity in subfam. Lobelioideae into account, this uniformity in the *L. tupa* complex seemed to offer some further support for the monophyly of this group. The herbaceous members of subfam. Lobelioideae are karyologically quite variable, show the entire range of chromosome numbers, and include numerous dysploids, but are predominantly diploid. The woody species, by contrast, are much less variable. Nearly all woody species are tetraploid, with only a few diploids and hexaploids and no dysploid numbers known. Lammers (1993) interpreted this fact as support for the hypothesis that woodiness is apomorphic within the family. Although the reviewers here agree with this hypothesis, we would not consider the karyological findings as unequivocal. However, a general trend of higher chromosome numbers at higher latitudes and higher elevations within the subfamily was demonstrated convincingly.

Garcia Jacas und Susanna (1992) investigated southwest Mediterranean species of *Centaurea* sect. *Acrocentron* (Compositae). They especially dealt with basic chromosome numbers of this section, and could dismiss $x = 9$ as a result of a previous miscount. Although direct chromosomal evidence could not be obtained, the authors could provide numerous convincing biogeographic arguments that chromosomal evolution in sect. *Acrocentron* went from $x = 11$ to $x = 10$.

The east African genus *Guizotia* was analyzed by Dagne and Heneen (1992) with special reference to races of *G. abyssinica*, a locally important oil crop.

Three *Crepis* species endemic to Greece were studied by Kamari (1992). They differed considerably in ploidy level and C-band properties. Tetraploids had more heterochromatin than was expected from possible diploid ancestors.

Studies in the genus *Lactuca* were performed with respect to the cultivated lettuce and its wild relatives (Koopman et al. 1993), and to the cytogeographic differentiation of *Lactuca* in the Iberian peninsula (Mejías 1993). In the latter study, chromosomal findings supported the present taxonomic grouping within the genus. Comparable to the results in *Centaurea* sect. *Acrocentron*, the eastern Mediterranean also in *Lactuca* appeared to be a center of origin for more modern western Mediterranean taxa. Koopman et al. (1993) concluded that the evolution of specific heterochromatin classes was not linked to the domestication of *L. sativa*. Therefore, the disparity of chromosomal and RFLP data in conclusions regarding a relationship in terms of genetic similiarity was assumed to rely on the different experimental approaches, and would reflect a nonparallel evolution of DNA classes.

b) Monocotyledoneae

In the perennial genus *Paris* (Liliaceae) Miyamoto et al. (1992) presented a comprehensive C-band study in eastern Asian species, the center of diversity of this genus. They showed that there is no distinct chromosomal disruption between tropical and temperate species of this genus. *Paris thibetica*, restricted to Tibet, had the lowest amount of heterochromatin. *Paris japonica*, a presumptive relict of the Ice Age, showed the most complicated C-band pattern and had the highest proportion of heterochromatin in the total genome.

The genus *Muscari* was treated by Valdés and Díaz Lifante (1992), who showed that *M. matritensis* differs from *M. comosum* in chromosomes and reproductive behavior, and by Steck-Blaser (1992) who dealt with the Swiss taxa. Polyploids could not always be derived from the generally acknowledged base chromosome number of x = 9 in *Muscari*, and it was discussed that either aneuploidy or the occurrence of other basic chromosome numbers would be of explanatory value.

In North African species of *Anthericum* and *Asphodelus*, the lowest chromosome number found was 2n = 28 (Díaz Lifante 1992). Díaz Lifante discussed that this chromosome number might be derived from x = 7, an ancestral chromosome number in Asphodelaceae and Anthericaceae.

In the karyotype of a clone of *Iris variegata*, four chromosomes were heteromorphic (Karihaloo et al. 1993). On the basis of somatic chromosome morphology and meiotic associations, it was etablished that the heteromorphy was due to translocations between the long and the short chromosomes of this karyotype. Bridge-fragment configurations at meiotic metaphase I also suggested an involvement of paracentric inversion heterozygosity. While the basic karyotype of *I. variegata* appears to be similar to that of the other species of bearded irises with 24 chromosomes, structural alterations, similar to that recorded by Karihaloo et al. (1993), have probably been responsible for the chromosomal variations observed in other clones previously.

Luceño and Castroviejo (1991) and Luceño (1992) studied the meiotic behavior in a number of *Carex* species (Cyperaceae) growing in the Iberian peninsula and Macaronesia. It was suggested that the different chromosome numbers obtained in *C. laevigata* (from 2n = 69 to 2n = 80) were distributed according to an increase in chromosome fission along a geographic gradient in a north-south direction. Most of the species investigated showed a regular meiosis, with the exception of *C. leporina* and *C. disticha*. Chromosomal fusion and reciprocal translocations were suggested as the mechanisms of chromosome evolution in *C. leporina*, whereas a high intraindividual variation in chromosome numbers was observed in *C. disticha*.

A number of interesting studies were published in the Poaceae. Pohl and Clark (1992) for the first time reported a diploid woody bamboo, *Chusquea talamancensis* from Costa Rica, with n = 10. The other species of *Chusquea* and *Aulonemia* investigated were tetraploid (n = 20). The base number of x = 10 differs from the common number of x = 12 found in most woody and some of the herbaceous Bambusoideae.

In the South African genus *Tribolium*, Spies et al. (1992) found diploid to hexaploid specimens with a base chromosome number of x = 6. Frequent irregularities in meiosis such as unequal chromosome segregation and the presence of univalents suggested a hybrid origin of the species.

C-band patterns were analyzed in *Phleum* (Cai and Bullen 1991), *Phalaris* (Viinikaa and Kallio 1993), *Aegilops* (Georgiou et al. 1992), *Oryza* (Hamoud et al. 1991), and *Hordeum* (Linde-Laursen et al. 1992). In the hexaploid *Phleum pratense* (2n = 42), two sets of seven chromosomes were indistinguishable from each other in both morphology and banding patterns; a third set of seven was found to be different from these. Two genomes, A and B, were tentatively established. The banded karyotype in diploid *P. alpinum* (2n = 14) was close to the A genome in *P. pratense*, and the karyotype in *P. bertolonii* (2n = 14) was analogous to the B genome in *P. partense*, which suggests that these species were the genome donors of *P. partense*.

C-banded karyotypes of the 32 species of *Hordeum* divide them into four groups. In combination with chromosome number, and the number of morphology of marker chromosomes, the C-banded karyotypes of a number of species were sufficiently distinct to be of use for identification and the study of relationships. Most species of *Hordeum* of the Americas had similar banding features at the corresponding ploidy level, questioning the biological relevance of referring them to different sections.

Meiotic investigations were carried out in a number of taxa of the Triticeae (e.g., Salomon and Lu 1992; Liu and Wang 1993; Torabinejad and Mueller 1993). The issue (1–2) of Heriditas (vol. 116) especially dealt with this tribe. In this issue, Anamthawat-Jónsson and Heslop Harrison (1992) reported on species-specific DNA sequences, and Heslop-Harrison (1992) reviewed the molecular cytogenetics, cytology, and genomic comparisions in the Triticeae.

In a cytogeographic study, Koul and Gohil (1991) analyzed numerous grasses from Kashmir. Dubcovsky and Martínez (1991, 1992) studied species of *Festuca* from South America. The *F. pallescens* alliance from Argentina and Chile contained only hexaploid species (2n = 42), which were probably allopolyploids. Since no diploid fescues have been found in South America or other continents of the Southern Hemisphere, the origin of the genus was probably in the Northern Hemisphere.

The circumboreal grass *Arctagrostis latifolia* is represented in mainland Alaska by three chromosome races (Mitchell 1992). A widespead distribution was indicated for the tetraploids (2n = 28) and octoploids (2n = 56) and a limited distribution for the hexaploids (2n = 42). Both tetraploids and octoploids appeared to be well represented through the boreal interior and arctic, unglaciated refugia of Alaska. In the heavily glaciated area of southcentral Alaska only tetraploids were found. Cytotype distribution was interpreted as indicating a greater mobility for the tetraploid race, better enabling it to occupy deglaciated and other disturbed areas.

In palms (Arecaceae) Palomino and Quero (1992) pointed out the occurrence of bimodal karyotypes in the genus *Sabal* which is considered to belong to a group of more primitive palms. Comparable cases of biomodality were found by Röser (1993) on a larger scale of closely related genera which in this respect, as in others, differed considerably from other members of subfamily Coryphoideae. Additonally, in fluorochrome and C-band patterns, it was shown that the subfamily was composed of karyologically diverse groups, although they shared, apart from a few exceptions, the same chromosome number of 2n = 36. The coryphoid group showed two strongly separated subgroups. Close relationships among coryphoid, phoenicoid, and borassoid palms, which are united in subfam. Coryphoideae, were not detected. Nuclear differentiation, karyotypic changes, and possible basic chromosome numbers in palms were discussed.

Karyotypes and meiotic configurations of *Anthurium andraeanum* (Araceae) and closely related taxa were analyzed by Marutani et al. (1993). All species have 2n = 30 chromosomes. Despite an overall similiarity of karyotypes, the variation observed in *A. nymphaeifolium* and *A. ochranthum* suggested chromosomal rearrangements. In meiosis all taxa and all hybrids showed regular formation of bivalents, which suggested a close genomic relationship between the parental taxa of the hybrids.

Acknowledgment. The authors wish to thank the Fonds zur Förderung der wissenschaftlichen Forschung in Österreich (P-09657 BIO) for support.

References

Akiyama S, Wakabayashi M, Ohba H (1992) Bot J Linn Soc 109:247–257. **Anamthawat-Jónsson K**, Schwarzacher T, Leitch AR, Bennett MD, Heslop-Harrison JS (1990) Theor Appl Genet 79:721–728. **Anamthawat-Jónsson K**, Heslop-Harrison JS (1992) Hereditas 116:49–54.

Bailey JP, Stace CA (1992) Plant Syst Evol 180:29–52. **Benko-Iseppon AM**, Morawetz W (1993) Bot Acta 106:183–191. **Bennett ST**, Thomas SM (1991) Genome 34:868–878. **Bennett ST**, Kenton AY, Bennett MD (1992) Chromosoma 101:420–424. **Berg C**, Greilhuber J (1992) Genome 35:921–930. **Berg C**, Greilhuber J (1993a) Plant Syst Evol 184:133–151. **Berg C**, Greilhuber J (1993b) Plant Syst Evol 185:259–273.

Cai Q, Bullen MR (1991) Genome 34:42–58. **Clot B** (1992) An Jard Bot Madr 49:215–229.

Dagne K, Heneen WK (1992) Hereditas 117:73–83. **Dahlgren G** (1991) Plant Syst Evol 177:193–211. **Díaz Lifante Z** (1992) Willdenowia 22:143–148. **Dickson EE**, Arumaganathan K, Kresovich S, Doyle JJ (1992) Am J Bot 79:1081–1086. **Diosdado JC**, Pastor JE (1993) Bot J Linn Soc 111:23–35. **Diosdado JC**, Pastor JE, Valdés B (1993) Bot J Linn Soc 112:75–87. **Dubcovsky J**, Martínez AJ (1991) Can J Bot 69:2756–2761. **Dubcovsky J**, Martínez AJ (1992) Can J Bot 70:1134–1140.

Garcia Jacas N, Susanna de la Serra A (1992) Plant Syst Evol 179:1–18. **Georgiou A**, Karataglis S, Roupakias D (1992) Plant Syst Evol 180:105–114.

Hamoud MA, Hassan YA, Nagl W, Selim EE (1991) Cytologia 56:319–325. **Heslop-Harrison JS** (1992) Hereditas 116:93–99. **Heslop-Harrison JS**, Leitch AR, Schwarzacher T, Anamthawat-Jónsson K (1990) Heredity 65:385–392.

Iwatsubo Y, Naruhashi N (1993) Cytologia 58:95–98.

Kamari G (1992) Plant Syst Evol 182:1–19. **Karihaloo V**, Karihaloo JL, Koul AK (1993) Caryologia 46:77–85. **Koopman WJM**, de Long JH, de Vries IM (1993) Plant Syst Evol 185:249–257. **Koul KK**, Gohil RN (1991) Cytologia 56:437–452. **Krügel T** (1992a) Beitr Phytotaxon 15:69–86. **Krügel T** (1992b) Beitr Phytotaxon 15:87–92.

Lammers TG (1993) Am J Bot 80:660–675. **Lammers TG**, Hensold N (1992) Am J Bot 79:585–588. **Larkka J**, Jokela P, Pietelä L, Viinikka Y (1992) Caryologia 45:229–235. **Le HT**, Armstrong KC, Miki B (1989) Plant Mol Biol Rep 7:150–158. **Leitch AR**, Mosgöller W, Schwarzacher T, Bennett MD, Heslop-Harrison JS (1990) J Cell Sci 95:335–341. **Linde-Laursen I**, von Bothmer R, Jacobson N (1992) Hereditas 116:111–116. **Liu Z-W**, Wang R-RC (1993) Genome 36:102–111. **Luceño M** (1992) Willdenowia 22:149–165. **Luceño M**, Castroviejo S (1991) Plant Syst Evol 177:149–159. **Luque T** (1992) Bot J Linn Soc 110:77–94.

Marutani M, Sheffer RD, Kamemoto H (1993) Am J Bot 80:93–103. **Maxted N**, Callimassia MA, Bennett MD (1991) Plant Syst Evol 177:221–234. **Mejías JA** (19939 Bot Helv 103:113–130. **Mitchell WW** (1992) Can J Bot 70:80–83. **Miyamoto J**, Kurita S, Zhijian G, Hen L (1992) Cytologia 57:181–194. **Moscone E**, Lambrou M, Hunziker AT, Ehrendorfer F (1993) Plant Syst Evol 186:213–229.

Nair RR, Sasikumar B, Ravindran PN (1993) Cytologia 58:27–31. **Neuenschwander A** (1991) Bot Helv 101:39–52.

Olin-Fatih M, Heneen WK (1992) Genome 35:583–589.

Palomino G, Quero JH (1992) Cytologia 57:485–489. **Pandit MK**, Babu CR (1993) Bot J Linn Soc 111:371–378. **Piovano MA**, Bernardello LM (1991) Syst Bot 16:89–97. **Pohl RW**, Clark LG (1992) Am J Bot 79:478–480. **Pringle GJ**, Murray BG (1992a) Plant Breed 108:132–138. **Pringle GJ**, Murray BG (1992b) Plant Breed 108:139–148. **Pringle GJ**, Murray BG (1993) Bot J Linn Soc 111:331–342.

Röser M (1993) Bot Acta 106:170–182.

Salomon B, Lu B-R (1992) Plant Syst Evol 180:1–13. **Samuel R**, Morawetz W (1992) Caryologia 45:353–358. **Schierenbeck KA**, Stebbins GL, Patterson RW (1992) Plant Syst Evol 179:187–195. **Schwarzacher T**, Leitch AR, Bennett MD, Heslop-Harrison JS (1989) Ann Bot 64:315–324. **Spiess JJ**, Davidse G, du Plessis H (1992) Am J Bot 79:689–700. **Steck-Blaser B** (1992) Bot Helv 102:211–227.

Torabinejad J, Mueller RJ (1993) Genome 36:147–151.

Valdés B, Díaz Lifante Z (1992) Bot J Linn Soc 110:125–130. **Vasil'eva MG**, Pimenov MG (1991) Plant Syst Evol 177:117–138. **Viinikka Y**, Kallio M (1993) Caryologia 46:47–52.

Dr. Martin Röser
Institut für Botanik und Botanischer Garten
Universität Wien
Rennweg 14
A-1030 Wien, Austria

Professor Dr. Wilfried Morawetz
Forschungsstelle für Biosystematik und Ökologie
Österreichische Akademie der Wissenschaften
Kegelgasse 27
A-1030 Wien, Austria

E. Geobotany

I. Plant Population Ecology

By Walter Seidling, Uwe Starfinger, and Jürg Stöcklin[1]

1. Introduction

In Central Europe plant population ecology as a relatively young science is receiving more and more interest among ecologists. The recently published book on population ecology by Schmid and Stöcklin (1991) and the textbook of Urbanska (1992), both in German, illustrate this. Studies with relevance for population ecology are being carried out with different contexts by scientists from different ecological fields. Inevitably, it was necessary in this review to observe certain limits. We treat important aspects of population ecology but do not pretend to cover all of its fields. The selection reflects our interest in how processes on the population level contribute to an understanding of processes on the vegetation level.

2. Sexual Reproduction and Its Effects on Population Dynamics
(U. Starfinger)

In order to understand plant populations in their environments, it is necessary to know processes connected with the reproductive cycle. Many studies focus on individual phases of the reproductive cycle such as germination, establishment, juvenile phase, flowering, pollination, seed set, dispersal, and seed banks.

Relatively rarely are the whole life cycle of a given taxon or of several species of a genus studied. Differences between closely related species are of special interest. Boutin and Harper (1991) investigated and compared five *Veronica* species in order to show how the different life history traits determine their demographies. Faensen-Thiebes (1992) has shown the importance of several aspects of the life cycle of the biennial *Melilotus alba* such as germination, flower production, and seed dormancy, and how these interact to give the species the ability to compete successfully with perennials. Vrieling et al. (1992) studied coexisting monocarpic perennials and showed that different factors determined the population dynamics of them, e.g., seed bank size, recruitment processes, herbivory, dispersal, and post-dispersal seed predation. Guitián and Fuentes (1992) studied the reproductive biology of *Crataegus monogyna* and found remarkable differences to published results from other European regions. Nicholas et al. (1992) studied seed

[1] Authors in alphabetical order.

Progress in Botany, Vol. 55

production, germinability, and seedling and sapling densities of fir and spruce over several years.

a) Production of Seeds

Female reproductive success may be limited by resources, pollination, or ovule and seed mortality. These factors may vary considerably between seasons, as shown by Ehrlén (1992) for *Lathyrus vernus*, in which the resource limitation is not always reached due to predation or low pollination. Pollinator limitation and resource limitation influenced seed production of two *Geranium maculatum* populations to different degrees, indicating that selection may have variable effects on fecundity traits (Ågren and Willson 1992). Seed production was pollen-limited in a study on *Ipomopsis aggregata* (Campbell 1991). In three dioeciuous tropical tree species, fruit set was limited by pollen supply and negatively correlated to the distance between male and female trees, so that reproductive failure increases with isolation of females (House 1992). The same trend was shown for a self-incompatible annual by Kunin (1992): widely spaced plants produced less seed than plants growing closely together. Both cases point out that selection in self-incompatible plants may work against over-dispersion. Pollen limitation of fruit set and seed set was also found in *Glechoma hederacea*, not only in female plants but also in hermaphroditic plants even though they are self-compatible (Widén 1992). In some cases, however, female reproductive success may decrease with excess pollen deposition; possible causes are discussed in Young and Young (1992). The interesting case of a pollinating seed parasite is described by Thompson and Pellmyr (1992): under certain conditions the plant-insect relationship is mutualistic, in other cases commensalistic.

De Jong et al. (1992a) studied the consequences of mass or extended blooming for pollination. They developed a model to explain how mass blooming can be selected for among self-compatible species, whereas it is more favorable for self-incompatible species to present flowers one at a time. That pollination success is greater if the pollinators visit a plant several times was also shown by Mitchell and Waser (1992). Self-pollination of self-incompatible plants (geitonogamy) negatively affects fitness, which is a problem for plants, as attractiveness to pollinators increases pollen removal but also geitonogamy (Klinkhamer and de Jong 1993). In *Ipomopsis aggregata*, geitonogamy was stronger on larger individuals in a field experiment by de Jong et al. (1992b). Deleterious effects of geitonogamy, however, were overruled by other factors such as size-dependent seed abortion in this species.

The direct cost of seed production consists of energy and nutrients invested by an individual plant, the indirect cost is the reduction in allocation to future reproductive organs. Elmquist et al. (1991) found that allocation to reproduction reduces future growth and fecundity, and also defense against herbivores. In an experiment, Ashman (1992) found that seed production leads to smaller biomass allocation to flowers in the current season and reduced allocation to inflorescences in the following year. She interprets this effect as indirect costs within and between flowering seasons. In *Sidalcea oregana*, the allocation of nutrients and energy to flowers and floral organs decreased over the season.

The proportional allocation to the individual organs differed with the currency used, biomass, nitrogen, or phosphorus (Ashman and Baker 1992).

Tapper (1992) showed that irregular fruit production in ash in Sweden was not related to fruiting in the previous year or to climatically positive conditions. So it did not seem to be resource-limited. A correlation between the fruiting tendency and the time of leafing in the previous year was interpreted as a synchronization within the ash population in order to escape predation through predator satiation.

The size of seeds is an important factor, as it may influence survival rates of seedlings under adverse conditions. Aizen and Patterson (1990, 1992) showed a correlation between acorn size and geographic range of oaks in eastern North America. Jensen (1992) questioned some of their methods and results and found lower correlations. He concludes that a linear model may be inappropriate for the relationship between acorn size and range. In species from arid Australia, the seed size was more important in determing the size of young seedlings than growth rate or speed of germination (Jurado and Westoby 1992). The same was found in a dune grass. In this grass, larger seeds also had higher germination rate (Zhang and Maun 1991). Another correlation, that between seed size and frequency, is the subject of Hedge et al. (1991): they offer an explanation for the observation that frequent plants have smaller seeds than rare ones based on fractal geometry. *Trapa natans*, an annual floating-leaved aquatic, produces very large seeds (mean fresh weight 7.3 g). In poor growing conditions the number of the seeds is reduced but the size distribution remains unaltered (Kurihara and Ikusima 1991).

Large seeds can have higher predation risks than smaller ones. Long-term rodent removal from a desert habitat led to a winter annual flora which was dominated by, normally rare, large-seeded species (Samson et al. 1992). In mediterranean grasslands, sheep eat large quantities of seeds, small seeds have a better chance to pass unharmed through the digestive tract (Russi et al. 1992a).

Different types of seed are produced by many species (seed heteromorphism). In species of *Spergularia*, the production of two seed morphs was related to surrounding vegetation, indicating that seed heteromorphism has an adaptive value in habitats where water acts as a secondary dispersal agent (Telenius and Torstensson 1991; Telenius 1992).

b) Dispersal of Seeds

Dispersal plays a key role in a plant's choice of habitat (Bazzaz 1991). Dispersal spectra of all species of a given vegetation rarely show clear trends. Guitián and Sánchez (1992) found biotic dispersal to be most common among mature stages of vegetation in a sample of 46 plant communities from Spain, but no clear differences between Mediterranean and Eurosibirian vegetation.

How different modes of dispersal influence the ability of species to colonize sites was demonstrated by Dzwonko and Loster (1992) using woodland species in Poland: hovering and flying anemochores and endozoochores are the best colonizers. The mode of dispersal influences spatial patterns of three seedlings in oldfields (Myster and Pickett 1992).

Lack of long-distance dipersal mechanisms was found to be a major cause for the rareness of *Geranium purpureum*, a ballautochorous species, in near-natural vegetation in S. Spain (Herrera 1991).

Ernst et al. (1992) studied dispersal modes and distances in savanna grasses and found that animals, wind, and water help to colonize new sites, but the bulk of the diaspores remains in the vicinity of the parent plants.

Dispersal may occur in more than one step in some species, as demonstrated for *Pinus jeffreyi* by Vander Wall (1992). The seeds of this tree are wind-dispersed and subsequently collected and cached by rodents.

Wind dispersal can produce different patterns of seedlings in closely related species due to structural differences of the seed capsule (Blattner and Kadereit 1991). Andersen (1991) describes different models to predict diaspore flight of wind-dispersed plants and compares the results with experimental data.

The effects of ants on seeds of myrmecorous species in Australian sclerophyll vegetation varies from predation to dispersal among different ant species. Collection rates depend on seed characters (Hughes and Westoby 1992a, b).

Birds and mammals that eat fleshy fruits not only disperse seeds but often also influence germinability of seeds. An additional effect was described by Murphy et al. (1993): seeds passed through the gut of a specialized frugivorous bird showed the same germinability but a better establishment rate than those eaten by unspecialized frugivores. Characters of bird- and mammal-dispersed fruits such as the shape (Herrera 1992) or toxicity of fruits (Ehrlén and Eriksson 1993) are not necessarily adaptive. Dispersal is beneficial to a plant's fitness only if seeds are viable after consumption and defecation *and* safe sites are achieved. Whereas the first condition was fulfilled by foxes, the second was not in a study by Bustamante et al. (1992). Birds that eat multi-seeded fruits defecate small quantities of seeds over a prolonged time, which increases spatial distribution and germination success (Barnea et al. 1992).

c) Germination

An important experiment that illustrated the concept of safe sites for germination was the well-known and often cited experiment by Sagar (1959) using *Plantago* seeds. Primack and Miao (1991) repeated the experiment and – contrary to Sagar – could not find microsite influence on germination.

Conditions affecting germination can influence a species' performance very drastically: *Chenopodium ficifolium* was found to have changed in its temperature requirements for germination from the 1950s to present and is much more abundant now than formerly (Otte 1991). On the other hand, species with lower demands on germination temperatures declined in frequency (Otte 1993a, b).

In mediterranean therophyte pastures, varying weather conditions in autumn strongly influence species composition due to different germination requirements (Espigares and Peco 1993). Timing of germination, germination requirements, and seedbank characteristics helped to explain the distribution of three *Rumex* species (Voesenek and Blom 1992; Voesenek et al. 1992). Vegetation and litter cover influence germination and emergence of tree seedlings more strongly than soil factors in North American forest tree species (Burton and Bazzaz 1991). Germination is rare in a salt marsh plant community in New England because most species are unable to germinate due to salt stress (Shumway and Bertness 1992). Germination studies in Ericaceae species led to the determination of their successional status in maquis vegetation (Mesleard and Lepart 1991). Fire increases germination in two species of *Cistus* that are dominant in mediterranean habitats with frequent fires (Roy and Sonié 1992). Germination and dormancy characteristics in a rare *Solidago* species could not explain its rarity, as they were similar to that of widespread congeners (Buchele et al. 1991a). Different germination response after fire explained the co-existence of species in South African Mediterranean type vegetation (Kilian and Cowling 1992).

d) Seed Banks

Seed banks can be an important factor in the demography of plants, especially annuals (Kalisz and McPeek 1992). In hardwood forests, many tree species form winter-transient seed banks or permanent seed banks with a high turnover rate, as shown for mid to late successional species by Houle (1992). In savannas, the composition of the seed bank can react strongly to grazing, as shown in South Africa (O'Connor and Pickett 1992) and in Patagonia (Bertiller 1992). Species in a mediterranean pasture show different seed bank behavior: legumes have much higher carryover of seeds from one season to another than grasses (Russi et al. 1992b). Seed banks seem to be more important for vegetation affected by massive disturbance (Jiménez and Armesto 1992). Spatial distribution of seeds in a seedbank is strongly influenced by the size of the seeds (Chambers et al. 1991). The size of soil particles is an important determinant of germination success (Johnson and Fryer 1992).

Cyclic dormancy changes are well knwon from annual weeds of arable fields. Meyer and Kitchen (1992) found that dormancy is induced by chilling and removed by high temperatures in the short-lived perennial *Penstemon palmeri*, which ensures a persistent seed bank and germination in autumn only.

Seed banks are not always soil seed banks. Canopy seed storage occurs mainly in fire-prone habitats (Lamont 1991). So-called aerial seed banks exist also in desert plants in the Namib, where this behavior is correlated to temporal and spatial availability of water (Günster 1992).

e) Establishment of Seedlings

After germination, many plants are still in a susceptible phase of their life cycle. Mortality of seedlings can be quite high before they are established.

The fate of red oak seedlings was followed for 6 years. Under a tree overstory, much less survived than in the open (Crow 1992). In forests, leaf litter is an important factor for the establishment of seedlings. Molofsky and Augspurger (1992) found in a tropical forest that the studied tree species reacted differently to varying depths of litter: in some species germination was influenced by litter, in others by seedling survival. Thus, leaf litter can affect diversity of tree seedlings. Recruitment can principally depend on seed or microsite availability. Seed limitation has been underestimated according to Eriksson and Ehrlén (1992); many species are probably both seed- and microsite-limited.

Existing vegetation can have positive or negative influence on the establishment of seedlings; examples for both cases are found in the current literature. The establishment of woody species in old fields in North America is most strongly limited by seed and seedling predation. Seeds and seedlings in the vicinity of herbs have a higher predation risk than those in the open, which is described as a novel mechanism of inhibition (Gill and Marks 1991). In low-productivity grassland, positive effects of neighboring plants on seedlings establishment were demonstrated by Ryser (1993). He concluded that gaps are of minor importance for recruitment in this habitat. Kellman and Kading (1992) found that red oaks facilitate pine seedling establishment. Salonen (1992) demonstrated facilitation and inhibition effects using plant models made of plastic in a bare peat succession ecosystem: of 13 species, 3 had higher numbers and biomass under the artificial plants, 1 had lower seedling numbers.

The study of growth and survival of seedlings can help to understand why a certain species can or cannot grow in a given environment. Quintana-Ascencio et al. (1992) could explain regeneration of a tropical highland oak in this way.

3. Plant-Plant Interference: Competition

(W. Seidling)

Competition, the use of resources by an individual in such a way that others will not be able to use the same resource, is still one of the main topics in plant population biology. There is a great body of literature on plant-plant interactions originating from experiments or field observations. A smaller but increasing number of papers is based on modeling, many times connected with theoretical considerations. A third field of literature refers to interspecific competition, often taking growth characteristics (e.g., Wilson 1991) into consideration. These traits may serve as a source for classification of species with respect to their life strategies. Recently, Leishman and Westoby (1992) classified 300 species from a semi-arid woodland, using 43 traits.

a) Mechanisms of Competition

Campbell and Grime (1992) tested the well-known triangular model of plant strategies on a stress-disturbance matrix with additive mixtures from seven grass species. Stress and disturbance lower the effects of competition, a finding which is in accordance with the CSR model. However, stress tolerators were, even under the poorest soil conditions, the weaker partners, indicating tolerance of low nutrient conditions and not a superior ability to compete for nutrients at low external concentrations. Aerts and van der Peijl (1993) consider nutrient uptake kinetics and nutrient loss rates to explain the observed dominance of low-productive species on poor sites in spite of their low competitive ability.

Campbell et al. (1992) relate the vertical force exerted by growing shoots to the ability of plants to penetrate overlaying layers and take it as an indicator for competitive ability.

Shainsky and Radosevich (1992) report on mechanisms of competition among 2-year-old tree seedlings in mixtures and monocultures. Moisture and light deficits were the most prominent causes for growth reduction under nutrient-rich circumstances. Growth response of *Quercus douglasii* seedlings to competition for water mediated by different annual species revealed dependencies not only on density but also on species, indicating functional groups of competitors (Gordon and Rice 1993). Hints on the geometry of resource capture by plants determined by neighbors are given by Sutherland and Benjamin (1993). Mortimer (1992) advocates the use of the ratio of fine root length to leaf area instead of the common root-shoot ratio to estimate competitive abilities of species, which should be less biased by other functions of roots and shoots. Maas (1992) presents allocational and allometric features of two graminoid species (*Schoenus ferrugineus*, *Molinia caerulea*) to explain their competitional ability, but includes external factors to explain their coexistence in wet pastures.

Lüscher et al. (1992) report about the phenomenon of neighbor specifity between pairs of *Lolium perenne* genotypes and *Trifolium repens* genotypes originating from the same or different locations. A relative increase of *Trifolium* biomass was observed, if *Lolium* originated from the same location. They discuss their results on the background of coevolutionary processes which includes aspects of facilitation.

Allelopathy between two desert shrubs was found by Mahall and Callaway (1992): the observed asymmetric responses of roots could not be explained by depletion of water or nutrients. Pardales et al. (1992) demonstrate the allelopathic effects of *Colocasia esculenta* residues in sandy soil, reducing root growth of subsequently sown *Sorghum bicolor*.

Dioecious species can serve as valuable objects for studying competition among individuals with small but well-defined differences concerning their reproductive effort. According to Vasiliauskas and Aarssen (1992), no competitive weakness of female individuals of *Juniperus* populations was observed compared to males, contradicting in this case the idea that the female function should be more costly. In a 50:50 replacement mixture within a range of nitrogen supply Pilbeam and Robson (1992) tested two variants of *Lolium perenne*, one fast- and one slow-respiring. The outcome – the slow-respiring population was the better competitor – calls for further competition experiments with clones differing in only one physiological property.

Common methods of plant arrangements used in experiments were recently evaluated by Snaydon (1991). He summarized measures of plant competition and criticized replacement approaches. Cousens and O'Neill (1993), on the contrary, plead for adapted planting designs by saying "replacement series is one of a number of designs which can match to our specific objectives".

b) Self-Thinning in Plant Stands

Problems of self-thinning in even-aged plant stands still receive much attention because competition and population dynamics (density-dependent mortality) are inevitably intermingled. Kataoka et al. (1991) observed density-dependent genet mortality and tillering of surviving plants in Rhodes grass (*Chloris gayana*) after sowing. Subsequent to the first cutting, continuous genet mortality took place, which was compensated by ongoing tillering leading to few but large genets. Shoot number per genet was finally almost equal between different cultivars, but tiller size and density varied significantly, indicating high heritability of shoot size.

Size structure of plant stands over time is modeled by Yokozawa and Hara (1992), using photosynthesis and respiration parameters. They stress the significance of asymmetry even in the situation of two-sided competition. Theoretical considerations on the mode of competition (one-sided vs. two-sided) in tree populations were conducted by Hara (1992) and Kohyama (1992). Clark (1992) presents a theoretical model on the interplay of density-dependent and density-independent mortality over time in plant cohorts on the basis of the self-thinning rule. He derives an explanation of observed stages in forest stand development.

Dewar (1993) discusses restrictions of the –3/2 power law with the aid of simulation of carbon balances within tree stands. He concludes that the rate of respiration, canopy structure, and foliage dynamics are the major physiological parameters which determine the slope of the thinning line. The integration of population dynamics into self-thinning via response surfaces of mixtures is realized by Puettmann et al. (1992) for *Pseudotsuga menziesii-Alnus rubra* stands.

c) Changes in Growth Characteristics by Competition

Besides mortality and birth rate, growth characteristics of plants can be altered by competition. Those responses are referred to as phenotypic plasticity and concern plant size as well as growth habit. Actual studies give an impression of the wide range of responses among different plant species.

Hara et al. (1991) studied growth patterns in size-structured tree populations: Suppressed *Abies* ssp. individuals show little height growth but some increment in stem diameter. The contrary is the case in suppressed *Betula ermanii* specimens. These responses are not controlled by competition. They are interpreted as genetically fixed life-form traits. In contrast to this, alterations of allometric relationships through competition within populations of annuals are confirmed by Weiner and Thomas (1992), while Schmid and Bazzaz (1992) state that number and morphology of rhizoms of old-field perennials are partly altered through nutrient addition.

Tillering in grasses depends primarily on leaf appearance rate. After canopy closure, tiller production of *Festuca arundinacea* is lowered secondarily by reduced leaf production; but a strict mechanism for the control of branching, like apical dominance in dicots, does not seem to exist, as Skinner and Nelson (1992) sum up. Gibson et al. (1992) relate shoot and leaf-lamina zenith angles to plant density and other variables. The age-dependent decrease of shoot erectness slowed down or stopped at higher densities while age-dependent widening of leaf-lamina zenith angles stayed uninfluenced by density. Molau and Prentice (1992) applied computer-aided image analysis of leaf shape to investigate population structure.

d) Competition and Community Processes

Silvertown and Dale (1991) question the principle of transitivity ($A > B \mathrel{\hat{=}} B > C \xrightarrow{?} A > C$) to act in plant communities. Wilson and Roxburgh (1992) apply the "community matrix" theory (Levins 1968) to plant competition data, to explain stable multi-species constellations only relaying on pair-wise interaction. Quite another course is followed by Stone and Roberts (1991) to cope with multiple competition in communities. They stress the role of indirect interaction and postulate that 20 to 40% of all interactions between partners should be beneficial. Grubb (1992) raises another important issue in connection with competition in communities: interference between plants should not simply be seen as features which change gradually with resource supply. Temporal as well as spatial decoupling should be considered. Since undisturbed plant populations reproduce with some degree of autocorrelation deterministic models can be developed to simulate community development based on species-specific plant traits. Wilson (1992) models functionally organized communities on the base of a mosaic structure (fragments termed metacommunities). His model suggests that species dynamics at the patch level can explain certain properties expected on the community level (e.g., maintenance of species richness). The cellular automaton model worked out by Silvertown et al. (1992b) used replacement probabilities of five grass species. One important result was that random distribution of species resulted in much faster extinction of poor competitors than aggregated arrangements, which is common in natural communities. The idea of functional groups of plant species ("rooty species", "leafy species") is used by Reynolds and Pacala (1993) to model succession under different nutritional conditions. They conclude that instability, which results from fundamental differences between light and nutrient consumption, seems to be a main force in succession. Küppers (1992) uses ecophysiological traits of woody species concerning biomass increment and architectural arrangement to explain their position in forest succession. Alvarez-Buylla and Martinez-Ramos (1992) conceive the pioneer-climax framework as a continuum of tree life history traits.

The often observed high stability of gaps in ground layers of forests is mainly caused by the demographic processes of the surrounding plant populations (here *Narcissus pseudonarcissus*), according to Barkham (1992), and not due to environmental causes. Alteration of ecological conditions can deliver valuable observations on plant population dynamics. Some recently published papers ask about the influence of a rise in atmospheric CO_2 level on plant-plant interaction and changes in vegetation. Johnson et al. (1993) stress the handicap of C_4 plants as against C_3 plants concerning their carbon assimilation efficiency with increasing atmospheric CO_2 content. According to Bazzaz et al. (1992), enhanced oscillations of population size in annuals can result from higher CO_2 levels in the ambient air.

Efforts are being made to connect both population biology and community ecology. Levin (1992) recently argued that a problem hampers the reciprocal integration of findings: it is the problem of scale which includes both functional as well as topographic aspects. It can still be viewed as a major challenge.

Recently, vegetation scientists become more aware of scale within plant communities, as papers by Agnew et al. (1993) and Wiegleb (1992) show. Processes on a small scale are often identical to dynamics of the populations which build up the plant cover. Soukupová (1990), Herben et al. (1993), and van der Maarel and Sykes (1993) give good examples of small-scale dynamics within grassland communities. Schmidt and Brübach (1993) attribute a great part of temporal and spatial variation in abundance of plant species on artificial protosoil to dispersal and life-form strategies of the invading species. The importance of life-history traits and the role of phenotypic plasticity in *Potamogeton* species to survive in a highly dynamic and steadily changing environment of rivers is emphasized by Wiegleb et al. (1991).

4. The Clonal Growth Form and the Structure of Populations

(J. Stöcklin)

The iterative production of structurally similar modules is a general feature of all plants. Clonal growth in a more narrow sense means the production of new, potentially independent, but genetically identical "ramets". Characteristic abilities of clonal species which are attributable to their growth strategy include: (1) the small-scale colonization of open habitats, (2) the occupation of space and the monopolization of resources by forming dense patches, (3) the avoidance of competition by fugitive growth and (4) the opportunistic exploration for resources which are heterogenously distributed. Statements about clonal plants in demographic and functional terms require an understanding of those developmental processes which control the morphology and growth of an organism.

a) Architectural Plasticity in Clonal Plants

An important aspect in understanding the adaptive significance of clonal growth concerns the degree of predictable plasticity in the architecture of clonal plants. Schmid and Bazzaz (1990, 1991, 1992) examined the consequences of contrasting growth form in *Solidago* (compact growth form) and *Aster* (spreading growth form) for plasticity in size and architecture. The two taxa were well characterized by their respective architecture. In contrast to the large plasticity in size, architecture was influenced by different soil types only in *Aster*, the species with a "guerrilla" growth form. The authors observed a larger phenotypic plasticity in plants grown from rhizomes as compared to plants grown from seeds, and suggest a weaker control of the architectural blueprint (i.e., genetically determined growth plan) in clonal progeny (Schmid and Bazzaz 1990). In comparing the growth of native and transplanted shoots, Schmid and Brazzaz (1991) found a stronger clonal integration and parental support in *Solidago* than in *Aster*, thereby confirming earlier findings that in clonal species with a clumped habit, clonal integration is more likely than in plants with a spreading habit. Finally, Schmid and Bazzaz (1992) tested the expectation that the character of the rhizome system would predictably respond to differences in the abundance of resources and to competition. They found only partial support for this prediction in *Aster* and *Solidago*. As de Kroon and Knops (1990) already proposed, based on their investigation with *Carex flacca* and *Brachypodium silvaticum*, Schmid and Bazzaz (1992) suggested that the foraging tactic of clonal plants with spreading architecture may not apply to species with compact genet architecture.

Architectural parameters or clonal growth form may not only vary due to environmental conditions but can vary with age of the clone. This was shown in *Salix setchelliana* (Douglas 1991), a species native to glacial river gravel bars. Clones of this species appear to change in time from a more spreading growth form and environmental exploration to a more compact growth from with increased branching. Meyer (1992) observed that clones of *Solidago altissima* did not change their morphological growth characteristics such as length of rhizomes or branching angle as they increase in size and get older. Only demographic parameters changed with age. Branching intensity of rhizomes decreased, but survivorship increased. Plasticity in size of clones due to competition by neighbors was a result of difference in the survivorship of single rhizomes, branching intensity, however, was not affected by competition.

Interpretations of morphological changes in clonal plants that appear as a consequence of changing environmental conditions are more easily made with a knowledge of plant development. Birch and Hutchings (1992) showed that the use of a plastochron index to record the morphological growth of *Glechoma hederacea* may help to interpret its morphological plasticity. Internode elongation is accomplished relatively quickly, but leaf expansion and petiole development require a longer time period and could simultaneously continue for several consecutive ramets. This allows the plant to make continuous adjustments to changes in its environment. Price and Hutchings (1992) used anatomical methods to better understand the pattern of integration in this species. Unfortunately, studies of the internal organization of clonal species are still rare although such investigations might help in the interpretation of their functional behavior.

b) Physiological Integration and Characteristics of Clonal Growth

Vascular connections between sister ramets provide the potential for the translocation of photoassimilates, water, and minerals, and as a consequence for resource sharing among ramets, which in turn may affect genet performance in heterogeneous environments. Physiological integration may explain the low mortality and high genetic diversity in clonal plant populations (de Kroon et al. 1992). Jónsdóttir (1991) could show that death rate of grass tillers within clones was not density-dependent, but rates of birth and death of tillers varied in a synchronized way, providing evidence that already established tillers regulate the production of new ones. Between ramets of single clones of *Brachypodium pinnatum* and *Carex flacca*, de Kroon et al. (1992) observed asymmetric competition, suggesting little evidence for effective physiological integration. Hara et al. (1993), on the other hand, found that in *Phragmites australis* growth patterns and size-structure dynamics of individual shoots are in striking contrast to those of nonclonal plants. Even in the extreme situation of asymmetric competition, shoot populations converged to the same size-structure in height and weight as they grew, suggesting a regulatory mechanism between shoots. These authors suggest that belowground physiological integration in the form of controlled allocation patterns of remobilized resources is a major characteristic of many clonal plant species and guarantees high growth rate of or little suppression on small shoots.

As a rule, the support of daughter ramets during early growth by older ramets predominates in colonizing clonal species with a preponderantly linear growth form (examples include *Carex arenaria*, *Cynodon dactylon*). Evidence for a basipetal transport of assimilates and nutrients is more frequently found in species with a more compact growth from (for an overview of older literature, see Marshall 1990). Evans and Whitney (1992) showed that the acropetal transport of water in clones of the colonizing dune perennial *Hydrocotyle bonariensis* is likely to ameliorate the heterogenous saline conditions associated with coastal environments. Alpert (1991) observed both acropetal and basipetal transport of nitrogen in the stoloniferous *Fragaria chiloensis*, but such nutrient transport tended mainly to promote the growth of younger ramets. The same was observed by Abrahamson et al. (1991) in *Solidago altissima*. Mother rhizomes freely shared nutrients with their developing daughter ramets, but any nutrient sharing that occurred in this species between sister ramets was too little to significantly affect their growth. The authors suggest a clonal growth strategy for goldenrod, in which nutrients are not shared among sister ramets and where the production of rhizomes therefore is reduced in nutrient-poor microsites. In the foraging *Glechoma hederacea*, transport of assimilates and xylem transport was also almost exclusively acropetal (Price et al. 1992). The pattern of integration was influenced by vascular constraints and became restricted to single stolon branches as clones became larger and structurally more complex. Welker and Briske (1992) showed that in the caespitose graminoid *Schizachyrium scoparium*, transport of resources is exclusively acropetal with established ramets supporting juvenile ramets during early development. Integration is restricted to three connected ramet generations caused by the disintegration of vascular connections among ramet sequences.

A remarkably high level of integration including basipetal transport of assimilates to storage organs was found in *Podophyllum peltatum*, a clonal forest herb with a linear, sympodial rhizome (Landa et al. 1992) and in *Trifolium repens*, a grassland perennial (Turkington and Klein 1991; Chapman et al. 1992; Kemball et al. 1992). In both species such high levels of clonal integration may support the maintenance of clones in heterogenous and stressful environmental conditions.

c) Clonal Growth Pattern: Insights from Models

Model approaches have frequently been utilized to interpret the growth patterns of clonal plant species or to compensate for the lack of long-term studies of their population dynamics. Recent models are nonstationary and try to incorporate natural variation for growth parameters using stochastic simulations. Carlsson and Callaghan (1991) incorporated field data on between-year variation in flowering in a projection matrix model and by doing this were able to simulate the cyclic behavior of tiller numbers in *Carex bigelowii*. Cain et al. (1991) showed by stochastic simulation of clonal growth in *Solidago altissima* that variation for clonal growth parameters may increase the efficiency of substrate exploration and decrease the potential for intraclonal competition.

Whilst the above-mentioned models incorporate field data from real plants, more theoretical models try to make predictions about the evolutionary significance of plant form. De Kroon and Schieving (1991a), using a general model, tried to explain resource allocation patterns as a function of clonal morphology. Sakai (1991) developed a game model to examine evolutionary stable strategies for the architecture of herbaceous plants in open and closed habitats and dependent on the density of clones. Caraco and Kelly (1991) modeled physiological integration as an adaptive process to predict patterns in resource translocation in clonal plants growing in a stochastic environment. They predict that a large mother ramet should always share resources with a daughter ramet, but a smaller mother ramet should often abandon a daughter and allocate all its available resources to its own growth.

d) Growth Strategies of Clonal Plants and Their Ecology

Quite frequently, plant communities or at least one layer of the vegetation (e.g., herbs in a forest) are structured or even dominated by vigorous clonal species. The small-scale colonizing ability by clonal growth is often superior to germination and establishment from seeds. The studies of Klimes (1992) with *Rumex alpinus* and Soukupovà (1992) with *Calamagrostis canescens* illustrate the invading strength of clonal species in the case of increased nutrient availability or after disturbance. Both species are examples for the competitive strength which is characteristic for clonal species with compact growth form and high productivity. Tybjerg and Vestergaard (1992) studied the structure and growth of the rhizomatous *Polygonatum verticillatum* in a forest site. To describe the phalanx-like growth pattern, they used an architectural model based on the location and fate of buds and the size and spatial orientation of the bud products. Geber et al. (1992) observed genetic differences in demographic parameters among clones of water hyacinth. They suggest that these differences are related to the ability of clonal colonization vs. competitive strength and the capacity of establishment by vegetative spread or by seeds. Several weeds that are able to invade undisturbed plant communities are clonal species (Thompson 1991; Weber and Schmid 1992; Sukopp and Schick 1993). The Eurosibirian *Linaria vulgaris* is a noxious weed in North America which propagates by the development of buds on roots. Nadeau et al. (1991) studied the lateral spread of this species in order to predict the best time for weed control.

Recently, Bazzaz (1991) reviewed the specific abilities of clonal plants to integrate several patch types or to choose habitats by clonal spread and fragmentation. De Kroon and Schieving (1991b) characterized the efficient resource utilization of slow-growing clonal species in inherently poor environments as a "conservative strategy". Clonal species which dominate the vegetation and possess the ability to monopolize the available resources in space and time are characterized by these authors as having a "consolidation strategy". Stöcklin (1992b) identified five clonal growth strategies considering differences in growth form, growth rate, and plasticity. Each strategy offers benefits under particular environmental conditions. With examples from alpine plant communities, it was possible to show that different strategies prevail in different habitats.

Biological differences among closely related species, and their performance within plant communities may be explained by genet architecture. Stöcklin (1992a) compared the ability of lateral clonal growth in closely related *Epilobium* species. Differences among species in the shoot regeneration from subcotyledonary buds result in contrasting plant architecture and colonization strategies in their respective habitats. The expanding clonal growth of *E. fleischeri* is a prerequisite for the dominant role of this species in early succession on the perimeter of glaciers, and may also explain the long persistence during later succession when establishment by seeds is unlikely (Stöcklin 1990). Kik et al. (1990) found a relationship between plant architecture of different biotypes and habitat in the grass *Agrostis stolonifera*. Colonizing populations had the highest reproductive capacities and colonizing by seeds was by far more efficient than by stolons.

e) Dynamics in Clonal Populations and Recruitment from Seedlings

The development of ramet populations in a colonizing clonal plant may be very dynamic (Meyer and Schmid 1991). However, small-scale and short-term studies of clonal plant populations more often suggest stability. Eriksson (1992b) used a stochastic model to infer from demographic data sampled over a period of 4 years the long-term dynamics of the clonal dwarf shrub *Linnaea borealis*. Simulations indicated a considerable temporal variation in population size. Although variation in flowering was pronounced, this was of little importance for growth rate variation. Eriksson (1992b) suggests that flowering might be a selectively neutral trait in reproductive dynamics of clonal species. Carlsson and Callaghan (1991), on the other hand, found in *Carex bigelowii* that cyclicity of flowering is necessary to maintain a positive growth rate of tiller populations. The often recorded increase of sexual reproductive effort in clonal species with plant size would also suggest that in clonal species, flowering is related to population dynamics and has an important influence on plant fitness (Rusterholz et al. 1992). In clonal dioecious plants, as a consequence of the high resource cost of fruiting, the sex ratio in flowering ramets may be male-biased (Lovett Doust and Laporte 1991).

Recruitment from seedlings in established populations is often considered to occur infrequently in clonal plants. Meyer and Schmid (1991) observed an extremely rare germination of seeds in an established population of *Solidago altissima*. However, clonal species differ in respect to the frequency of repeated seedling recruitment. In an overview, Eriksson (1989) found in 40% of the 68 species for which information was available evidence for seedling recruitment in established populations. Repeated establishment from seeds was more frequent in grassland than among woodland species, in species growing clonally aboveground than in species growing clonally belowground, and it was more common in species lacking features that promote long distance seed dispersal than among species that possess such features. This last point is explored by Eriksson (1992a) from an evolutionary point of view and it is suggested that clonal propagation may have been augmented in lineages where long distance seed dispersal predominates in order to make genet fitness less dependent on local dispersal by seeds. Stöcklin and Zoller (1991), based on the comparison of the life cycle of *Epilobium* species, emphasize the close relationship between clonal growth strategy and the long-term consequences of seedling establishment for the understanding of population structure in clonal plants. However, field studies of the dynamics of clonal plants which include the seedling stage are rare, and in most cases information on the frequency of seedling establishment is only anecdotal. A close look might reveal that establishment from seeds is more frequent than assumed. Even in a species with predominant clonal growth such as the alpine pioneer *Geum reptans*, 5% of the new plants established in a population came from seeds (Rusterholz et al. 1993). Maze and Whalley (1992) studied the dynamics of seedlings in a clonal foredune grass. They found that seedling emergence occurs mainly on the front of the foredune, as this is an area of active sand accretion which is necessary for germination. Hossaert-McKey and Jarry (1992) studied investment in clonal growth and sexual reproduction in two "phalanx"-type *Lathyrus*-species. They found low seed production, limited seed dispersal, and low seedling establishment, but no evidence of competition between established clones and seedlings. A stock of dormant seeds in the soil is probably a counterbalance to the restricted gene flow in these species.

5. Impact of Herbivory on Plant Populations

(W. Seidling)

In ecosystem approaches it its common to regard bottom-up forces first, like Hunter and Price (1992) or Power (1992), who investigate food web relations (see also Grellmann and Schneider 1992 advocating a multi-leveled model). But Hunter and Price stress that top-down processes are important for the regulation of populations of low trophic level like plants as well: herbivory is a dynamic process influencing both the herbivore and the plant. For ecologists working with populations in aquatic systems (e.g., Elser 1992) it is a quite common perspective. For terrestrial ecosystems, these processes have only recently attracted attention. We will concentrate on top-down forces influencing plant fitness (survival, growth, reproduction).

a) Growth Responses of Plants After Herbivore Attack

After herbivore attack, plants have always lost some tissue (or sap), but normally do not die mainly because of their modular organization. A great variety of physiological and morphological responses after herbivore attack occur: changes in dry matter assimilation, redistribution of carbohydrates and minerals, delayed plant senescence, changes in leaf area ratio, plant architecture, and root development, as Prins and Verkaar (1992) recapitulate. To gain further insight into (over)compensation after tissue loss, they recommend physiological investigations under complex situations as well as field studies. Responses of woody species after browsing, with special reference to African savannas, were summarized by Bergström (1992).

Several defoliation and branch-clipping experiments demonstrate that the consequences for the individual fitness of a plant depend on the intensity to tissue loss, on the temporal pattern of damage, on the parts of a plant attacked, on the relative age (stage) of a plant, and on the local availability of resources. That the leaves remaining after herbivore attack increase their photosynthetic capacity to a degree related to the amount of tissue loss has been found for many species. Ovaska et al. (1992) attribute this primarily to the improved competitive status of the remaining leaves. Oesterheld (1992) tested the compensatory growth response of two grass species after different defoliation intensities. His results are consistent with an exponential increase in aboveground relative growth rate (cf. Hilbert et al. 1981) with increasing tissue loss. Root growth is meanwhile restricted through preferential allocation of currently produced carbohydrates towards leaf growth.

Compensatory growth of *Pinus sylvestris* under different stand densities was found by Edenius et al. (1993). Compensation was most delayed in the case of severe (100% annual shoot biomass) clipping and, suprisingly, was best in densely grown stands. Individuals of a tropical pioneer tree species in forest gaps were defoliated to 25, 50 and 75% by Núñez-Farfán and Dirzo (1991). While the control performed the best growth, no clear difference between the treatments were found, suggesting that compensatory growth depends only to a minor extent on tissue loss. Reichman and Smith (1991) found that loss of root tissue may generally be more harmful (here in *Tragopogon dubius*) in terms of compensatory growth than leaf loss.

Price and Hutchings (1992) investigated the response to two degrees and different patterns of leaf removal in a glasshouse experiment with *Glechoma hederacea*. While compensation for tissue loss was never complete, a 50% leaf loss applied to consecutive pairs of leaves along the axes with the next two pairs left intact resulted in only minor growth reduction. This and other results are in conformity with resource transport patterns (Price et al. 1992) and support the idea of decreasing integration of modules with plant age or size. According to Marquis (1992), the fitness of a subtropical shrub was reduced by a 10% defoliation of a single branch before flowering, but was not reduced if the same treatment was employed during flowering. A scattered 10% leaf loss had no effect at either time. Besides compensatory growth, these results indicate restricted resource movement in woody plants.

Doak (1992) investigates the impact of a principal herbivore on its perennial host *Epilobium latifolium* and modeled its effect over the whole plant lifespan (37 years). According to his results, seed production was generally more influenced than vegetative growth, but interaction between intensity and frequency of feeding mediated the net damage in a nonlinear way.

Graminoids are known for their high tolerance to defoliation. For the clonal sedge *Carex bigelowii*, a higher density of shoots and biomass per area was ascertained after grazing, in spite of decreased tiller size and a lower rate of flowering (Jónsdóttir 1991). Pyšek (1992), who applied insecticides for herbivore exclusion, obtained similar results: reduction in seed number (36.5%) was the most pronounced difference by the end of the growing season in the perennial *Senecio ovatus*. Differences in weight and leaf area existing earlier had almost ceased by that time.

Marriott and Haystead (1992) investigated N uptake of defoliated and intact *Trifolium repens* plants. They found reduced N content in defoliated plants, but the mineral N uptake from soil was 50% higher.

Regrowth after herbivore attack is often not merely compensation for lost tissue, but alteration in morphological features may also occur. Tolvanen et al. (1992) studied the reaction of dwarf shrubs (*Vaccinium myrtillus, V. vitis-idea*) after defoliation or current-year branch clipping. A more vigorous regrowth of new branches was observed after the removal of whole branches than after entire defoliation of branches, because apical dominance is broken only in the first case. Against general expectation, compensatory growth on the ramet level was better in the evergreen *V. vitis-idea.*

Strong evidence for overcompensation and enhanced plant fitness in terms of higher seed mass in given by Paige (1992). Changed architectural features and probably induced chemical defense of *Ipomopsis aggregata* after tissue loss are interpreted as a switch from mutualistic to an antagonistic interaction towards ungulates (because secondary herbivory is greatly restricted).

Karban and Strauss (1993) report high mortality among *Erigeron glaucus* seedlings and herbivore-specific morphological aberrations after gopher attack, but especially spittlebugs reduced the number of flower heads and, as a consequence, seed production. The number of seeds per head was comparatively invariable. The authors guess that there is a non-neglectable influence of herbivory on life-time fitness of *E. glaucus* clones. The iteroparous *Yucca elata* responds to cattle feeding on its flowers with a high degree of clonality (Kerley et al. 1993). This type of herbivory has a strong effect on the demography of this clonal species, resulting in few but large old genets and little generative regeneration. Studies which address problems connected with pathogens and plant population processes generally earn little attention among ecologists. Impacts on sexual reproduction as well as on vegetative growth occur and intermingle. Wennström and Ericson (1992) investigate systemic infection by a rust fungus of clonal plants. They point out that rhizome length and its fragmentation ability can be a function of infestation avoidance.

b) Induced Anti-Herbivory Defense

In regrown parts of attacked plants, as well as in tissue grown prior to damage, changes of morphological or chemical qualities are often observed, which may offer protection against repeated herbivory. Investigations concern the question whether plants are enforced by basic physiological conditions (e.g., N deficiency see Bryant et al. 1983) or and to what degree extra resource investment for protection has evolved to lower the attractiveness of a plant.

Leather (1993) shows that early damage within a season can prevent herbivory later: artificial defoliation of *Prunus padus* during summer resulted in less damage by late-season defoliators. Bryant et al. (1991) investigated long-term-induced defense against defoliating insects in African woody species. According to the "carbon/nutrient balance model" (Bryant et al. 1983), they conclude that inherently slow-growing species (like evergreens) tend to increase leaf quality for insect herbivores in the following season, due to a comparatively higher loss of carbon, while in rapid-growing trees, palatability of leaves decreases (reduced N and P concentrations, increased total phenol and condensed-tannin concentrations). Likewise, on the basis of the "carbon/nutrient hypothesis", Lincoln (1993) is concerned with the susceptibility of plants through insect herbivores after atmospheric CO_2 enrichment, while Lambers (1993) concludes that phenolic compounds generally accumulate under such conditions mainly due to a greater limitation of nutrients.

Experiments by Rausher et al. (1993) with the annual *Ipomoea purpurea* give evidence that changes after herbivore attack, in spite of being systemic, do not lead to a general unpalatability. Only parts of plants reduced their susceptibility to some of their herbivore insects.

Besides the defoliation regime, climatic conditions influence the attractiveness of *Betula pubescens* leaves for insect herbivores, as is shown by Senn et al. (1992). They consider this fact to be an evolutionary source of the observed high degree of leaf polymorphism within mountain birch. Ernst et al. (1991) report about short-term waves (1–3 months) of total phenolic content of leaves of savanna tree species which are overlaid by a general decrease with leaf age. They suppose a climatic agent as cause for the observed fluctuations. The authors call for caution in interpreting chemical compounds as pure defense mechanisms.

c) Permanent Defense in Plants

Permanent resistance to herbivores can be achieved by chemical as well as morphological properties (including structures like extrafloral nectaries which attract potential predators of herbivores, e.g., Rashbrook et al. 1992). Recent research seems to underline strong genetic control over defense traits. This can be taken as indirect evidence that herbivory has acted and still acts as a force in plant population processes and plant evolution. In grasses, silica acts as an effective defense against mammals as Gali-Muhtasib et al. (1992) report. Palo et al. (1992) focus on twigs eaten by mammals during winter. They state a trade-off between high concentration of phenolics in the apical part of twigs reducing digestability for browsers and the higher proportion of fiber in the more proximal parts. Since pressure from browsers should decrease with tree height, it would be of interest whether a within-crown gradient exists, besides the general age-/size-dependent decrease in phenolic content. A similar approach has been applied by Hjältén and Palo (1992), offering winter dormant branches of four tree species. In this experiment, voles and hares did not prefer N-rich twigs, but the total concentration of phenolics seemed to be the factor best explaining their feeding preferences.

Rousi et al. (1993) suppose substantial genetic variation among *Betula pendula* families with regard to their attractiveness to hares and voles. Their results did not confirm the postulated trade-off between expenditures for defense and growth. Bowers et al. (1992) show that in *Plantago lanceolata* glycoside content and relative palatability varies according to season, habitat, and genotype.

The "hybrids-as-sinks hypothesis", stating that hybrids are more susceptible to herbivory, could not be substantiated by Aguilar and Boecklen (1992), who investigated patterns of herbivory in the *Quercus grisea* x *gambelii* complex. Paige and Capman (1993) use a recombinant DNA technology to investigate the relationship between hybridization (*Populus angustifolia* x *fremontii* complex) and pest resistance. Aphid resistance was found to be polygenic. Strong correlation between morphological and resistantial traits do not exist. Moreover, resistance trait expression seems to vary with environment (elevation) and could therefore be covered in terms of phenotypic plasticity. They warn against generalizations regarding effects of hybridization and call for more genetic studies.

An example of permanent chemical defense is given by Zangerl and Berenbaum (1993): *Pastinaca sativa* specimens with high concentration of furanocoumarins are not attacked by parsnib webworm. Since concentrations of these secondary metabolites are partly under genetic control, subpopulation-specific utilization patterns occur.

In *Pinus ponderosa* β-pinene and β-phellandrene content is under strong genetic control. Both impede feeding activity of Albert's squirrel (Snyder 1992). Due to dramatic reduction of the host's fitness after squirrel attack, this type of selective herbivory is supposed to be an important agent of natural selection in ponderosa pine populations.

Investigations on three dioecious *Salix* species by Danell et al. (1991) attribute the observed female-biased sex ratio in boreal areas at least partly to selective vole herbivory. The application of this relation is extended to local environmental gradients by the authors in the sense that poor sites tend to have a higher proportion of female shrubs.

Since seeds represent a valuable resource of nutrients for consumers, they may be equipped with different defense mechanisms. Siemens et al. (1992) studied two congeneric Fabaceae. The species with a seed coat-based chemical defense was less successful than the other, which protected its seeds mechanically by early abscission. The study concludes that quite different defense systems can evolve under almost the same predator spectrum.

d) Costs of Anti-Herbivory Defense

Costs of anti-herbivory measures of plants (Simms 1992) must be differentiated in expenses for mechanical structures and those for chemical repellants. Skogsmyr and Fagerström (1992) calculate physiological costs for production and maintenance of tannins, terpenes, nicotine, and mechanical defense structures under different innate growth rates and nutrient availabilities. Under N shortage, mechanical defense seems to be the most expensive measure, while under C shortage tannins in higher concentrations are most expensive. Nicotine is always comparatively cheap.

Salomonson et al. (1992) state that the maintenance of a toxic defense system requires the high availability of potassium in the soil, and call this a key factor for the occurrence of toxic species.

Briske and Anderson (1992) could not confirm a suggested trade-off between anti-herbivore defense measures of plant species evolved under a grazing regime (like *Schizachyrium scoparium*) and growth of species lacking such experience.

Karban (1993) asks for the costs of induced defense in comparison to untouched plants of *Gossypium thurberi*. He could not find support for the idea that induced resistance is advantageous over constitutive resistance. The question of whether selection for high constitutive defense against spider mites in cotton may lower the ability of the plant to produce induced defenses against mites or verticillium wilt could also not be corroborated by Brody and Karban (1992).

e) Stress and Herbivory

According to Price (1991), the attractiveness of plants to certain herbivores does not concur well with the "plant stress hypothesis" (herbivores achieving greatest gain from stressed plants, e.g., Louda and Collinge 1992, who found increased herbivory on *Cardamine cardifolia* plants with water stress) and only partly with the "climatic release hypothesis" (insect outbreaks after a series of favorable seasons). He calls for diversification of hypotheses on plant-herbivore relationships, and suggests the "plant vigor hypothesis" to explain repeated utilization of "vigorous modules". In this context, it is interesting that grasshoppers, but also elk, fed almost exclusively on vigorous willow shoots regrown after burns and not on the unburned controls (Stein et al. 1992). The specimens most attacked by a weevil in the biennial *Cynoglossum officinale* are the roots of large and not of small plants (Prins et al. 1992).

On the contrary, findings of Rousi et al. (1991) do not support the "plant vigor hypothesis": the tallest *Betula* seedlings were most resistent against hare browsing. Schmitz et al. (1992) found that dried twigs of the chemically defended *Populus balsamifera* (defense is effective only in vigorous condition) were prefered to fresh ones of the less defended *Salix glauca*. These findings suppose that the "plant vigor hypothesis" does not hold for mammalian herbivore plant interaction in general.

Masters et al. (1993) suggest a model which is partly in accordance with the "plant stress hypothesis": root feeding enhances foliar feeding via stress response of the plant, while foliar feeding has a reducing effect on belowground herbivory by reducing root growth. A long-term study (up to 12 years) on leaf demography in eucalypts was done by Lowman (1992). Peaks of consumption occurred on one side during emergence (vigorous parts), while in older leaves measures of defense seem to be more effective. On the other side, generally higher damage occurred in shady parts of the crowns (weak parts).

Since the above-mentioned hypotheses are subject to debate, a unifying concept is still lacking. To explain different incidences of anti-herbivory measures in general, Grubb (1992) suggests the "scarcity-accessibility hypothesis", which he himself calls a "broad-brushed" approach. The main idea behind it is that degree and mode of defense cannot be explained by only one dimension. He considers at least six properties simultaneously.

f) Herbivory and Community Processes

Defoliation experiments refer mostly to short-term effects on single plants, while impacts on plant populations under field conditions or within communities are mostly studied by exclosure experiments. Since herbivory occurs always in combination with other forces, experimental design is of great importance. An approach with certain relevance for competition experiments in general has been carried out by Reader (1992a). He showed that reduced herbivory can be responsible for the improved survival or growth of target plants, which were relieved from their neighbor plants, and not – as usually supposed – reduced competition for resources.

According to the results of a combined competition-exclosure experiment, Reader (1992b) concludes that hollow-ridge variation of three perennial forbs is caused by relief-orientated herbivory (probably mainly mollusk herbivory) and not due to competitive or other environmental interaction. Rees and Brown (1992) refer to the question of whether herbivory and competition are purely additive or whether herbivory acts on growth rates of competing plants as well. Results of a comparative approach with four species of Brassicaceae deny the last assumption. Furthermore, the importance of mollusk herbivory for succession is emphasized.

Plant populations within communities rarely show simple proportionality of responses according to different degrees of influencing factors: as Müller-Schärer (1991) demonstrates, belowground herbivory in *Centaurea maculosa* applied without competition can leave biomass or fecundity unchanged; but applied together with grass competition, high mortality and reductions of fitness parameters were observed under certain soil conditions. An even more complex situation is given in field studies on direct and indirect impact of grazing: disturbance, enhanced resource supply, and tissue loss through herbivory can be related to different life-cycle stages of various species. Fernández et al. (1992) focus on two dominant shrub species in the Patagonian steppe, one (*Senecio filaginoides*) facilitated, one (*Mulinum spinosum*) slightly hindered by sheep grazing on the surrounding grass matrix. Population structure of *Senecio* was characterized by a threefold increase of the smallest height class and a higher rate of seedling recruitment under grazing conditions. Silvertown et al. (1992a) focus on an annual (*Geranium dissectum*) influenced directly and indirectly by sheep grazing. The dependence of plant performance on season of grazing and seeding should be underlined. Lowman and Heatwole (1992) tried to evaluate leaf loss of eucalypt tree populations by insects. More than half of over 5000 leaves monitored senesced from other causes. In *Eucalyptus nova-anglica*, loss of leaf surface area reached 300%. In this species, insect herbivory is supposed to contribute to tree mortality and dieback phenomena to a higher degree. They found indications that less vigorous trees are attacked more heavily (see Sect. 5.e).

6. Applied Aspects

(U. Starfinger)

A major field of application for population ecology is species conservation (Cornelius 1991; Cornelius and Meyer 1991). Various aspects concerning animals and plants were presented in Seitz and Loeschke (1991).

Long-term demography of endangered species as a basis for conservation activities is rarely available. An example is given by Cinquemani-Kuehn and Leopold (1992). Spiess (1991) developed conservation programs for bog species on the basis of demography and strategy types. Hegazy (1992) gives recommendations for the conservation of a threatened salt marsh species on the basis of surviorship curve, fecundity schedule, and life table studies. The life cycle of an endangered *Solidago* species was studied by Buchele et al. (1991b), and was not found to be different from that of congeneric widespread species. To predict the fate of the population of the rare *Banksia cuneata*, Burgman and Lamont (1992) develop quasi extinction risk curves under different weather and fire conditions. This case study is of general importance because it shows that modeling can focus attention on key conditions for long-term survival.

Special attention must be paid to the seed banks of endangered species (Otte 1992; Albrecht 1993). In a rare wetland plant community in Canada, rare plant species were represented in the seed bank in even lower precentages than in the vegetation cover, so disturbance can soon lead to their replacement through common species (Wisheu and Keddy 1991). The role of the seed bank is crucial in the restoration of vegetation (Skoglund 1992). Gerhardt (1993) studied tree seedling growth and survival in order to understand ways to restore tropical dry forest.

In many species, instability on a local scale is contrasted by high stability on a higher, metapopulation scale (cf. Hanski 1991). Van der Meijden et al. (1992) investigated fugitive biennials. On the metapopulation level, they survive in spite of short-lived local populations, comparatively rare safe sites and lack of long-distance dispersal by large and long-lasting seed banks relying on spatially dependent patch (disturbance) distribution. Ouborg (1993) used the meta-population concept and the classical island biogeography concept to study extinction and colonization processes with respect to species conservation. The dynamic meta-population approach seems to be a better tool to explain effects of isolation. O'Connor (1991) moves on the same level of regional integration when modeling processes of local extinction in semiarid grasslands.

Processes which are important for nature and species conservation, such as herbivore-plant interactions (Tscharntke 1991) or the immigration of alien species (Starfinger 1991a) can be better understood with population ecological approaches, but few exist as of now. The success of immigrating species has been studied with population ecological methods by D'Antonio (1993), Kowarik (1992), Meyer and Schmid (1992), Starfinger (1991b), and Weber and Schmid (1992). A common result of these studies is that invasions can only be understood from a combination of factors relating to the invader, the invaded community, and abiotic factors.

A further aspect is the reaction of plant populations to anthropogenic pollution or devastation of the environment. Bornkamm (1991) gives a size-orientated demography of *Typha angustifolia* and *Phragmites australis* shoots influenced by different levels of boron load. Tschurr (1992) and Hasler (1992) studied the regeneration capacity of clonal alpine species after damage in the context of the renaturation of ski runs.

References

Abrahamson WG, Anderson SS, McCrea KD (1991) Am J Bot 78(11):1508–1514. **Aerts R**, van der Peijl MJ (1993) Oikos 144–147. **Agnew ADQ**, Collins SL, van der Maarel E (1993) J Veg Sci 4:146–148. **Ågren J**, Willson MF (1992) Oecologia 92:177–182. **Aguilar JM**, Boecklen WJ (1992) Oikos 64:498–504. **Aizen MA**, Patterson WA (1990) J Biogeogr 17:327–332. **Aizen MA**, Patterson WA (1992) J Biogeogr 19:581–582. **Albrecht H** (1993) Schriftenr Schutz gefährdeter Pflanzenarten 5 (in press). **Allen RB**, Platt KH (1990) Oikos 57:199–206. **Alpert P** (1991) Ecology 72(1):69–80. **Alvares-Buylla ER**, Martinez-Ramos M (1992) J Ecol 80:275–290. **Andersen M** (1991) Am Nat 137:476–497. **Ashman T-L** (1992) Oecologia 92:266–272. **Ashman T-L** (1992) Ecology 73:1237–1243.

Barkham JP (1992) J Ecol 80:797–808. **Barnea A**, Yom-Tov Y, Friedman J (1992) Acta Oecol 13:209–219. **Bazzaz FA** (1991) Am Nat 137 Suppl:116–139. **Bazzaz FA**, Ackerly DD, Woodward FI, Rochefort L (1992) J Ecol 80:643–651. **Bergström R** (1992) J Veg Sci 3:315–324. **Bertiller MB** (1992) J Veg Sci 3:47–54. **Birch CPD**, Hutchings MJ (1992) Oikos 63:387–394. **Blattner F**, Kadereit JW (1991) Flora 185:55–64. **Bornkamm R** (1991) Verh Ges Ökol 20:927–934. **Boutin C**, Harper JL (1991) J Ecol 79:199–221. **Bowers MD**, Collinge SK, Gamble SE, Schmitt J (1992) Oecologia 91:201–207. **Briske DD**, Anderson VJ (1992) Vegetatio 103:41–49. **Brody AK**, Karban R (1992) Oikos 65:301–306. **Bryant JP**, Chapin FS, Klein DR (1983) Oikos 40:357–368. **Bryant JP**, Heitkonig I, Kuropat P, Owen-Smith N (1991) Am Nat 137:50–63. **Buchele DE**, Baskin JM, Baskin CC (1991a) Bull Torrey Bot Club 118:288–291. **Buchele DE**, Baskin JM, Baskin CC (1991b) Bull Torrey Bot Club 118:281–297. **Burgman MA**, Lamont BB (1992) J Appl Ecol 29:719–727. **Burton PJ**, Bazzaz FA (1991) Am J Bot 78:131–149. **Bustamante RO**, Simonetti JA, Mella JE (1992) Acta Oecol 13:203–208.

Cain ML, Pacala SW, Silander JA Jr (1991) Oecologia 88:477–485. **Campbell BD**, Grime JP (1992) Ecology 73:15–29. **Campbell BD**, Grime JP, Mackey JML (1992) J Ecol 80:633–641. **Campbell DR** (1991) Am Nat 137:713–737. **Caraco T**, Kelly CK (1991) Ecology 72(1):81–93. **Carlsson BA**, Callaghan TV (1991) Oikos 60:231–240. **Chambers JC**, MacMahon JA, Haefner JH (1991) Ecology 72:1668–1677. **Chapman DF**, Robson MJ, Snaydon RW (1992) Oecologia 89:338–347. **Cinquemani-Kuehn DM**, Leopold DJ (1992) Bull Torrey Bot Club 119:65–76. **Clark JS** (1992) Theor Popul Biol 42:172–198. **Cornelius R** (1991) Verh Ges Ökol 20:905–915. **Cornelius R**, Meyer G (1991) In: Henle K, Kaule G (eds) Arten- und Biotopschutzforschung für Deutschland. Forschungszentrum Jülich, Ber ökol Forsch 4:119–130. **Cousens R**, O'Neill M (1993) Oikos 66:347–352. **Crow TR** (1992) Oecologia 91:192–200.

Danell K, Hjältén J, Ericson L, Elmqvist T (1991) Oikos 62:145–152. **D'Antonio CM** (1993) Ecology 74:83–95. **de Jong TJ**, Klinkhamer PGL, van Staalduinen MJ (1992a) Funct Ecol 6:606–615. **de Jong TJ**, Waser NM, Price MV, Ring RM (1992b) Oecologia 89:310–315. **de Kroon H**, Knops J (1990) Oikos 59:39–49. **de Kroon H**, Schieving F (1991a) J Ecol 79:519–530. **de Kroon H**, Schieving F (1991b) In: van Groenendael J, de Kroon H (eds) Clonal growth in plants. Academic Publ, The Hague, pp 113–130. **de Kroon H**, Hara T, Kwant R (1992) Oikos 63:410–419. **Dewar RC** (1993) Ann Bot 71:147–159. **Doak DF** (1992) Ecology 73:2086–2099. **Douglas DA** (1991) Can J Bot 69:590–596. **Dzwonko Z**, Loster S (1992) J Biogeogr 19:195–204.

Edenius L, Danell K, Bergström R (1993) Oikos 66:286–292. **Ehrlén J** (1992) Ecology 73:1820–1831. **Ehrlén J**, Eriksson O (1993) Oikos 66:107–113. **Elmquist T**, Cates RG, Harper JK, Gardfjell (1991) Oikos 61:65–72. **Elser JJ** (1992) Ecology 73:887–902. **Eriksson O** (1989) Oikos 55:231–238. **Eriksson O** (1992a) Oikos 63:439–448. **Eriksson O** (1992b) J Veg Sci 3:61–69. **Eriksson O**, Ehrlén J (1992) Oecologia 91:360–364. **Ernst WHO**, Kuiters AT, Nelissen HJM, Tolsma DJ (1991) Acta Bot Neerl 40:63–74. **Ernst WHO**, Veenendaal EM, Kebakile MM (1992) Vegetatio 102:1–11. **Espigares T**, Peco B (1993) J Veg Sci 4:189–194. **Evans JP**, Whitney S (1992) Am J Bot 79(12):1344–1347.

Faensen-Thiebes A (1992) Flora 186:377–391. **Fernández RJ**, Nuñez AH, Soriano A (1992) Oecologia 91:39–46.

Gali-Muhtasib HU, Smith CC, Higgins JJ (1992) Ecology 73:1724–1729. **Geber MA**, Watson MA, Furnish R (1992) J Ecol 80:329–341. **Gerhardt K** (1993) J Veg Sci 4:95–102. **Gibson D**, Casal JJ, Deregibus VA (1992) Ann Bot 70:69–73. **Gill DS**, Marks PL (1991) Ecol Monogr 61:183–205. **Gordon DR**, Rice KJ (1993) Ecology 74:68–82. **Grellmann D**, Schneider M (1992) Verh Ges Ökol 21:71–77. **Grubb PJ** (1992) J Ecol 80:585–610. **Gruber F** (1992) Ber Foschungszentr Waldökosyst A 86/I+II. **Guitián J**, Fuentes M (1992) Acta Oecol 13:3–11. **Guitián J**, Sánchez JM (1992) Vegetatio 98:157–164. **Günster A** (1992) J Biogeogr 19:563–572.

Hanski I (1991) Biol J Linn Soc 42:17–38. **Hara T** (1992) Ann Bot 69:509–513. **Hara T**, Kimura M, Kikuzawa K (1991) J Ecol 79:1085–1098. **Hara T**, van der Toorn J, Mook JH (1993) J Ecol

81:47–60. **Hasler AR** (1992) Veröff Geobot Inst ETH Stift Rübel 111:1–104. **Hedge SG**, Lokesha R, Ganeshaiah KN, Uma Shankar R (1991) Oikos 62:100–101. **Hegazy AK** (1992) J Appl Ecol 29:549–557. **Herben T**, Krahulec F, Hadincová V, Kovárová M (1993) J Veg Sci 4:171–178. **Herrera CM** (1992) Ecology 73:1832–1841. **Herrera J** (1991) Oikos 62:209–215. **Hilbert DW**, Swift DM, Detling JK, Dyer MI (1981) Oecologia 51:14–18. **Hjältén J**, Palo T (1992) Oikos 63:477–483. **Hossaert-McKey M**, Jarry M (1992) J Ecol 80:555–565. **Houle G** (1992) J Ecol 80:611–623. **House SM** (1992) J Ecol 80:57–69. **Hughes L**, Westoby M (1992a) Ecology 73:1285–1299. **Hughes L**, Westoby M (1992b) Ecology 73:1300–1312. **Hunter MD**, Price PW (1992) Ecology 73:724–732.

Jensen RJ (1992) J Biogeogr 19:573–579. **Jiménez HE**, Armesto JJ (1992) J Veg Sci 3:579–586. **Johnson EA**, Fryer GI (1992) J Ecol 80:823–836. **Johnson HB**, Polley HW, Mayeux HS (1993) Vegetatio 104/105:157–170. Jónsdóttir GÁ (1991) Acta Bot Neerl 40:311–318. **Jurado E**, Westoby M (1992) J Ecol 80:407–416.

Kalisz S, McPeek MA (1992) Ecology 73:1082–1093. **Karban R** (1993) Ecology 74:9–19. **Karban R**, Strauss SY (1993) Ecology 74:39–46. **Kataoka M**, Ibaraki K, Tokunaga H (1991) J Appl Ecol 28:842–854. **Kellman M**, Kading M (1992) J Veg Sci 3:679–688. **Kemball WD**, Palmer MJ, Marshall C (1992) Oikos 63:366–375. **Kerley GIH**, Tiver F, Whitford WG (1993) Oecologia 93:12–17. **Kik C**, van Andel J, van Delden W, Joenje W, Bijlsma R (1990) J Ecol 78:949–961. **Kilian D**, Cowling RM (1992) J Veg Sci 3:637–646. **Klimes L** (1992) Oikos 63:402–409. **Klinkhamer PGL**, de Jong T (1993) Oikos 66:180–184. **Kohyama T** (1992) Ann Bot 70:451–460. **Kowarik I** (1992) Einführung und Ausbreitung nichteinheimischer Gehölzarten in Berlin und Brandenburg und ihre Folgen für Flora und Vegetation. Verh Bot Ver Berlin Brandenb Beih 3:1–188. **Kunin WE** (1992) Oecologia 91:129–133. **Küppers M** (1992) In: Teller A, Mathy P, Jeffers JNR (eds) Responses of forest ecosystems to environmental changes. Elsevier, London, pp 628–630. **Kurihara M**, Ikusima I (1991) Vegetatio 97:117–124.

Lambers H (1993) Vegetatio 104/105:263–271. **Lamont BB** (1991) Oikos 60:266–268. **Landa K**, Benner B, Watson MA, Gartner J (1992) Oikos 63:348–356. **Leather SR** (1993) Oikos 66:43–47. **Leishman MR**, Westoby M (1992) J Ecol 80:417–424. **Levin SA** (1992) Ecology 73:1943–1967. **Levins R** (1968) Evolution in changing environments. Princeton Univ Press, Princeton. **Lincoln DE** (1993) Vegetatio 104/105:273–280. **Louda SM**, Collinge SK (1992) Ecology 73:153–169. **Lovett Doust L**, Laporte G (1991) J Ecol 79:477–489. **Lowman MD** (1992) J Ecol 80:433–447. **Lowman MD**, Heatwole H (1992) Ecology 73:129–142. **Lüscher A**, Connolly J, Jacquard P (1992) Oecologia 91:404–409.

Maas D (1992) Verh Ges Ökol 21:413–420. **Mahall BE**, Callaway RM (1992) Ecology 73:2145–2151. **Marquis RJ** (1992) Ecology 73:143–152. **Marriott CA**, Haystead A (1992) Ann Bot 69:429–435. **Marshall C** (1990) In: van Groenendael J, de Kroon H (eds) Clonal Growth in Plants. Academic Publishing, The Hague, pp 23–41. **Masters GJ**, Brown VK, Gange AC (1993) Oikos 66:148–151. **Maze KM**, Whalley RDB (1992) Austr J Ecol 17(2):189–194. **Mesleard F**, Lepart J (1991) J Veg Sci 2:155–164. **Meyer AH** (1992) Dissertation, Uni Basel. **Meyer AH**, Schmid B (1991) In: Schmid B, Stöcklin J (eds) Populationsbiologie der Pflanzen. Birkhäuser, Basel, pp 123–146. **Meyer AH**, Schmid B (1992) Verh Ges Ökol 21:285–293. **Meyer SE**, Kitchen SG (1992) J Ecol 80:115–122. **Mitchell RJ**, Waser NM (1992) Ecology 73:633–638. **Molau U**, Prentice HC (1992) J Ecol 80:149–161. **Molofsky J**, Augspurger CK (1992) Ecology 73:68–77. **Mortimer SR** (1992) J Veg Sci 3:665–672. **Müller-Schärer H** (1991) J Appl Ecol 28:759–776. **Murphy SR**, Reid N, Yan Z, Venables WN (1993) Oecologia 93:171–176. **Myster RW**, Pickett STA (1992) Bull Torrey Bot Club 119:145–151.

Nadau LB, Dale MRT, King JR (1991) Can J Bot 69:2539–2544. **Nicholas NS**, Eagar C, Bonner FT (1992) Bull Torrey Bot Club 119:289–299. **Núñez-Farfán J**, Dirzo R (1991) J Veg Sci 2:459–464.

O'Connor TG (1991) Am Nat 137:753–773. **O'Connor TG**, Pickett GA (1992) J Appl Ecol 29:247–260. **Oesterheld M** (1992) Oecologia 92:313–316. **Otte A** (1991) In Mahn EG, Tietze F (eds) Agro-Ökosysteme und Habitatinseln in der Argrarlandschaft. Wiss Beitr MLU Halle Wittenberg 6:38–48. **Otte A** (1992) Landw Jahrb 69:837–860. **Otte A** (1993) Proc 35th Symp Int Assoc Veg Sci, Shanghai (in press). **Otte A** (1993) Schriftenr Schutz gefährdeter Pflanzenarten 5 (in press). **Ouborg NJ** (1993) Oikos 66:298–308. **Ovaska J**, Walls M, Mutikainen P (1992) J Exp Bot 43:1301–1307.

Paige KN (1992) Ecology 73:2076–2085. **Paige KN**, Capman WC (1993) Evolution 47:36–45. **Palo RT**, Bergström R, Danell K (1992) Oikos 65:450–454. **Pardales JR**, Kono Y, Yamauchi A, Iijima M (1992) Ann Bot 69:493–496. **Pilbeam CJ**, Robson MJ (1992) Ann Bot 69:79–86. **Power ME** (1992) Ecology 73:733–746. **Price EAC**, Hutchings MJ (1992) J Ecol 80:39–47. **Price EAC**, Marshall C, Hutchings MJ (1992) J Ecol 80:25–38. **Price PW** (1991) Oikos 62:244–251. **Primack RB**, Miao S (1991) Bull Torrey Bot Club 118:154–160. **Prins AH**, Verkaar JH (1992) In: Ayres PG (ed) Pests and pathogens. Plant responses to foliar attack. Bios, Oxford, pp 13–31. **Prins AH**, Nell HK, Klinkhamer PGL (1992) Oikos 65:409–413. **Puettmann KJ**, Hibbs DE, Hann DW (1992) J Ecol 80:449–458. **Pyšek P** (1992) Oecologia 91:596–628.

Quintana-Ascencio PF, Gonzalez-Espinosa M, Ramirez-Marcial N (1992) Bull Torrey Bot Club 119:6–18.

Rashbrook VK, Compton SG, Lawton JH (1992) Ecology 73:2167–2174. **Rausher MD**, Iwao K, Simms EL, Ohsaki N, Hall D (1993) Ecology 74:20–29. **Reader RJ** (1992a) Ecology 73:373–376. **Reader RJ** (1992b) Oikos 65:414–418. **Rees M**, Brown VK (1992) J Ecol 80:353–360. **Reichman OJ**, Smith SC (1991) Ecology 72:116–124. **Reynolds HL**, Pacala SW (1993) Am Nat 141:51–70. **Rousi M**, Tahvanainen J, Uotila I (1991) Am Nat 137:64–82. **Rousi M**, Tahvanainen J, Henttnonen H, Uotila I (1993) Ecology 74:30–38. **Roy J**, Sonié L (1992) J Appl Ecol 29:647–655. **Russi L**, Cocks PS, Robers EH (1992a) J Appl Ecol 29:763–771. **Russi L**, Cocks PS, Roberts EH (1992b) J Appl Ecol 29:772–778. **Rusterholz HP**, Stöcklin J, Schmid B (1992) Mesogee 52:16–17. **Rusterholz HP**, Stöcklin J, Schmid B (1993) Verh Ges Ökol 22:337–346. **Ryser P** (1993) J Veg Sci 4:195–202.

Sagar GR (1959) The biology of some sympatric species of grassland. Ph D Thesis, Univ Oxford. **Sakai S** (1991) J Theor Biol 148:535–544. **Salomonson A**, Ohlson M, Ericson L (1992) Oikos 65:493–501. **Salonen V** (1992) J Veg Sci 3:109–112. **Samson DA**, Philippi TE, Davidson DW (1992) Oikos 92:61–80. **Schmid B**, Bazzaz FA (1990) Ecology 71(2):523–535. **Schmid B**, Bazzaz FA (1991) Flora 185:335–344. **Schmid B**, Bazzaz FA (1992) Oikos 65:13–24. **Schmid B**, Stöcklin J (eds) (1991) Populationsbiologie der Pflanzen. Birkhäuser, Basel, 351 pp. **Schmidt W**, Brübach M (1993) J Veg Sci 4:247–254. **Schmitz OJ**, Hik DS, Sinclair ARE (1992) Oikos 65:295–300. **Seitz A**, Loeschke V (eds) (1991) Species conservation: a population-biological approach. Birkhäuser, Basel. **Senn J**, Hanhimäki S, Haukioja E (1992) Oikos 63. **Shainsky LJ**, Radosevich SR (1992) Ecology 73:30–45. **Shumway SW**, Bertness MD (1992) Oecologia 92:490–497. **Siemens DH**, Johnson CD, Ribardo KJ (1992) Ecology 73:2152–2166. **Silvertown J**, Dale P (1991) Oikos 61:441–444. **Silvertown J**, Watt TA, Smith B, Treweek JR (1992a) J Veg Sci 3:35–40. **Silvertown J**, Holtier S, Johnson J, Dale P (1992b) J Ecol 80:527–534. **Simms EL** (1992) In: Fritz RS, Simms EL (eds) Plant resistance to herbivores and pathogens: ecology, evolution, and genetics. University of Chicago Press, Chicago, pp 392–425. **Skinner RH**, Nelson CJ (1992) Ann Bot 70:493–499. **Skoglund J** (1992) J Veg Sci 3:357–360. **Skogsmyr I**, Fagerström T (1992) Oikos 64:451–457. **Snaydon RW** (1991) J Appl Ecol 28:930–946. **Snyder MA** (1992) Ecology 73:1730–1741. **Soukupová L** (1990) In: Krahulec F, Agnew ADQ, Agnew S, Willems JH (eds) Spatial processes in plant communities, Academia, Praha, pp 145–159. **Soukupová L** (1992) Oikos 63:395–401. **Spiess R** (1991) Arch Naturschutz Landschaftsforsch 31:79–94. **Starfinger U** (1991a) In: Henle K, Kaule G (eds) Arten- und Biotopschutzforschung für Deutschland. Forschungszentrum Jülich, Ber ökol Forsch 4:225–233. **Starfinger U** (1991b) In: Seitz A, Loeschke V (eds) Species conservation: a population-biological approach. Birkhäuser, Basel, pp 171–184. **Stein SJ**, Price JW, Abrahamson WG, Sacchi CF (1992) Oikos 65:190–196. **Stöcklin J** (1990) Verh Ges Ökol 19(2):30–43. **Stöcklin J** (1992a) Bot J Linn Soc 108:35–47. **Stöcklin J** (1992b) Bot Helv 102:3–21. **Stöcklin J**, Zoller

H (1991) In: Schmid B, Stöcklin J (eds) Populationsbiologie der Pflanzen. Birkhäuser, Basel, pp 147–164. **Stone L**, Roberts A (1991) Ecology 72:1964–1972. **Sukopp H**, Schick B (1993) Diss Bot 196:163–174. **Sutherland RA**, Benjamin LR (1993) Ann Bot 71:131–134.

Tapper P-G (1992) J Veg Sci 3:41–46. **Telenius A** (1992) Acta Bot Neerl 41:305–318. **Telenius A**, Torstensson P (1991) Oikos 61:216–222. **Thompson JD** (1991) BioScience 41:393–01. **Thompson JN**, Pellmyr O (1992) Ecology 73:1780–1791. **Tolvanen AK**, Laine K, Pakonen T, Saari E, Havas P (1992) Acta Oecol 13:607–615. **Tscharntke T** (1990) J Appl Ecol 27:679–692. **Tscharntke T** (1991) In: Henle K, Kaule G (eds) Arten- und Biotopschutzforschung für Deutschland. Forschungszentrum Jülich, Ber ökol Forsch 4:201–207. **Tschurr P** (1992) Veröff Geobot Inst ETH Stift Rübel 108:1–121. **Turkington R**, Klein E (1991) Can J Bot 69:226–228. **Tybjerg H**, Vestergaard P (1992) Oikos 65:395–408.

Urbanska KM (1992) Populationsbiologie der Pflanzen. Fischer, Stuttgart, pp 374.

van der Maarel E, Sykes MT (1993) J Veg Sci 4:179–188. **van der Meijden E**, Klinkhamer PGL, de Jong TJ (1992) Acta Bot Neerl 41:249–270. **Vander Wall SB** (1992) Ecology 73:614–621. **Vasiliauskas SA**, Aarssen LW (1992) Ecology 73:622–632. **Voesenek LACJ**, Blom CWPM (1992) Acta Bot Neerl 41:319–329. **Voesenek LACJ**, de Graaf MCC, Blom WCPM (1992) Acta Bot Neerl 41:331–343. **Vrieling K**, de Jong T, Klinkhamer PGL, van der Meijden E, Prins A (1992) Verh Ges Ökol 21:295–301.

Weber E, Schmid B (1992) Diss Bot 196:209–227. **Weiner J**, Thomas SC (1992) Ecology 73:648–656. **Welker JM**, Briske DD (1991) Oikos 63:357–365. **Wennström A**, Ericson L (1992) J Ecol 80:71–77. **Widén M** (1992) Oikos 63:430–438. **Wiegleb G** (1992) Verh Ges Ökol 21:327–338. **Wiegleb G**, Brux H, Herr W (1991) Vegetatio 97:161–172. **Wisheu IC**, Keddy PA (1991) J Veg Sci 2:181–188. **Wilson DS** (1992) Ecology 73:1984–2000. **Wilson JB**, Roxburgh SH (1992) Oikos 65:343–348. **Wilson SD** (1991) Oikos 62:292–298.

Yokozawa M, Hara T (1992) Ann Bot 70:305–316. **Young HJ**, Young TP (1992) Ecology 73:639–647.

Zangerl AR, Berenbaum MR (1993) Ecology 74:47–54. **Zhang J**, Maun MA (1991) Bull Torrey Bot Club 118:141–153.

Dr. Walter Seidling
Dr. Uwe Starfinger
Institut für Ökologie
Technische Universtität Berlin
Schmidt-Ott-Straße 1
D-12165 Berlin, FRG

Dr. Jürg Stöcklin
Botanisches Institut
Universität Basel
Schönbeinstraße 6
CH-4056 Basel, Switzerland

Subject Index

SPRINGER NATURE

GPSR Compliance

The European Union's (EU) General Product Safety Regulation (GPSR) is a set of rules that requires consumer products to be safe and our obligations to ensure this.

If you have any concerns about our products, you can contact us on ProductSafety@springernature.com

In case Publisher is established outside the EU, the EU authorized representative is:

Springer Nature Customer Service Center GmbH
Europaplatz 3
69115 Heidelberg, Germany

FSC
www.fsc.org
MIX
Papier | Fördert gute Waldnutzung
FSC® C083411

Kolibri 360 GmbH
Ferdinand-Jühlke-Straße 7
99095 Erfurt, Deutschland
produktsicherheit@kolibri360.de